# THE SCIENCE
# AND WONDERS
# OF THE
# ATMOSPHERE

# THE SCIENCE AND WONDERS OF THE ATMOSPHERE

STANLEY DAVID GEDZELMAN
City College of New York

John Wiley & Sons
New York
Chichester
Brisbane
Toronto

Photo research by Shari Segel and
Rosemary Eakins/Research Reports

Photo Editor, Stella Kupferberg

**Library of Congress Cataloging in Publication Data**

Gedzelman, Stanley David, 1944-
    The science and wonders of the atmosphere.

    Bibliography: p.
    Includes index.
    1.  Meteorology.  I.  Title.
QC861.2.G43        551.5        79-23835
ISBN: 0-471-02972-6

Printed in the United States of America

10 9 8 7 6 5 4 3 2 1

# PREFACE

Not long ago meteorology was such an obscure subject that many people thought it was the study of meteorites. Then, in the wake of social change, meteorology was brought into the limelight. First, attention was focused on the pollution and desecration of the environment. Second, as the old, strict college curricular requirements were relaxed, meteorology began to serve as a "relevant" and popular alternative to physics or chemistry.

As a result of these changes, the number of people taking introductory meteorology has greatly increased and I, like many other meteorology instructors, have found myself devoting a much larger percentage of my time teaching basic-level courses. When I first began lecturing in the large introductory class given at City College, I had no intention of writing a textbook. However, after several semesters I realized that I was not satisfied with any book on the market and so decided to write one myself.

With my class in mind, I prepared an extensive set of lecture notes and naively thought that transforming them into a textbook would be a relatively easy job. I now have much more respect for all authors of textbooks. However, I feel even more strongly than before that there are important fundamental differences that set *The Science and Wonders of the Atmosphere* apart from all other elementary meteorology textbooks. It is now appropriate to tell you just what my approach has been.

*The Science and Wonders of the Atmosphere* contains many accounts and descriptions of extreme and sometimes disastrous weather events as well as explanations of beautiful phenomena of the atmosphere such as rainbows, halos, the colors of the sky, and the many shapes of clouds. I believe it is these *wonders* that attract most people to the atmospheric sciences in the first place. Furthermore, far from distorting people's viewpoints, the observations of these phenomena provide excellent examples of many basic scientific laws and principles. Moreover, making astute observations constitutes an essential prerequisite for scientific reasoning. Thus, right after the Introduction you will find treatments of the cloud forms (Chapter 2) and the various optical phenomena (Chapter 3) so that you may quickly become inveterate sky watchers.

Although I have tried to make *The Science and Wonders of the Atmosphere* as easy and pleasurable to learn from as possible, there are nonetheless many difficult concepts in meteorology. One principal difficulty lies in visualizing the three-dimensional processes of the atmosphere when looking at figures on a flat page. For this reason, many figures have been given the appearance of depth. Color has also been used to advantage; for example, in many figures red is used to represent warm air and gray has been used to signify cold air. Important sections designed to help you read weather maps, contour maps, and meteorological charts have been placed in Chapter 1.

Throughout the book, I have presented the difficult concepts together with the phenomena they explain. Often, analogies from everyday life are used to help render things more comprehensible. Thus, for example, when you look at how raindrops grow, you will also read how some insects can walk on water, and why flyswatters have holes. The important scientific laws are often presented in the historical context in which they were discovered, partly for the sake of interest, but largely because the understanding of these concepts will frequently develop in your minds in a way similar to their historical development.

The scientific aspects of this book are virtually self-contained, and all scientific laws and principles are presented as if you have never seen them before. This is important enough if you intend to specialize in meteorology or some related topic, but I believe it is absolutely essential if this is to be your only exposure to a physical science. In meteorology we make use of findings from all the sciences but most of all from physics, which represents one of the profound intellectual achievements of humanity and a cornerstone of the modern technological world. This is why you will find that so many fundamental laws of physics have been treated at length, and why I have placed such emphasis on the *science* of the atmosphere. To do any less might make it appear that progress in meteorology has come independently of progress in other areas of science; in fact, nothing could be further from the truth.

You might wonder why mathematics has been included to such an extent in *The Science and Wonders of the Atmosphere,* especially since I have stressed my attempts to facilitate learning. However, to avoid mathematics is to give a distorted view of meteorology, which is a *quantitative* science. You may find mathematics difficult, and consequently unpleasant, but there are many instances in which *qualitative* information alone is almost worthless. For example, you cannot get major accumulations of snow unless the temperature remains below 0°C (Celsius or centigrade). Once the temperature rises above that critical *number,* the snow begins to melt.

The mathematics is presented in a straightforward, practical manner. Every mathematical formula in the book is illustrated by simple, completely worked out examples that require no more than a knowledge of elementary algebra. I very strongly recommend that you try to read these examples, because where they appear in the discussions they form an integral part of the subject under consideration. And, if you read slowly and carefully, you may find that the examples are not as difficult as you had expected. In fact, I have found that despite the tearing out of hair and gnashing of teeth among my nonscience students the first time they see me put an equation on the blackboard each semester, they inevitably perform above average on the mathematical questions of their examinations; and a few even revel in their new-found mathematical talent!

I have included a problem section at the end of each chapter. Naturally some of the problems are fairly simple, but many are challenging or present important additional information. A few projects and experiments are also suggested. Hints or solutions to the more difficult problems are given at the back of the book.

Once you begin reading and solving the problems you will quickly find that temperatures are not given in degrees Fahrenheit, distances are not given in feet or miles, and weights are not given in pounds or ounces. In short, I have used metric units as much as possible—first, because it is a more rational system of weights and measures and, second, because it is slowly becoming the standard worldwide. You will therefore be forced to get used to the metric units a little sooner than otherwise, but if you have any trouble with them or with any symbols or abbreviations promptly turn to the list of abbreviations, symbols, and units that follow the Table of Contents. Definitions are also quite important in science, and you will see technical terms written in boldface where they are defined. A more extensive list of definitions is given in the Glossary at the back of the book.

*The Science and Wonders of the Atmosphere* covers a wide range of topics. This was done for a number of reasons. First, there is enough material on various aspects of climatology so that it can also serve as an introductory climatology textbook for those students who are without a prior course in atmospheric science. The book is also designed to allow the instructor greater freedom in choosing the extra topics, which may vary from semester to semester. The book can also be used as a general, nontechnical reference by professionals or by anyone who might want to learn something about weather and the atmosphere. Material not regularly included in basic meteorology textbooks includes optical phenomena (Chapter 3); consequences of the molecular nature of air (Chapter 5); the evolution and composition of the planets and their atmospheres (Chapter 6; some of this material is still highly conjectural); agricultural meteorology and climate near the ground (Chapter 9); details of the growth of raindrops (Chapter 12); lightning (Chapter 19); and much of the material on climates of the past and present and their various effects on the earth and on us (Chapters 22 to 25). However, if you cannot find a topic in this book, or wish to investigate something in greater detail, you will probably be able to find a good reference in the Bibliography at the back of the book.

Now I will ask *you* for a favor. I am always on the lookout for good photographs (if in black and white, they should be glossy prints; if in color, 35 mm slides are best), especially of clouds or optical phenomena. If you feel you have an outstanding picture, send a *duplicate* (because I may not be able to return it) to me at City College with your name and address, including details of where and when the photo was taken. If that photograph is used in any future edition of this book, you will of course receive credit for it.

No atmospheric science book can be written in a vacuum, and I owe thanks to many. My parents, brothers, friends, and good teachers helped to inspire and prepare me long before the book was conceived. My wife, Bernice, was a constant source of encouragement. I am also grateful for the time, facilities, and complete academic freedom afforded me at City College of New York and Tel Aviv University. The doors of countless libraries and museums were also open to me.

Throughout the book I have refrained from giving credit to scientists who are still active, but I certainly feel indebted to them for their enormous contributions. The content of this book is due in large part to their many remarkable accomplishments and discoveries that have advanced meteorology tremendously in the years since World War II.

Many colleagues helped me along the way. My fellow teachers and friends in the Department of Earth and Planetary Sciences at City College increased my appreciation and understanding of geology, while members of various other departments greatly enhanced my knowledge of other fields. The entire manuscript was thoroughly and critically reviewed by Professors John M. Wallace, of the University of Washington, and Lance Bosart, of the State University of New York at Albany. Specific chapters or sections were read by Professors Jerome Spar and Albert Ehrlich, of the City College of New York; Zev Levin, George Ohring, and Akiva Bar-Nun, of Tel Aviv University; Dr. Charles Tony Gordon, of the Geophysical Fluid Dynamics Laboratory; and Dr. Israel Tzur, of Tel Aviv University; as well as Robert Salzman, Susan Salzman, and Ann Gordon. The comments of all of these people eliminated errors and led to improvements in the manuscript. I also was afforded great hospitality and much helpful information by Alva Wallis, of the National Climatic Center, and Vince Olivier's entire staff of satellite meteorologists at the World Weather Building.

Large sections of the manuscript were typed (and read) by my wife Bernice, my mother Rita, my brother Robert, Helen Goldberg, and Margaret Sullivan. Their comments helped me to simplify many explanations. And, of course, I thank my students; most of the explanations you will read here were first tried successfully on them. You do not see the many they "refused" to understand.

Final thanks go to the highly professional staff at John Wiley & Sons. Particular thanks go to Butch Cooper who coordinated it all, John Balbalis who helped in the conception of the figures, Stella Kupferberg for photograph research, Joan Knizeski who transformed the entire manuscript into readable English, Kenny Beck for his design, and Malcolm Easterlin and Elizabeth Doble who managed to transform a crude manuscript into a finished product. It is apparent to me that excellence is their first motive.

Stanley Gedzelman

# CONTENTS

# ABBREVIATIONS, SYMBOLS, AND UNITS USED IN THIS BOOK

Science textbooks, such as this one, are often full of abbreviations, symbols, and units, so it is helpful to see them all together before you begin reading. Here are the abbreviations commonly used in this book for *units of measure*.

| | |
|---|---|
| Length | centimeter (cm); meter (m); kilometer (km) |
| Mass | gram (g); kilogram (kg) |
| Time | second (s); minute (min); hour (h) |
| Energy | calories (cal) |
| Temperature | Kelvin or absolute (°K); Fahrenheit (°F); Celsius or centigrade (°C) |
| Pressure | millibar (mb) |

There are also various combinations of these abbreviations; for example, meters per second is abbreviated m/s and calories per square centimeter per minute is abbreviated cal/(cm²)(min).

Here are the symbols commonly used for the variables and concepts that appear in equations.

Absolute temperature *(T)*  Mass *(m)*
Acceleration *(a)*  Mixing ratio *(W)*
Acceleration of gravity *(g)*  Molecular weight *(M)*
Area *(A)*  Pressure *(P)*
Coriolis acceleration *(C)*  Pressure gradient *(PG)*
Density *(ρ)*  Radiation *(R)*
Dew point temperature *(T_d)*  Radius *(r)*
Distance *(d)*  Relative humidity (RH)
Energy *(e)*  Saturated mixing ratio *(W_s)*
Force *(F)*  Speed *(s or v)*
Height *(H)*  Velocity *(v)*
Intensity of radiation *(I)*  Volume *(V)*
Kinetic energy (KE)  Wavelength *(L)*
Length *(L)*  Weight *(w)*
 Wet bulb temperature *(T_w)*

Abbreviations of the *cloud types* are given on page 23, abbreviations of the various *air-mass types* are given on page 274, and abbreviations of the *climate types* are given on page 387–389 and in Appendix 2.

Other commonly used abbreviations include:

Clean Air Turbulence (CAT)  National Weather Service (NWS)
East (E)  North (N)
Eastern Standard Time (EST)  Northern Hemisphere (NH)
General Circulation Model (GCM)  Showalter Stability Index (SSI)
Greenwich Mean Time (GMT)  South (S)
Intertropical Convergence Zone (ITCZ)  Southern Hemisphere (SH)
Latitude (lat.)  Temperature Humidity Index (THI)
Lifting Condensation Level (LCL)  West (W)

Most people get used to units such as feet, pounds, and degrees Fahrenheit, but in the sciences it is best to standardize and convert to units such as meters, kilograms, and degrees centigrade (these are called *metric* units). Here are some frequently used conversions.

| Length | | | | | |
|---|---|---|---|---|---|
| | 1 | million microns | = | 1000 | millimeters |
| | 1000 | millimeters | = | 100 | centimeters |
| | 100 | centimeters | = | 1 | meter |
| | 1000 | meters | = | 1 | kilometer |
| | 0.394 | inch | = | 1 | centimeter |
| | 2.54 | centimeters | = | 1 | inch |
| | 1 | meter | = | 3.28 | feet |
| | 1 | foot | = | 0.305 | meter |
| | 1 | kilometer | = | 0.621 | mile |
| | 1 | mile | = | 1.61 | kilometers |
| | 1 | kilometer | = | 0.54 | nautical mile |
| | 1 | nautical mile | = | 1.85 | kilometers |
| | 60 | nautical miles | = | 1 | degree of latitude |
| | 1 | degree of latitude | = | 111 | kilometers |

| Mass | | | | | |
|---|---|---|---|---|---|
| | 1000 | grams | = | 1 | kilogram |
| | 1 | kilogram | = | 2.20 | pounds |
| | 1 | pound | = | 0.454 | kilogram |

*Note:* The pound conversion is true only on earth; a pound is really a unit of weight (force).

| Speed | | | | | |
|---|---|---|---|---|---|
| | 1 | meter per second | = | 2.24 | miles per hour |
| | 2.24 | miles per hour | = | 1.94 | knots |
| | 1.94 | knots | = | 3.60 | kilometers per hour |
| | 1 | knot | = | 1.15 | miles per hour |
| | 1.15 | miles per hour | = | 0.515 | meter per second |
| | 0.515 | meter per second | = | 1.85 | kilometers per hour |

| Energy | | | | | |
|---|---|---|---|---|---|
| | 1 | joule | = | 0.239 | calorie |
| | 0.239 | calorie | = | 6,250,000,000,000,000,000 electron volts | |
| | 1 | calorie | = | 4.186 | joules |

| Power | | | | | |
|---|---|---|---|---|---|
| | 1 | joule per second | = | 1 | watt |
| | 1 | watt | = | 14.3 | calories per minute |
| | 14.3 | calories per minute | = | 0.00134 | horsepower |
| | 1 | horsepower | = | 746 | watts |

| Force | | | | | |
|---|---|---|---|---|---|
| | 1 | newton | = | 1 | kilogram meter per second per second |
| | 1 | kilogram meter per second per second | = | 0.225 | pound |

| Pressure | | | | | |
|---|---|---|---|---|---|
| | 1 | pascal | = | 1 | newton per square meter |
| | 1 | inch of mercury | = | 33.86 | millibars |
| | 1 | millibar | = | 100 | pascals |
| | 100 | pascals | = | 0.0295 | inch of mercury |
| | Average atmospheric pressure | | = | 1013.25 | millibars |
| | 1013.25 | millibars | = | 14.7 | pounds per square inch |
| | 14.7 | pounds per square inch | = | 29.92 | inches of mercury |
| | 29.92 | inches of mercury | = | 760 | millimeters of mercury |

Temperature conversions are presented on page 81.

*Time Zones*
Greenwich Mean Time is 5 hours ahead of Eastern Standard Time, but only 4 hours ahead of Eastern Daylight Time. Eastern Standard Time is one hour ahead of Central Standard Time, two hours ahead of Mountain Standard Time, and three hours ahead of Pacific Standard Time. For example,
1200 GMT = 0700 EST = 0600 CST = 0500 MST = 0400 PST

  ABBREVIATIONS, SYMBOLS, AND UNITS USED IN THIS BOOK

# INTRODUCTION

There is a story about a man who lived near the shore of Long Island. A barometer that he had purchased arrived in the mail one morning. When the man opened the package, the barometer needle was pointing to "hurricane." He shook the barometer, but the needle didn't budge. Assuming that it was broken, he went to the post office to send it back to the manufacturer. When he returned home a day later his beach house was gone—it had been destroyed by the hurricane of 1938.

This storm, which struck Long Island and New England on September 21, 1938, was not predicted by meteorologists. For more than a hundred years prior to 1938, no severe hurricane had come near Long Island, and people began to think that no hurricane ever could. When this particular hurricane was located off the east coast of Florida it was moving slowly northward up the east coast of the United States (straight toward Long Island). Everyone assumed that it would soon turn toward the east and die over the Atlantic Ocean like all "good" hurricanes were supposed to. This hurricane didn't. It suddenly accelerated and raced at an unprecedented speed straight toward Long Island and New England.

People in the soon-to-be-struck area did receive some warning (from nature) that a violent storm was approaching. Clouds high in the sky came racing from the south; the tide began to rise sharply along the beach; and huge waves piled onto the shore. However, these early warning signs were provided by the elements—not by meteorologists—so almost no one paid much attention to them.

The situation deteriorated rapidly. Coastal areas suffered incredible damage when the hurricane struck (see Figures 1.1 and 1.2). As the center of the storm smashed into the south shore of Long Island, a giant wave or surge of water passed right over the dunes along the beach and crumbled many houses. The tide rose so high that a wave knocked a man into the top of a telephone pole. Wind gusts reaching 160 knots uprooted or snapped millions of trees across Long Island and New England. Almost 700 people were killed by the storm—many from drowning, many when houses or buildings collapsed right on them, and many when they were struck by objects picked up by the wind and accelerated to speeds far greater than a human can throw a baseball.

The only fortunate thing was that this hurricane struck about a week after the beach crowds had left for the winter. Otherwise, the death toll would have been far, far higher.

After the storm had passed the atmosphere grew quiet once again, but one part of the earth had been severely mangled. Why hadn't the meteorologists given the people a fair warning? What were the reasons that the storm's motion and fury were not anticipated?

The storm was not predicted partially because of its unprecedented speed. There were few ships at sea in the vicinity of the storm's path so that almost no information was available. Furthermore, in 1938 meteorologists had

FIGURE 1.1
Flooding in downtown Providence, Rhode Island, after the hurricane of 1938.

FIGURE 1.2
Trees downed by the hurricane of 1938.

no radar or satellites to warn of an approaching storm.

By November 1970, the situation had changed. Satellites enabled meteorologists to keep an almost constant watch of the weather anywhere on earth. Meteorologists were watching when a small hurricane developed in the Bay of Bengal on November 9, 1970, and they followed its course as it moved straight for the coast of East Pakistan (now Bangladesh; see Figure 1.3). More than 24 hours in advance they correctly predicted that the hurricane would intensify and strike the coastal lowlands on November 12. This time what followed was not their fault.

First, some background information on the geography of the region may be necessary. Much of what is now Bangladesh consists of the Ganges-Brahmaputra delta, which lies only a few meters above sea level. The coastal waters teem with fish, and the land is extremely fertile; therefore, the area is one of the most densely populated of any region on earth. In addition, November is harvest time (in this area), so the population swells even more with the influx of migrant workers.

Because of the primitive means of communication and transportation, almost no attempt was made to evacuate the people. When the hurricane struck a wall of water as much as 4 meters (m) higher than the normal high tide surged inland—almost instantly drowning 100,000 people. The death toll from the storm finally reached an estimated 300,000. The fishing industry and harvest were destroyed, and starvation was rampant. Within a year a bloody revolution in East Pakistan was successful; the new nation of Bangladesh was created.

Such extraordinary events cause people to wonder and ask, "What could possibly have produced such a disaster?" These disasters are produced by the same forces and laws that produce our every-day weather. Perhaps science was born after a storm.

Before reading any further, why not attempt to write briefly what *you* think causes the wind and the rain. Perhaps you don't even know where to begin. In that case, give your impressions if you have any. Do not be afraid to write down something that may later appear to be naive or even stupid, for we all must start somewhere, and finding the correct approach is half the problem.

Now that you have tried (have you?), permit me to help you.

## SOME ANCIENT METEOROLOGISTS

Let us see first what others in the past have thought about the elements. People living in precivilized society or early civilizations were generally more exposed to the ele-

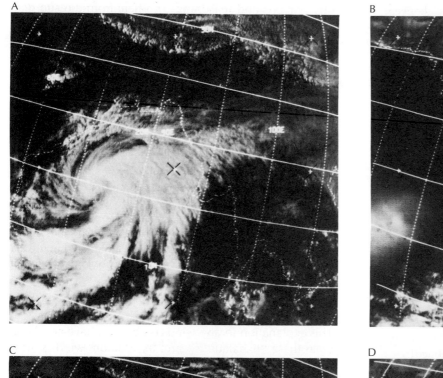

A

B

C

D
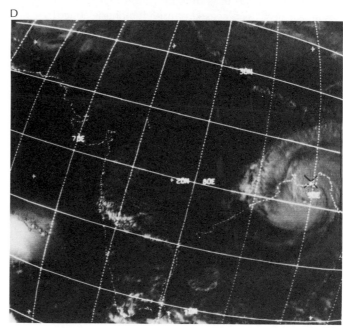

FIGURE 1.3
Sequence of satellite pictures of a cyclone that struck East Pakistan, November 12, 1970. This cyclone may have produced the greatest death toll of any single storm. Clouds appear white and clear areas dark in almost all satellite photos.

ments, and hence more aware of the weather than we are. The occasionally terrifying nature of the weather prompted attempts at explanations that were given more in terms of the imagination, desires, and fears than on any rational scheme. In their approach to natural phenomena,

these people believed that the weather represented the expression of the moods of the gods and spirits. In *The Odyssey,* Poseidon, the god of the sea, was angry at the Greek hero, Odysseus, who was floating on a raft at sea. When he saw that Odysseus was getting near land

. . . he gathered the clouds and troubled the waters of the deep, . . . and he roused all storms of all manner of winds, and shrouded in clouds the land and sea: and down sped night from heaven. The east wind and the south wind clashed, and the stormy west and the north, that is born in the bright air, rolling onward a great wave. Then were the knees of Odysseus loosened and his heart melted, and heavily he spake to his own great spirit.

*The Odyssey,* Book V

In this paragraph Homer was essentially answering the question, "Why did this happen *to me?*" There are times when we may all feel like this, but ask yourself if you are satisfied with such an explanation for wind and rain. Personally, I am not. Neither were some of the later Greeks who grew skeptical of these poetic but childlike attempts at explaining natural occurrences. In the *History of the Persian Wars,* Herodotus reports that

The storm lasted three days. At length the Magians, by offering victims to the winds, and charming them with the help of conjurers, while at the same time they sacrificed to Thetis and the Nereids, succeeded in ending the storm four days after it first began; *or perhaps it ceased of itself* [emphasis added].

*History of the Persian Wars,* Book 7

Can you detect a note of irreverence? People began to realize that a more rational approach to such problems was called for. Nature is not the whimsical expression of the moods of the gods and, despite what you may feel at times, there is nothing "personal" in the events of nature. Instead, nature proceeds by basic principles or laws.

The Greeks never did find the correct laws, but they did start the search. The great philosopher Aristotle wrote a book on meteorology. This was perhaps his worst book because most of his explanations in it are worthless, but occasionally he came out with a gem. Read his explanation of rain.

The efficient, controlling, and first cause is the circle of the sun's revolution. . . . The earth is at rest, and the moisture about it is evaporated by the sun's rays and the other heat from above and rises upwards; but when the heat which caused it to rise leaves it . . . the vapor cools and condenses again as a result of the loss of heat and the height and turns from air into water: and having become water falls again onto the earth.

*Meteorologica,* Book 1, Chapter 9

At least one mistake in the above paragraph is easy to spot. We know now that the earth goes around the sun. Nevertheless, the above quotation contains many good observations and reasoning. The heat from the sun *does* evaporate water from the earth's surface. When the air and vapor rises in the atmosphere, its temperature *does* fall (although it doesn't lose heat!) and the vapor *does* often condense back to its liquid form.

You must remember that this is one of the best quotes from Aristotle's *Meteorologica.* Now read his explanation of the wind.

Now there is in the earth a large amount of fire and heat, and the sun not only draws up the moisture on the earth's surface but also heats and so dries the earth itself; and this must produce exhalations which are of the two kinds we have described, namely vaporous and smoky (air). . . . The dry exhalation is the origin and natural substance of the winds. That this must be the case is evident from the facts. . . . Yet it is absurd to suppose that the air which surrounds us becomes wind simply by being in motion.

*Meteorologica,* Book 2, Chapter 4

This explanation for the winds is absurd. I have no idea of what these "exhalations" are. In addition, wind *is* air in motion. In this quote there is too much "logic" and too little observation. The Greeks were excessively enchanted with the power of words. Good scientific work involves gathering accurate observations and then using reason to connect them. Unfortunately, many of the Greeks and their successors thought that they could do without the facts. After the heyday of the Greek culture, intellectual interest in the world of nature declined in the West, and people grew more interested in the Rock of Ages than in the age of rocks.

If you think that the last quote from Aristotle was bad, here is one by Isidore:

The clouds are to be understood as holy evangelists, who pour the rain of the divine word on those who believe. For the air itself, empty and thin, signifies the empty and wandering minds of men, and then, thickened and turned into clouds, typifies the confirmation in the Faith of minds chosen from among the empty vanity of the unfaithful. And just as rainy clouds are made from the empty air, so the holy evangelists are gathered to the faith from the vanity of this world.

*On the Nature of Things*

Isidore, Bishop of Seville, was one of the great intellectuals of the Middle Ages! To the intellectuals of his day, the natural world was nothing more than a demonstration and proof of the existing religious dogma. These views of nature were absurd; yet, they were accepted as gospel for more than a thousand years.

Then, starting around the year 1600, various instruments such as the thermometer and barometer were invented, and scientists began looking very carefully at the atmosphere and *measuring* its properties and behavior. Since then, we have made enormous progress in our understanding of the atmosphere and in our ability to predict the weather. You may say that the weather forecasters still do a rotten job, but read on and you will find out why.

## THE ULTIMATE CAUSE OF THE WIND AND RAIN—WEATHER IN A NUTSHELL

This section contains the key to the subject of meteorology in a simplified, skeletal form. It is such an important section that I suggest you read it over just before beginning each new chapter in this book—read it until you get sick of reading it.

**1.** The sun's heating varies over the earth and with the seasons.

The *directness* of the sun's rays (sun overhead means direct) and the length of the daylight period determine the total heat input to the earth and atmosphere. This heat input is far greater near the equator where the sun is nearly overhead than at the poles where the sun is low in the sky. In addition, the sun is higher in the sky in summer (July for the Northern Hemisphere; January for the Southern Hemisphere) than in winter (January for the Northern Hemisphere; July for the Southern Hemisphere). Finally, the sun's heat also varies from place to place because of irregularities on the surface of the earth. (From this point on NH and SH will be used in place of Northern Hemisphere and Southern Hemisphere, respectively.)

**2.** The differences of air temperature over the earth cause the winds.

Temperature varies over the globe as a result of the unequal solar heating. Generally, air at the equator is far warmer than air at the poles. Warm air is light and tends to rise; cold air is heavy and tends to sink. The winds,

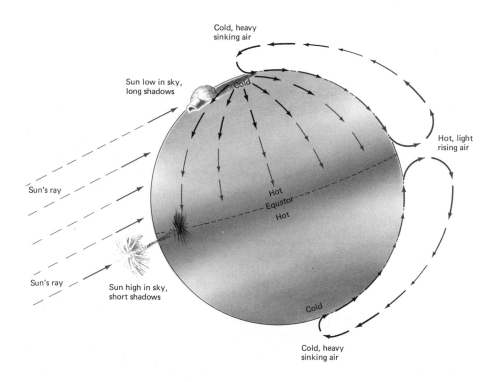

Cold, heavy sinking air

Sun low in sky, long shadows

Cold

Hot, light rising air

Sun's ray

Hot
Equator
Hot

Sun's ray

Sun high in sky, short shadows

Cold

Cold, heavy sinking air

FIGURE 1.4
Idealized wind system on a uniform earth which does not rotate. To produce this wind system, the sun would have to revolve around the earth. Notice the long shadows near the pole.

which are largely horizontal, serve to "fill the void." You might envision this simple, ideal pattern of global winds: the heated, light air near the equator rises and spreads out toward the poles, while the cold, heavy polar air sinks and spreads out along the ground, moving back toward the equator to fill the void (see Figure 1.4).

**3.** The rotation of the earth destroys this simple wind pattern, twisting the winds and producing great wind spirals that are known as high and low pressure areas.

Once anything moves on the rotating earth it is subjected to a twisting force to the right of *its* direction of motion in the NH and to the left of *its* direction of motion in the SH (see Figure 1.5). This effect is known as the **Coriolis force.** The air that rises in the equatorial regions and spreads poleward is so twisted by the Coriolis force that it cools, grows heavy, and sinks in subtropical latitudes, long before it reaches the poles. In the midlatitudes the winds are so distorted by the Coriolis force that a series of spiraling high and low pressure areas are always found (see Figure 1.6).

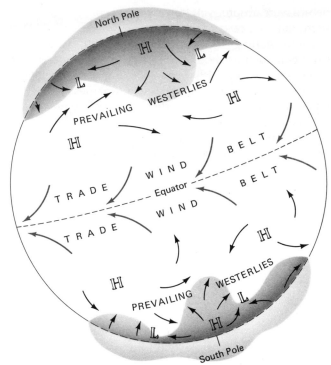

FIGURE 1.6
A more realistic picture of the global winds. Notice that the tropical latitudes are characterized by the reasonably steady trade wind belt. Air still rises near the equator. Other features show that the middle latitudes are dominated by large swirling high (H) and low (L) pressure areas. Near the poles are large blobs of frigid air. The boundaries of this frigid air are wavy and are called *fronts*.

FIGURE 1.5
All *moving* objects on the rotating earth are forced to their right in the northern hemisphere and to their left in the southern hemisphere. This is called the Coriolis force.

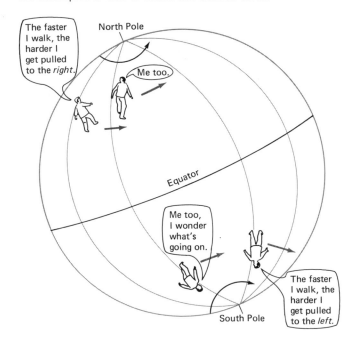

In the NH, air in low pressure areas spirals counterclockwise and inward to the center near ground level, whereas air in high pressure areas spirals clockwise and outward from the center near ground level. For the SH, air in low pressure areas spirals clockwise, but still inward near ground level. Guess which way the air spirals in high pressure areas in the SH! The high and low pressure areas are typically 1000 kilometers (km) or more in diameter.

**4.** Since cool air can "hold" less water vapor than warm air, rain and other forms of precipitation are caused by cooling the air.

When the air is warm, the water on the ground will evaporate. Water vapor is a gas and is every bit as "dry" and invisible as the air. A great deal of vapor can remain in the air if the air is warm, but only a small amount if the air is cold. Therefore, precipitation (rain or snow or hail, etc.) occurs when air is cooled sufficiently to make some of the vapor condense back to its liquid (water) or sublime to its solid (ice) form and fall to the ground.

This is basically the same process as that which causes freezers to need defrosting periodically. Warm air holding much vapor rushes into the freezer when the door is opened. Inside the freezer the air is cooled considerably, and the vapor *sublimes* to ice or frost on the freezer walls.

**5.** Pressure in the atmosphere decreases with increasing height.

The average pressure of the atmosphere at sea level is approximately 1000 **millibars** (hereafter written as mb); this is almost equal to the weight of 10,000 kilograms (kg), for every one square meter column of air extending to the top of the atmosphere. The atmospheric pressure is therefore due to the weight of the air above you. As you ascend, there is less air above you and hence lower pressure.

**6.** Decreasing the air pressure causes the temperature to drop.

When you pump up a tire or a basketball the pump gets very hot. A general rule is that when air pressure is increased the temperature rises. Conversely, when air pressure is decreased the temperature drops. However, you must be careful.

**Warning**
This does *not* mean that the air in high pressure areas is warm and the air in low pressure areas is cold. In fact, some of the coldest air is found in high pressure areas! High or low pressure implies almost nothing about temperature; but the *process* of increasing pressure implies that the temperature of a particular blob or **parcel** of air is rising, and the *process* of decreasing pressure implies that its temperature is falling.

**7.** Clouds and precipitation are caused by rising air; clear weather is caused by sinking air.

The most efficient way to cool a large mass of air rapidly is to lower its pressure. This is done by lifting the air since as it rises the pressure decreases. As a result the temperature decreases and so does the amount of vapor that the air can hold. The vapor rapidly condenses and clouds (which are composed of billions of tiny water droplets or ice crystals), and sometimes precipitation, result.

*Virtually all clouds and precipitation are due to rising air.* On occasion, the air can be cooled sufficiently by other processes to form clouds or fog, but it rarely leads to precipitation.

Conversely, when air sinks its temperature rises so that its capacity for holding vapor increases. Then any cloud droplets would tend to evaporate and the clouds would disappear.

Now you are ready to learn a little bit about some of the world's climates. Since air generally rises near the equator, there should be plenty of clouds and rain. Since the air in the subtropics generally sinks, the weather should be dry and even desertlike (see Figure 1.7). These facts are confirmed by the world rainfall maps (see Figures 12.2 and 12.3).

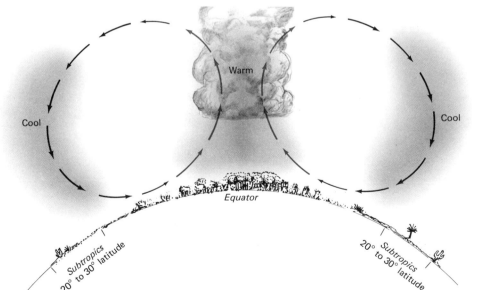

FIGURE 1.7
The rising air near the equator produces abundant rainfall so that the world's tropical rain forests (jungles) are located there. The sinking air of the subtropical latitudes causes dry conditions so that many of the world's deserts are located between latitudes 20 and 30°.

**8.** Rising air in low pressure areas causes clouds and precipitation; sinking air in high pressure areas causes clear weather.

Since the air near the ground spirals inward toward the center of low pressure areas, the pressure would rapidly increase if the air didn't leave the center by rising. When the pressure remains lower than that of the surroundings, you can rest assured that the air which spiraled in toward the center of the low also rises and therefore cools. This leads to clouds and precipitation.

On the other hand, air near the ground spirals out of high pressure areas and, if the high pressure is to be maintained, air from above must sink to take its place. The sinking air warms and the air clears (see Figure 1.8).

Now that you know the "secret" of meteorology, never forget it.

# TV WEATHER REPORTS AND THE WEATHER MAP

Millions of people watch the weather reports on the TV news program each day. Many of us watch a particular news program simply because we feel that it gives the best weather presentation. Yet few people really seem to understand what the TV meteorologists are talking about. In this section we will look at the main elements of a TV weather report.

TV meteorologists generally talk about four things, besides giving the current weather conditions and the forecast. They talk about (1) high pressure areas, (2) low pressure areas, (3) cold fronts, and (4) warm fronts. These topics are illustrated by showing weather maps and satellite pictures.

Since you will almost be a professional meteorologist by the end of this book, you will have to learn how to read weather maps. You must begin by learning how to read the information at each weather station; the **station model** is presented in Figure 1.9; it tells the wind direction and speed, the amount of clouds, the pressure and pressure change in the last three hours, the type of weather (i.e., rain, snow, etc.), the temperature, and something called the dew point temperature. In Figure 1.9 all of the information is presented in a standard format, with the temperature always on the upper left and pressure always on the upper right. Each weather map may contain 100 or more weather stations—which is a considerable amount of information. At first it may appear overwhelming, but as soon as you are familiar with the station model the weather map will no longer look so strange.

What about the main features on the weather maps?

You already know a little about the high and low pressure areas. You know that high pressure areas imply sinking air which implies clear weather and that low pressure areas imply rising air which implies cloudy, wet weather. You also know that these pressure systems are typically a thousand or so kilometers in diameter.

Many people ask just how high the pressure has to be in

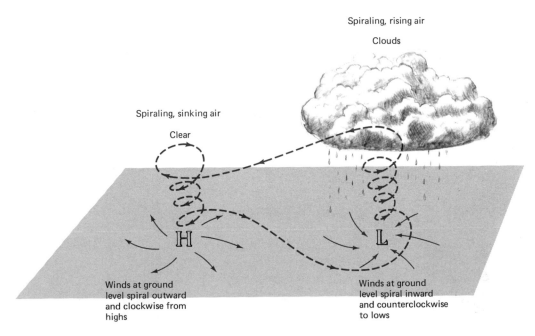

FIGURE 1.8
A simplified drawing of the spiraling winds and vertical motions in high and low pressure areas. Air near the ground spirals counterclockwise, converging into low pressure areas where it then rises, producing clouds and rain or snow. Air near the ground spirals clockwise, diverging outward from highs. Thus, sinking air and mostly clear weather dominates in highs.

Spiraling, rising air

Clouds

Spiraling, sinking air

Clear

Winds at ground level spiral outward and clockwise from highs

Winds at ground level spiral inward and counterclockwise to lows

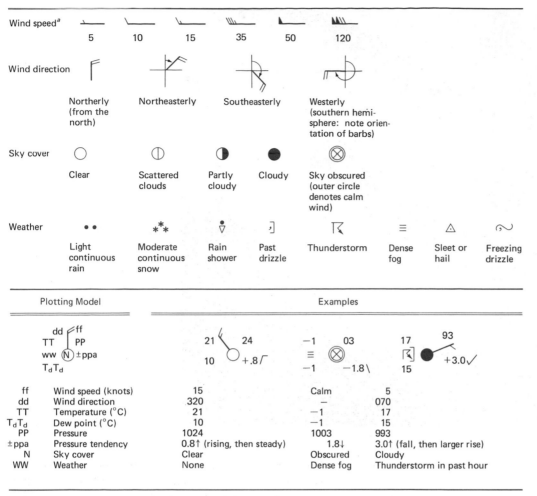

| Plotting Model | Examples | | |
|---|---|---|---|
| ff | Wind speed (knots) | 15 | Calm | 5 |
| dd | Wind direction | 320 | — | 070 |
| TT | Temperature (°C) | 21 | −1 | 17 |
| $T_dT_d$ | Dew point (°C) | 10 | −1 | 15 |
| PP | Pressure | 1024 | 1003 | 993 |
| ±ppa | Pressure tendency | 0.8↑ (rising, then steady) | 1.8↓ | 3.0↑ (fall, then larger rise) |
| N | Sky cover | Clear | Obscured | Cloudy |
| WW | Weather | None | Dense fog | Thunderstorm in past hour |

[a]Units: knots (1 knot = 1 nautical mile per hour; 1.95 knots = 1 m/s). 60 nautical miles = 1 deg of latitude.

FIGURE 1.9
A simplified weather station model, complete with interpretation and some standard weather symbols. A more complete list of weather symbols is given in Appendix 1.

order to have a high pressure area. The answer is that there is *no* specific number. The average pressure of the atmosphere at sea level is 1013 mb (millibars are used by meteorologists as their favorite pressure unit). Typical high pressure areas will have central pressures anywhere between 1020 and 1050 mb, while typical low pressure areas will have central pressures anywhere between 970 and 1010 mb.

High and low pressure areas are determined in a relative manner. Any time the pressure in a certain region is higher than the pressure of the surrounding areas, we have a high pressure area. Any time the pressure in a certain region is lower than that of the surroundings, we have a low pressure area. I have even seen low pressure areas with central pressures of 1025 mb! They were called low pressure areas simply because the atmospheric pressure all around them was higher than 1025 mb.

You must now learn something about fronts. The term **front** originated during World War I when soldiers on both sides of the war spent months in trenches along the battle front. The word "front" therefore implies a line of conflict. In meteorology, *a front is a narrow zone across which the temperature, humidity, and wind change abruptly.* Although the air on each side of the front will eventually mix and the contrasts will be destroyed, the very process of mixing can be quite violent. Weather at or near fronts can be quite severe. Much of our rain, snow, thunderstorms, and tornadoes occurs near fronts (see Figure 1.10).

As you travel from the equator to the poles the weather will naturally become colder. However, the temperatures will *not* drop steadily, but rather will fall in an abrupt and irregular manner. For instance, you might travel poleward for 2000 km and the temperature will drop by only 5°C (C

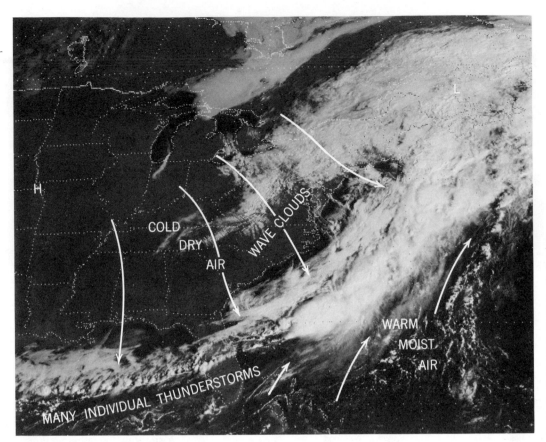

COLD
DRY
AIR

WAVE CLOUDS

WARM
MOIST
AIR

MANY INDIVIDUAL THUNDERSTORMS

H

L

FIGURE 1.10
The satellite picture of the cold front of 1330 GMT, October 3, 1977. The small round cloud blobs over Florida and the Gulf of Mexico are actually thunderstorms. Notice the cloud bands over the Appalachian Mountains. These often form when strong winds blow across and over the mountain ridges and are known as mountain wave clouds.

stands for Celsius or centigrade), but in the next 50 km it might drop by as much as 20°C or more.

How sharp does the temperature contrast have to be before the transition zone is called a front? Temperature contrasts vary dramatically from one front to another so there is no simple answer to the question. Over land in midlatitudes a typical temperature difference might be 5°C across 200 km. Temperature differences tend to be much smaller across fronts in the subtropics or anywhere out at sea.

Since fronts are narrow zones separating warm air from cold air, the following question is nonsensical—"Which is colder, a cold front or a warm front?" Fronts are merely *boundaries* separating warm and cold air masses and therefore have no temperatures of their own.

What then is the difference between a warm front and a cold front? The answer is simple. A cold front is a front in which the cold air is advancing, whereas a warm front is a front in which the warm air is advancing. This means that the temperature will drop when a cold front passes by and rise when a warm front passes by. (For the moment we will ignore the occasional exceptions to this rule.) The temperature changes are usually more abrupt at cold fronts.

Fronts invariably run through low pressure areas. Almost all low pressure areas found outside the tropics (the so-called **extratropical cyclones**) have fronts running right through their centers. Fronts also tend to skirt between high pressure areas. Meteorologists therefore often

FIGURE 1.11
Symbols for the different types of fronts. Double arrows indicate the direction of motion of the fronts.

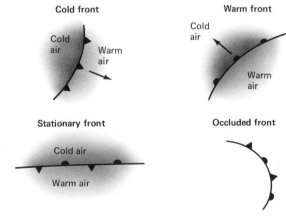

Cold front

Cold
air

Warm
air

Warm front

Cold
air

Warm
air

Stationary front

Cold air

Warm air

Occluded front

speak of frontal **troughs** (i.e., valleys) since the fronts are essentially located in "valleys" of low pressure.

Fronts are depicted on weather maps by using heavy lines. Cold fronts have spikes, while warm fronts have semicircles on the lines. These features always protrude in the direction toward which the fronts are moving. The depiction (see Figure 1.11) for stationary fronts (guess what that means) and occluded fronts (which you will learn about later) are also shown.

A simplified version of a classical weather situation for the United States is shown in Figures 1.12 and 1.13. These show the weather at 0300 GMT (GMT stands for Greenwich Mean Time, which is 5 hours ahead of EST or Eastern Standard Time) on December 5 and 6, 1977. On each map a cold front and warm front are seen as meeting in the center of the low pressure area. The warm front and cold front on each map together act to separate one large mass of cold air in the north from another large mass of warm air in the south.

As is typical, a large area of rain or snow, several

Surface Weather Map 0300 GMT December 5, 1977

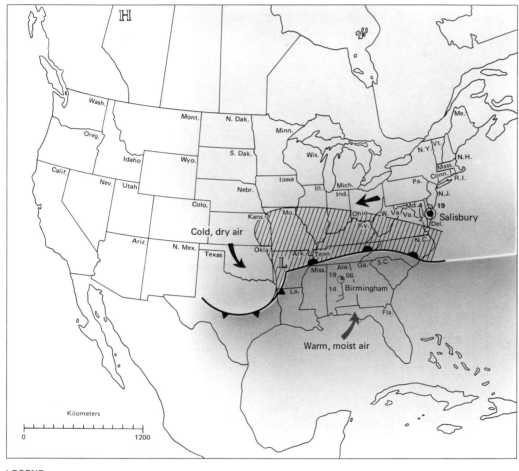

FIGURE 1.12
Surface weather map at 0300 GMT, December 5, 1977. At this time Birmingham, Alabama, is warm while Salisbury, Maryland, is cold. The hatching always indicates the general region of precipitation for surface weather maps.

LEGEND

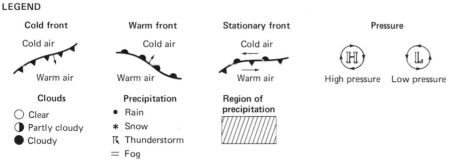

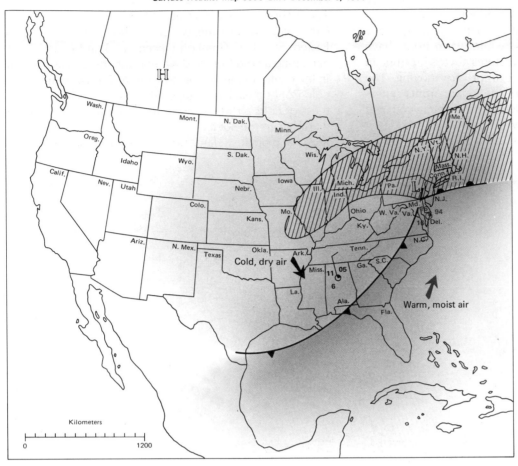

## LEGEND

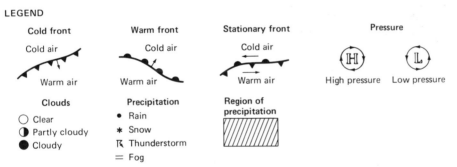

**Cold front**

Cold air

Warm air

**Warm front**

Cold air

Warm air

**Stationary front**

Cold air

Warm air

**Pressure**

High pressure    Low pressure

**Clouds**

○ Clear
◐ Partly cloudy
● Cloudy

**Precipitation**

● Rain
∗ Snow
℞ Thunderstorm
= Fog

**Region of precipitation**

FIGURE 1.13
The surface map 24 hours later. Birmingham has become cold, whereas Salisbury has become warm. Note that the low (L) has moved to the northeast, while the high (H) has moved to the southeast.

hundred kilometers wide occurs to the north of the warm front. The precipitation associated with the warm front typically lasts 12 to 24 hours. Along the cold front there is often a narrower band (50 to 200 km wide) in which showers or thundershowers may occur.

By comparing Figures 1.12 and 1.13 you should notice that sometime between December 5 and 6 the warm front passed through Salisbury, Maryland, and the temperature rose, while a cold front passed through Birmingham, Alabama, and the temperature fell.

Here is a question for you. If the fronts continue to move at the same rate, when do you think the cold front will pass through Salisbury? Perhaps the satellite picture (Figure 1.14) for later that day will help!

You should also notice that the low pressure area has moved toward the northeast. The high pressure area also moved, but it moved toward the southeast. In the United States low pressure areas typically move toward the northeast, whereas high pressure areas typically move toward the southeast.

12    INTRODUCTION

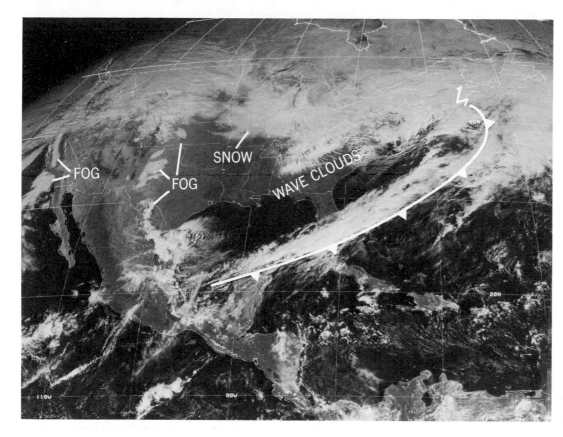

FIGURE 1.14
The satellite picture of 1700
GMT December 6, 1977.

On the average, low and high pressure areas move with speeds of approximately 40 km/h (per hour) or 1000 km/day. During the winter they move somewhat faster than this, and during the summer they move somewhat slower.

In summary, not only do the winds blow in spiraling patterns around high and low pressure areas, but the pressure areas themselves also move. The areas of rain and snow move with the low pressure system as a whole and not with the winds.

## Warning

The speed and direction in which high and low pressure areas move bear *no* simple relationship to the speed and direction of the wind itself. You may find that this sounds confusing at first because tricky concepts are involved. Fortunately, there is a simple analogy to this situation that can help to clarify matters. If you spin a top on the floor, the top may move across the floor in any direction and at any speed. On the other hand, the top may not move at all, no matter how hard you spin it.

For the moment envision a low pressure area as a top that (1) spins counterclockwise when you look down on it and (2) tends to move toward the northeast. The speed

of rotation of the top can be thought of as the wind speed, while the rate at which the top moves across the floor can be thought of as the speed at which the low pressure area as a whole moves (see Figure 1.15).

Many have found this subtle distinction somewhat confusing. In 1743, Benjamin Franklin and his brother exchanged letters in which they described the weather during an eclipse. Benjamin complained that it became cloudy shortly before the eclipse; yet, his brother noted that clouds did not roll in until several hours afterward. Benjamin was living in Philadelphia, while his brother was living in Boston. Now, it is a well-known fact that during stormy weather in Philadelphia and Boston the wind usually blows *from* the northeast. Therefore, some people at that time thought that storms also came from the northeast. In fact, so little was known about the nature of storms at that time that many people thought that storms arose spontaneously out of nowhere. You can imagine how surprised Benjamin Franklin was when he subsequently found out that storms almost always hit Philadelphia first and then moved *toward* the northeast and hit Boston almost a day later (see Figure 1.16). The storms actually moved in the opposite direction of the winds!

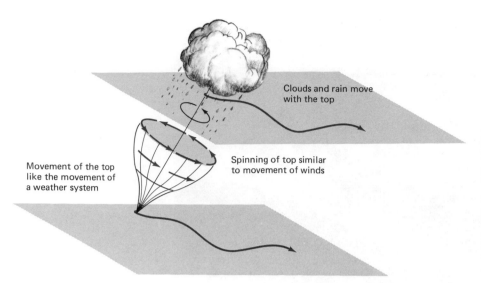

FIGURE 1.15
Think of a storm (a low pressure area) as a counterclockwise spinning top that moves across the floor. The spinning of the top represents the winds, while the movement of the top represents the motion of the storm and clouds. Then you can see that the movement of the storm is independent of the wind direction.

# CONTOUR MAPS

In the study of meteorology it is absolutely necessary to use maps. Unfortunately, many people have trouble reading maps. Their difficulty probably stems from the fact that while a scene is three-dimensional, the map used to represent the scene is two-dimensional. A simple map will depict some of the features of the ground such as rivers, lakes, and mountains. In the study of meteorology, maps called contour maps are often used to depict quantities such as temperature and pressure. These maps are even more abstract than regular road maps, and most people find them almost impossible to understand at first. I hope some of the mystery involved in understanding regular and contour maps will be removed in the following paragraphs.

The secret about most maps is that they depict what the ground features look like from *above*. In other words, maps give you a bird's eye view of the terrain. A map therefore shows you what you would see if you looked down on the earth. This is precisely why pictures taken from satellites look just like maps. Maps generally conform to one additional *convention*—the top of the page usually points northward.

Now, imagine that you are searching for a hidden gold mine. The map in Figure 1.17 will help you find the mine (marked ✱). You can see that the mine is located somewhere to the southwest of Gold Hill. If you lived in Fools Gold City, then you would have to cross Rapid River in order to get to the mine. The map is worth at least a hundred words of description, but a picture is still worth a thousand words. A picture indicates details and features in the landscape in such a way that almost anyone can recognize them.

FIGURE 1.16
Ben Franklin's amazing discovery that storms tend to move *toward* the northeast even though the winds during storms tend to blow *from* the northeast.

Day #1

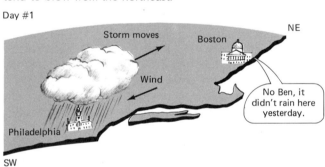

Day #2

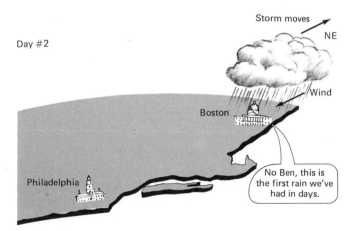

**Map**

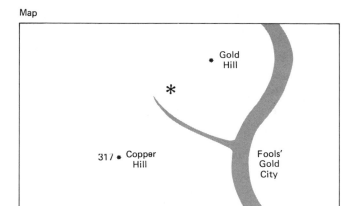

Gold
Hill

*

31 / * Copper
Hill

Fools'
Gold
City

**Scene**

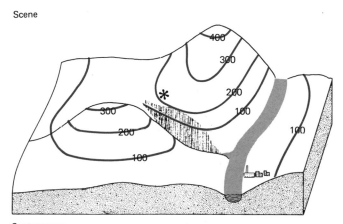

**Contour map**

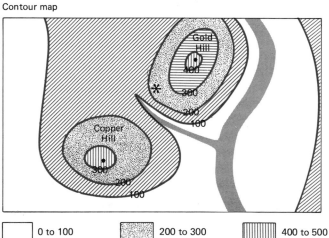

| | 0 to 100 | | 200 to 300 | | 400 to 500 |
| | 100 to 200 | | 300 to 400 | | |

FIGURE 1.17
A regular map, a drawing, and a contour map, all depicting the same situation. The contour map contains lines called contour lines, which are lines of equal height. For example, the 400 meter (m) contour line shows where the land is exactly 400 m above sea level. Everywhere inside the line, the land is more than 400 m above sea level; while everywhere outside the line, it is less.

If pictures are so much better than maps, why don't we use pictures instead? We use maps in the cases where pictures have disadvantages. No picture could accurately show the rivers and mountains of the United States, simply because the United States is too large. A picture may be better for a local scene, but a map is often better for a large and complex scene. Moreover, say that you wanted to depict an abstract quantity such as temperature at various places in the United States. No one has yet been able to draw a picture of temperature because it is abstract and therefore invisible.

The height of the land or the temperature of the land is conveniently depicted by using **contour maps.** At first a contour map makes little sense; but once you have learned how to read it, it shows all the details of the landscape. As I have mentioned, contour maps are essential in the study of meteorology. They provide the meteorologist with an accurate picture of the pressure patterns, the temperature patterns, and the patterns of all the other quantities that might be needed.

How do you read a contour map? A contour map, like a regular map, is what the land looks like when viewed from above. The contour lines shown on the map are lines of equal height.

**Rule 1:** Every point on a given contour line has the same value (of height above sea level).

When you are drawing a weather map you will not draw lines for height above sea level; instead, you will draw lines for pressure. These lines are called **isobars,** and every point on a given isobar has the same value of pressure. Similarly, every point on a given **isotherm** has the same value of temperature.

Contour lines are drawn at specific intervals. For instance, when height contours are drawn the height interval is often 100 m (meters). Therefore, a contour line is drawn from 100 m above sea level, another for 200 m above sea level, and so on. Isobars are often drawn at pressure intervals of 4 mb, whereas isotherms are often drawn at temperature intervals of 10 degrees.

**Rule 2:** Every contour line separates regions with greater values than on the line itself from regions with smaller values than on the line itself.

In the drawing and contour map of Figure 1.17, the land elevation on Gold Hill is greater than 100 m above sea level at all points within the 100 m contour and less than 100 m above sea level at all points outside the 100 m contour. The hidden gold mine ✱ is located between the 200 m contour and the 300 m contour. We therefore know that the mine lies at an elevation between 200 and 300 m above sea level. Can you figure out the elevation of Fools Gold City?

Another feature that contour maps show is the slope of the land. Rapid River lies in a small canyon that is about 200 m deep. The canyon walls rise steeply from the river banks. On the contour map we see that along the river the contour lines are close together. This leads us to Rule 3.

**Rule 3:** The closer the contour lines, the steeper the slope or the larger the gradient.

The word **gradient** means almost the same thing as the word **slope,** but gradient is a more general term. The slope of the land is the difference in elevation between two places divided by the *horizontal* distance between them. The gradient of a quantity is defined as the difference of the quantity at two points divided by the distance between them. When expressed in terms of an equation, we can write

$$\text{Gradient of a quantity} = \frac{\text{difference of the quantity between two points}}{\text{distance between the two points}}$$

---

Example 1.1

The pressure in Boston is 1000 mb when the pressure in New York City is 1010 mb. Boston is 320 km away from New York City. What is the pressure gradient?

The solution of this difficult problem involves no more than simply substituting numbers into the above equation.

Substitute:

$$\text{Pressure gradient} = \frac{10 \text{ mb difference}}{320 \text{ km}}$$

Solution
0.031 mb/km

---

The pressure gradient is extremely useful in meteorology because it is closely related to the wind speed. Generally, the larger the pressure gradient, the larger the wind speed. The pressure gradient in the example above is quite large and would be associated with wind speeds of approximately 80 km/h. The gradients of many other quantities also provide the meteorologist with much useful and important information.

It is also informative to depict the temperature gradient by drawing isotherms. When isotherms are drawn for the weather map of December 6, 1977, you can see how they help to locate the fronts (see Figure 1.18). The fronts are found at the southern border of the region where the isotherms are closely spaced. The isotherms are far apart in the warm air to the south of the fronts, simply because there is hardly any temperature contrast there.

**Rule 4:** The shape of the contours indicates the shape of the map features.

Round contours indicate either round hills or bowllike depressions in the land, while V-shaped contours indicate either valleys or ridges. The isobars for December 6, 1977 are shown in Figure 1.19. Let us examine several features that are indicated by the isobars. The center of the low pressure area has a pressure that is lower than 992 mb, but higher than 988 mb; otherwise, the 988 mb isobar would have been drawn. The high pressure area has a central pressure that is higher than 1040 mb but lower than 1044 mb; otherwise, the 1044 mb isobar would have been drawn.[1]

The isobars surrounding the cold front are distinctly V-shaped, as you should expect, because most cold fronts are located in low pressure troughs.

Notice also that the isobars make the high pressure area look like a large drop of molasses about the drip off the bottom of a spoon. Indeed, it depicts a giant "drop" of cold air oozing southward.

---

[1] On *all* surface weather maps, the pressure values you read are the so-called sea-level pressure values. The actual pressure at each station depends both on its altitude above sea level and the weather. Standard formulas are used to "adjust" the station pressure to an expected value at sea level, thereby eliminating the altitude dependence.

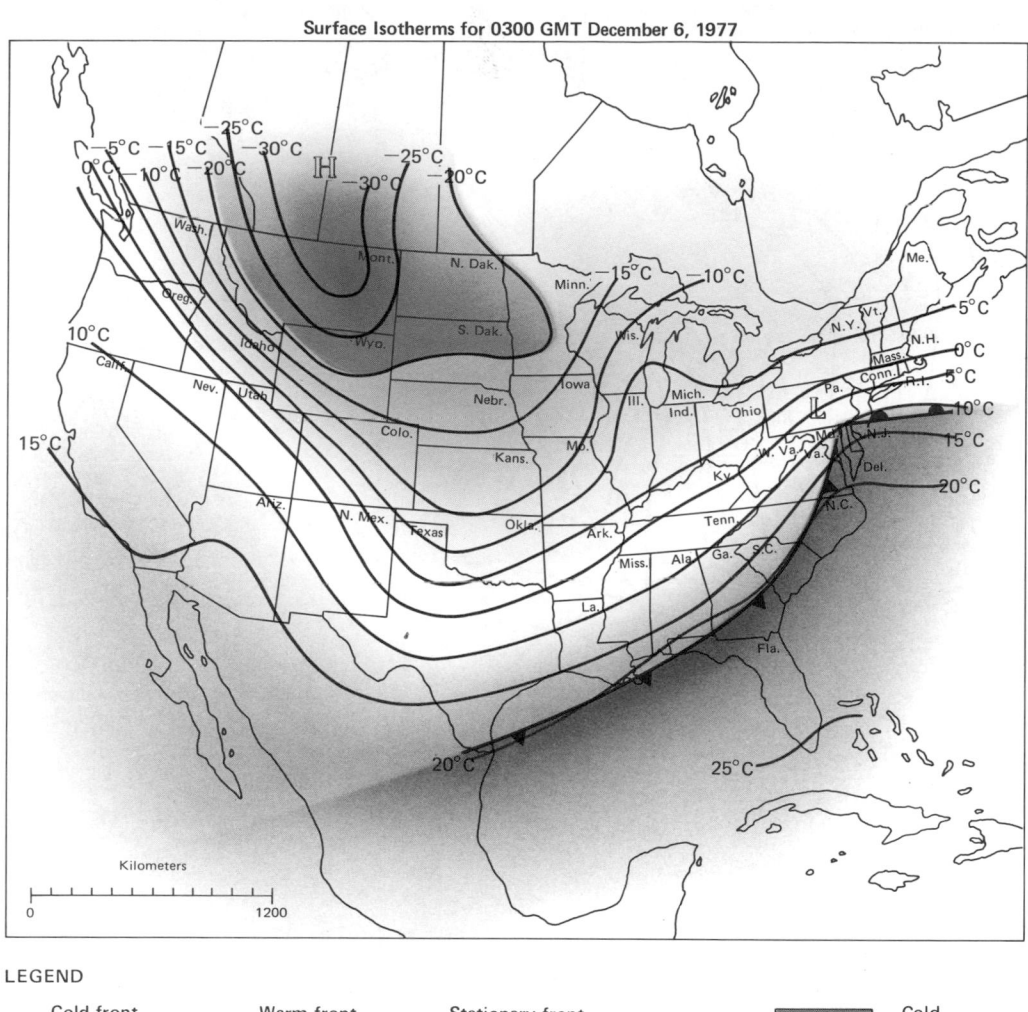

Surface Isotherms for 0300 GMT December 6, 1977

## LEGEND

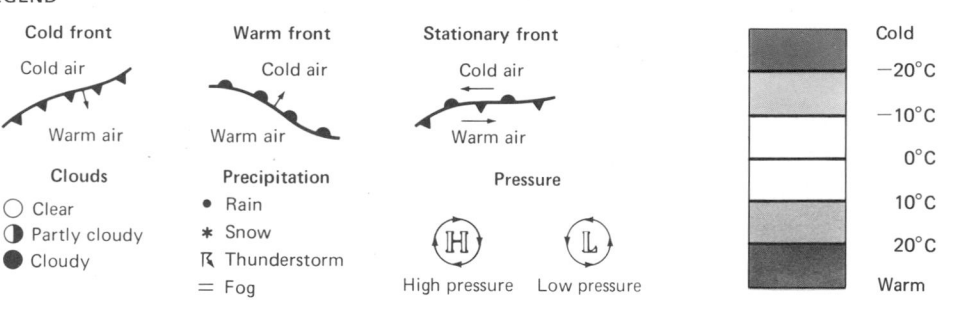

**Cold front**

Cold air

Warm air

**Warm front**

Cold air

Warm air

**Stationary front**

Cold air

Warm air

**Clouds**

○ Clear
◐ Partly cloudy
● Cloudy

**Precipitation**

• Rain
＊ Snow
Ⓡ Thunderstorm
＝ Fog

**Pressure**

Ⓗ
High pressure

Ⓛ
Low pressure

Cold
−20°C
−10°C
0°C
10°C
20°C
Warm

FIGURE 1.18
The surface weather map of
0300 GMT, December 6,
1977, showing the isotherms
at intervals of 5°C. Notice that
the isotherms are very helpful
in finding the fronts.

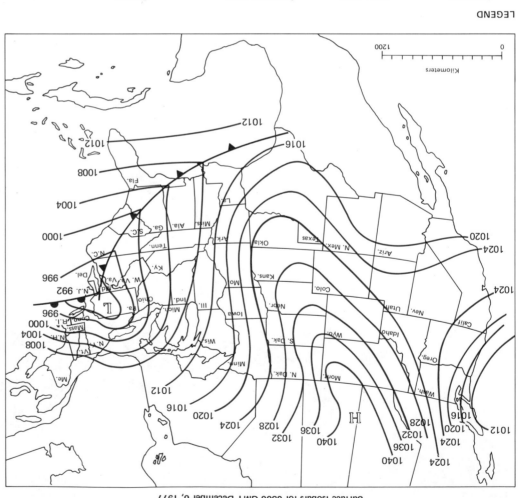

Surface Isobars for 0300 GMT December 6, 1977

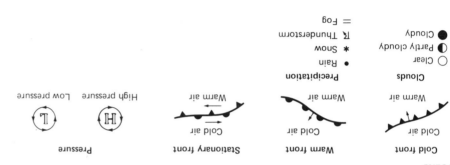

**LEGEND**

| Cold front | Warm front | Stationary front | Pressure |
|---|---|---|---|
| Cold air ▲▲▲ Warm air | Cold air ●●● Warm air | Cold air ▲●▲● Warm air | High pressure Ⓗ Low pressure Ⓛ |

**Clouds**
○ Clear
◐ Partly cloudy
● Cloudy

**Precipitation**
● Rain
✳ Snow
℞ Thunderstorm
≡ Fog

**FIGURE 1.19**
The surface weather map of
0300 GMT, December 6,
1977, showing the isobars at
intervals of 4 mb.

## PROBLEMS

**1-1** Describe the most severe weather event you have ever experienced.

**1-2** Look at Figure 1.3. Compare the size of the hurricane to the size of India. What is the hurricane's approximate diameter in kilometers?

**1-3** Search through the literature written before A.D. 1600 and find an "explanation" for some meteorological phenomenon. Compare it with the explanation given in this book.

**1-4** Explain why a flame on a candle always points upward no matter how the candle is turned.

**1-5** Look at Figure 1.6 and describe how air spirals around high pressure areas in the SH.

**1-6** Will the air coming out of an automobile tire feel cold or hot? Explain and then perform this experiment (without letting out too much air).

**1-7** Can you see water vapor?

**1-8** Explain why there is no contradiction between the two statements: (1) warm air rises, and (2) rising air cools.

**1-9** Air blowing from the northeast passes over a mountain range. Where will most of the rain fall: (a) on the northeast side, (b) on the southwest side, (c) on the top. Explain.

**1-10** Watch the TV weather reports several days in a row and see how fronts, highs, and lows move. Compare the weather patterns of the TV weather map with Figure 1.12 or Figure 1.13.

**1-11** By reading the station model of St. Louis, Missouri, in Figure 17.31, give the temperature, dew point temperature, pressure, cloudiness, weather, wind direction, and wind speed.

**1-12** Does the air rise more rapidly (a) to the north or (b) to the south of warm fronts? Explain.

**1-13** By using the isotherms of Figure 1.18, what do you estimate the temperature to be at Little Rock, Arkansas?

**1-14** Using Figure 1.19, estimate the pressure gradient in Arkansas in units of millibars per kilometer (mb/km).

**1-15** Considering that fronts lie in pressure troughs, explain why rising pressure is one indication that a cold front has just passed.

**1-16** Explain how the wind can be blowing from the northeast at the same time that the storm is moving toward the northeast.

# CLOUDS

At 5:30 A.M. on July 16, 1945, the Atomic Age was blasted into existence. The first atomic bomb was exploded in the desert of New Mexico on that morning, although the world was not to find out about the atom bomb for another few weeks when the second bomb was dropped on Hiroshima, Japan.

The few scientists and soldiers permitted to witness that first secret explosion were astounded by what they saw. The enormous energy of the atom was released in a flash of light, and a giant fireball about 2 km in diameter rose and expanded into the sky. Finally, the giant mushroom-shaped cloud took form—the cloud that represents death and destruction to so many people. Even so, it is not quite so distinctive as you may think (see Figure 2.1).

Each cloud represents the signature of the winds that produced it. Rising air produces the majority of clouds and the resulting cloud shapes follow directly from the *patterns of rising and sinking air.* It is probable that no two clouds have ever been exactly alike but, since there are only a small number of basic patterns of air motion, there can be only a few *different basic cloud forms.*

Many people are not fully aware of the variety of cloud forms and shapes, and the notion that clouds inevitably must make a day gloomy is more cliché than fact. By watching the sky day after day you will soon see that while clouds sometimes do appear ominous and gloomy, they are also capable of exhibiting the most delicately beautiful shapes and forms. Some cloud forms occur in bad weather while, oddly enough, some indicate good weather to come! A knowledge of the state of the sky (i.e., the cloud forms) is actually very useful in weather predic-tion. In the days before weather was routinely reported, farmers, sailors, and other people who worked out-of-doors were able to forecast changes in the weather with considerable skill simply by watching the clouds.

In this chapter I will describe and explain the character-istic forms and shapes of clouds, and how they can be used to help predict the weather. When you read this

FIGURE 2.1
A nuclear explosion.

book will have become "clearer."

chapter you will probably find that the cloud pictures are not very meaningful at first. Therefore, make it a point to look at the sky each day and compare what you see to the pictures in this book. If the weather is cooperative, within about two weeks you will be able to identify most cloud types rather easily, and by then the cloud pictures in this

# CLOUD COMPOSITION AND CLOUD FORMS

Clouds are not solid objects. If you have ever been in a fog (literally), then you have been inside a cloud that touched the ground. Clouds are composed of countless millions of tiny water droplets or, if the air is cold enough, tiny ice crystals. The tiny droplets average about 10 microns in diameter (a **micron** is one millionth of a meter and each letter on this page is about 2000 microns high), and there are about 100 million of them in each cubic meter. Ice crystals tend to be much larger but are much less concentrated.

The droplets and crystals are carried along by the winds and therefore can easily pass over obstacles such as mountains. Therefore, if a cloud is seen on only one side of a mountain, it is *not* because the mountain has blocked the cloud; the tiny droplets or crystals simply evaporate once they sink down the other side of the mountain.

Why do clouds appear white when ice and water are both clear? This is easy to answer. Take an ice cube that is clear and smash it into tiny pieces; then notice that it has turned white. Smash glass into a pile of fragments and the same thing will happen. If you shine a red light on it, it will appear pink. If you put it in the shade, it will appear gray. Since the cloud is composed of millions of tiny droplets or crystals, it acts like the crystals of a smashed ice cube and reflects light in a confused irregular manner. Thus sunlight reflected from clouds is white most of the day, but near sunset or sunrise when the sun turns red the clouds bathed in direct sunlight turn pink or even red.

Around the year 1800 two men noticed that there were a limited number of cloud forms, and so decided to classify clouds. One of these men was Baptist Lamarck (1744–1829), who is better known for his work in biology. Lamarck's cloud classification was upstaged by the classification of Luke Howard (1772–1864), who was also trained as a biologist, and today we still use a modified form of Howard's original scheme. One advantage of Howard's classification was that he used Latin words.

Clouds form in three basic patterns (see Figure 2.2, 2.3, and Plate 1).

FIGURE 2.2
A cumulus cloud as seen from Longs Peak, Colorado.

FIGURE 2.3
A stratus cloud over a small island.

1. Curly or fibrous clouds are known as **cirrus** clouds.
2. Layered or stratified clouds are known as **stratus** clouds.
3. Lumpy or heaped clouds are known as **cumulus** clouds.

Clouds are also distinguished by the heights above ground level at which they form. Thus we have the following classifications:

1. High clouds whose bases usually lie more than 6 km above ground level (prefix **cirro**).

2. Middle clouds whose bases lie between 2 and 6 km (prefix **alto**).
3. Low clouds whose bases lie below 2 km (no prefix).
4. Clouds of vertical development (no prefix).

Several combinations can occur (see Figure 2.4). Often clouds appear in a distinct layer, but still appear lumpy. These are called stratocumulus. Then we also distinguish clouds that produce precipitation by adding the term **nimbo** or **nimbus**. Thus in table form we have the following descriptions:

|  | High | Middle | Low | Vertical |
|---|---|---|---|---|
| Curly | Cirrus (Ci) | — | — | — |
| Layered | Cirrostratus (Cs) | Altostratus (As) | Stratus (S), nimbostratus (Ns) | — |
| Lumpy | — | — | Cumulus (Cu) | Cumulonimbus (Cb) |
| Partly layered, partly lumpy | Cirrocumulus (Cc) | Altocumulus (Ac) | Stratocumulus (Sc) | — |

Adding to the list of Latin prefixes and suffixes, we find terms like **fracto** (fractured or broken) and **lenticularis** (lens-shaped), and other terms which are generally self-explanatory (that is, if you know Latin).

FIGURE 2.4
A schematic diagram of the different cloud types.

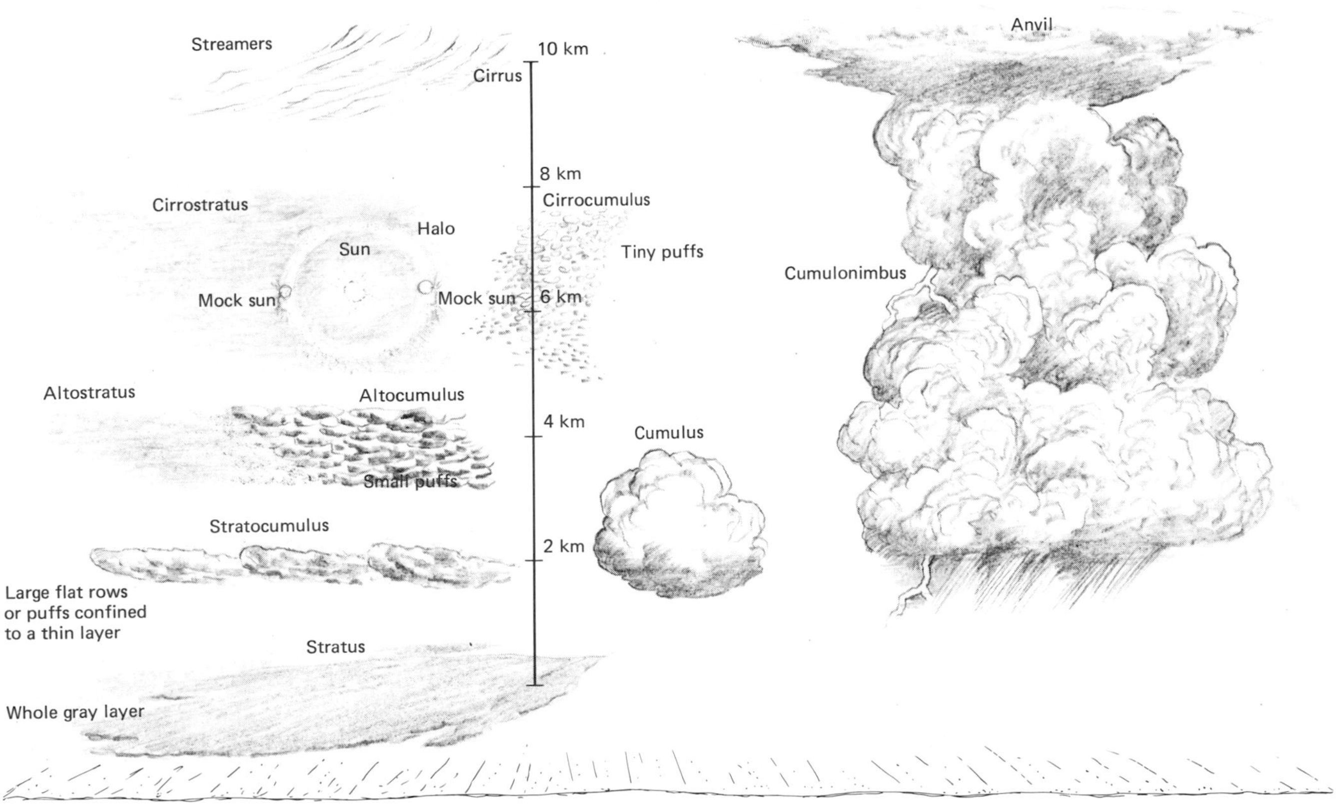

Different cloud forms

# CUMULUS AND CUMULONIMBUS

The mushroom-shaped cloud of the atom bomb is an extreme example of a cumulus cloud. A local blob of exceptionally hot air can form from an explosion. Since hot air is light, this blob will begin to rise and force its way through the surrounding, relatively quiescent atmosphere. While the air at the top of the blob is slowed down by resistance with the quiescent atmosphere, the air in the center of the blob meets less resistance and rises quickly. This results in the mushroom shape since the air near the top of the blob cannot move very rapidly upward, so that it is squeezed outward to form the cap, while the rapidly rising air below forms the stem (see Figure 2.5).

The cap of the mushroom can become very wide if the rising blob of air happens to rise into a layer of warmer air. The air near the top of the blob loses its buoyancy once the surroundings are just as warm and light but, since the air from the lower part of the blob will still be rising for a while, the top of the blob will be forced to spread out.

It is fairly common these days to see movies in which bombs explode, so next time you see one be sure to watch for the mushroom shape and notice how the smoky air moves outward and downward near the cap and

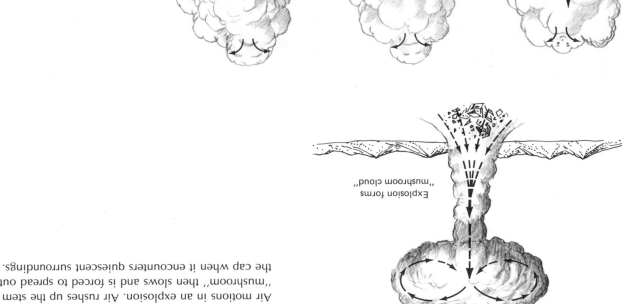

Explosion forms "mushroom cloud"

upward in the stem of the mushroom. For small bombs, the mushroom shape lasts only a few seconds.

After all this talk about mushrooms you may notice that most cumulus clouds more closely resemble a cauliflower (see Figure 2.6). This is because large cumulus clouds consist not merely of one blob of warm air (known technically as a **thermal**) but of several thermals. These thermals are familiar to both glider pilots and birds who search for such blobs of warm rising air to keep them aloft longer. If you watch the individual thermals or cauliflower (florets) of a large cumulus you will notice that as the flower itself moves upward, the lumps on the outside of the flower move outward and downward just as in the mushroom clouds of explosions (see Figure 2.7).

Thermals are, of course, easier to find if they become visible or, in other words, if they produce clouds. An essential question for us to answer is, "When will a thermal become visible as a cloud?" You should recall from the first chapter that as air rises its temperature drops. A blob of air will continue to rise as long as it is warmer than the surrounding air. However, as the temperature of the air in the thermal drops, it is less able to hold all its water vapor. Once the thermal reaches a certain height the water vapor in the air will begin to condense into millions of tiny droplets, and a cloud will form. The name for this critical height is the lifting condensation level or LCL, and you

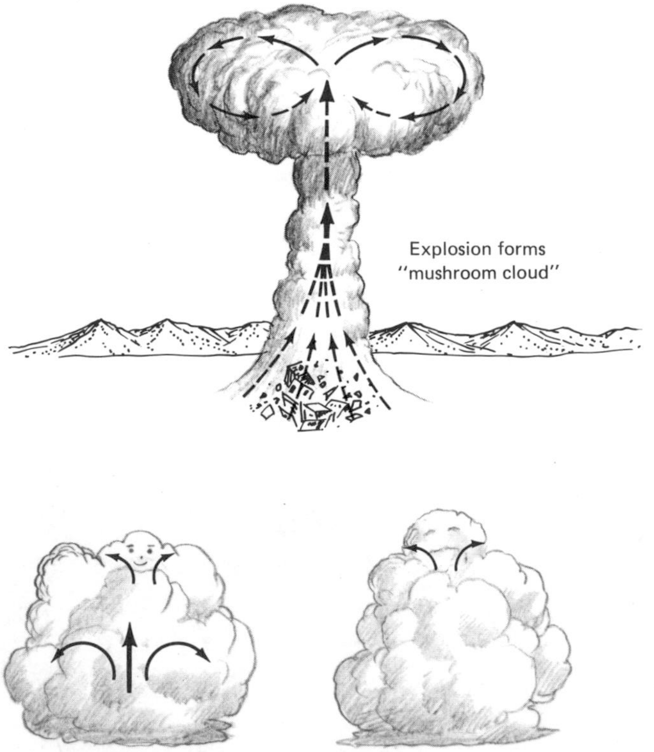

FIGURE 2.5
Air motions in an explosion. Air rushes up the stem of the "mushroom," then slows and is forced to spread out to form the cap when it encounters quiescent surroundings.

FIGURE 2.6
Air motions in a rising thermal. As the cauliflower grows the air moves upward, outward, and finally downward.

FIGURE 2.8
A cumulonimbus cloud with a large anvil extending downwind. Note the elevated dome where air rising has overshot like a fountain into the stratosphere. Notice also that there are no small cumulus clouds near the large cumulonimbus cloud!

will notice that cumulus clouds generally have quite flat bases (which occur at the LCL). If a thermal peters out before it reaches the LCL, there will be no cloud. Once a thermal reaches the LCL the cloud base will form, and the cloud will extend as high as the thermal is able to rise.

If a cumulus begins to grow quite large its appearance may suddenly change. The entire top may flatten and spread out. The cloud top will then smooth out and resemble a giant mushroom or anvil (see Figure 2.8). Whenever a cumulus cloud reaches this stage it produces rain, thunder, and lightning and is known as a **cumulonimbus** cloud or a thunderstorm. Cumulonimbus clouds are the castles of the sky. The fact that the top has spread out indicates that the cloud air has risen into surroundings that are noticeably warmer. This always occurs by the bottom of the stratosphere (see Chapter 7).

The anvil of the cumulonimbus cloud in Figure 2.8 is obviously lopsided. This is simply due to the fact that strong winds at that level are blowing the cloud particles

FIGURE 2.7
Time lapse photographs of a growing cumulonimbus cloud.

much excess heat near the ground, virtually all the rain falls from cumulus or cumulonimbus clouds.

At most inland locations cumulus clouds tend to form during the daytime when hot air is generated by sunlight. Notice how dramatically out of place cumulus clouds are in Magritte's surrealistic painting *The Empire of Light* (see Figure 2.10).

It is possible to write a "biography" of a typical day with cumulus clouds. In the morning the sky is clear and the ground is cold. It may even be foggy. As the sun rises the ground gradually grows warmer, the fog soon clears, and thermals begin to rise from the ground. At first they do not produce any clouds (because they do not reach the LCL), but generally by 10:00 or 11:00 A.M. small cumulus clouds begin to appear all over the sky (Plate 1). As the heat of the day continues to build up, some of these small clouds grow significantly larger; but in the competition for the hot air supply, most of the small clouds are not successful and die (i.e., sink and evaporate).

The few successful clouds (the survival of the fittest) continue to grow and reach maturity as proud cumulonimbus clouds in midafternoon. In their pride and youthful excessiveness they bring about their own destruction. First, they block the sunlight from reaching the ground, cutting off the necessary heat supply. Second, the very rain (and sometimes hail) they produce creates a weight that offsets the buoyancy of the air. The air stops rising, and the clouds begin to shrink and die. Then it is often too late to start again, because by this time the sun is setting and therefore the heat source is gone. The day ends with a sky of flattened, stooped, aged versions of the once proud cumulonimbus clouds known as **stratocumulus** clouds (see Figure 2.11). Then, soon after sunset even these last remnants disappear.

Thunderstorms are not always the end product of a day that begins with small cumulus clouds. The later in the day the cumulus clouds first appear, the less chance there is that they will grow very large. The appearance of small daytime cumulus clouds is a sign that the weather will most likely continue to be good for another day, since cumulus clouds tend to dissipate once the sun goes down. For this reason cumulus clouds are generally considered to be fair weather clouds, but you must always watch to see that they do not grow too large.

There are days when the sky is filled with a layer of very flattened cumulus clouds, that is, *stratocumulus*. Stratocumulus clouds also form when the hot air rising from the ground is not permitted above a certain level because of the overall sinking tendency of the air in a high pressure area, or because above that layer there is a layer

FIGURE 2.9
Mamma clouds at the base of a cumulonimbus cloud. This often indicates severe weather.

downwind. Often, in fact, entire cumulus or cumulonimbus clouds are tilted, because the wind speed varies with height.

There is one additional feature of cumulonimbus clouds that deserves mention at this point. In severe thunderstorms the underside of the cloud sometimes develops huge and ominous protuberances, appropriately enough called **mamma** (see Figure 2.9), that are most easily seen when the sun is near the horizon. The mamma are often a warning sign that all hell may soon break loose, that is, that large hail or even tornadoes may soon follow. Occasionally, the mamma are not seen until after the storm is over.

The next time that a thunderstorm occurs you should notice that it is invariably quite warm before it arrives, but the temperature drops suddenly once the storm begins. The thunderstorm is merely exchanging warm air for cold—as the warm air rises, cold heavy air from above sinks to take its place.

The details of this process are somewhat involved and will be discussed later; but the net effect of all the air movements is that cumulus and cumulonimbus clouds play an essential role in the heating of the atmosphere between 2 and 15 km and, conversely, in relieving the intense heat that would otherwise build up near ground level. It is estimated that if cumulus clouds were "turned off," and if the hot air near the ground were not permitted to rise, then the earth's surface would be about 10°C warmer.

In the tropical regions of the earth where there is so

of very warm air into which the rising thermals cannot penetrate (see Figures 2.12 and 2.13).

Clouds of similar shape that form at higher levels are known as **altocumulus** or *cirrocumulus* clouds (see Figure 2.14 and Plate 2). At first it may be difficult for you to distinguish between these three categories, but the experienced eye has little trouble. Generally, if you hold your hand at arms length, the stratocumulus cloud elements will be fist sized, the altocumulus cloud elements will be thumb sized, and the cirrocumulus puffs will be the size of your pinky nail.

Cumulus clouds are small clouds and generally range from a few meters to a few kilometers in height and width. Cumulonimbus clouds start from about 3 km in diameter and may on rare occasion exceed 30 km in width and 20 km in height—which is truly monstrous. The largest cumulonimbus cloud I have ever seen reached a height of 19 km above sea level, and this was the cloud that caused the Big Thompson River in Colorado to flood on August 1, 1976 (see Figure 2.15). More than 130 people drowned when the tremendous rains from this thunderstorm poured down the Big Thompson River Canyon at speeds

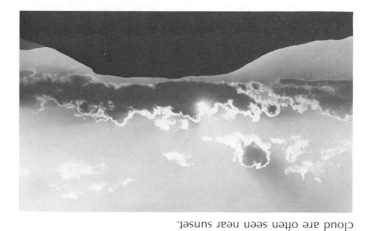

FIGURE 2.11
Stratocumulus clouds settling from afternoon cumulus. These cloud are often seen near sunset.

FIGURE 2.10
*The Empire of Light* by René Margritte (1950). Daytime cumulus clouds with a nighttime street scene. (Collection, The Museum of Modern Art, New York. Gift of D. and J. de Menil. Oil on canvas, 31 × 39.)

FIGURE 2.13
Stratocumulus clouds confined to a thin layer. These often appear in deep blue skies.

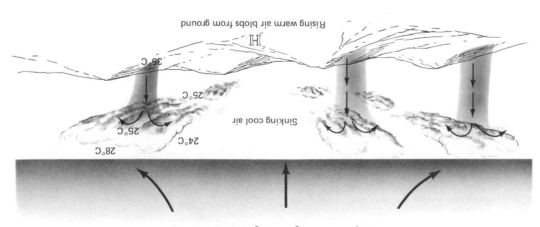

FIGURE 2.12
Stratocumulus clouds confined to a thin layer. This occurs when the warm sinking air of a high pressure area acts like a lid, just above the level at which the cloud forms.

Rising warm air blobs from ground

Sinking cool air

Very warm sinking air of high acts like a lid

common on tropical islands to see lightning over the ocean at night (see Figure 2.17 and 2.18). There is also an increased tendency for cumulus and cumulonimbus clouds to form over cities because of the extra heat produced in the cities.

Finally, cumulonimbus clouds are quite common over volcanoes and even over forest or grassland fires. We probably have all seen at least one movie in which a desperate farmer's crop was saved from fire at the last moment by a thunderstorm. It may very well have been that the fire itself provided the heat which triggered the thunderstorm, so that such last minute salvation is not necessarily completely fictional (see Figure 2.19). Unfortunately, reality often goes one step further, and it is quite possible that the same thunderstorm would also produce enough large hail to destroy the farmer's crop after all. Why don't they make a movie about that?

## STRATUS CLOUDS

The sunlit tops of *cumulus* clouds glisten bright white, while their shaded bottoms appear gray. With *stratus* clouds we see no such color variations for all we can see is the dull, gray bottom of the clouds. Whereas cumulus clouds are typically a kilometer wide, stratus clouds generally cover the sky from horizon to horizon. A single stratus cloud may literally be hundreds of kilometers in length and width. From the vantage point of space it is quite possible to see the entire cloud and think of it as beautiful (see Figure 1.14); but from the ground our view is limited, and we often think of stratus as ugly.

You will recall that clouds are produced when air is cooled sufficiently. In general, the most efficient way to cool the air is to make it rise; almost all clouds are produced in this manner. On occasion, low stratus clouds are *not* caused by rising air. These somewhat exceptional clouds are produced when the air is cooled by contact with a cold surface. Such low stratus clouds are actually all too common in certain regions of the world where the earth's surface is significantly colder than the air layer directly above. Thus, wherever cold ocean currents are found in the tropical and subtropical regions of the earth (such as coastal California), low-lying stratus clouds or even fog are common. This type of stratus cloud is also common over the sea ice in polar regions during the summer when the sun tries to heat the air but the sea ice cools it down. Stratus clouds produced in this manner rarely *lead to precipitation and ironically are often found in some of the driest regions on earth!*

exceeding 80 km/h and raised the water level more than 4.5 m in a few minutes. I saw this particular cloud from a distance of almost 80 km, and it still looked large.

Cumulus clouds may form anywhere in the sky but because they are caused by rising hot air blobs they will tend to form over those places where the ground is slightly warmer. Thus, during the daytime they tend to form over mountains (see Figure 2.16) rather than over nearby lowlands or valleys; they also tend to form over land (which is quickly heated by sunlight) rather than over water. At night these tendencies reverse, so that it is quite

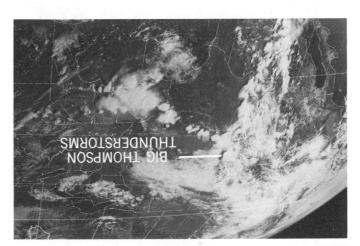

**FIGURE 2.15**
Satellite picture of the very large thunderstorm over Colorado which was responsible for the Big Thompson flood. Notice that this thunderstorm was one of several that formed along a front.

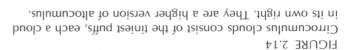

**FIGURE 2.14**
Circumcumulus clouds consist of the tiniest puffs, each a cloud in its own right. They are a higher version of altocumulus.

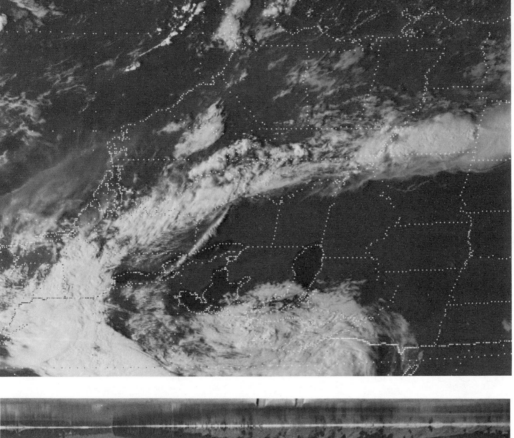

FIGURE 2.17
Satellite photo of 1231 GMT,
July 1, 1977, showing early
morning cumulonimbus
clouds over warm water
around Florida.

FIGURE 2.16
Cumulus clouds over the
Grand Tetons. Sunlight strikes
each mountain, heating the
rocks. This warms the air
which then rises up the
slopes. Thus, each mountain
has its own cloud.

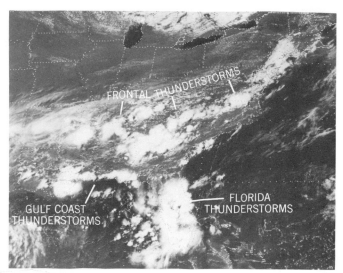

FIGURE 2.18
The satellite photo at 2030 GMT later the same afternoon. Even larger cumulonimbus now lie over the hot land of Florida, while over the water it is mostly clear.

Most other stratus clouds are produced when an entire layer of air is lifted. This does *not* imply that the entire atmosphere is rising but merely that one sloping layer of air is sliding upward. This can happen when the ground is sloped as in mountainous regions. More commonly it occurs when a warm air mass meets a cold air mass at a warm front. As the warm air mass pushes against the cold air mass, the cold air mass retreats slowly (and grudgingly) (see Figure 2.20). Some of the warm air, being light, slowly slides upward over the cold air mass. As it rises the excess vapor condenses into droplets or ice crystals, and soon snow and rain begin to fall from a single sheet of clouds that has spread across the sky. As the warm air slides higher and higher up over the dome-shaped mass of cold air, more and more vapor is wrung from the air. Not only do the clouds get higher but they also grow thinner (since as the air rises it is getting "wrung" dry like a sponge). Thus, the dark **nimbostratus** clouds that form when the air first begins to rise become graded—almost imperceptably to **altostratus** clouds and then to **cirrostratus** clouds. Finally, cirrus clouds form as the air nears the top of the cold air dome. By this point virtually all the vapor in the once warm, humid air has condensed and fallen toward the ground as either rain or snow.

What distinguishing features enable us to identify stratus clouds? The answer is—*no distinguishing features*. When the day is overcast and gloomy be assured that stratus clouds lie above. Surprisingly enough, it is easier to identify altostratus and cirrostratus clouds. Altostratus clouds have noticeably higher bases than stratus clouds

FIGURE 2.19
Lightning from a large cumulus cloud started a forest fire in the Bitterroot National Forest, Montana, on August 5, 1961. The heat from the forest fire then made the cloud become gigantic.

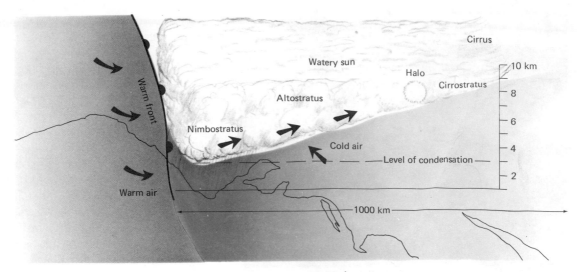

FIGURE 2.20
Schematic view of the clouds associated with a warm front. Compare this to the weather map in Figure 1.12

and are lighter gray. Often the sun can be vaguely seen through altostratus clouds as a round disk that is generally not bright enough to cast shadows. This is known as a "watery" sun (see Figure 2.21). Sometimes a corona (see next chapter) is visible directly around the sun or moon if the altostratus clouds are thin enough.

Cirrostratus clouds are even higher and thinner and are generally accompanied by halo phenomena (see next chapter), which constitutes proof that these clouds consist of ice crystals and not droplets. Cirrostratus clouds are transparent enough so that the sunlight can still cast weakened shadows (see Figure 2.22).

It has been known for countless centuries that a halo around the sun or the moon is a reasonably sure sign that rain or snow will follow within 24 hours. The halo itself has nothing to do with the oncoming precipitation, but the cirrostratus cloud (which produces the halo) does. Cirrostratus clouds usually appear in the cloud sequence shown in Figure 17.11. The entire weather pattern moves from west to east. Look at Figure 17.10 carefully and no-

FIGURE 2.21
Altostratus clouds can often be identified by "watery" suns or moons. They are favorites in horror movies.

FIGURE 2.22
Cirrostratus clouds are the clouds that produce halos.

tice that over Philadelphia at first there are cirrus clouds, then cirrostratus clouds (and a halo), then altostratus clouds, and finally rain from nimbostratus clouds. Therefore always remember the following rule.

**Rule:** If you see a halo around the sun or the moon expect rain or snow to begin within 12 to 24 hours. If the clouds seem to get thicker and lower after seeing the halo, be even more confident that rain or snow will follow soon.

This is perhaps the most reliable outdoor weather forecasting rule you can use.

# CIRRUS CLOUDS

The delicately fibrous clouds that appear high in the sky are the cirrus clouds. Cirrus clouds like almost all clouds are produced by rising air, but the shapes of the cirrus clouds are a result of the swirling winds beneath the cloud tops. Perhaps the best way to picture how the shapes of these delicate, silky clouds are produced is to think of them as military bombers. For if you look carefully (see Plate 1), you will see that each strand can be traced back to one small parent cloud (or kernel). The parent cloud produces ice crystals, which then fall out into air below.

The major part of the "cloud" is actually just a trail of falling ice crystals that has been twisted by the wind. The bomber is driven by jet engines, while the bombs that fall are gradually slowed down (in their forward motion) by

air resistance. Thus, the trail of bombs lags behind the plane. The cirrus clouds are driven by the enormous winds (100 knots or more) of the atmosphere at heights of 10 km, and as the crystals fall into the air below (where the wind is almost invariably slower) they lag behind the parent cloud (see Figure 2.23). If the wind direction also varies with height, the trails can get twisted and become quite complicated. Unfortunately, when there are too many long or twisted trails, or when the air below the clouds is hazy enough, the appearance of the clouds gets smudged and you can no longer see their delicate beauty.

Unlike bombs that eventually reach the ground, the ice crystal from cirrus usually evaporate and the trails end. It is interesting that trails of cirrus frequently begin with the condensation trails (**contrails**) made by jets (see Figure 2.24). The exhaust from the jets contains a number of pollutants and water vapor. When the water vapor is introduced into the surrounding frigid air, it often condenses (or rather sublimates) into ice crystals that then fall a considerable distance before evaporating.

Cirrus clouds are not merely beautiful; they may also have some interesting meteorological significance. They may actually "seed" clouds below them with ice crystals and thereby help to produce precipitation. This process will be explained in detail in Chapter 12. Meteorologists have artificially seeded some clouds in similar manner and have had some success in producing rain.

# CLOUD WAVES AND BANDS

Have you ever seen a flying saucer? Perhaps it would be more appropriate to ask if you have ever seen anything that looked more like a flying saucer than the cloud in Figure 2.25. How many people, do you think, have seen clouds like this one and then run raving like lunatics to report "proof" the earth is being invaded. Such flying saucer-shaped clouds are quite common over Mt. Rainier where, appropriately enough, the first modern "sighting" of a flying saucer was made.

The cloud in Figure 2.25 is an example of a mountain wave cloud (altocumulus lenticularis). Mountain wave clouds are in many ways the most fantastic of all cloud types in appearance. They generaly appear very smooth and lenslike. Mountain wave clouds are generated when strong winds blow across a mountain ridge, but they generally do not move. *This means that the air blows straight through mountain wave clouds!*

Does this seem mysterious and paradoxical to you? Doesn't it seem as if clouds should move with the wind? This apparent paradox makes it difficult for most people

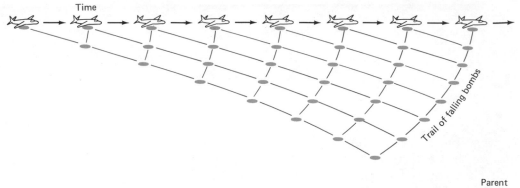

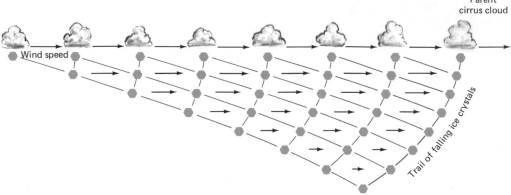

FIGURE 2.23
The streamers of cirrus clouds form when ice crystals fall out of the parent cloud and then are twisted by the wind. Since the wind is usually faster as you go higher, the falling crystals lag behind the parent cloud much like falling bombs lag behind the bomber.

FIGURE 2.25
Mountain wave cloud or flying saucer? These clouds formed over Mt. McKinley, Alaska.

FIGURE 2.24
X marks the spot where contrails from jets crossed. The contrails have spread somewhat and have become cirrus clouds.

to understand mountain wave clouds, but they are really quite simple to understand. Again, an analogy may be helpful. If you have ever seen water flow down a stream or even down the gutter of a road, you should notice that when the water flows over a large rock or other obstacle, foam is produced. The foam stays in the same place even though the water is moving downstream (see Figure 2.26). The foam is composed of constantly changing bubbles that are formed when waves break. The bubbles *do* move, but they burst before they get far downstream—this is why the foam itself does not appear to move. The foam is not a "solid" object but merely a *pattern* of bubbles. Similarly, the mountain wave cloud is not a solid object, but only a *pattern* of cloud droplets that literally flow through the cloud.

When air flows across a mountain ridge it is forced upward into a region where it is too cold and heavy. Once it crosses the ridge it sinks back because of its weight, but like a spring that has been stretched too far it overshoots; so, as the air moves downstream, it continues to oscillate (bounce) up and down (see Figure 2.27).

FIGURE 2.26
The foam **pattern** stays in one place even though the water and the individual bubbles flow downstream. Similarly, mountain wave clouds remain stationary even though the wind blows across the mountain ridges.

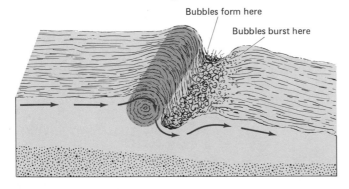

Bubbles form here

Bubbles burst here

A pattern of standing (not moving) waves forms in the air downstream from the mountain. The clouds form when the air in a moist layer rises high enough for condensation to occur. But as soon as the air sinks, the droplets evaporate and the cloud disappears.[1] The clouds are attached to the wave pattern, and a cloud is located in the crest of each wave while the trough of each wave is clear. The waves do not move, so the clouds cannot move.

Mountain wave clouds are used by glider pilots who know that strong updrafts can be found when mountain wave clouds are seen. Glider pilots have reached heights of over 11 km in the mountain waves of the Rockies. However, great caution and experience are necessary on these days for there are also strong downdrafts. Many inexperienced glider pilots have been killed by getting caught in downdrafts that have forced the glider to smash into a mountainside. A region of extreme danger and turbulence is called the **rotor** (see Figure 2.27). This is like a gigantic, breaking ocean wave and has been responsible for numerous crashes of gliders and engine planes as well. Under proper cloud and wind conditions, time-lapse photography has dramatically shown the incredible power of rotors.

Wave clouds (billows) can also form in the air far from mountains. Waves can be produced in the air when the wind speed varies with height (**wind shear**). Known as **gravity waves,** they are generated in much the same manner as the waves produced when wind blows over water. If wind shear occurs in a sufficiently moist atmospheric layer then clouds will form in the crests of the waves, and clear gaps will form in the troughs of the waves.

The clouds (or wave crests) tend to line up in a direction perpendicular to the wind (see Figures 2.28 and 2.29). These waves will move with the average speed of

[1] The clouds do not disappear so quickly when ice crystals form. The ice crystals do not evaporate as quickly as the droplets and therefore leave a trail (Plate 3).

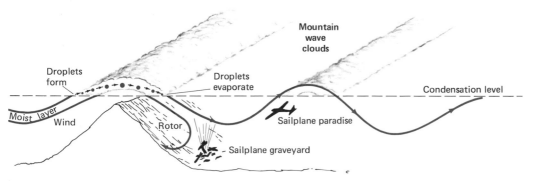

Mountain wave clouds

Droplets form

Droplets evaporate

Condensation level

Moist layer

Wind

Rotor

Sailplane paradise

Sailplane graveyard

FIGURE 2.27
A schematic picture of the airflow that produces mountain wave clouds. Notice the rotor that is a region of turbulence and danger to pilots.

FIGURE 2.28
Wind shear can produce
waves in midair that are made
visible when clouds form at
wave crests.

FIGURE 2.29
An example of wave clouds
formed by wind shear.

the layer. As with mountain wave clouds, these clouds are attached to the waves. Wavelengths vary considerably but typically are less than a kilometer. When they occur low in the sky they produce stratocumulus bands. While at greater heights, they produce either altocumulus bands (higher) or cirrocumulus bands (highest), known as a mackerel sky (see Figures 2.30 and 2.31).

Pilots avoid flying through such wave clouds because they know it makes for rough flying. Sometimes these waves exist in layers of the atmosphere that are too dry to produce clouds so that without warning planes move into regions of clear air turbulence (CAT). There are times when this turbulence is so severe that it causes structural damage to planes.

FIGURE 2.30
Cirrocumulus bands perpendicular to the wind.

FIGURE 2.31
Altocumulus bands at Cardiff.

Fortunately, it is possible to minimize the risk of encountering CAT. This is done by looking at upper-level weather maps and avoiding the regions of strongest winds and wind shear.

Wave clouds (mountain waves included) not only serve a useful purpose to pilots, they also indicate that an important process is taking place in the atmosphere. These gravity waves help to keep the high-level winds in the atmosphere from growing too strong. In 1888 Herman von Helmholtz (1821–1894) first suggested this, but at that time there was absolutely no knowledge of the winds above ground level. Only recently has the importance of Helmholtz's idea been realized.

Stratocumulus or its higher relatives, altocumulus and

cirrocumulus clouds, can take on a cellular appearance. The cellular appearance can, under idea conditions, resemble honeycombs in a beehive or the so-called Bénard convection cells that can be plainly seen at the top of a cup of hot cocoa that is just beginning to cool. Often the cells are so large that they can only be seen from a satellite.

Convection cells can either be closed, which means that rising air with clouds form in the center while the fringes are clear, or they can be open. Closed convection cells tend to form in the stratus cloud layer over cold

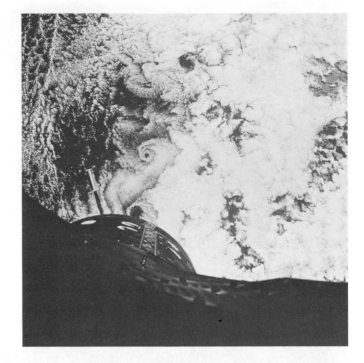

FIGURE 2.32
Closed convection cells forming over cold waters off Baja California can be seen on the right side of the figure. Notice also the eddies that form when the wind blows past Guadaloupe Island and the open convection cells over the warmer waters further offshore.

FIGURE 2.33
Open convection cells forming over warm waters of the Atlantic around 20°N, 20°W latitudes.

waters (see Figure 2.32). Open convection cells tend to form over warm ocean waters, and sometimes the rising air at the fringes of each cell produces a ring of thunderstorms. (See Figures 2.32 and 2.33.)

If you stir the cup of hot cocoa very gently, the convection cells will stretch out into bands. In the atmosphere the winds sometimes stretch out the convection cells into lines called cloud **streets** or bands, which form almost parallel to the wind direction. Cloud streets are quite common when cold winter air from the land pours over the warmer ocean. The surface air heats and rises to form clouds that are then aligned by the winds into streets. At first cloud streets may appear like wave clouds, but they are notably different (see Plate 4 and Figure 2.34). With cloud streets, the principal role of the wind is to align the already existing clouds. Cloud streets are therefore parallel to the wind, whereas wave clouds are perpendicular.

Sometimes both streets and waves are seen in the same cloud layer. In this case the streets are generally much wider, and the sky looks magnificent (see Figure 2.35).

FIGURE 2.34
Satellite pictures over Lake Michigan showing contrasting views. One picture is taken in early winter and shows cold air streaming over the lake. The other is taken in the late spring when the lake is colder than the land. Guess which picture is which! Also, guess the wind direction in each.

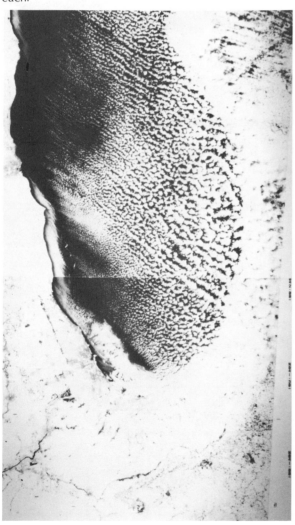
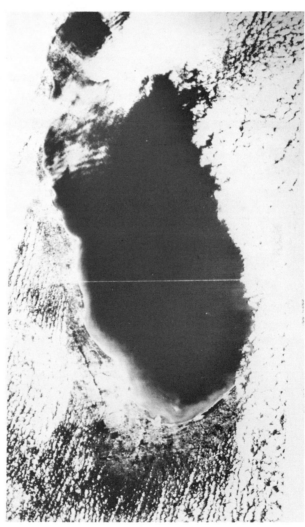

FIGURE 2.35
Altocumulus streets and waves.

# PROBLEMS

**2-1** Keep a daily cloud journal for two weeks: column 1, date; column 2, cloud type; column 3, weather that day. Look at the clouds each day at breakfast and again at lunch. Relate the clouds to the weather changes. Give yourself 1 point for each different cloud type you see and 2 points for each cloud observation that enabled you to predict the weather to come.

**2-2** Explain why the bottoms of large cumulus clouds are so flat while the sides and tops are so fluffy.

**2-3** Why is it often clear for several kilometers around a large cumulonimbus cloud? (See Figure 2.8.)

**2-4** Why does it tend to be cloudy during the day in the mountains?

**2-5** Clouds that are thin or have high bases rarely, if ever, produce precipitation. Name these types of clouds.

**2-6** Sometimes the sides of a cumulus cloud will be sinking. Explain why this occurs.

**2-7** Along which coasts do you expect to find low stratus clouds and fog similar to those that form along the coast of California?

**2-8** Look at Figure 2.20 and compare it with Figure 1.12. Then check to see if you explained Problem 1-12 correctly.

**2-9** What are the distinguishing features of altostratus clouds and cirrostratus clouds?

**2-10** How can you distinguish among stratocumulus, altocumulus, and cirrocumulus clouds?

**2-11** Judging from Figure 2.20 and from the rule for forecasting precipitation after seeing a halo, estimate how fast lows move.

**2-12** Draw rough sketches of a typical cumulus cloud, a stratus cloud, and a cirrus cloud. Then explain how each gets its basic shape.

**2-13** By looking at cirrus clouds, what can you tell about the winds high in the atmosphere?

**2-14** Explain why mountain wave clouds hardly move even though the wind is constantly blowing through them.

**2-15** Explain why mountain wave clouds form over mountains, or downwind from them, but never far upwind.

**2-16** Explain why birds soar in circles on relatively calm days, but in straight lines almost parallel to the winds on windy days.

**2-17** Explain why glider pilots love to see mountain wave clouds, but why they must be careful when they do.

**2-18** Why can sunlight help build up cumulus clouds, but also help to dissipate altocumulus, cirrocumulus, and cirrus clouds (and occasionally stratocumulus clouds, too)?

**2-19** Look at Figure 17.30. Judging from the cloud streets you see over the Gulf of Mexico, in what direction do you think the wind is blowing? Is the air cold or warm compared to the water below?

**2-20** Explain how a cellular appearance of altocumulus clouds at night is similar to Bénard convection cells that sometimes form at the top of a cup of hot cocoa.

**2-21** See if you can find any open or closed convection cells in Figure 16.6 or Figure 18.14.

**2-22** Tell a friend about the most beautiful cloud you have ever seen, and explain how it got its shape.

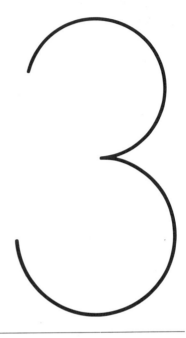

# OPTICAL PHENOMENA

As each day comes to an end, the sky near the setting sun turns slowly from blue to red. Those of us who rise early can witness the same process in reverse order at dawn.

If you watch the sunset carefully you will notice that near the horizon the sky may first become green, then yellow, then orange, and finally red. Each point in the sky is colored somewhat differently so that it is not easy to paint an accurate sunset. Furthermore, there are some days when the red color is much more intense and other days when it is pale. It is safe to assume that no two sunsets are ever *exactly* alike.

This chapter is devoted to describing and explaining the beautiful and occasionally weird optical phenomena —the vivid colors of the sky, rainbows, halos, and mirages—that are produced within our atmosphere. Although it is true that some optical phenomena can be used as crude predictors of weather, they have little direct "practical" value. However, understanding them often adds to our sense of beauty.

## THE NATURE OF LIGHT

The optical phenomena observed in the atmosphere result when the path that light takes becomes distorted by obstacles in the air. The obstacles may be raindrops or even the molecules of air. In order to understand these phenomena we must first understand something about the nature of light.

Isaac Newton (1642–1727) was perhaps the greatest scientist who ever lived. He was the inventor of calculus (invented independently by Gottfried W. Von Leibnitz, 1646–1716) and was the first to explain the orbital motions of the planets and the comets. He was also the first to understand something of the nature of color, but in his great work on optics he argued that light is not a wave phenomenon but rather consists of streams of corpuscles or particles.

Almost everyone today knows that light does consist of waves that travel at a speed of 300 million m/s (meters per

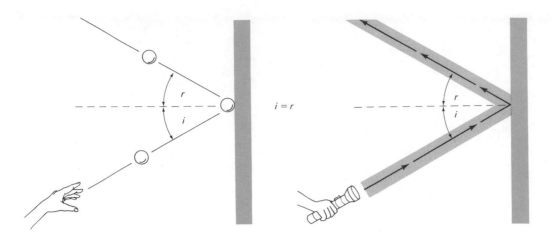

second) through a vacuum. So why did Newton reject the notion of light waves?

Newton used two reasons to reject the hypothesis of light waves. First, *sound* waves cannot travel through a vacuum since they are merely alternate compressions and rarefactions of air. Similarly, water waves need to travel through water. Newton felt that all waves must travel through a medium such as air or water and, since the light of the sun had to travel through empty space to reach the earth, he argued that it could not consist of waves but rather had to consist of some type of particle.

Newton's second argument is a little more technical, but scientists felt it to be far more impressive. It had long been known that light travels in straight lines so long as it is not disturbed. In addition, when light strikes a mirror it will be reflected the same way a ball will bounce after striking a wall (see Figure 3.1). This simple and common behavior is known as the law of reflection.

**Law of Reflection:** The angle of incidence equals the angle of reflection.

This law might lead us to believe that light is like a solid ball or particle, except for the fact that waves also often travel in straight lines and also can be reflected by walls, such as docks (or the sides of a bathtub). Newton's argument was more subtle. He knew that waves will bend around a corner. Light, however, does not seem to bend at all when passing by a corner, because the shadows of walls *seem* quite straight (see Figure 3.2).

The fact is that light actually does bend around corners! Yet the bending effect is so small that under normal circumstances it is not visible to the naked eye. A scientist named Francesco Grimaldi (1618–1663) had actually noticed this small bending effect (known as **diffraction**) when he had observed bands of color at the edge of a shadow. Newton knew of the work of Grimaldi but some-

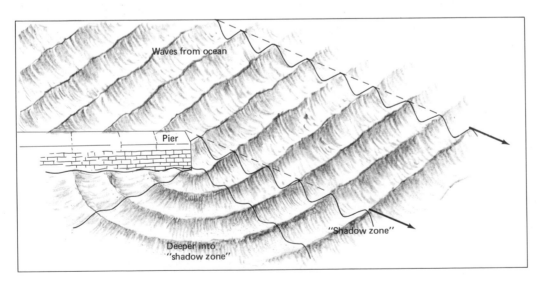

FIGURE 3.2
Waves from the open ocean curve around and into the calmer waters behind piers where they are much smaller than before. The arrows indicate the direction in which the waves are moving. Light waves also have this property, but their intensity is so low in the "shadow zone" that it appears dark.

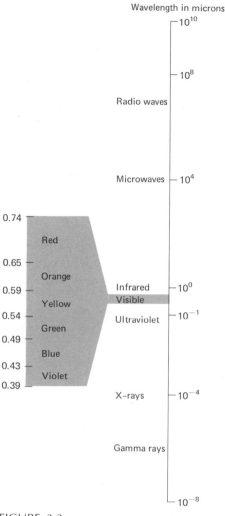

Wavelength in microns

FIGURE 3.3
The spectrum of electromagnetic radiation.

the shortest (see Figure 3.3). All visible light is confined to a narrow band of wavelengths from 0.39 to 0.74 microns. You might wonder why all the colors of light fit into such a small range of wavelengths. Isn't it possible for these waves to have longer or shorter wavelengths? The answer is yes.

Longer and shorter electromagnetic waves most certainly do exist, but our eyes are not sensitive to these other waves (i.e., we cannot see them). Among the shorter waves are ultraviolet rays, which cause sunburn, and X rays. Among the longer waves are infrared waves (often thought of as heat waves), which are emitted from the earth and from our bodies, as well as microwaves and radio waves.

Why then is it that we can only see a very small range of electromagnetic waves? The sun emits visible light at maximum intensity. Approximately 45% of all energy in the electromagnetic radiation emitted by the sun is concentrated in the extremely narrow band occupied by visible light. Nine percent is ultraviolet, while 46% is spread over the infrared wavelengths. It is partially for this reason that our eyes are sensitive to visible light and not to shorter or longer waves. There is simply not enough energy in the longer and shorter wavelengths. Because there is so little microwave energy radiated by the sun, eyes sensitive only to microwaves would reveal a very dark earth.

A matter of philosophical importance must be mentioned at this point. Many people hate mathematics or quantitative subjects, but enjoy the beauty of the arts or qualitative subjects. This hatred of numbers cannot be as complete as many people feel for it can be said that all the beautiful colors of the spectrum—red, orange, yellow, green, blue, and violet—differ by nothing other than wavelength (which is a number). Beauty may well be inseparable from a mathematical plan.

Even though Newton did not know that light consists of waves, he nevertheless made a fundamental discovery about the nature of color. Newton took an ordinary glass prism and allowed one beam of sunlight to go through it. He was certainly not the first person to perform this interesting and beautiful experiment.

I procured me a triangular glass prism to try therewith the *celebrated* phenomena of colours.

*Optics*

For at least 2000 years people had known of this effect, but Newton was the first to realize that *colors are merely components of white light!* In this experiment a narrow beam of normal (white) light is aimed at a glass prism. On

how its great significance escaped him. Its significance also escaped Grimaldi and, oddly enough, for almost 150 years it even escaped the advocates of the viewpoint that light is a wave phenomenon.

I am sure that you have seen the beautiful colors produced by oil slicks on wet roads or in river or lakes. This is the phenomenon that led to the discovery of the wave nature of light. Although it was Newton who first discovered them (they are technically known as Newton's rings), his explanation for the rings is wrong. In 1801, Thomas Young (1773–1829) realized that the color in oil slicks (as well as the phenomenon of diffraction) necessarily proves that light consists of waves.

After this fundamental discovery it was a simple matter to find the wavelengths of light waves. Of all the colors in the spectrum, red has the longest wavelength and violet

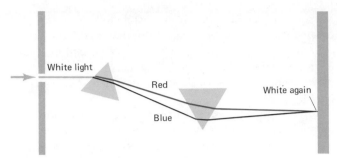

**FIGURE 3.4**
Newton's famous experiment rejoining the colors of the spectrum by using a second prism to recover the original white light.

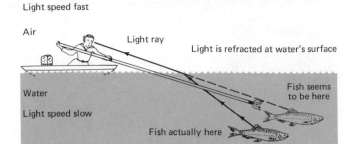

**FIGURE 3.5**
Light bends sharply at the water surface due to refraction, making the fish appear to be somewhere else.

passing through the prism, the light is bent (*refracted*) and split (*dispersed*) into all the colors of the spectrum. Newton was the first to realize that the major features of the colors of the rainbow are due to such **dispersion.** To prove this he "procured him" a second prism to put the colors back together again and, lo and behold, when the colors combined the light became white again!

Figure 3.4 and Plate 5 show that light bends when it enters the prism and bends again when it leaves the prism. This bending of light is known as **refraction.** Refraction is not the same as the slight bending of light *around* obstacles, which is known as diffraction. Refraction can occur when light passes from one medium to another (say from air to water); therefore, light will bend sharply at the interface, or it can occur within a given medium (e.g., air) when density differences cause the light to bend gradually.

Many amusement parks have a house of mirrors. A house of mirrors is essentially a labyrinth whose walls consist of mirrors. It is very difficult to find your way out of this labyrinth, because it is almost impossible to distinguish between reflected and nonreflected light rays.

Refraction can also cause confusion. Optical illusions such as mirages are caused by rather unexpected refraction. From all appearances it is not possible (except by experience) to tell if a light ray has been refracted. If you put a stick halfway into the water and view the stick obliquely, it will seem to be bent right at the surface of the water. It is not the stick but the light rays that are bent as they leave the water.

Let us say that you try to spear a fish in the water. Looking down in the water, you spot a fish. If you are inexperienced you will not realize that the fish is not in the place it appears to be. This is simply the result of the fact that the light rays are bent at the water surface (see Figure 3.5).

*Bending of light rays by refraction always results when one part of the wave moves faster than another part (see*

Figure 3.6). Light rays will bend sharply at an interface where the speed of light changes suddenly or will bend gradually in a continuous medium such as air, where the speed of light changes gradually.

The direction of motion of the light (or any wave) is indicated by the rays in Figure 3.6. The rays are always perpendicular to the wave crests, and as the wave crests bend the light rays do also.

An excellent example of refraction of waves can be seen at the beach. As waves approach the beach they slow down because waves travel more slowly when the water gets shallow. If the waves approach the beach at an oblique angle, they will bend in toward the shore and face the shore more directly as they near the beach. The bending is caused by refraction. As part of the wave crest nears the beach it slows down; the part of the wave crest further from the beach begins to catch up.

From Figures 3.6 and 3.7 you can see the following rule at work.

**FIGURE 3.6**
Refraction occurs when one part of the wave moves faster than another part. Waves always bend *toward* the region where they move more *slowly*.

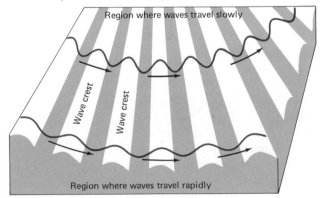

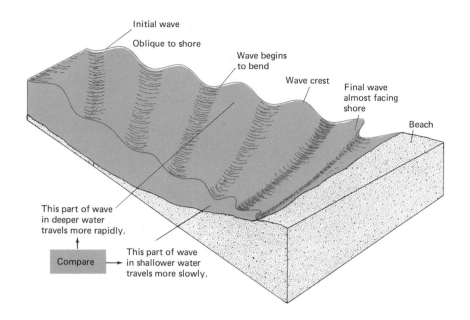

Initial wave

Oblique to shore

Wave begins
to bend

Wave crest

Final wave
almost facing
shore

Beach

This part of wave
in deeper water
travels more rapidly.

Compare →

This part of wave
in shallower water
travels more slowly.

FIGURE 3.7
Ocean waves refract directly in toward the beach, because they travel more slowly where the water is shallow.

**Rule:** Refraction causes waves (rays) to bend *toward* the region in which they move the slowest and to bend away from the region in which they move the fastest.

In air the speed of light slows down as the air becomes denser. Generally the air becomes compressed and denser as you descend in the atmosphere because of the great weight (pressure) due to the air above it. Air also becomes denser when it is colder.

On March 21 and September 21 (the equinoxes), there are supposed to be exactly 12 hours of sunlight. This means that on these days the sun should rise at 6:00 A.M. and set at 6:00 P.M. If you view the sunrise and sunset on these days from a point where there are no obstructions, you will find that the sun will rise at about 5:56 A.M. and set at about 6:04 P.M. The precise time of sunrise and sunset depends slightly on the latitude, but you can expect about eight extra minutes of sunlight each day of the year.

Eight extra minutes of sunlight each day? Yes! This is partly due to the finite size of the sun, but is mainly due to the refraction of sunlight in the atmosphere. Although you may think that the sun's rays come straight to you from the sun, you are in for a surprise. The sun's rays are bent slightly by atmospheric refraction. At the precise moment when you think that the bottom of the sun is just touching the horizon, the sun has actually just dipped below the horizon (see Figure 3.8).

The light rays of the sun are bent by refraction, because the air near the ground is denser than the air above. Light waves therefore travel more slowly near the ground and therefore bend in the same manner as the curvature of the earth. In a certain sense the setting and rising of the sun is an optical illusion.

A mirage is an optical illusion that is the result of the refraction of light by air with unusual density patterns. The desert mirage makes you think that you have found water when you are actually seeing blue sky (see Figure 3.9). It occurs on the sunniest of days when the ground becomes so hot that it superheats the air lying just above it. Several meters above the ground the air is colder and heavier. Therefore, light waves travel faster near the

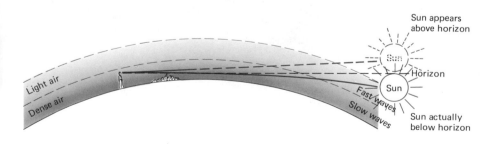

Sun appears
above horizon

Horizon

Sun

Fast waves

Slow waves

Sun actually
below horizon

Light air

Dense air

FIGURE 3.8
Because of refraction, you can still see the sun just after it has actually set.

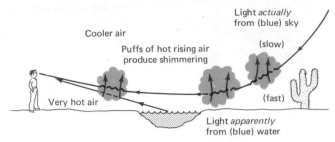

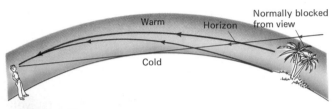

FIGURE 3.9
A desert mirage is caused when light waves from the sky are refracted so that they appear to come from the ground. Shimmering is produced by rising blobs of heated air. This appears as waves.

FIGURE 3.10
When air near the ground is very cold, compared to the air a few meters higher, you can actually see places normally below the horizon.

ground than several meters up, and so they bend away from the ground.

What makes the optical illusion even more convincing is that the ''water'' seems to shimmer. The shimmering results because the superheated, light air at the ground is rising in irregular puffs or eddies. These constantly chang-

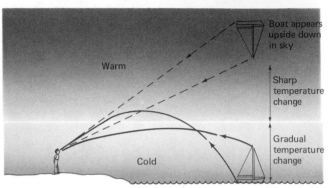

FIGURE 3.11
Temperature profile that could produce an inverted image.

FIGURE 3.12
A mirage of the sun. Double images of the sun or moon near the horizon are common whenever the water temperature is much different from the air temperature directly above it. This photo was taken near dawn when the air was much colder than the water.

PLATE 1
A cirrus cloud viewed from Wildcat Mountain, New Hampshire. Notice the long streamers, which indicate the clouds are moving toward the right. There are also some small cumulus clouds in the lower right corner.

PLATE 2
Altocumulus clouds are merely a higher version of stratocumulus. They often produce the most spectacular sunsets and sunrises.

PLATE 3
Mountain wave clouds with ice crystal tails, seen near Denver.

PLATE 4
Satellite view of stratocumulus cloud streets over Georgia. You can tell the wind is blowing from the south because otherwise the clouds would have extended some distance over the water.

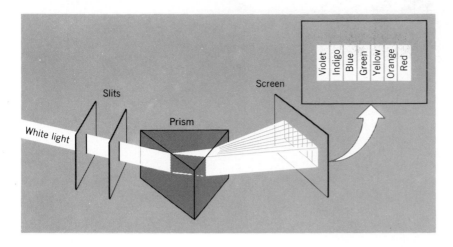

Violet — Indigo Blue　Green　　Yellow　　Orange　　Red

PLATE 5
The prism and the spectrum of
colors.

PLATE 6
A double rainbow.

PLATE 7
A sun pillar.

PLATE 8
A red sky can be used to forecast
the weather in much the same way
that a rainbow can. This sunset picture was taken facing west, where it
is clearing.

PLATE 9
An iridescent mountain wave cloud.

PLATE 10
The solar flare of December 1973.

PLATE 11
The aurora borealis in Fairbanks,
Alaska.

PLATE 12
Hurricane Gertrude.

(a)

PLATE 13
(a) India as seen from space. Note the cumulus clouds over the land and the clear corridor just offshore. (b) Florida as seen from space. Again clouds are mainly over land, but not over the rivers and lakes.

(b)

PLATE 14
The annual spectacle of fall foliage
in Vermont.

ing temperature irregularities cause the light waves to take irregular, constantly changing paths that result in shimmering.

If you look at a tar-covered road at a glancing angle on a hot, sunny day, you can probably see a similar mirage.

On other days when the surface of the earth is very cold and the air touching the surface is colder than the air several meters up, light rays will bend in the opposite way from that of the desert mirage. Then, small objects on earth may appear as "castles" in the sky. This is sometimes called the *fata morgana* after the magician–sister of King Arthur. This type of situation may also be the origin of many "sightings" of UFOs.

On occasion, people at sea have also spotted islands or other objects in the distance that were normally blocked from view by the curvature of the earth as a result of this type of refraction (see Figure 3.10).

Depending on the precise temperature structure of the air, the images of objects in the distance may appear either right side up or upside down and may appear either magnified or compressed. Several images of the same object may be visible. Some distortion is to be expected in

FIGURE 3.13
The rainbow is produced when sunlight strikes raindrops. Spots of light come from countless millions of raindrops so that it appears to be a continuous arc. Each raindrop at a 42½° angle from your shadow sends a spot of red light, while each at a 40½° angle sends a spot of blue light to your eyes.

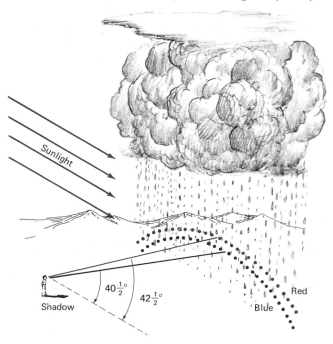

these unusual situations, and this tends to reinforce feelings that something supernatural is happening (see Figures 3.11 and 3.12). It may indeed be "super," but it is entirely natural.

# RAINBOWS

And it shall come to pass, when I bring a cloud over the earth, that the bow shall be seen in the cloud: and I will remember my covenant . . . and the waters shall no more become a flood to destroy all flesh.

Exodus, Chapter 9, 14–15

According to the bible, the rainbow is the sign of God's promise to mankind that he will never again flood the earth. Indeed, rainbows often indicate that the rain has passed. Generally, it will be sunny when you see a rainbow, but rain clouds (usually cumulonimbus) will be just a short distance away.

Rainbows are simple to find. They only occur when sunlight passes through gaps in the clouds and strikes the raindrops beneath the clouds (see Figure 3.13). Thus, when there are showers nearby, simply look in the part of the sky opposite the sun at a 42° angle from your shadow, and if there is a rainbow, that is where it will be. This is the *primary rainbow* and it always forms a circular arc at a 42° angle from the shadow of your head. The red color always appears on the outside or top of the bow.

Whenever the primary rainbow is bright enough, another bow (the secondary bow) can be seen a short distance above it. The **secondary rainbow** forms at a 51° angle from your shadow, but the color sequence is the reverse of the primary bow's, with red on the bottom of the secondary bow. The primary rainbow is always brighter than the secondary bow.

Outside of the tropics or the dry summer subtropics (see Chapter 22) rainbows are seen more during the spring and summer, since there are more showers and thunderstorms then. In these regions during the spring and summer, rainbows will be seen early in the morning or late in the afternoon, but not around noon. This is due to the fact that at noon the sun is quite high in the sky so that 42° from your shadow you can see only the ground.

Sailors have long known that rainbows can be used to predict the weather. Generally, thunderstorms move from west to east in the midlatitudes; this verifies the adage

*Rainbow in morning, sailor's warning*
*Rainbow at night, sailor's delight.*

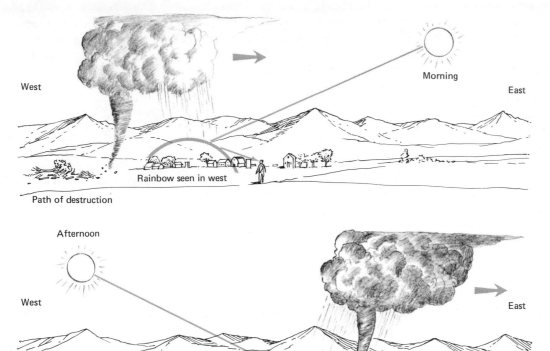

West

Morning

East

Rainbow seen in west

Path of destruction

Afternoon

West

East

Path of destruction          Rainbow seen in east

FIGURE 3.14
The rainbow as a tool in weather forecasting. This rule works in the midlatitudes where thunderstorms generally move from west to east. In the tropics, thunderstorms generally move from east to west, so the rule must be changed.

FIGURE 3.15
Path of light producing the primary rainbow. There must be two refractions and one reflection. Furthermore, only one ray enters the drop at the correct angle to be a rainbow ray.

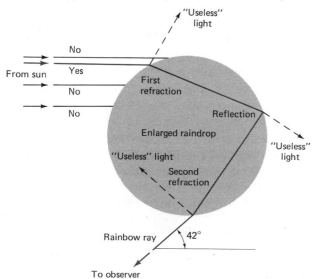

In the morning the sun is in the east; to see a rainbow you must be facing west where it is raining. Since thunderstorms come from the west take warning from the morning rainbow. At night (actually late in the afternoon, but afternoon doesn't rhyme well) the sun is in the west; to see a rainbow—well, you can figure that one out for yourselves (see Figure 3.14).

You can make your own rainbow with nothing more than a garden hose. Just shoot a fine spray of water droplets in the air and look away from the sun at a 42° angle from your shadow. This experiment (which should be performed in summer so that you don't freeze) might indicate to you that the bible is not quite correct about the rainbow. The rainbow is not produced within the cloud but is produced by sunlight striking the raindrops beneath the cloud.

Sunlight striking raindrops produces rainbows. Each raindrop acts like a combination of a prism and a mirror when sunlight strikes it. In order to produce the *primary* rainbow, a beam of light must first enter the raindrop where it is refracted; next, it must penetrate to the back of the drop where it is reflected; and finally, it must leave the drop where it is refracted a second time (see Figure 3.15).

Newton's discovery of dispersion then explains the colors of the rainbow—but there is still one main problem. If light hits the center of the raindrop, it will bounce straight back; instead of a 42° angle, there will be a 0° angle. Depending on the precise part of the drop that the ray of sunlight hits, the returning ray will come back at a variety of angles, from 0 to 42°. Why then is the rainbow only seen at 42°?

René Descartes (1600–1657) had solved that problem several years earlier in an appendix to his famous work, *A Discourse On Method*. To prove how wonderful his method was, Descartes proceeded to solve a mathematical problem that had eluded scientists for 2000 years. In Descartes' own words:

The principal difficulty still remained, which was to determine why, since there are many other rays which can reach the eye after two refractions and one or two reflections when the globe [raindrop] is in some other position, it is only those of which I have spoken which exhibit the colors.

I then took my pen and made an accurate calculation of the paths of the rays which fall on different points of a globe of water to determine at what angles after two refractions and one or two reflections they will come to the eye and then I found that after *one reflection and two refractions* there are many more rays which come to the eye at an angle of 41° to 42° than any *smaller* angle, and none which come at any *larger* angle [emphasis added].

*A Discourse on Method*

Of all the rays that strike the raindrop, the rainbow ray spreads out the least and is therefore the most intense. Also, because the rainbow ray occurs at the largest angle, the colors do not get smudged out (see Figure 3.16). Clearly the rainbow is an amazing phenomenon. Descartes' explanation of the rainbow came within 20 years after Willebrod Snell van Royen (d. 1626) had finally discovered the correct law of refraction (people had been trying for 2000 years prior to Snell's discovery). It is said that in making his proof Descartes computed the deviation angle for 10,000 rays. In fact, he did not do that much work, but he still had to compute many deviation angles to make a strong case. A mere 30 years after Descartes' laborious mathematical proof, Newton invented calculus (which is like the assembly line of mathematics) and was able to solve the problem in less than an hour.

In reading Descartes' quote you should realize that while the primary rainbow is produced by two refractions and one reflection; the secondary rainbow is produced by two refractions and two reflections (see Figure 3.17). You should also be able to explain the fact that when you can

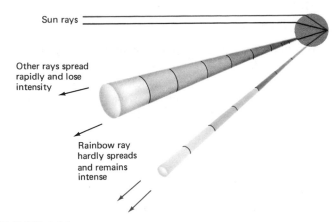

FIGURE 3.16
The rainbow ray is the most deflected of all the rays. It is also the most intense because it spreads out the least.

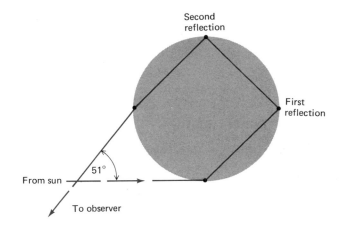

FIGURE 3.17
Path of light producing the secondary rainbow. There must be two refractions and two reflections. As with the primary rainbow, only one ray enters the drop at the correct angle to be a rainbow ray.

see both rainbows, the sky is darkest between the two rainbows (see Plate 6)!

There are some rather unusual or subtle features about rainbows that might be of interest. On occasion, sunlight is reflected from a lake or river in such a way that the reflected sunlight will produce its own rainbow (reflected rainbow). If there are waves on the surface of the lake, the reflected rainbow will actually quiver. A second feature is even more subtle. Occasionally, several bands of color can be seen just within the primary rainbow. These extra bands are known as **supernumerary** rainbows, and they were explained by Thomas Young once he realized that light consisted of waves. They are due to the diffraction of light.

# HALOS

Very often there will be beautiful halos—rings and arcs and spots of light—in the sky above, while entire populations of unseeing people walk below. To the careful observer of the sky the various halos may prove to be even more fascinating sights than rainbows.

The most common halo forms a circular ring around the sun or moon. This you will find by stretching out your arm and fingers; when your thumb covers the sun or moon, your pinky will touch the halo. (When looking for halos you must, of course, be careful to shield your eyes from the direct sunlight.)

Halos are produced when sunlight strikes ice crystals and then penetrates to the ground. Halos are therefore produced mainly when a rather thin, uniform veil of cirrostratus clouds covers the sky. They may be seen at any time of the year, for even on the hottest days the air temperature at the level of the cirrostratus clouds (near 10 km) is low enough for ice crystals to form. The practical value of halos for weather forecasting (see Chapter 2) follows from the fact that cirrostratus clouds often precede warm fronts and precipitation.

Just as the precise shape of the rainbow is due to the geometry of the circle, the precise shapes of the halos, arcs, and so on, are due to the shapes of the ice crystals. All ice crystals are basically six sided (hexagonal). The differences in crystal shapes account for some of the variety of phenomena that are observed. Plates and columns are most commonly observed, and some ice crystals are even bullet shaped. Figure 3.18 contains a small sampling of the possible shapes of ice crystals.

The crystal shape determines the kind of halo or arc that will be produced. Interestingly enough, the crystal shape depends on the temperature at which it forms so that by looking at halos we can tell something of the temperature at cloud level. For instance, in crudest terms plates tend to form when the temperature is either between 0 and $-4°C$ or between $-10$ and $-22°C$, whereas columns and needles tend to form when the temperature lies between $-4$ and $-10°C$ or between $-22$ and $-50°C$.

Let us see first how the normal halo is produced. Light from the sun enters one side of the crystal and is refracted because light travels more slowly through ice than through air. This light leaves the crystal through another side and is refracted again. The light can be bent by as little as 22° from its initial direction (see Figure 3.19). From geometrical considerations it turns out that not only is 22° the smallest angle at which the light can be bent, but under normal conditions more rays are bent at this angle than at any other angle. This last sentence should sound similar to the discussion of the primary rainbow. Thus, the halo appears at 22° angle from the sun or moon, red light is bent slightly less due to dispersion, and the inside of the halo appears dull red.

For the sake of convenience, only the six rectangular sides are called sides, but the hexagonal surfaces of the ice crystal are called the top and bottom. The 22° halo results

**FIGURE 3.18**
Common shapes of ice crystals.

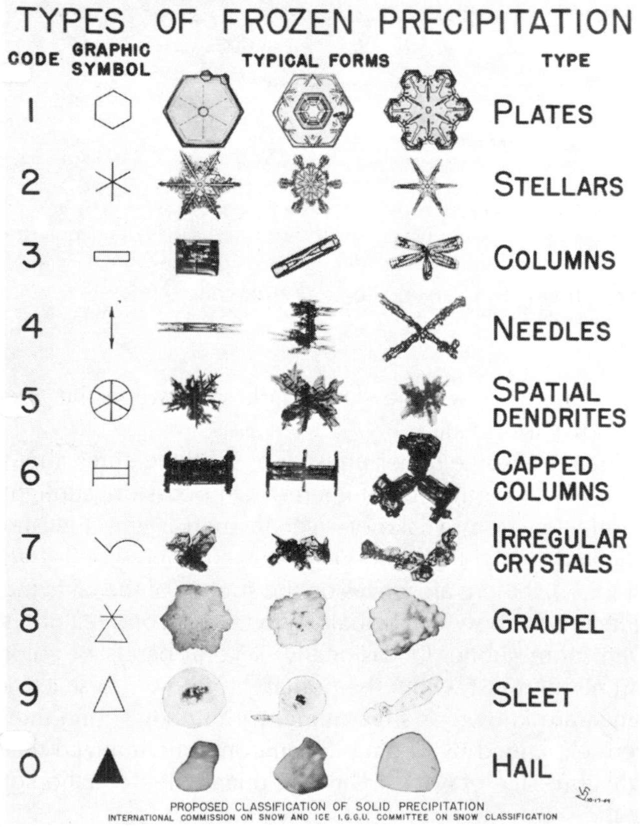

## TYPES OF FROZEN PRECIPITATION

| CODE | GRAPHIC SYMBOL | TYPICAL FORMS | TYPE |
|---|---|---|---|
| 1 | | | PLATES |
| 2 | | | STELLARS |
| 3 | | | COLUMNS |
| 4 | | | NEEDLES |
| 5 | | | SPATIAL DENDRITES |
| 6 | | | CAPPED COLUMNS |
| 7 | | | IRREGULAR CRYSTALS |
| 8 | | | GRAUPEL |
| 9 | | | SLEET |
| 0 | | | HAIL |

PROPOSED CLASSIFICATION OF SOLID PRECIPITATION
INTERNATIONAL COMMISSION ON SNOW AND ICE I.G.G.U. COMMITTEE ON SNOW CLASSIFICATION

**FIGURE 3.19**
Paths of light through ice crystals to produce 22° and 46° halos.

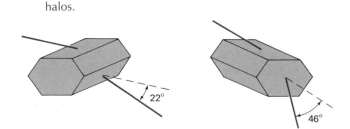

when light enters one of the six sides and leaves another side.

There is another halo called the *great halo* which is not often seen. It appears at an angle of 46° from the sun or moon and forms on the same basic principle as the normal 22° halo. In the great halo, light either enters the *top* of the crystal and then emerges from one of the sides, or it enters one side and then passes out the bottom.

The very seldom observed 8° halo is due to the "bullet" ice crystals that are less common. Bullets also produce halos of 17° and 90°, which are extremely unusual.

Basically, the size of the halos depends on the angle between the faces of the ice crystals. What is it, then, that causes the **sun dogs** (also known as **mock suns** or **parhelia**)? These are bright spots of light at the same height as the sun and just outside the normal halo. What causes the various other arcs?

If you throw an ice cream stick in the air you will observe that it falls almost the same way every time. The long side of the stick remains almost horizontal. Single bullet-shaped crystals tend to fall vertically. Many of the unusual arcs form as a result of the fact that a large percentage of the crystals tend to "line up" or, in other words, orient themselves in a particular manner as they fall, depending on the crystal shape. Circular halos form when the crystals have no preferred orientation in space. The parhelia or "sun dogs" form when crystals fall in such a manner that the six sides are all vertical. This can happen for falling bullets, plates, and umbrella-shaped capped columns larger than about 25 microns. Some of the possible halos and arcs are shown in Figures 3.20, 3.21, Plate 7 and in Table 3.1.

FIGURE 3.20
Halo complex in Antarctica.

# THE COLORS OF THE SKY

Why is the sky blue? This is a problem that has puzzled people for at least as long as the rainbow problem. Newton had thought that the color was due to the reflection of sunlight off hollow water droplets, but his theory was completely wrong (hollow water droplets or bubbles do not even occur in the atmosphere). In 1815, Brucke performed an experiment with a cloud of smoke; smoke consists of tiny particles. When the cloud of smoke was illuminated with white light it appeared bluish (like the color of smoke coming directly from a cigarette). Brucke realized that this simple experiment might have something to do with the color of the sky, but he couldn't prove it. (Leonardo da Vinci had made a similar observation while watching smoke from a fire sometime around 1500!)

FIGURE 3.21
A variety of the more common halo phenomena.

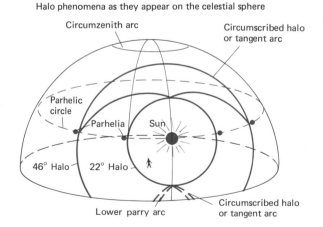

Halo phenomena as they appear on the celestial sphere

TABLE 3.1
Description of Some Halo Phenomena

| Name | Appearance | Crystal Features | Orientation | Path of Light | Comments |
|---|---|---|---|---|---|
| 8° and 17° halos | Circular arc around sun or moon | Bullet shaped | Random | Refraction, light through tapering bullet faces | Very rare and hard to see |
| 22° halo | Circular arc around sun or moon | Small columns or plates | Random | Refraction, light through crystal sides | Very common, often brighter on top and bottom |
| 46° halo | Circular arc around sun or moon | Small columns | Random | Refraction, light through top or bottom and one side | Rare |
| Parhelia to 22° halo | Bright-colored spots of light at same elevation as sun or moon and just outside halo | Medium-sized bullets, capped columns or plates | 6 sides, vertical | Refraction, light through crystal sides | Cannot form when sun or moon above about 60°, common |
| Parhelic circle | Horizontal circle that runs through sun or moon | Medium-sized plates or columns | Some faces, vertical | Reflection, light off vertical faces | Not colored |
| Tangent arcs to 22° halo (circumscribed halo) | Arcs tangent to 22° halo at top and bottom lying outside halo | Medium-sized columns | Hexagonal, faces vertical | Refraction, light through sides | Shape changes with sun's altitude, merging with halo when sun is above 60° |
| Circumhorizontal arc | Horizontal arc below altitude of sun or moon | Same as for parhelia | 6 sides, vertical | Refraction, light in side and out bottom | Cannot form when sun or moon below about 60° |
| Circumzenithal arc | Horizontal arc around zenith above altitude of sun or moon | Same as for parhelia | 6 sides, vertical | Refraction, light in top and out the side | Cannot form when sun or moon above about 60° |
| Light pillar | Vertical arc usually above sun or moon | Large-sized plates | 6 sides, almost vertical | Reflection, light usually off crystal bottom | Not colored, except when sun is at horizon; usually seen when sun or moon is low in sky |
| Subsun | Spot of light as far below as sun or moon is above horizon | Medium-sized plates | 6 sides, vertical | Reflection, light off crystal top | Must be seen from above—in plane or on mountain |

In 1881, Lord Rayleigh finally provided the explanation. The blue color of the sky is due to the air molecules which can scatter (reflect) the light waves that pass by. If there were no air molecules, the sky would appear as black as night, and the stars would be visible even during the day. The sun would appear even brighter than it now is.

*Sky* light is due to millions upon millions of air molecules and other minute dust particles (**aerosols**) that scatter sunlight so that it can reach our eyes from all directions—sky light is scattered sunlight (see Figure 3.22). When a puffy cumulus cloud floats by, the sides of the cloud are bright white when they are in the direct sunlight. This is because each of the tiny cloud droplets reflects sunlight, and the reflected sunlight makes the sides of the cloud bright white. The cloud appears white rather than blue because the cloud droplets reflect all colors of light equally well. *The sky appears blue because air molecules scatter blue light more easily.*

Why is this so? The air molecules are so small that not only are they smaller than the cloud droplets, but they are also far smaller than light waves. Most light waves can easily bypass these tiny molecules and continue on their way. However, the *shorter* the wavelength of light, the more difficulty it has bypassing the molecules and the more likely that it will strike the molecules and be scattered in some other direction. The precise law discovered

by Rayleigh is that, if the scattering object is small compared to the wavelength of light, then if the wavelength of light were suddenly cut in half, 16 times as much light would be scattered. This process is known as **Rayleigh scattering.** Rayleigh scattering does not apply to cloud droplets. Cloud droplets are large compared to the wavelength of light; therefore, they reflect all colors equally well because no wavelengths of light can bypass them.

Since blue has a short wavelength, it seems reasonable that the sky is blue. Nevertheless, violet has the shortest wavelength of any visible color so you might wonder why the sky isn't violet instead of blue. This is so for a number of reasons.

1. There is more blue light than violet light in direct sunlight.
2. Physiologically, our eyes are more sensitive to blue light than to violet light.
3. There is enough light of other wavelengths to ''smudge'' the color and make the overall effect appear blue.

Near sunrise and sunset the sun and horizon sky turn red (see Figure 3.23). You must have noticed that the sun is not so bright at sunset time as it is at midday. This is because when the sun is near the horizon the sun's rays must pass obliquely through the atmosphere and, thus, far more light is scattered than when the sun is high in the sky. Near sunset, as the direct sunlight approaches the observer, more and more of the shorter wavelengths (violet, blue, green) get scattered by the air and are removed from the direct sunbeam. After a while the only light left in the direct sunbeam in appreciable amounts is the long or red. This behavior can be generalized to state:

**Rule:** Any distant light in the atmosphere appears red because the blue light has been scattered.

This rule explains a number of phenomena. When the moon is at the horizon it, too, appears reddish or orangish. When a dark thunderstorm comes overhead but it is still clear at the horizon, the horizon sky turns orange. This appearance takes on ominous overtones, but the color is nothing more than the natural color of a distant light. In this case the distant light is the illuminated part of the sky.

Because sky light is scattered sunlight, it is mainly blue at first. As this light approaches the observer it gets scattered again. This secondary scattering progressively depletes the blue color as it penetrates the darkness under the cloud (see Figure 3.24). By the time it reaches the ob-

FIGURE 3.22
Sky light is scattered light. Air molecules are so small that they scatter shorter blue waves much more easily than longer red rays so that the sky appears blue. If there were no air molecules, the sky would appear pitch black and stars could be seen all day.

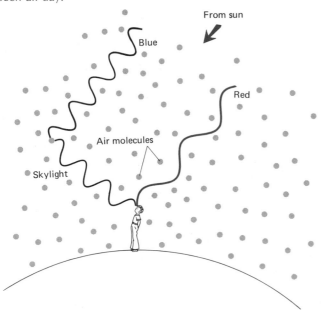

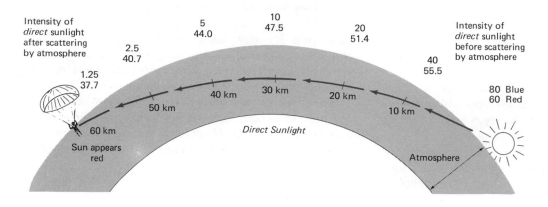

Blue light depleted
more rapidly by scattering

FIGURE 3.23
The sun (or moon) looks red near the horizon because almost all the blue light has been scattered.

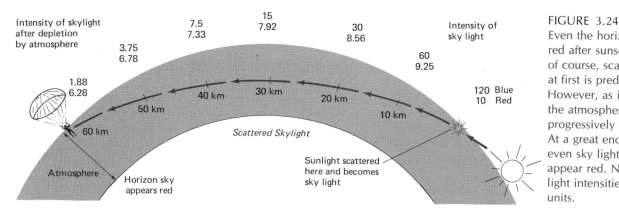

FIGURE 3.24
Even the horizon sky appears red after sunset. Sky light is, of course, scattered light that at first is predominantly blue. However, as it passes through the atmosphere, the blue gets progressively scattered again. At a great enough distance, even sky light will, therefore, appear red. Numbers give light intensities in relative units.

server under the cloud, so much of the blue has been depleted that the light is mainly orange.

The same thing happens for the sky near the horizon after sunset time, but the red color is far deeper since at sunset the light is coming from a very great distance. However, the most dramatic reddening of the horizon sky occurs during a total solar eclipse when the observer is cast into the darkest shadow, while 100 km away it is light.

The entire sky can appear deep red when clouds overhead reflect the red light of the rising or setting sun. It is under these conditions that the following saying makes sense,

*Red sky in the morning, sailor take warning.*
*Red sky at night, sailor delight.*

The logic is much the same as for the rainbow. If you see a red sky at night, it must be clear in the west to allow the sunlight through. Then, since weather moves from the west, it should clear soon (see Plate 8).

## CORONAS AND LIGHT WAVES

We are used to saying that one plus one equals two. But this simple rule of addition does not work so simply for light waves. If the crest (top) of one wave meets the trough (bottom) of another wave of equal amplitude, the two waves will **interfere** with each other or cancel each other. Scientists would say that these two canceling waves are one-half of a wavelength out of **phase.** Only when two waves are exactly in phase can they add up simply. With light waves brightness is due to the change from wave crest to trough. So when two waves cancel, there is no brightness.

Now imagine that light waves are permitted into a box through one very narrow slit. Some light will reach all portions of the opposite wall. If we then allow light to enter the box from a second very narrow slit, then at certain points in the box it will interfere with the light from the first slit, while at other points the light will add. The points of cancellation of light are the points at which the light is exactly one-half of a wavelength out of phase, and at these points it will be dark (see Figure 3.25).

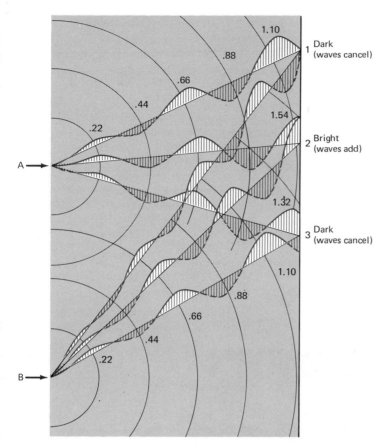

FIGURE 3.25
Interference of light waves. At point 1, the crest of the wave from *A* is canceled by the trough of the wave from *B*. Thus, it will be dark at point 1. Similarly, it will be dark at point 3. At point 2, it will be bright because the crests of both waves arrive together and hence add.

Consider light with as wavelength of 0.44 microns (blue light). Now at point 1 in Figure 3.25, one path is 0.22 microns longer than the other path. This is half a wavelength; therefore, the two waves cancel, and there is no light. However, at point 2, one path is 0.44 microns longer than the other, so the waves are back in phase (crest lines up with crest), and point 2 is bright. At point 3, the two paths differ by 0.66 microns, and you can see that the waves are out of phase again.

This results in a pattern of alternating light and dark bands, known as an interference or diffraction pattern. As mentioned earlier, this was discovered by Grimaldi and explained 150 years later by Thomas Young.

Interference patterns are often beautifully colored and not merely light and dark. It is simple to explain the colors. A point 3, in Figure 3.25 the blue light from the two paths is completely out of phase, but the red light

with a wavelength of 0.66 microns would be completely *in* phase, since the two paths differ by 0.66 microns. Therefore, if white light was shined through the slits, point 3 would appear red. On the other hand, there will be certain points where red light will interfere but blue light will not; these points will then appear blue.

This is the effect that produces the supernumerary rainbows; it also produces the coronas that form around the sun or moon. In the introduction to his work *Optics,* Isaac Newton admitted that he could not explain the coronas around the sun or moon. This was because he did not realize that light consists of waves.

Coronas generally occur when thin, semitransparent altostratus or altocumulus clouds are in the sky. The tiny droplets act in much the same manner as the slits in the box and produce a series of colored rings immediately around the moon or sun. Coronas around the moon are usually much easier to see; you often can be blinded looking for coronas around the sun unless you look through tinted glass.

Sometimes the thin altocumulus clouds appear iridescent if the direct sunlight is blocked by a lower cloud. Iridescence is also due to interference and consists of parts of coronas. Thin mountain wave clouds often produce the most spectacular iridescent displays, because the drop sizes are all similar (see Plate 9).

# OTHER OPTICAL PHENOMENA

There are many other optical phenomena that can be seen in the atmosphere but, unfortunately, there is not room enough to discuss them all. Perhaps it is appropriate to describe just two more phenomena.

### Crepuscular Rays

Sometimes you will see beams of sunlight passing around the edges or through chinks in the clouds. These light beams appear similar to searchlight beams near airports or in theaters and, indeed, they are rendered visible because of the scattering of light off air molecules and dust particles in the air. Usually the sunbeams appear to diverge or spread out after they bypass the clouds; these "spreading" rays are called **crepuscular rays.** Occasionally the sunbeams appear to narrow and they are then called **anticrepuscular rays.**

But how can sunbeams appear to diverge or converge when we know that they are basically parallel? The explanation is simple — the diverging or converging is merely a result of perspective, as with the apparent converging of railroad tracks in the distance. When you look in the direction of the sun the rays will appear to diverge as they approach you from the cloud level; when you look away from the sun the rays will appear to converge as they pass from cloud level.

### The Heiligenschein

When you wake up in the morning and the dew is on the grass, rush outside before the sun has a chance to evaporate the dewdrops. Stand on the grass and look at your shadow. You will see a bright glow directly around the shadow of your head. This is known as the **heiligenschein** (see Figure 3.26). The dewdrops resting on the grass act to reflect and focus light precisely back from the direction it came. Reflectors used on street signs and bicycles (and cat's eyes) are based on the same principle.

Benvenuto Cellini (1500–1571), the famous Italian Renaissance artist and goldsmith, had a different explanation for the heiligenschein.

There is one thing I must not leave out — perhaps the greatest that ever happened to any man — and I write this to testify to the divinities and mysteries of God which he deigned to make me worthy of. From the time I had my vision till now, a light — a brilliant splendor — had rested above my head . . . . It can be seen above my shadow, in the morning, for two hours after the sun has risen; it can be seen much better when the grass is wet with soft dew.

*Autobiography of Benvenuto Cellini*

Which explanation do you prefer?

FIGURE 3.26
The heiligenschein always forms around the shadow of the observer even if it be the camera.
Notice there is no heiligenschein where footsteps have crushed dewdrops.

# PROBLEMS

**3-1** Prove the law of reflection by shining a searchlight obliquely at a mirror in a dark room.

**3-2** Construct the following experiment to simulate diffraction. Take a flat pan and fasten a divider securely in the middle. The divider should have a narrow vertical slit to allow water to pass through Then make waves on one side of the pan only. Watch how the waves spread out on the other side.

**3-3** The thickness of colored oil slicks is some multiple of one-half the wavelength. What is the thinnest oil slick that will appear blue?

**3-4** Perform the following experiment to demonstrate refraction and dispersion. Allow a *narrow beam* of sunlight (a searchlight in a dark room can also be used) to strike a glass prism (a chandelier or vase can be substituted for a prism, so long as the glass has sharp angles).

**3-5** Demonstrate refraction by placing a stick halfway in water and viewing it obliquely.

**3-6** From a high vantage point watch waves refract as they approach the shoreline.

**3-7** Explain why refraction adds to the length of the day.

**3-8** Explain why mirages are more common when there is a sudden, drastic change in air temperature.

**3-9** When the sun or moon is near the horizon, double images are most common if the air temperature is either much colder or much warmer than the surface below. Explain.

**3-10** To see a rainbow in the sky, you must look in the opposite direction from the sun. Why?

**3-11** Can a rainbow be produced by moonlight?

**3-12** Why don't you see rainbows near noon in the summer (this is not the case in polar regions)?

**3-13** Simulate a rainbow in the following way. Obtain a spherical glass beaker and fill it with water. Then in a piece of cardbcard cut a circular hole equal in size to the beaker. Move your equipment to a dark room. Now shine a searchlight through the hole and onto the beaker. When the beaker is close to the cardboard, you should see a circular ring of light on the cardboard.

**3-14** Produce a rainbow in broad daylight using the water spray from a hose (or watch the spray from a fountain in sunlight).

**3-15** Exactly how do rainbows form?

**3-16** Why is the sky darkest between the primary and secondary rainbows?

**3-17** How would a rainbow be used to forecast weather near the *equator*? Explain.

**3-18** Keep on the lookout for all halos and rainbows. See how many of the phenomena listed in Table 3.1 you can see, and give yourself an extra point for each rare phenomenon you see.

**3-19** Explain why the sky is blue.

**3-20** Explain why the sky near the horizon is not so blue (and why it turns red near sunrise or sunset).

**3-21** Explain why the blue rings of coronas are closer to the sun or moon than the red rings.

**3-22** When you see a ring of light that closely surrounds the moon is it a halo or a corona?

**3-23** Why are halos produced by cirrostratus clouds and not altostratus or altocumulus clouds, while with coronas the reverse is true?

**3-24** Considering that aerosols tend to scatter light by only small angles, explain whether it is easier to see crepuscular rays or anticrepuscular rays.

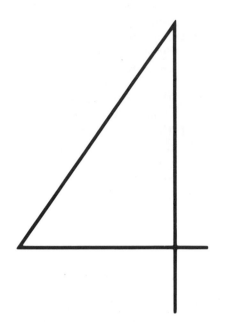

# THE EARTH
# IN SPACE

Just as it is the source of life on earth, the sun is the source of the weather on earth. Virtually 100% of all the light and heat on earth comes from the sun. Because this light is spread unevenly over the earth's surface, some places on earth are colder than others. The resulting temperature differences cause the winds and drive the atmosphere like a giant engine.

In the first chapter you read that the polar lands receive the indirect (oblique) rays of the sun, while the equatorial regions receive direct (overhead) rays, with the result that the equatorial regions are warmer than the polar regions. You also read that during July the sun's rays in the NH are more direct than during January so that the warmest time of the year, summer, occurs around July in the NH.

There are at least two questions that we have yet to ask:

1. Why are the sun's rays overhead at the equator and during summer, and why are they low in the sky at the poles and during winter?

2. Why are the sun's rays more effective when the sun is higher in the sky?

I will begin this chapter by answering these questions.

To answer the first question fully it is necessary to discuss a little bit of astronomy, that is, to see what the earth and the sun look like when viewed from space. After this, we will look at the precise law (known as the sine law) that explains why overhead rays are more effective heaters than rays, which are lower in the sky and strike the earth obliquely.

This chapter will also lead you into further intricacies. Here is a third question that deserves close attention.

3. Why are temperatures on earth generally so amenable to life?

For example, the planet, Venus, lies somewhat closer to the sun than we do, but Venus is too hot to support life as we know it. The temperature on the surface of Venus was recently (1975) determined to be a sizzling 500°C by a Russian lander. The planet Jupiter lies about five times farther from the sun than we are and is a frozen wasteland —far too cold to support life as we know it. The temperature on the top of the thick clouds that cover Jupiter is estimated to be roughly −110°C. Mars is the only planet (besides the earth, of course) on which there is even a remote chance for life to exist. Why are we on earth so lucky to have almost perfect temperatures? In the last part of this chapter we will begin to approach an answer to this question. We will also look at the solar system so that you can become familiar with the planets, our closest neighbors in space. Our understanding of the solar system will also set the stage for later discussions concerning the origin, structure, and temperatures of the planets and their atmospheres.

To begin, you should know that the earth is a giant ball-shaped rock about 12,700 km in diameter and 40,000 km in circumference. Even so, when you compare the earth to the sun you may be surprised for the *sun is more than 100 times wider than the earth.* The sun is a giant ball of gas 1,400,000 km in diameter. The sun is so large that if it were placed where the moon is now, the earth would be embedded, only a little more than halfway from the sun's center, since the moon averages only 385,000 km from the earth. The only reason that the sun looks so small to us on earth is that it is so far away. The sun is 149.5 million km from the earth on the average. If you decided to drive your car to the sun (and it didn't overheat on the way), then it would take you 200 years if you drove 24 hours each day at a speed of 90 km/h.

## THE EARTH AND SUNLIGHT

Christopher Columbus (1446?–1506) was definitely not the first person to realize that the earth is round (spherical). All the educated people of his time were well aware that the earth is a sphere. Why then did Columbus encounter so much opposition when he presented his expedition plans to the king and queen of Spain?

Columbus made quite a mistake! He actually knew less about the size of the earth than did his scholarly contemporaries. Columbus thought that the earth was much smaller than it is; he therefore calculated that Japan and China lay only about 5000 km to the west of Spain. Scholars living at the time had much better estimates of the size of the earth and knew that Japan and China lay at least 16,000 km to the west of Spain. At the time, supplies would never have lasted during a sea voyage of 16,000 km; so, it is quite understandable that people did not want to finance Columbus. Columbus persisted for over 10 years, and only by constantly appealing to the religious sentiments of the king and queen of Spain was he finally rewarded with a small fleet of three ships. By sheer coincidence, he reached land after sailing nearly 5000 km westward, becoming more convinced than ever that Japan lay nearby. We know that Columbus was wrong, but his persistence changed the course of history.

Everyone now knows that the earth is round, but how did the people living during Columbus' time know it? No one was large enough to see that the earth is round, and no one up to that time had ever traveled completely around the world.

The discovery that the earth is round was a result of the study of geometry. It was probably the ancient Greeks who first realized that the earth is round since they were masters of geometry and astronomy. The Greeks made up for their almost total lack of success in meteorology by excelling in these other fields. When they discovered that they could use geometrical principles when thinking about the earth and the heavens, they quickly determined a number of interesting facts. Besides proving that the earth is round, they also computed the sizes of the earth, moon, and sun as well as the distances from the earth to the sun and to the moon. Some of their calculations were remarkably accurate.

One way to prove that the earth is spherical is to view it from space. At first you could not be sure if the earth were a sphere or if it were merely a flat disk (like a plate). However, as the earth turned (or as you went around the earth) you would soon see only an edge *if the earth were a disk.* We know that the earth is spherical, because no matter how it is turned it always appears circular.

But we know the ancient Greeks had no satellites or rockets so they could never see the earth from space. Yet there were times that they saw the shadow of the earth! This occurs about twice a year—whenever there is an eclipse of the moon (a lunar eclipse). During a lunar eclipse the earth gets directly inbetween the sun and the moon. The earth therefore blocks the sunlight from reaching the moon. The moon is then in the earth's shadow and gets dark (see Figures 4.1 and 4.2).

At the beginning and the end of each lunar eclipse the edge of the earth's shadow crosses the moon. The Greeks noticed that the edge of the earth's shadow *always* formed part of a circle. They then reasoned correctly and said that this fact proves that the outline of the earth is round in every direction and therefore must be spherical.

Once the Greeks realized that the earth is spherical, they wanted to find out exactly how large it is. A Greek named Eratosthenes found the answer to this question somewhat before 200 B.C. He noticed that at noon on June 21 (which is the first day of summer and is also known as the summer solstice) the sun was directly overhead at Syene (now Aswan), since the sunlight penetrated straight to the bottom of a vertical well. We will never know how Eratosthenes first realized this. Perhaps he had been thrown into the well by an angry mob for teaching that the earth is round.

Exactly one year later Eratosthenes was safely back in Alexandria, Egypt. Alexandria is located just about 800 km north of Syene. At noon on June 21, Eratosthenes observed that the sun was *not* overhead but instead was at an angle of 7° from the zenith (vertical). What does this mean?

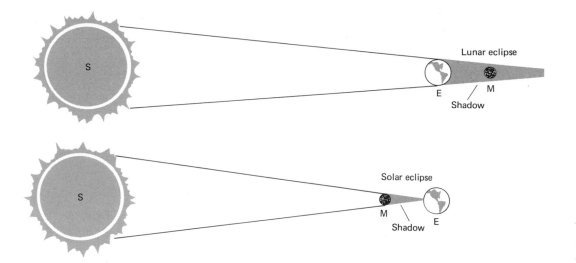

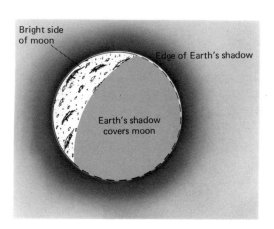

FIGURE 4.1
The positions of the earth, moon, and sun for a lunar and solar eclipse.

FIGURE 4.2
The earth's shadow creeping across the moon. The edge of the shadow is always circular. During a lunar eclipse, surprisingly, the moon never gets completely dark, but rather is dull red. Can you explain this using the principles of refraction and the scattering of sunlight as it passes through the earth's atmosphere?

There are at least two possible explanations. If the earth were flat the sunlight would have to be coming at a different direction at each of the cities. This would also imply that the sun is only a little more than 6400 km away from the earth, and the Greeks knew that the sun was much further away than that (see Figure 4.3). They also knew that the earth is not flat but rather is spherical.

Eratosthenes gave the correct explanation. Because the sun is so far away, all sunbeams striking the earth at a given time are virtually parallel. The 7° difference in the elevation of the sun between the two cities (Alexandria and Syene) must then be due to the fact that the earth's surface is curved. Since every circle has 360°, an angle of 7° represents almost one-fiftieth (1/50) of the entire circle (see Figure 4.4). This change of angle by 7° takes place in a distance of 800 km. How many kilometers, asked Eratosthenes, would it take to go completely around the earth (i.e., change the angle by 360°)? This is a relatively simple mathematical problem.

Example 4.1
Problem in sentence form:
    800 km is about 1/50 of the earth's circumference.

Problem in equation form:
    800 km = (1/50)C       C = earth's circumference

Solve this equation by multiplying each side of the equation by 50.

Solution
    40,000 km = C

You can see that the calculation of Eratosthenes was remarkably accurate.

**Note:** Always remember that in mathematics the word *is* often means *equals,* the word *of* often means *multiplied by,* and the word *per* means *divided by.*

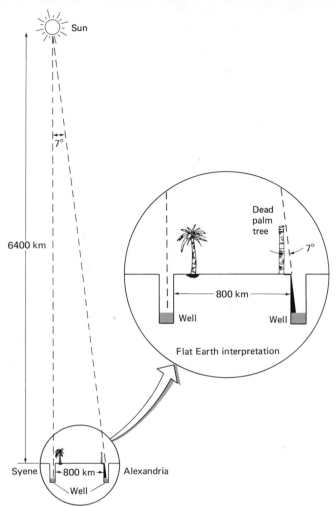

FIGURE 4.3
At noon on June 21, the sun appears at a different angle in the sky at Syene than at Alexandria. *This is the interpretation if the earth were flat.* The sun would be only 6400 km away by this interpretation, which of course is wrong.

Light emitted from the sun spreads out uniformly in all directions and only a small fraction of the sun's rays strike the earth. Even though the sunlight spreads out as it moves away from the sun, we know that all rays that strike the earth at a given time are virtually parallel. You might think that this statement contradicts the first sentence of the paragraph. After all, spreading rays cannot be parallel. On the other hand, some of the rays can be *almost* parallel to others.

It is actually quite easy to show that all rays striking the earth are almost parallel. Imagine that a candle is held close to a person in a dark room. On the wall opposite the candle there will be a large shadow of the person. This shows that the light rays hitting the edges of the person are

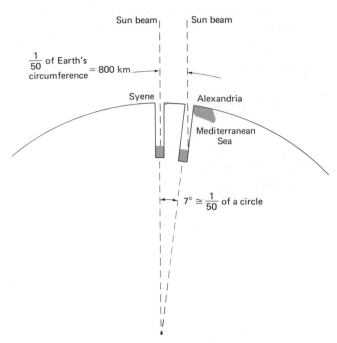

FIGURE 4.4
The correct interpretation shows the sun at a great distance. Then the 800 km between Alexandria and Syene is about 1/50 of the earth's circumference.

not parallel at all. However, as the candle is moved back away from the person, the shadow on the wall gets smaller. This shows that the light rays striking the person are more nearly parallel when the source of light is farther away (see Figure 4.5).

The same reasoning works for the light reaching the earth from the sun. If the earth were reduced to the size of a pea, as it is in Figure 4.6, then you can see that the farther that the sun is located from the earth, the more closely parallel the light rays striking the earth become. This diagram unfortunately does not even begin to show how closely parallel all the sun's rays are on earth. If it were drawn to scale, then the sun would have to be 75 m away! The angle between the two rays striking the opposite ends of the earth are then a mere one-two hundreth (1/200) of a degree!

**Note:** Because the sun itself is so large, rays coming from opposite ends of the sun actually differ by an angle of ½°. A sunbeam is therefore not technically a ray, but instead is a very thin cone of light. For most purposes, in meteorology this slight angle is unimportant and is therefore neglected; but as you can see in Figure 4.1 the angle *is* important for such events as eclipses.

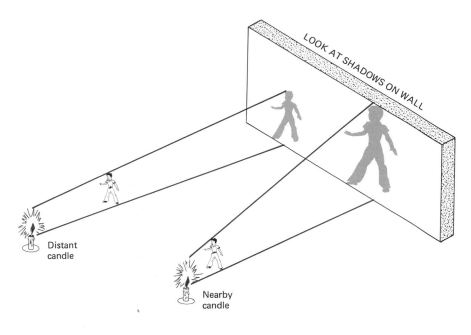

LOOK AT SHADOWS ON WALL

Distant candle

Nearby candle

FIGURE 4.5
The closer the light, the more it spreads out. Compare the sizes of the two shadows.

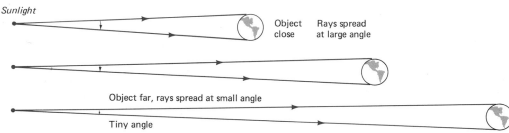

Sunlight

Object close — Rays spread at large angle

Object far, rays spread at small angle

Tiny angle

FIGURE 4.6
Since the sun is so far away, sun rays all over the earth are virtually parallel. For an earth this size, the sun would actually be 75 m away!

The fact that sunbeams are virtually parallel all over earth means that, *above* the atmosphere and clouds, *direct* sunlight is equally intense all over earth. Worldwide differences in temperature must therefore be due to some other factor such as the directness (directness = overhead) with which the sun's rays strike the earth and atmosphere. This point has already been mentioned in Chapter 1. We have also seen that sunbeams, since they are virtually parallel, are also important when dealing with many of the optical phenomena such as rainbows.

## THE SEASONS: THE EARTH'S ORBIT

This is not a book about the intellectual exploits of the ancient Greeks but I feel compelled once again to give credit where credit is due. Shortly after 1500, the great scientist Nicolaus Copernicus (1473–1543) discovered that the earth revolves around the sun. Prior to that time every-

one had thought that the sun revolved around the earth. Everyone, that is, except for Aristarchus of Samos who lived earlier than 200 B.C. Aristarchus attempted to measure the distance from the sun to the earth. Although he grossly underestimated the distance, he realized that the sun is very, very far away from the earth. In order for the sun to appear as large as it does, considering its great distance, it must be far larger than the earth.

Aristarchus then reasoned that, since the sun is far larger than the earth (he thought that it was 7 times larger while we know that it is actually about 110 times larger), it is more reasonable to assume that the small earth revolves around the larger sun.

Aristarchus stood alone and later came to doubt his own work. Many others countered his argument by saying that if the earth moved then it should pass by some of the stars (technically this is known to astronomers as **parallax**). Since the earth does not seem to move past any of the stars, said the opponents of Aristarchus, then the earth must not be moving. It would be ridiculous, they argued further, to say that the reason that the earth does not seem to pass the stars is due to the great distance of the stars. No stars, they said, could be that far away. But they were wrong, and the fact is that the stars are very, very far away.

The nearest star is the sun. It is an average of 149.5 million km away from the earth. The next nearest star is almost a million times farther than that! Within the past 100 years extremely sensitive telescopes have actually shown that we do slightly pass by a few of the closest stars in the sky. Neither Copernicus nor any of the other ancients could possibly have seen this minute amount of parallax. Copernicus assumed that the earth moved around the sun because the mathematical scheme of the motions of the planets becomes far simpler and more rational once the sun is placed in the center of the solar system. Mathematics may be very difficult at times, but it is a very powerful tool for scientific problems.

When the earth lies at its average distance from the sun, the intensity of *direct* sunlight above the atmosphere is 1.95 cal/(cm²)(min) (i.e., calories per square centimeter per minute) or 1.38 kw/m² (i.e., kilowatts per square meter). This is known as the **solar constant** because, to the best of our knowledge, the total amount of light coming from the sun varies very little, if at all.

Once every year the earth revolves around the sun in an almost circular orbit. The orbit is actually an ellipse — which is a slightly flattened circle. While the earth is 149.5 million km from the sun on the average, around January 1 it comes as close as 147 million km; around July 1 it recedes as far as 152 million km from the sun. Think about this for a minute. Do you realize that *we have winter in the NH when the earth is closest to the sun, and summer when the earth is farthest from the sun?*

This may seem confusing since, when the earth is closer to the sun in January, *direct* sunlight is actually more intense. You might ask why this doesn't lead to warmer weather in January. Direct sunlight is about 7% more intense on January 1 than it is on July 1.

However, two other factors are far more important in determining the difference between winter and summer. The temperature of a region on earth strongly depends on the directness with which the sun's rays strike the earth (remember that direct means overhead) and also on the number of hours of daylight each day. These last two factors decrease by as much as 100% from summer to winter, depending on your location on earth. The fact that the earth is closer to the sun during January is therefore of secondary importance in determining the temperatures on earth. Its effect is merely to make the winters in the NH a little warmer and the summers a little cooler than they would otherwise be. Here we are implying two things; one that during July in the NH the sun climbs higher in the

sky, and second that there are more daylight hours than during January. This is indeed the case.

In the middle latitudes of the NH the sun rises each morning in the eastern part of the sky. If you face the sun it moves toward your right hand, ascending until noon (forgetting about daylight savings time) when it is due south and is at its highest point in the sky for that day. After noon it begins to descend, but continues moving toward your right hand and finally sets in the western part of the sky.

Each year at noon on June 21, the sun reaches its highest point for the year. June 21 is also the longest day of the year and is called the summer solstice. On December 21 of each year the *noon* sun is at its lowest point in the sky. December 21 is also the shortest day of the year and is called the winter solstice (see Figure 4.7).

These differences are related to the fact that, if you travel far enough north, the sun will never set on June 21 but will never rise on December 21. Any point from 66½° latitude to the North pole will have 24 hours of daylight on June 21, but 24 hours of darkness on December 21 (see Figure 4.8).

The time has now come to explain all these facts. As the earth revolves around the sun, it also rotates like a spinning top about its axis. The earth's axis is an imaginary line running from the North Pole through the center of the earth to the South Pole. Each day and night on earth takes 24 hours and is slightly more than one rotation. Each year has almost exactly 365¼ days and nights.

You should notice that on June 21, the North Pole is tilted toward the sun at an angle of approximately 23½°. June 21 is the first day of summer in the NH, but the first day of winter in the SH. Figure 4.8 shows that on June 21,

FIGURE 4.7
Paths of the sun across the sky at various times of the year at about 55°N latitude.

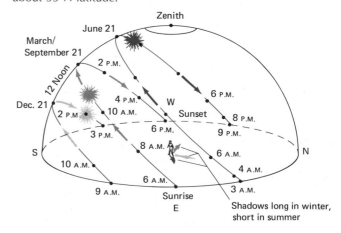

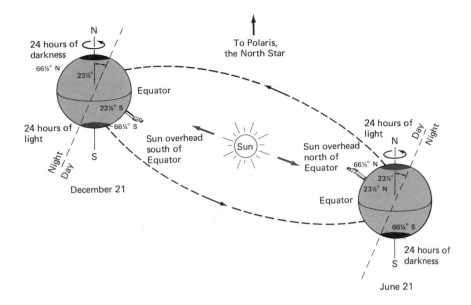

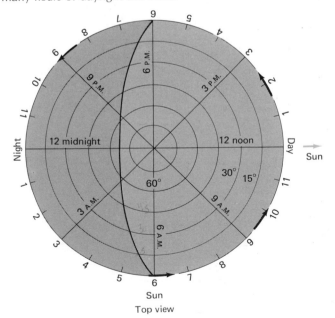

FIGURE 4.8
The earth in its orbit around the sun. Notice for instance, on December 21, the sun is directly overhead at 23½° S latitude. At this time, the sun is low in the sky in the NH, and there is more darkness than light. North of 66½° N latitude, there is 24 hours of darkness; south of 66½° S latitude, there is 24 hours of light.

the sun at noon is nearly overhead throughout much of the NH, but is low in the sky in the SH.

You should also be able to see from Figure 4.8 that the sun never sets near the North Pole on June 21, because even the midnight line there is bathed in sunlight. On the other hand, near the South Pole there is not a hint of the sun all day, because even the noon line there lies in darkness.

The chart of the earth's orbit shows that the earth's axis always aims to the same point in space—near the North Star, Polaris. Because the earth goes around the sun, half a revolution (or half a year) after the North Pole has tilted toward the sun, we know that it must tilt away from the sun simply because the earth is now on the other "side" of the sun.

December 21 is half a year after June 21. The situation on December 21 is therefore the exact opposite of the situation on June 21. On December 21, the sun is high at noon in the SH where it is the first day of summer, and low in the NH where it is the first day of winter. The situation at the poles has also reversed. Now the South Pole has 24 hours of sunlight, whereas the North Pole has 24 hours of darkness.

On two days of the year, March 21 and September 21, the earth's axis tilts neither toward nor away from the sun. These two days are therefore called the equinoxes and, except for effects due to refraction of sunlight by the atmosphere, the day and night everywhere on earth are each exactly 12 hours.

A top view of the earth shows how many hours of sunlight can be expected in the NH. Since a complete ro-

tation is made in 24 hours, we can tell how long daylight is in the following way. In one rotation you complete a circle around a *given* latitude. The latitude lines are circles (see Figure 4.9). To compute the number of hours of

FIGURE 4.9
The number of hours of sunlight at any latitude in the NH on June 21. The heavy solid line divides night and day. Three-fourths of the 60° latitude line is in daylight, so day is 18 hours long. When does the sun rise at 30° latitude, and how many hours of daylight are there?

daylight, simply find the fraction of the latitude circle that is on the daylight side of the earth. For instance, you might ask how many hours of daylight do you get on June 21 at 60° N latitude?

---

Example 4.2
How many hours of daylight are there at 60° N latitude on June 21?

Information: From Figure 4.9 we see that 3/4 of the latitude circle lies in daylight, and we know that each latitude circle takes 24 hours to complete. Therefore
　　3/4 of 24 hours is daylight

Solution
In equation form:
　　(3/4) (24 hours) = daylight = 18 hours

---

Table 4.1 gives daylight information for a number of world cities and locations. Hardly anyone looks at the tables contained in textbooks, but they do often contain useful information. I now ask you to notice some strange information contained in Table 4.1. Over the equator the sun is not highest on June 21, but rather at two other times —March 21 and September 21. At Caracas, Venezuela (latitude 10°), the sun is highest around May 21 and July 21. You can see that the sun behaves somewhat differently in the tropical regions than it does in the midlatitude regions. This leads to certain differences between tropical and midlatitude climates. For example, many places around 10° N latitude are warmest in May. Notice also that in Buenos Aires, Argentina, which is south of the equator, the sun is low and the day is short on June 21, but on December 21 the sun is high and the day is long. Finally, look at the figures for the North Pole where the sun never gets high in the sky but both day and night last for six months apiece.

At this point, you should be aware that the earth's orbit actually changes slightly from year to year. These changes have a potential impact on climate changes and therefore will be discussed in Chapter 24.

TABLE 4.1
Length of Day (in Hours and Minutes) and Elevation Angle of the Sun at Noon[a]

| | Latitude and Representative City | | | | | | | |
|---|---|---|---|---|---|---|---|---|
| Date | 90° N North Pole | 66½° N Salehkard | 60° N Bergen | 50° N Frankfurt | 23½° N Havana | 10° N Caracas | 0° S Kisangani | 40° S Valdivia |
| December 21 | 0:00 — | 0:17 0° | 6:22 7° | 8:30 17° | 10:47 43° | 11:51 57° | 12:07 67° | 15:17 74° |
| January 21 and November 21 | 0:00 — | 4:28 3° | 7:15 10° | 8:53 20° | 10:57 46° | 11:53 60° | 12:06 70° | 14:52 71° |
| February 21 and October 21 | 0:00 — | 8:27 12° | 9:24 19° | 10:22 29° | 11:46 55° | 11:57 69° | 12:06 79° | 13:29 62° |
| March 21 and September 21 | 24:00 0° | 12:15 24° | 12:12 30° | 12:09 40° | 12:07 67° | 12:06 80° | 12:06 90° | 12:08 50° |
| April 21 and August 21 | 24:00 12° | 16:03 35° | 15:00 41° | 13:56 51° | 12:24 78° | 12:15 89° | 12:06 79° | 10:47 38° |
| May 21 and July 21 | 24:00 20° | 20:04 44° | 17:11 52° | 15:27 62° | 13:17 87° | 12:20 80° | 12:06 70° | 9:24 30° |
| June 21 | 24:00 24° | 24:00 47° | 18:04 55° | 15:50 65° | 13:27 90° | 12:37 77° | 12:07 67° | 9:01 27° |

[a] To the nearest whole degree.

# THE GENERAL LAW FOR THE INTENSITY OF RADIATION

In the following few pages you will learn precisely how the intensity of sunlight depends on how high the sun is in the sky and how far you are from the sun.

Take a searchlight in a dark room, stand close to the wall, and shine it directly at the wall. You will see a small circle of bright light. Then, aim the searchlight obliquely at the wall, and you will get a larger ellipse of duller light. Then shine the light directly at the wall once again and back away from the wall. The circle of light will get larger and duller (see Figure 4.10).

In each case the same total amount of light strikes the wall, but the light is always less intense when it has spread over a larger area. On this basis we can write a general intensity law. The total radiation ($R$)—which remains constant—is equal to the intensity ($I$) of the radiation times the area ($A$) over which it has spread. In equation form this appears as

**The General Intensity Rule:**

$$R = (I_a)\,(A_a) = (I_c)(A_c)$$

or rearranging,

$$I_c = (I_a)\left(\frac{A_a}{A_c}\right)$$

where the subscripts stand for areas $a$ or $c$.

This law can now be used to show how sunlight intensity depends on the angle of the sun in the sky or the distance from the sun.

**The Sine Law of Sunlight Intensity**

The precise amount that the sun's rays spread out on the ground depend on the angle of the sun in the sky. Thus, in Figure 4.11 you can see that when the sun is low in the sky, as it is near the poles, the sunlight spreads over a large ellipse and so is not very intense. On the other hand, near the equator where the sunlight is almost directly overhead you get a smaller circle of more intense light.

Now we use the general intensity law and Figure 4.12 to find precisely how sunlight intensity depends on angle. In Figure 4.12 sides $a$, $b$, and $c$ form a right triangle in which side $c$ is the hypotenuse and side $a$ is opposite the sun's elevation angle, $a'$. The sine of angle $a'$ is therefore given by

$$\sin a' = \frac{\text{opposite side}}{\text{hypotenuse}} = \frac{a}{c}$$

but this is also equal to

$$\sin a' = \frac{A_a}{A_c}$$

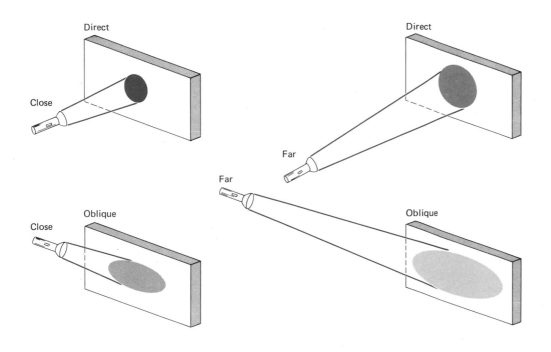

FIGURE 4.10
Illustration of the general intensity law. The light is most intense when the searchlight is close to the wall and aimed directly at it. The light is least intense when the searchlight is far from the wall and aimed obliquely at it.

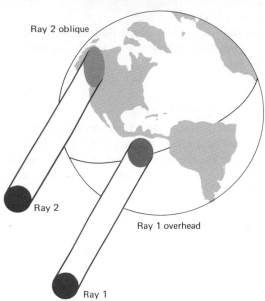

**FIGURE 4.11**
Sunlight always strikes the high latitude obliquely, and so it spreads out more and is, therefore, less intense. Shadows are also long (refer back to Figure 1.4).

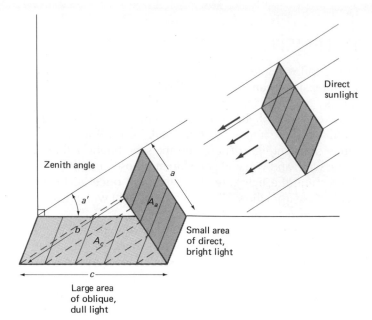

**FIGURE 4.12**
Illustration for the sine law applied to light intensity.

Now simply substitute this into the general intensity rule, and you arrive at

**The Sine Law:**

$$I_c = (I_a)(\sin a')$$

---

Example 4.3
Compare the intensity of sunlight at noon in New York City on June 21 with the intensity on December 21.

Information:
At noon on June 21 in NYC, $a' = 73°$.
At noon on December 21 in NYC, $a' = 26°$.
The intensity of direct sunlight, $I_a \cong 2.0$ calories per square centimeter per minute. (In all calculations we use 2.0 rather than the more accurate 1.95, for computational ease.)

Equation:
$$I_c = (I_a)(\sin a')$$

Solution
Substitute:

| June 21 | December 21 |
|---|---|
| $I_c = (I_a)(\sin 73°)$ | $I_c = (I_a)(\sin 26°)$ |
| $= (2)(0.956)$ | $= (2)(0.438)$ |
| $= 1.912$ | $= 0.876$ |

You can see that the sun's rays at noon in New York City are more than twice as intense on June 21 than they are on December 21.

Example 4.4
Compare the intensity of sunlight at noon in Bergen, Norway, on June 21 with the intensity at noon on December 21.

Information:
Use Table 4.1 to find angle $a'$.
At noon on June 21 in Bergen, $a' = 53°$
At noon on December 21 in Bergen, $a' = 6°$.
The intensity of direct sunlight, $I_a \cong 2.0$ calories per square centimeter per minute.

Equation:
$$I_c = (I_a)(\sin a')$$

Solution
Substitute:

| June 21 | December 21 |
|---|---|
| $I_c = (I_a)(\sin 53°)$ | $I_c = (I_a)(\sin 6°)$ |
| $= (2)(0.805)$ | $= (2)(0.105)$ |
| $= 1.61$ | $= 0.21$ |

You can see that the sun's rays at noon in Bergen, Norway, are almost 8 times more intense on June 21 than on December 21. No wonder winter is colder than summer! Did you ever imagine that a silly little angle could be that important?

TABLE 4.2
Values of Sines

| Angle in Degrees | Sine | Angle in Degrees | Sine | Angles in Degrees | Sine |
|---|---|---|---|---|---|
| 0 | 0.000 | 31 | 0.515 | 61 | 0.875 |
| 1 | 0.018 | 32 | 0.530 | 62 | 0.883 |
| 2 | 0.035 | 33 | 0.545 | 63 | 0.891 |
| 3 | 0.052 | 34 | 0.559 | 64 | 0.899 |
| 4 | 0.070 | 35 | 0.574 | 65 | 0.906 |
| 5 | 0.087 | 36 | 0.588 | 66 | 0.914 |
| 6 | 0.105 | 37 | 0.602 | 67 | 0.921 |
| 7 | 0.122 | 38 | 0.616 | 68 | 0.927 |
| 8 | 0.139 | 39 | 0.629 | 69 | 0.934 |
| 9 | 0.156 | 40 | 0.643 | 70 | 0.940 |
| 10 | 0.174 | 41 | 0.656 | 71 | 0.946 |
| 11 | 0.191 | 42 | 0.669 | 72 | 0.951 |
| 12 | 0.208 | 43 | 0.682 | 73 | 0.956 |
| 13 | 0.225 | 44 | 0.695 | 74 | 0.951 |
| 14 | 0.242 | 45 | 0.707 | 75 | 0.966 |
| 15 | 0.259 | 46 | 0.719 | 76 | 0.970 |
| 16 | 0.276 | 47 | 0.731 | 77 | 0.974 |
| 17 | 0.292 | 48 | 0.743 | 78 | 0.978 |
| 18 | 0.309 | 49 | 0.755 | 79 | 0.982 |
| 19 | 0.326 | 50 | 0.766 | 80 | 0.985 |
| 20 | 0.342 | 51 | 0.777 | 81 | 0.988 |
| 21 | 0.358 | 52 | 0.788 | 82 | 0.990 |
| 22 | 0.375 | 53 | 0.799 | 83 | 0.993 |
| 23 | 0.391 | 54 | 0.809 | 84 | 0.995 |
| 24 | 0.407 | 55 | 0.819 | 85 | 0.996 |
| 25 | 0.423 | 56 | 0.829 | 86 | 0.998 |
| 26 | 0.438 | 57 | 0.839 | 87 | 0.999 |
| 27 | 0.454 | 58 | 0.848 | 88 | 0.999 |
| 28 | 0.470 | 59 | 0.857 | 89 | 1.000 |
| 29 | 0.485 | 60 | 0.866 | 90 | 1.000 |
| 30 | 0.500 | | | | |

## The Inverse Square Law

Humankind has taken a giant step! We have reached the moon and are reaching for the planets. It is quite possible that by the year 2000 astronauts will have landed on the planet Mars. What will they find when they get there? Will there be life? If not, will the planet at least be hospitable to life?

We know that Mars presents certain problems for maintaining life. First, there is an atmosphere on Mars but it is very thin (the earth's atmosphere is almost 200 times denser). Second, Mars is 228 million km from the sun on the average. This is about 1.5 times the distance between the sun and the earth. Because Mars is so far from the sun, the sun's rays are much weaker on Mars than they are on earth. As a result, Mars is a rather cold planet.

Is Mars too cold to support life? In this section, I will

begin to answer this question. The temperature of an object such as a planet depends on the intensity of sunlight it receives, which depends on its distance from the sun.

Remember how the circle of light from the searchlight spread out and grew less intense when you backed away from the wall? Obviously, sunlight gets less intense as distance increases, but what is the precise law?

Your first impression might be that if the light is twice as far away, then it should be half as intense. If this is your impression then you are wrong! The correct answer is that a light twice as far away will be only 0.25 as intense! Scientists call this relationship the **inverse square law.**

The general intensity law and Figure 4.13 should help you understand the inverse square law. As the distance ($d$) doubles (or triples), the light spreads out over 4 (9) times the area. In general, the area varies with the distance squared, or in mathematical form

$$\frac{A_a}{A_c} = \left(\frac{d_a}{d_c}\right)^2$$

When you substitute this into the general intensity law you find the famous inverse square law

**Inverse Square Law:**

$$I_c = I_a \left(\frac{d_a}{d_c}\right)^2$$

**Note:** The inverse square law is one of the basic and most important laws in science. Not only is it true of the intensity of light, but it is also true for gravitational attraction (as Isaac Newton discovered when the apple fell on his head) and for the attraction or repulsion of electrical charges.

FIGURE 4.13
Illustration for the inverse square law applied to light intensity.

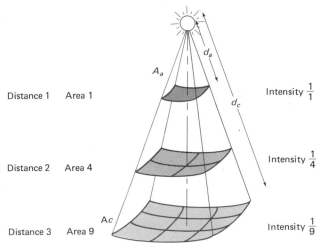

## Example 4.5

What is the intensity of direct sunlight on the planet Jupiter?

Information: We know (1) that the intensity of direct sunlight on the earth is $I_a = 2.0$ calories per square centimeter per minute, and (2) from Table 4.3 that Jupiter is 5.2 times farther from the sun than the earth is. In terms of equations this information takes the form shown below.

1. $I_a = 2.0$
2. $\dfrac{d_a}{d_c} = \dfrac{1.0}{5.2}$    or    $\left(\dfrac{d_a}{d_c}\right)^2 = \dfrac{1.0}{27.0}$

Equation: The inverse square law, naturally

$$I_c = I_a \left(\frac{d_a}{d_c}\right)^2$$

Solution
Substitute:

$$I_c = \frac{1}{27.0}(2.0)$$
$$= 0.074$$

In short, we find that the intensity of direct sunlight on the planet Jupiter is about 1/27 of the intensity on earth, and therefore we can expect that Jupiter is a far colder planet than the earth. A sunny day on Jupiter would appear about as dark as a cloudy, rainy day on earth.

See if you can figure out how direct sunlight on the planet Pluto compares with direct sunlight on the earth. From Table 4.3 we find that Pluto is roughly 40 times farther from the sun than we are (on the average). Using the inverse square law you should find that the intensity of direct sunlight on the planet Pluto is only 1/1600 that of earth. Bright sunlight on the planet Pluto therefore appears no brighter than twilight on earth.

# THE REST OF THE SOLAR SYSTEM

The sun, the planets and their moons, and other heavenly bodies such as the asteroids and comets constitute the solar system. We find that the earth and the other planets all formed at about the same time, some 4.6 billion years ago. All the planets except for Mercury and Pluto have appreciable atmospheres. The chemical makeup of the planets and their atmospheres, in fact, depends on the size of the planet and the distance from the sun. Each planet and each atmosphere is therefore unique, but all are close relatives.

All bodies in our solar system revolve around the sun. The sun is a small-to-medium sized star, but it still dwarfs all the planets. The sun is actually a giant ball of gas held together by its own gravity. It rotates on its axis, turning once in about 27 days.

TABLE 4.3
Some Properties of the Sun and the Planets

| | Mass— Relative to Earth[a] | Distance from Sun | | Diameter | | Time to Orbit Sun in Years | Mean Surface Temperature (°C) |
| | | In Millions of Kilometers | In Units Relative to Earth-Sun Distance | In Thousands of Kilometers | Relative to Earth | | |
|---|---|---|---|---|---|---|---|
| Sun | 329,390 | — | — | 1390.6 | 109 | — | 5570° |
| Mercury | 0.055 | 58 | 0.39 | 5.14 | 0.40 | 0.24 | 260°[b] |
| Venus | 0.807 | 108 | 0.72 | 12.62 | 0.99 | 0.62 | 500° |
| Earth | 1.0 | 150 | 1.0 | 12.76 | 1.0 | 1.0 | 15° |
| Mars | 0.107 | 228 | 1.52 | 6.86 | 0.54 | 1.88 | −60° |
| Jupiter | 314.5 | 778 | 5.20 | 143.60 | 11.2 | 11.9 | −110° |
| Saturn | 94.1 | 1426 | 9.52 | 120.60 | 9.5 | 29.5 | −190° |
| Uranus | 14.4 | 2869 | 19.2 | 53.40 | 4.18 | 84.1 | −215° |
| Neptune | 16.7 | 4495 | 30.0 | 49.70 | 3.90 | 165 | −225° |
| Pluto | ? | 5900 | 39.4 | 3.48 | 0.27 | 249 | −235° |

[a] The mass of the earth is $(6)(10)^{24}$ kilograms.
[b] On the sunlit side.

On the next clear, moonless night go out and look up at the stars. In the city we can see little of the beauty of the heavens, because the city lights drown out all but the brightest stars. But in the country the heavens appear in all their glory and grandeur, and the sky we see there is almost as it was when Moses, Jesus, and Mohammed walked the earth.

The relative positions of all the stars are virtually fixed with respect to all the other stars. The paths they take across the sky are so regular that they have been accurately known for well over 4000 years. When the ancients looked carefully at the sky they noticed that a few of the so-called "stars" did not remain in the same relative position but rather moved with respect to all the fixed stars. In-deed, *planetes* is the Greek word for wanderer. The ancients correctly reasoned that the planets are much closer to the earth than the fixed stars, but that is all they knew about them.

We now know that the planets are our closest neighbors in space and are very different from the stars. Stars produce their own light and heat, whereas the planets merely reflect the light of the sun.

The planet Mercury is the most rapid wanderer across the skies because it is closest to the sun. A year on Mercury is only 88 of our days long. Mercury is a small, hot, cratered planet with a diameter less than half that of the earth. (See Figure 4.14.)

Venus is the brightest object in the sky other than the

FIGURE 4.14
The planet Mercury. Like the moon, it is covered by craters and has no atmosphere.

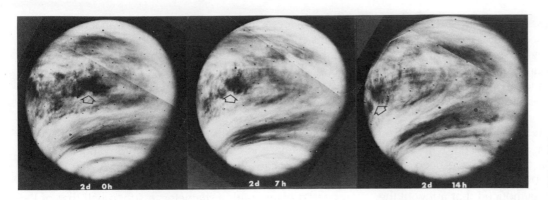

FIGURE 4.15
The planet Venus with its cloud bands, as seen in ultra-violet light.

sun or the moon. Venus is nearly as large as the earth and comes closest to the earth of all the planets. It circles the sun in about 225 days but takes 240 days to turn on its axis, so that on Venus the sun actually rises in the west. (See Figure 4.15.)

Mars is a small planet and like the earth has icecaps near the poles. Mars is much colder than the earth but still seems to be the most likely candidate of all the other planets to support life. (See Figure 4.16.)

These four planets—Mercury, Venus, Earth, and Mars—are called the inner planets because they are close to the sun. They are all basically large balls of rock and iron. This is not true of the remaining planets, which are called the outer planets. The outer planets (except for Pluto) are all far larger than any of the inner planets and are mainly composed of gases and ices that remain frozen because they are so far from the sun (see Figures 4.17 and 4.18).

The first of the outer planets is Jupiter, the largest of the planets. Jupiter is more than 11 times wider than the earth and is more massive than all the other planets combined. If Jupiter had been only a few times larger, it would have evolved into a small star instead of a planet. As it is, Jupiter produces more heat internally than it gets from the sun. It takes Jupiter 12 years to circle the sun, but only 10 hours to complete a day.

Saturn, another giant among the planets, comes after Jupiter. Saturn is one of the most beautiful objects in the sky because of the rings which circle it. Like the moons of Jupiter, the rings of Saturn can only be seen with a telescope. Saturn is almost ten times farther from the sun than we are, and it takes almost 30 years to circle the sun. Saturn is so light that it would float if it were placed in a large enough ocean.

Uranus and Neptune follow, but there is not much that we know about them yet. For example, the rings around Uranus were just discovered in 1977. Both planets are about 4 times wider than the earth and are extremely cold due to their great distance from the sun.

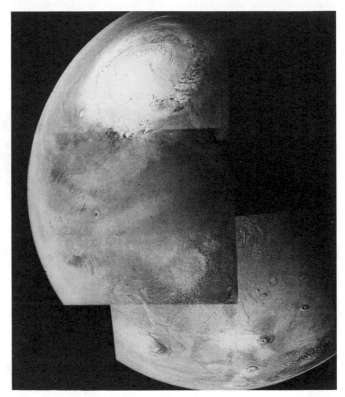

FIGURE 4.16
The planet Mars with ice caps.

Beyond Neptune lies Pluto, a small barren chunk of rock or ice with a large moon, just discovered in 1977. Pluto lies more than 40 times farther from the sun than the earth does and takes almost 248 years to circle the sun. Pluto was discovered in 1930 after years of effort. The existence of Pluto had long been suspected because something was disturbing the orbit of Neptune around the sun. Even so the discovery of Pluto has turned out to be a fluke. It *is* true that something is disturbing the orbit of Neptune, but Pluto is far too light to be responsible! Could there be another planet? We don't know yet.

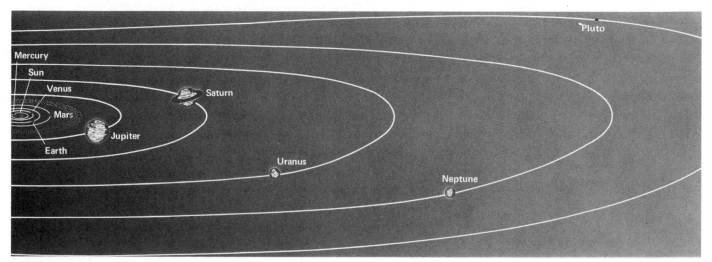

FIGURE 4.17
Relative distances of the planets.

FIGURE 4.18
Relative sizes of the planets.

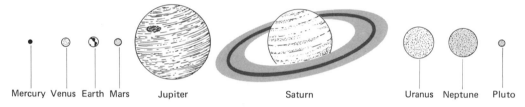

Mercury  Venus  Earth  Mars    Jupiter         Saturn         Uranus  Neptune  Pluto

# PROBLEMS

**4-1** How did the ancient Greeks know that the earth is spherical?

**4-2** Why is sunlight virtually parallel all over the earth?

**4-3** Why are lunar eclipses much more common than solar eclipses?

**4-4** During a total lunar eclipse the moon never gets completely dark but instead is a very dull red. What does the earth's atmosphere have to do with this?

**4-5** Using the concept of parallax, explain why cumulus clouds *seem* to move by faster than cirrus clouds.

**4-6** When is the earth closest to the sun? What season is it then in the NH?

**4-7** Explain why it is winter in one hemisphere at the same time that it is summer in the other.

**4-8** Explain why the sun rises in the northeast and sets in the northwest during July, but rises in the southeast and sets in the southwest during January.

**4-9** Why does the sun never set on June 21, north of the Arctic Circle?

**4-10** Explain why the sun circles the sky from left to right at 40° N latitude, but from right to left at 40° S latitude.

**4-11** Using Figure 4.9, find the time of sunrise and sunset and the length of daylight at 30° N latitude on June 21.

**4-12** Demonstrate to yourself that in a room with only one weak light it is easiest to read a book when the book faces the light directly and when the book is brought near the light.

**4-13** Compute the intensity of sunlight striking the ground at noon in Valdivia, Chile, on (a) December 21 and (b) June 21. Assume no light has been depleted by the atmosphere (refer to Table 4.1).

**4-14** What is the intensity of direct sunlight on the planet Mercury (use Table 4.3)?

**4-15** Which planet intercepts the greatest *total* amount of sunlight?

**4-16** How far does a planet have to be from the sun to have a solar constant 10 times as great as on earth?

# AIR MOLECULES

The air is composed of millions upon millions of tiny molecules that are rapidly moving around and constantly colliding with one another. Even though we are not able to keep track of any single molecule, the laws of probability and statistics enable us to predict the average behavior of large numbers of molecules. This is all that we need to be able to make fairly good weather forecasts and to understand how air behaves.

By thinking of the air (mathematically) as essentially a bunch of bouncing billiard balls, it becomes possible to explain air pressure and temperature and such complicated phenomena as diffusion, viscosity, heat conduction, and condensation and evaporation. The purpose of this chapter is to explain these features and properties of the air.

## PRESSURE OR IMPACTS OF MOLECULES

Near the end of the movie, *Goldfinger,* James Bond (the hero) and Goldfinger (the villain) have a final fight in a high-flying jet. The fight ends when James Bond shoots a bullet through a window in the plane. The window shatters and Goldfinger gets "sucked" out of the plane like a piece of putty.

The cabins of all jets are pressurized, because at heights of 10 km or more the air pressure is so low that human beings would quickly suffocate. The difference between the air pressure inside the cabin and outside is so great that it can easily "suck" a person out of the plane if a door or window should accidentally open. Actually, the person would not be "sucked" out of the plane but would be pushed out by the escaping air molecules!

*The pressure of the air is due to the impacts of air molecules.* When you catch a medicine ball, the impact of the ball against your body invariably knocks you backward. Individual molecules are far, far too small to be felt, but air pressure is the result of millions of impacts every second. Normally, molecules hit you equally on all sides so that you are not pushed anywhere, but when a pressurized cabin is suddenly opened many more molecules rush out than in. Then, any object that "gets in the way" simply gets pushed out of the opening by the escaping air molecules (see Figure 5.1).

Since pressure is the result of impacts, it might seem strange that air pressure should feel continuous rather than sporadic. For instance, when you change altitude by climbing a mountain or skin diving or riding in an elevator, you often feel a continuous pressure in your ears until your ears "pop." Similarly, when one person applauds, you hear distinct, sporadic claps, but when ten thousand people applaud the sound of all the impacts seems almost continuous. The larger the number of impacts of hands or molecules, the more continuous the sensation.

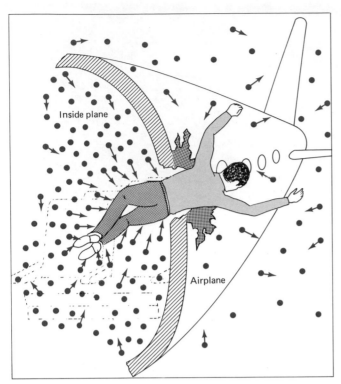

FIGURE 5.1
Molecules pushing a man out of a pressurized airplane cabin. It is incorrect to say the man gets "sucked" out.

TABLE 5.1
Probability for Heads Showing When *10* Coins Are Tossed

| Number of Heads | Probability |
| --- | --- |
| 0 | 1/1024 |
| 1 | 10/1024 |
| 2 | 45/1024 |
| 3 | 120/1024 |
| 4 | 210/1024 |
| 5 | 252/1024 |
| 6 | 210/1024 |
| 7 | 120/1024 |
| 8 | 45/1024 |
| 9 | 10/1024 |
| 10 | 1/1024 |

flip 100 coins. These bounds change to 54% and 46% heads with 1000 coins and 50.008% and 49.992% heads with 1 billion coins.

Generalizing, you can see that the *more coins* you flip, the *less perceptible* are the relative variations in the outcome. When large numbers are involved everything seems to average out to a more or less constant value, and the coin example demonstrates a general rule of statistics. The rule works equally well for molecules and, since there are millions upon millions of molecular impacts against your body each second, the pressure variations from second to second are completely insignificant.

Since air pressure results from the impacts of molecules, the magnitude of the pressure on any object depends on the following three factors:

**1.** The speed of the molecules.
**2.** The mass of the molecules.
**3.** The frequency of the impacts.

When molecules move faster and are more massive, they exert a larger pressure. The pressure also is larger when the impacts occur more frequently.

The last factor, the frequency of the impacts, is actually somewhat complicated. The frequency of the impacts depends both on the concentration and the speed of the molecules. Summarizing these three factors, we find that the equation for pressure due to air molecules is as follows

$$P = 1/300 \; \rho v^2 = 1/300 \frac{m}{V} v^2$$

where $P$ = pressure, $v$ = velocity or speed, $m$ = mass, $V$ = volume, and $\rho$ = density. The factor 1/300 is a constant

This property can be explained by the laws of statistics. In the following argument think of the molecules as coins. Imagine that in each second a molecule has a 50% chance of colliding against you. When a coin lands heads up it represents a collision, but when it lands tails up it represents no collision.

Start by flipping 10 coins each second. On the average trial (or second), 5 of the coins will land heads up (50% heads), and on more than 2% of the trials there will either be 9 or 10 heads or tails. Thus, extreme events are unusual but not too unusual (see Table 5.1).

If air pressure were the result of only 10 molecules, pressure would feel quite sporadic and irregular. As with the coins there would be some seconds in which 90% of the molecules collided with you, while in other seconds only 10% of the molecules would collide. However, air pressure is due to countless millions of molecules in each thimbleful of air. What does that imply?

Now, flip a larger number of coins. This could be very time consuming, but the point is the more coins you flip the more likely that the number of heads and tails will be nearly equal. In fact, there is only a 1% **probability** that you will get more than 63% or less than 37% heads if you

designed to make the units work out when $P$ is given in millibars, $m/V \equiv \rho$ in kilograms per cubic meter, and $v$ in meters per second.

Since we can measure the pressure of the air with a barometer and the density with the aid of a scale, we can use this law to tell how fast a typical molecule of air moves.

---

Example 5.1
What is a typical velocity for an air molecule under normal atmospheric conditions?

Information: Under normal atmospheric conditions at sea level, $P = 1000$ mb and density, $\rho = 1.2$ kilograms per cubic meter.

Equation:

$$P = 1/300 \; \rho v^2$$

Rearrange to solve for velocity, $v$:

$$v^2 = \frac{300P}{\rho}$$

Solution
Substitute:

$$v^2 = \frac{(300)(1000)}{1.2} = (25)(10^4)$$

Take the square root of each side:

$$v = (5)(10^2) = 500 \text{ m/s}$$

**Note:** This is close to the speed of sound in air. Sound vibrations are transmitted by colliding molecules and, therefore, it is natural that sound waves move with roughly the same speed as the molecules do.

---

# AIR PRESSURE AND THE BAROMETER

Near the beginning of the seventeenth century, large pumps were built to suck water up from mine shafts. Unfortunately, the pumps could never suck the water higher than about 10 m and therefore were useless since the mine shafts were inevitably much deeper than 10 m. Even Galileo could not build a pump that would suck the water any higher.

What these miners did not know was that they had created a barometer. Their pumps freely sucked air mole-

cules out of the tube; but even after all the air had been sucked out and a vacuum was created, the water did not follow for any more than 10 m (see Figure 5.2). Why?

What really is happening inside the tube is that the water is being *pushed* upward by the outside air pressure. When the air is sucked out of the tube it exerts no pressure on the water. Therefore, the water will rise as high as the atmospheric pressure can force it.

Thus, the water will rise until the pressure caused by its weight balances the pressure of the surrounding atmosphere. At sea level under normal atmospheric conditions (1013 mb) the water column will rise to a height of 10.4 m.

The first barometer was invented by Evangelista Torricelli in 1643 and it used water. Within a year mercury was used because it only takes a column of mercury 0.76 m high to balance the pressure of the atmosphere.

In 1647, Pascal (some say Descartes got the idea in the same year) made one of the first completely scientific predictions. He reasoned as follows: since the air pressure is due to the weight of air above you, there should be less pressure when you climb a mountain, and so the mercury in the barometer should not rise as high.

The following year, Pascal got his brother-in-law and some helpers to climb Puy du Dome, and when they got to the top the mercury did not rise as high in the barometer as it had at the base. The news of this successful prediction caused great jubilation all across Europe, because it represented a triumph of the new scientific method.

For this triumphant prediction we have recently hon-

FIGURE 5.2
A suck pump at sea level cannot "pull" water more than 10.4 m higher than the surrounding water level. Actually, once the pump sucks the air out of the tube, the surrounding atmospheric pressure forces the water up the tube.

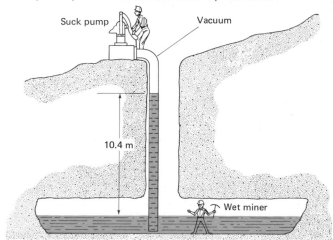

ored Pascal by naming a unit of pressure after him. Thus, average atmospheric pressure is 101,325 **pascals.** One hundred pascals are equal to 1 millibar.

# TEMPERATURE AND ENERGY

Take a piece of paper and try to write down a definition of temperature. You will probably find that this is not so easy to do. You might be tempted to write that when it is hot the temperature is high and when it is cold the temperature is low. Everyone knows what you mean, but that still does not make a very good definition.

Air temperature is determined by the speed of the molecules. The more rapidly the molecules move, the higher the temperature.

**Definition:** Temperature is proportional to the average kinetic energy of the molecules.

Perhaps this does not sound very exciting, but it is quite useful and informative. First, it is essential to define the **kinetic energy,** KE. The kinetic energy is the energy of motion and is given by the following equation

$$KE = \tfrac{1}{2}mv^2$$

The symbols $m$ and $v$ stand for the mass and velocity, respectively. In science it is important to be consistent with units so that when $m$ is given in kilograms (kg) and $v$ in meters per second (m/s), the energy is given in units of joules.

---

Example 5.2
A snowball that has a mass of 10 kg is moving with a velocity of 5 m/s. What is its kinetic energy?

Equation:
  $KE = \tfrac{1}{2}mv^2$

Substitute:
  $KE = \tfrac{1}{2}(10)(5)^2$

Solution
  $KE = 125$ joules

---

This may not seem very significant to you, but the example will take on meaning when the subject is switched to power. **Power** is defined as the rate at which energy is generated. If you generate one joule of energy in one second then that is equal to one **watt.** Therefore, if a person throws a 10 kg snowball at a velocity of 5 m/s once every

second, he is generating 125 watts (w) of power (746 watts equal one horsepower). Anyone who does this is working at quite a pace (i.e., like a horse)!

One of the great discoveries in science is the principle that energy is conserved. As a ball rolls down a hill it moves faster and faster. This means that it is gaining kinetic energy. Does this mean that it is gaining energy? The answer is no.

There are many different forms of energy. A ball at the top of a hill has a lot of **potential energy.** Potential energy is due to the attraction of gravity; the higher an object is, the more potential energy it has. When the ball descends the hill this potential energy is converted into kinetic energy.

In the atmosphere when cold, heavy air lies on top of warm, light air there is a lot of potential energy. When the cold, heavy air begins to sink and the warm, light air begins to rise this potential energy is converted directly into kinetic energy—which in everyday talk means that winds are produced.

Another common form of energy is heat energy, more properly called **internal energy.** After you apply the brakes and bring your car to a stop where has the kinetic energy gone? Just feel the brakes. They will be extremely hot. The car has lost its kinetic energy, but the molecules in the brakes feel hot because they have gained that kinetic energy and are moving very rapidly. Remember that temperature is proportional to the kinetic energy of the molecules.

Heat energy is often expressed in units called **calories.** Technically, a calorie is the amount of heat energy it takes to raise the temperature of one gram of water by one degree centigrade. One calorie is equal to 4.186 joules.

**Note:** The calories commonly used in connection with food are 1000 times larger than the calories mentioned above. This often causes confusion.

Since calories are units of energy, sunlight provides us with a form of power. After all, the solar constant is given in units of calories per square centimeter, or watts per square meter. In fact, since direct sunlight provides 1360 watts per square meter, it is no wonder that we have such high hopes for solar power.

Temperature is proportional to the average kinetic energy of the molecules and is given by the following equation

$$T = (0.00004)Mv^2$$

where $M$ is the molecular weight of the molecules (see Chapter 6), and $T$ is the **absolute temperature.**

Unfortunately, most people are not accustomed to the

absolute temperature scale. In fact, people are just beginning to get used to the Celsius (or centigrade) scale after years of using the Fahrenheit temperature scale. Why then, are you being forced to learn yet another temperature scale?

Scientists use the absolute temperature scale in all scientific laws because the absolute temperature (also called the Kelvin temperature) is proportional to kinetic energy. You should notice that according to the temperature equation, a temperature of absolute zero means that the molecules are absolutely still ($v = 0$). What this means is that when a gas is cooled to absolute zero there can be no wind and even no air, because all the molecules will be lying in a heap on the ground!

It is easy to switch from one temperature scale to another.

**Rule:** To change from absolute to centigrade, simply subtract 273° from the absolute temperature.

Example 5.3
Express 300° absolute in degrees centigrade.

Equation:

$$°C = T - 273$$
$$= 300 - 273$$
$$= 27°C$$

**Note:** Naturally, to change from centigrade to absolute simply *add* 273 to the centigrade temperature.

**Rule:** To change centigrade to Fahrenheit, first multiply the centigrade temperature by 1.8 and then add 32°.

Example 5.4
Express 35°C in degrees Fahrenheit.

Equation:
$$°F = (1.8)(°C) + 32$$
$$= (1.8)(35) + 32$$
$$= 95°F$$

Now see if you can find the rule used to change Fahrenheit to centigrade. A graph of the different temperature scales is shown in Figure 5.3.

Now let's return to the scientific definition of temperature. It can be used to tell why there is so little hydrogen and helium in our atmosphere, even though these two elements are so common through most of the universe. Air is a mixture of gases of different molecules and all coexist

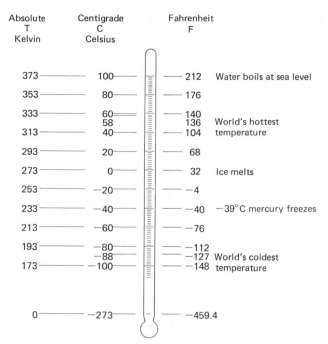

FIGURE 5.3
A thermometer showing the different temperature scales.

at the same temperature. Then, since $T$ is the same regardless of the type of molecule the product, $Mv^2$ also has to be the same.

Since hydrogen and helium have such small molecular weights (see Chapter 6), they must have correspondingly large velocities. Therefore, for any given temperature, hydrogen and helium molecules travel much faster than heavier molecules such as nitrogen and oxygen. This makes it possible for hydrogen and helium molecules to completely escape from the gravitational field of the earth (just as rockets do), while the more massive $O_2$ and $N_2$ are retained. Over the course of the earth's history, most of the hydrogen and helium have escaped from the earth.

## THE IDEAL GAS

Why is cold air heavy and warm air light? Thinking of molecules as billiard balls will help to answer this question. As the air is heated up, the molecules (or billiard balls) move faster and push outward. Imagine that you have blown up a balloon halfway. Then put the balloon in the freezer for a few minutes. When you take the balloon out again you should notice that it has shrunk. Then, rub the balloon to heat it up (don't use the stove), and you will see the balloon expand again.

As the air is heated up the molecules speed up and push harder against the walls of the balloon, forcing it to expand. Thus, warming leads to expansion or lighter air.

Children at zoos and country fairs always want you to buy them balloons. When they get helium filled balloons (helium is lighter than air) sooner or later you will see the balloon rising into the sky. As the balloon rises it will expand and eventually burst. Why does the balloon expand as it rises? To get more general, why does air expand as it rises through the atmosphere?

Once again, it is helpful to think of the air as a bunch of molecules. As air rises the pressure decreases. When the pressure outside the balloon decreases the molecules inside are free to push the balloon walls outward. Eventually, the balloon rises high enough so that there is very little pressure outside the balloon; then, the air inside the balloon may force the balloon to stretch past the bursting point.

All these facts can be explained mathematically by using an equation called the **ideal gas equation.** The ideal gas equation results when the pressure law is combined mathematically with the definition of temperature. For air in our atmosphere ($M = 29$) the ideal gas equation is given by

$$P = (2.87)\rho T$$

or equivalently

$$P = (2.87)\frac{m}{V}T$$

where $P$ is given in millibars, $\rho$ in kilograms per cubic meter, and T in degrees absolute.

The ideal gas law can be used to show *why* cold air is dense (or heavy) and why warm air is light. For the sake of argument hold the pressure constant. Then the product, $\rho T$ also must remain constant. This means that when the temperature is low, the density will be high and vice versa.

---

Example 5.5
A child "drops" an air-filled balloon. The balloon rises into the sky. Will it burst before it reaches an altitude of 10 km?

Information: The atmospheric pressure at 10 km is 300 mb, and the temperature there is $-33°C$. The air in the balloon has a mass, $m = 0.01$ kg, and the balloon will burst if it is stretched past a volume, $V = 0.02$ m³ (cubic meters).

Equation:
The ideal gas equation:

$$P = (2.87)\frac{m}{V}T$$

Rearrange to solve for volume, $V$:

$$V = (2.87)\frac{m}{P}T$$

Substitute:

$$V = (2.87)\frac{(0.01)}{300}(-33) = -0.003157 \text{ m}^3\text{???}$$

**Error**
Volume cannot be a negative number so something must be *wrong*. What is wrong? The answer is simple.

In all scientific laws you must use the absolute temperature and not some artificial scale such as centigrade. Thus, we must use $T = -33 + 273 = 240°$.

Solution
Substitute:

$$V = (2.87)\frac{(0.01)}{300}(240) = 0.023 \text{ m}^3$$

This is past the bursting point.

**Note:** When high-altitude balloons are released at the ground, they are quite flaccid and have much room for expansion as the balloon rises. Otherwise, the balloons would burst long before they reached the desired height.

---

You should recall that one of the most important secrets of the weather is that rising air leads to precipitation and clouds. This is true because rising air cools. You can now easily understand why rising air cools. Once again, it is helpful to think of air in the balloon.

As the balloon rises we know that it expands. This means that the balloon walls are slowly retreating from the molecules inside the balloon. When a ball is thrown against a retreating wall it will bounce off more slowly than it struck. This also happens to the molecules inside the expanding balloon. But since the rebounding molecules have lost some speed, they have also lost some temperature (see Figure 5.4).

# DIFFUSION

Diffusion is one of the basic physical processes of nature. If diffusion did not take place we would die within minutes; but what is more important, there could not be any rain or snow.

A number of years ago a friend of mine was traveling cross country in a military transport flight. As the plane was flying over the Rocky Mountains it encountered some

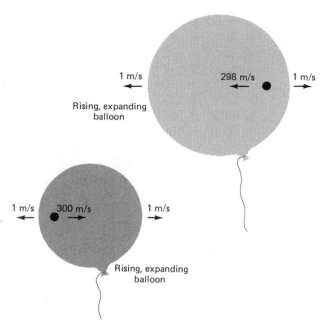

FIGURE 5.4
Molecules always bounce off a retreating wall more slowly than they hit. As molecules slow down, temperatures fall.

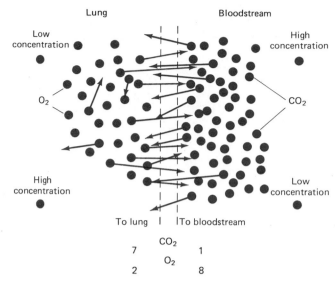

FIGURE 5.5
The principle of diffusion can be illustrated by the exchange of $O_2$ and $CO_2$ in the lungs. $CO_2$ diffuses from the bloodstream, where it is more concentrated, to the lungs where it is less concentrated, whereas oxygen diffuses in the opposite way. Numbers involved are schematic. You can see that this process works molecule by molecule and its reliability (and our lives) depend on the laws of statistics.

turbulence, and the soldiers inside were getting quite shaken up and weren't feeling too well. My friend happened to be hungry at the time and unwrapped a limberger cheese sandwich. Slowly, the aromatic molecules of limberger cheese diffused through the plane; as they reached the soldiers' noses, the poor men all vomited. This community effort finally inspired my friend to do likewise, possibly because of other aromatic molecules that diffused toward him.

**Diffusion** is a molecule by molecule spreading or averaging process. Right above the limberger cheese sandwich there was a large concentration of aromatic molecules. These molecules moved about in all directions at an average speed determined by the air temperature. Gradually, they spread throughout the plane's cabin.

Diffusion always acts in the same manner. On the average, molecules will always spread out from a region where they are concentrated to a region where they are sparse. This process will continue until the concentration is completely uniform everywhere. Diffusion is generally a slow process, but it is totally reliable because so many molecules are involved (see Figure 5.5).

An excellent example of diffusion takes place inside our lungs. The air entering the lungs is rich in oxygen, $O_2$, but poor in carbon dioxide, $CO_2$. Conversely, the air in the bloodstream has been depleted of $O_2$ and enriched in $CO_2$. Molecule by molecule, $O_2$ diffuses from the lungs into the bloodstream where it is less concentrated, while

$CO_2$ diffuses out of the bloodstream where it is concentrated.

This diffusion process would fail to work if the $O_2$ content of the air were seriously reduced, as it is several kilometers above sea level.

Evaporation and condensation are also examples of diffusion. Imagine a drop of water falling through the desert air. At the surface of the drop there is a high concentration of vapor molecules, while in the surrounding air there is a low concentration. Vapor molecules will then diffuse into the surrounding air. Unfortunately for the drop, as each vapor molecule leaves the immediate vicinity of the drop it is replaced by another molecule from within the drop. By this process the entire drop soon evaporates into thin air (see Figure 5.6).

**Heat conduction** is really diffusion of heat. An example of heat conduction is simple to give. Consider a volume of air divided into two regions. In region 1 the air is hotter than in region 2. Therefore, the molecules in region 1, on the average, are faster moving. When the molecules are allowed to move freely between the two regions, the faster molecules will spread out from region 1 while the slower molecules will spread out from region 2. The end result is a uniform temperature and, technically speaking, heat has been conducted or diffused from the warm region to the cold region.

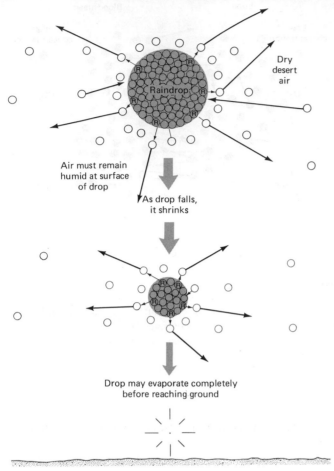

Dry desert air

Raindrop

Air must remain humid at surface of drop

As drop falls, it shrinks

Drop may evaporate completely before reaching ground

**FIGURE 5.6**
When a raindrop falls through dry desert air beneath the cloud it quickly evaporates. Vapor molecules at the surface of the drop quickly diffuse into the surroundings and then must be replaced by molecules from inside the drop.

Theoretically, it is possible that diffusion acts in the opposite sense, that is, to increase contrasts rather than to reduce them. You could envision a situation in which an ice cube is put into a glass of water, and the ice cube becomes colder while the water gets hotter and begins to boil. Ridiculous, you might say, but that is precisely what could happen if diffusion failed to work in the normal sense. Fortunately, this is only a hypothetical situation that is mathematically possible but so unlikely that it never will happen. It is far, far more likely to break the bank at Monte Carlo starting with one penny!

Recently, the wind-chill factor has become a frequently used term. Meteorologists are fond of saying, "outside the temperature is 0°C, but the wind-chill factor makes it feel like −20°C." What does that mean?

The **wind-chill factor** must be explained in terms of heat conduction. When it is calm, on a cold day your body will heat a thin layer of molecules near your skin. These heated molecules will diffuse slowly away from your body and be replaced by other cool, slow-moving molecules. Since diffusion is a somewhat slow process, your body will only contact a relatively small number of slow-moving molecules at any one time (see Figure 5.7).

But as soon as a wind starts blowing, it will blow away this protective layer of warmer molecules and constantly keep bringing new, cold molecules against your skin. Up to a certain wind speed, the faster the wind the more cold molecules will be brought against you and the greater the cooling effect. (See Table 5.2.)

**FIGURE 5.7**
The wind-chill factor. With no wind, a protective layer of warm air molecules forms around your body and only slowly diffuses away. On the other hand, the wind acts to constantly bring new cold molecules against your body. On a very hot day, simply reverse the hot and cold molecules, and you have the wind-scorch factor.

No wind: Cold molecules slowly diffuse toward body

Wind: Organized drift of cold molecules against body (notice "goose bumps")

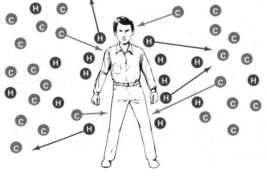
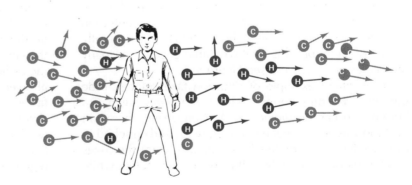

**TABLE 5.2**
The Wind-Chill Factor

| Wind Speed (Knots) | Temperature (°C) | | | | | | | | | |
|---|---|---|---|---|---|---|---|---|---|---|
| | 5 | 0 | −5 | −10 | −15 | −20 | −25 | −30 | −35 | −40 |
| 0 | 7 | 2 | −2 | −7 | −12 | −17 | −22 | −26 | −31 | −36 |
| 4[a] | 5 | 0 | −5 | −10 | −15 | −20 | −25 | −30 | −35 | −40 |
| 10 | −3 | −9 | −16 | −22 | −28 | −34 | −41 | −47 | −53 | −59 |
| 15 | −6 | −13 | −21 | −27 | −34 | −40 | −48 | −55 | −61 | −68 |
| 20 | −8 | −16 | −24 | −30 | −37 | −44 | −52 | −60 | −66 | −74 |
| 25 | −10 | −18 | −26 | −32 | −39 | −47 | −55 | −63 | −70 | −78 |
| 30 | −11 | −19 | −28 | −34 | −41 | −49 | −57 | −65 | −73 | −81 |
| 35 | −12 | −20 | −29 | −35 | −42 | −50 | −58 | −66 | −74 | −82 |
| 40 | −12 | −20 | −29 | −35 | −42 | −50 | −58 | −66 | −74 | −82 |

[a] By definition there is no wind-chill factor when the wind speed is 4 knots.

The largest wind-chill factor occurs when the wind speed is about 25 m/s. However, once the wind speed rises above this it means that the molecules themselves are moving faster, and faster molecules mean warmer air. This is why meteors and rockets entering the atmosphere turn white hot and often burn up.

When the temperature rises above about 33°C (this also depends on the humidity), a protective layer of cool molecules forms around your body in still air. Then, any wind will expose you to hotter molecules and will therefore heat your body. Therefore, above about 33°C we should speak of a "wind-scorch" factor.

**Viscosity** is friction for fluids. Air is a fluid that has some viscosity, but certainly not as much as molasses. Viscosity is another example of a diffusion process. When there is no wind, molecules move about randomly in all directions. Wind occurs when the molecules have a definite tendency to move in a particular direction. Viscosity is felt when the wind is stronger in one place than in another (i.e., when there is wind shear). Then the molecules from the faster wind region gradually spread into the slower region, and vice versa. As a result the winds tend to average out. It is one of the ironies of nature that velocity differences will even out more rapidly in a fluid where the molecules are freer to interchange, or in other words when the fluid is viscous and sticky.

## SOLID, LIQUID, AND GAS

Matter normally appears either as a solid, liquid, or gas (see Figure 5.8). In a solid the molecules are stuck together but can vibrate slightly. In a liquid the molecules are free to slide over one another, but none can wander away from the mass. In a gas the molecules are not connected and they move about space only occasionally colliding with one another.

Chemical bonds (which are really electromagnetic forces) try to hold the molecules together in solid form. However, as the molecules of any solid are heated, they begin to vibrate more and more rapidly until finally they have enough energy to slide away from their neighbors. At this point the solid is said to melt.

If you continue heating the liquid, the molecules will continue to speed up until they have enough energy to completely break free from all other molecules. The liquid then evaporates or boils, and it disintegrates into a gas.

These processes occur in reverse order as well, but not so easily. For instance, when the temperature rises above 0°C ice automatically melts, because the molecules become too energetic to stay in line. On the other hand, it is possible to cool a tiny cloud droplet of liquid water down as low as −40°C before it must freeze!

It may take you some time to get used to this fact, but it is a fact, nonetheless. Small droplets of water do not freeze automatically at 0°C simply because the molecules do not automatically line up in the proper order to form an ice crystal. This phenomenon, called **supercooling,** was discovered by Gabriel Fahrenheit (1686–1736) in 1724. Supercooling of *cloud droplets* was observed by Horace DeSaussure (1740–1799) before 1783 during one of his many explorations in the mountains. This was then forgotten for almost one hundred years.

In meteorology we are deeply concerned with these so-called **phase changes**—the changes between ice, liquid water, and water vapor. All the phase changes involve enormous amounts of heat. For instance, it takes 80 calo-

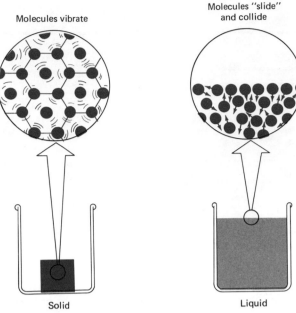

Molecules vibrate

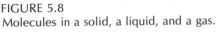

Molecules "slide" and collide

Molecules move freely and occasionally collide

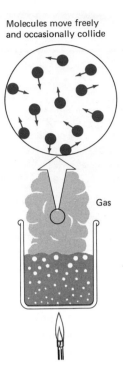

Gas

Solid

Liquid

FIGURE 5.8
Molecules in a solid, a liquid, and a gas.

ries of heat to melt one gram of ice, and it takes an additional 100 calories to heat that gram of water to the boiling point (100°C). It then takes 540 calories to boil the gram of water. No wonder it takes water so long to boil away completely.

The heat involved in the phase changes is called **latent heat.** Thus, the latent heat of fusion is 80 calories per gram, and the latent heat of vaporization for water is 540 calories at 100°C and 600 calories at 0°C.

After swimming, when you get out of the water you usually feel quite cool (unless the air is extremely warm and humid). You feel cool because water is evaporating from your skin into "thin air." In order to "free" the water molecules, latent heat energy had to be supplied to the water molecules. The question is, where did this energy come from?

The answer is that the energy was supplied by your skin and the air. Thus, as heat was added to the water molecules, it was taken from your body and from the air. This brings us to a sometimes confusing point.

Heat must be supplied from the surroundings to melt ice or to evaporate water. Therefore, as far as the surroundings are concerned, melting and evaporation are cooling processes! Conversely, freezing and condensation are heating processes for the surroundings (see Figure 5.9). You will see several other examples of this later in the book, but one more should suffice for now.

The reason that human beings sweat is to provide cooling for the body. However, the body is only cooled when the sweat evaporates. Humid days are the most uncomfortable because the air is already filled with vapor and little evaporation can take place. You can sweat all you want on a humid day, but there is little cooling power since there is little evaporation.

FIGURE 5.9
The different phase changes. The heat involved is the number of calories per each gram of $H_2O$ at 0°C. The plus sign indicates that heat is released to the surroundings; the minus sign indicates that heat is extracted from the surroundings.

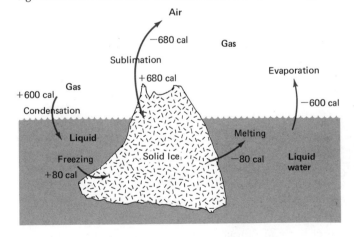

# PROBLEMS

**5-1** Explain why air pressure is proportional to the square of the speed of the air molecules.

**5-2** If a hole is suddenly made in the wall of a high-flying plane, you cannot get sucked out, but only pushed out of the plane by the air molecules. Why?

**5-3** Can a giant drink out of a 30 m straw at sea level?

**5-4** Does a column of mercury in a mercury barometer rise more near the Dead Sea or on the top of Mount Everest? Explain.

**5-5** Explain why atmospheric pressure decreases with altitude.

**5-6** A person weighing 50 kilograms climbs a 2 km high mountain in 20,000 seconds (5 hours, 33 minutes) and in the process acquires 1 million joules of potential energy. (a) What was the average rate of power expended? (b) If the person then fell down the mountain and all the potential energy converted to kinetic energy, what would be his or her velocity at the bottom? (Neglect air resistance so that this is not a terminal velocity.) (c) How many food calories is this equal to?

**5-7** Using the appropriate formulas, convert 25°C to the equivalent Fahrenheit (°F) and Kelvin (°K) temperatures. Then check your answer by looking at Figure 5.3.

**5-8** In like manner convert (a) 14°F to degrees centigrade and degrees Kelvin and (b) 148°K to degrees centigrade and degrees Fahrenheit.

**5-9** Suppose a molecule initially needs an upward speed of 11,000 m/s to escape from the earth, at what temperature will an average molecule of (a) hydrogen and (b) oxygen be able to escape to space, assuming it doesn't collide with any other molecules on the way? (*Hint:* The molecular weight of hydrogen is 2 and the molecular weight of oxygen is 32.)

**5-10** Do the same computations for the planet Mercury, where a molecule only needs an initial velocity of 4,100 m/s. Do you think Mercury can hold an atmosphere?

**5-11** Explain why a balloon expands when it is heated and also when it rises.

**5-12** Explain why your ears "pop" when you change altitude quickly (as in an elevator).

**5-13** As a balloon rises it expands and the air in it cools. Explain why these two facts are not contradictory.

**5-14** A balloon rises from sea level, where the pressure is 1000 mb and the temperature is 300° absolute, until it reaches the level where the pressure is only 500 mb. What is the final temperature at 500 mb? If the ideal gas equation can help, use it. If not, explain why not.

**5-15** An aerosol can, at an atmospheric pressure of 1000 mb, is thrown in a fire and heated from 300 to 900° absolute. The can cannot expand (i.e., change volume) but will burst when the pressure inside exceeds the air pressure outside by 1000 mb. Will the can burst?

**5-16** Do not allow the temperature of an ideal gas to change. If its volume doubles, what will its pressure be?

**5-17** Do not allow the pressure of an ideal gas to change. If its volume doubles, what will its pressure be?

**5-18** Demonstrate the validity of the statistical nature of diffusion in one of the two following ways. (a) Separate red and black cards. Then shuffle them and keep on shuffling them until all the red and black cards have separated again. (b) Take a teaspoonful of salt and a teaspoonful of sugar. Then put them in a closed jar and shake. Keep on shaking until you are able to take out a teaspoonful of pure sugar or pure salt.

**5-19** In which of the following situations would you rather be thrown outdoors stark naked: (a) $T = 0°C$, wind speed = 30 knots or (b) $T = -15°C$, wind speed 0?

**5-20** A pot of water at 0°C is put on the stove. It takes 10 minutes to *start* boiling. If you have to leave the house on a short errand, roughly how much time do you have before all the water will boil away?

**5-21** Explain why the melting of ice and the evaporation of water both act to cool the surroundings.

**5-22** Make a bet with a friend that pure water can be cooled below 0°C and still not freeze. Why is this a sure bet?

**5-23** Explain why hot days are more uncomfortable when it is humid.

**5-24** Direct sun rays are used to evaporate a layer of water 1 cm deep at 0°C (each square centimeter column then weighs 1 gram). How long will it take to evaporate the water completely? (Use the solar constant.)

**5-25** What do the following two facts have in common? (a) Fires do not burn as brightly at high altitudes. (b) Pregnant women living high in the Andes Mountains have always come closer to sea level to give birth.

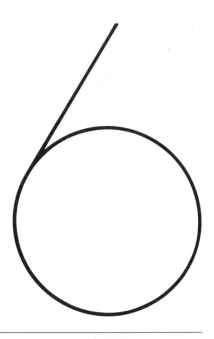

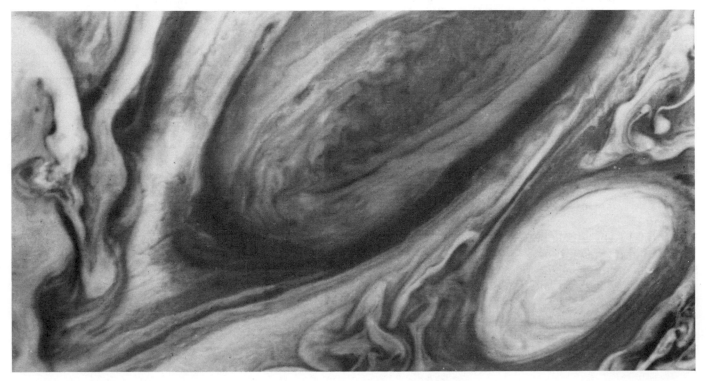

# THE COMPOSITION
# AND EVOLUTION OF
# OUR ATMOSPHERE
# AND THE REST OF
# THE SOLAR SYSTEM

In the course of respiration and digestion our bodies produce and emit gases. The chemical nature of those gases depends on the chemical composition of our bodies. In a similar manner the earth's atmosphere originated when volcanoes ejected gases from the earth's interior in a process known as **outgassing.** Once the gases reached the surface of the earth they were chemically altered by sunlight and by reacting with the rocks, water, plants, and animals on the surface.

Near active volcanoes the air usually stinks. The gases emitted by volcanoes are much different from the colorless and odorless air we breathe. Similarly, while most of the other planets also have atmospheres none have the oxygen we need to breathe in sufficient quantities. It is therefore quite an intriguing problem to explain why our atmosphere is what it is.

This chapter begins with a brief discussion on the nature of atoms and molecules. We will then look at the chemical composition of the sun, the planets, and their atmospheres, with emphasis on our own earth, of course. Finally, we will take a reasonable view (although not the only one) of the history of our atmosphere and the solar system.

## ATOMS AND MOLECULES

Every object whether solid, liquid, or gas is composed of atoms or groups of atoms called molecules. In a gas or liquid the atoms and molecules are free to move around, but in a solid they are locked into place and often lined up in a precise order to form crystals.

Each atom consists of a central nucleus composed of tightly packed protons and neutrons. Electrons whirl around the nucleus in much the same manner that the planets revolve around the sun (see Figure 6.1). The number of neutrons can vary somewhat but is usually close to the number of protons for any atom.

The atomic weight of an atom is almost exactly equal to the sum of the number of protons plus neutrons in that atom. Since it takes about 1836 electrons to weigh as much as one proton or neutron, the weight of electrons does not add much to the total weight of the atom.

The lightest atom is hydrogen. The normal form of the hydrogen atom consists of one proton and one electron. Its atomic weight is therefore 1. This means that you would need approximately

$$602,000,000,000,000,000,000,000$$

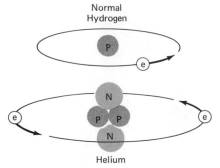

Normal
Hydrogen

Helium

FIGURE 6.1
Atoms are like miniature solar systems. Electrons, held by electrical forces, circle the nucleus, composed of protons and neutrons. Hydrogen and helium and shown.

atoms of normal hydrogen to weigh 1 gram!

The normal form of the oxygen atom has eight protons and eight neutrons. Its atomic weight is therefore 16. This means that it takes approximately

$$602,000,000,000,000,000,000,000$$

atoms of normal oxygen to weigh 16 grams. Atoms are very tiny, indeed!

One can easily get tired of writing such large numbers. There *is* an easier way to express large numbers. They should be written in a form called **exponent form.** Here are several examples of numbers written both in normal and exponent form.

$6,020 = (6.02)(10^3)$      $5,000 = (5)(10^3)$
$6,020,000 = (6.02)(10^6)$      $5,000,000 = (5)(10^6)$
$0.602 = (6.02)(10^{-1})$      $0.5 = (5.0)(10^{-1})$
$0.0602 = (6.02)(10^{-2})$      $0.05 = (5.0)(10^{-2})$
$0.00000602 = (6.02)(10^{-6})$      $0.000005 = (5.0)(10^{-6})$

The easy way to write the number of hydrogen atoms in 1 gram is, therefore,

$$602,000,000,000,000,000,000,000 = (6.02)(10^{23})$$

This is called **Avogadro's number.**

Almost all hydrogen atoms have one proton but no neutrons. A few hydrogen atoms have one neutron and fewer still have two neutrons. Different forms of the same atom are called **isotopes.** Oxygen always has eight protons. It usually has eight neutrons but may have nine or ten. Most atoms have a few isotopes. Table 6.1 contains a list of some common elements and their isotopes.

Isotopes have a large number of uses. In the earth sciences they are used mainly to tell the age of rocks and fossils. They are also used to tell what the temperature was on the earth in the distant past. You can read more about isotopes in Chapter 24.

The planets are kept in their orbits around the sun by gravitational attraction. Similarly, the electrons are kept in their orbits around the nucleus by electrical attraction. Protons and electrons are electrically charged particles. Each proton contains the same amount of electrical charge. According to convention, we say that the proton has a positive charge. Every electron contains the exact opposite charge of a proton; we therefore say that the electron has a negative charge. The neutron is neutral (has no charge) and seems to be a combination of a proton and an electron.

It is a fundamental law of electricity that unlike charges attract each other, whereas like charges repel each other (see Figure 6.2). This, of course, explains why the electrons are bound to the nucleus, but it can't explain how all the protons in the nucleus are held together. Special forces called nuclear forces (which will not be mentioned again in this book) hold the protons and neutrons of the nucleus together. The nuclear forces do not work so effectively with very large atoms, which is why many of them break apart radioactively.

Many of the atoms in our body and other chemical substances are found in groups called molecules. Every water molecule consists of two atoms of hydrogen with one atom of oxygen, and all are bound together by electrical forces. Symbolically, water is therefore written $H_2O$. The oxygen in our atmosphere also consists of molecules—

TABLE 6.1
Some Common Elements and Their Isotopes

| Element | Number of Protons or Electrons | Possible Number of Neutrons[a] |
|---|---|---|
| Hydrogen | 1 | *0, 1, 2* |
| Carbon | 6 | *4, 5,* 6, *7,* 8, *9, 10* |
| Oxygen | 8 | *5,* 6, *7,* 8, *9,* 10, *11, 12* |
| Iron | 26 | *26, 27,* 28, *29,* 30, *31, 32, 33, 34, 35* |

[a] Radioactive isotopes appear in italic. Most of these are extremely rare.

THE COMPOSITION AND EVOLUTION OF OUR ATMOSPHERE AND THE REST OF THE SOLAR SYSTEM

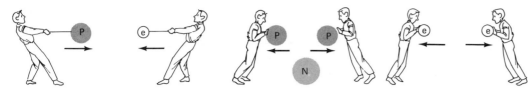

FIGURE 6.2
Electrical forces of electrons and protons. Unlike charges attract, like charges repel.

that is, molecules with two oxygen atoms bound together. Molecules will break apart into their individual atoms at extremely high temperatures of several thousand degrees centigrade, but they are usually stable at room temperature. A list of some common molecules is presented in Table 6.2.

TABLE 6.2
Some Common Molecules and Their Molecular Weights

| Name | Symbol | Common Molecular Weight |
|---|---|---|
| Oxygen | $O_2$ | 32 |
| Ozone | $O_3$ | 48 |
| Hydrogen | $H_2$ | 2 |
| Nitrogen | $N_2$ | 28 |
| Water | $H_2O$ | 18 |
| Carbon dioxide | $CO_2$ | 44 |
| Carbon monoxide | $CO$ | 28 |
| Quartz | $SiO_2$ | 60 |
| Rust (iron oxide) | $Fe_3O_4$ | 176 |
| Sulfuric acid | $H_2SO_4$ | 98 |
| Sugar | $C_6H_{12}O_6$ | 170 |
| Salt (sodium chloride) | $NaCl$ | 58 |

It is possible to think of a chemical reaction as an interchange of the atoms between molecules. For instance, when a lump of coal burns the carbon atoms in it combine with the oxygen molecules in the air to form molecules of carbon dioxide. Chemists write reactions in the symbolic form shown in Figure 6.3. Unfortunately, coal is not pure carbon because it contains a variety of impurities such as sulfur, which also reacts chemically. When the coal is burned sulfur also combines with oxygen to form sulfur dioxide ($SO_2$), which is a poisonous gas. You will read more about this in Chapter 14.

Look carefully at table salt and you will see that it consists of little cubic crystals. Almost all minerals and gems and many other substances as well (including ice) have a crystal structure. This simply means that the atoms of these substances align in a specific order into columns and rows. The very shape of the crystals often comes from the order in which the atoms line up. Thus, table salt crystals are cubes because the sodium and chlorine atoms of which salt is composed line up in a cubic lattice (see Figure 6.4). The different clay minerals consist of loosely connected sheets—which is why they feel slippery. When you slide your hand over wet clay you are ripping off a few of the thin sheets.

The crystals that meteorologists are most familiar with are ice crystals. As you have seen in the discussion of halos, ice crystals are basically hexagonal. This is due to the structure of the $H_2O$ molecules and the positions of the hydrogen atoms at 120° angles (each angle of a hexagon is 120°). Unfortunately there is no simple lattice structure for ice crystals as there is for salt.

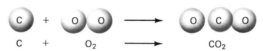

FIGURE 6.3
The chemical reaction of burning carbon and oxygen to form carbon dioxide. The arrow indicates the direction of the reaction. The direction of most reactions can be reversed under appropriate conditions.

FIGURE 6.4
The arrangement of atoms in crystals of table salt, NaCl. This is why salt crystals are cubic.

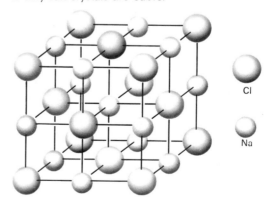

# THE CHEMICAL COMPOSITION OF THE EARTH, SUN, AND PLANETS

Of all the atmospheres in the solar system only our own contains life-sustaining oxygen, $O_2$. Yet oxygen is not one of the gases emitted by volcanoes in significant quantities. When we look into space we see that 99% of the mass of the universe is made up of hydrogen or helium. Yet in our atmosphere less than ten molecules in a million is hydrogen or helium. Truly, our atmosphere is a thing of wonder.

We must begin with the chemical composition of the earth and the rest of the solar system, since that is the ultimate source of our atmosphere. Table 6.3 contains a list of the most common elements that are found on and in the earth and in the rest of the solar system.

Consider first Column C in Table 6.3. This gives the relative abundances of different elements in the crust or outer regions of the earth. Oxygen and silicon are by far the dominant elements. This is not strange at all when you consider that most rocks and minerals are silicates, which means that they contain oxygen and silicon combined with certain metals such as iron, aluminum, sodium, calcium, potassium, or magnesium. For instance, sand is largely silicon dioxide, $SiO_2$, while clay contains aluminum in addition to silicon and oxygen.

Now look at Column B of Table 6.3 and compare it with Column A. Look at all the iron in our earth. There is easily enough iron to make the presidents of U.S. Steel and Bethlehem Steel happy. Unfortunately for them, most of this iron is buried deep within the earth. By now you

should realize that the inside or core of the earth is quite different from the crust or outer regions of the earth.

If you look at Table 6.3 once again, you will notice that the sun has a drastically different constitution from the earth. Almost 99% of the mass of the sun (and hence the solar system) consists of hydrogen and helium. In short, the sun is a tremendous sphere of gas. Astronomers have shown that hydrogen and helium also constitute about 99% of the rest of the universe. Yet on earth, there are only insignificant amounts of these two elements. Why is the earth, and, for that matter, the rest of the planets so different from the rest of the universe? The answer will become evident in the next few pages.

All of the planets exhibit a layered nature. The core of every planet, so far as we can tell, is composed of iron. This iron core is surrounded by a layer or layers of silicate rocks. For the four inner planets, this is basically all that there is, because the masses of the atmospheres or oceans are insignificant in comparison with the masses of the planets. The outer planets have an entirely different structure. The iron and rocky layers are buried by thick coverings of hydrogen, helium, and some water, ammonia, and methane. For instance, both Jupiter and Saturn are about 85% hydrogen and helium!

## TABLE 6.3
The Abundance of the More Common Elements on Earth and on the Sun (Percent of Total Mass)

| Element | (A) Sun | (B) Whole Earth | (C) Earth's Crust |
|---|---|---|---|
| Hydrogen | 77.3 | 0.0053 | Rare |
| Helium | 21.4 | Rare | Rare |
| Oxygen | 0.84 | 29.5 | 46.6 |
| Carbon | 0.35 | 0.05 | Rare |
| Silicon | 0.07 | 15.2 | 27.7 |
| Magnesium | 0.06 | 12.7 | 4.09 |
| Nickel | 0.007 | 2.4 | Rare |
| Sulfur | 0.04 | 1.9 | Rare |
| Calcium | 0.007 | 1.1 | 3.63 |
| Aluminum | 0.006 | 1.1 | 8.13 |
| Iron | 0.115 | 34.6 | 5.0 |
| Sodium | Rare | Rare | 2.83 |
| Potassium | Rare | Rare | 2.59 |

# THE ATMOSPHERES OF THE EARTH, SUN, AND PLANETS

The lowest 80 km of our atmosphere is a very well-homogenized mixture of gases composed of roughly 78% nitrogen ($N_2$), 21% oxygen ($O_2$), and 1% Argon (Ar) by volume, *after* you have removed the water vapor. $CO_2$ is a very minor constituent of the atmosphere—only about one-thirtieth (1/30) of 1%. Table 6.4 is a little more accurate (but just as dull) and, although it is accurate for dried air, the real air is never completely dry. The amount of water vapor varies considerably from day to day and may contain as much as 4% vapor on a hot, muggy day. Generally, it contains around 1% vapor. Water vapor, by the way, is completely different than liquid water in that it is as dry and invisible as any other gas in the atmosphere.

Ozone, $O_3$, is also variable and, at the surface of the earth, the air may contain up to 0.00007% of $O_3$. Even in the ozone layer, ozone concentrations usually vary from 0.0001 to 0.0002%. Therefore, ozone only constitutes a very tiny percentage of the atmosphere. However, because of its ability to absorb ultraviolet radiation and thus protect us from lethal sunburn, rarely has so little meant so much. Ozone concentrations are also larger in cities

TABLE 6.4
Composition of Dried Air[a]

| Constituent | Percent of Total Molecules | Percent of Mass |
|---|---|---|
| Nitrogen ($N_2$) | 78.08 | 75.51 |
| Oxygen ($O_2$) | 20.95 | 23.14 |
| Argon (A) | 0.93 | 1.28 |
| Carbon dioxide ($CO_2$) | 0.0325 | 0.049 |
| Neon (Ne) | 0.0018 | 0.0012 |
| Helium (He) | 0.0005 | 0.0001 |
| Krypton (Kr) | 0.0001 | 0.0003 |
| Hydrogen (H) | 0.00005 | 0.000002 |
| Ozone ($O_3$)[b] | 0.0006 | 0.0010 |

[a] In actual air the amount of water vapor can range up to about 4%, and then the other constituents will be correspondingly reduced. Water vapor is not included in this table because of its high variability.
[b] The concentration of ozone is quite variable, this is just an average figure.

where automobile exhaust fumes react with $O_2$ and sunlight to produce $O_3$ and generally smelly, corrosive air.

These are some of the dull facts, except — if there were no $O_2$, we would die of suffocation; if there were no $O_3$, we'd die of sunburn; if there were no $H_2O$, we'd die of thirst; and if there were no $CO_2$ or $N_2$, we'd die of starvation.

Since the sun is a large ball of gas we must be rather arbitrary when we define the boundaries of its atmosphere. Therefore we will say that the bottom of the sun's atmosphere occurs in the layer known as the **photosphere,** because the photosphere contains the visible surface of the sun. Like the interior of the sun, the solar atmosphere is composed almost entirely of hydrogen and helium.

Light and heat produced deep within the sun slowly work their way to the photosphere. Once reaching the photosphere, the light and heat are suddenly released and head out into space. The photosphere is by far the coolest part of the sun, but still has temperatures about 6000°K, except in certain regions known as sunspots. Sunspots were reported by the ancient Chinese, but seem to have been unknown in the western world until the time of Galileo. When Galileo discovered them with his telescope and reported them to the world, he once again upset a lot of people. According to some theologians, the sun was supposed to be a perfect sphere. When Galileo reported that the sun had spots (or blemishes), needless to say he did not make himself very popular.

Since measurements have begun we have found that the sun puts out virtually the same total amount of light and heat all the time, even though the number of sunspots

varies dramatically from year to year (see Figure 6.5). In fact, the sunspots seem to come in cycles of roughly 11 and 22 years, and these cycles also may have some effect on climate. For example, there is some evidence of a relationship between the Little Ice Age and the almost complete disappearance of sunspots (known as the Maunder Minimum) in the decades around 1700 (see Figure 6.6). This will be discussed more fully in Chapter 24.

The outer layer of the sun is called the **corona** and is usually about 1 million or more degrees. During a time of active sunspots, the corona is also affected and it becomes far hotter — perhaps 2 million or more degrees. Giant flares and prominences burst out of the photosphere into the corona, and the sun suffers what might be called a mild "stomachache." (see Plate 9)

Even during normal times, hot hydrogen gas is constantly being ejected from the sun because of the extreme heat and temperatures of the corona. The ejected gas whizzes out into space and passes by the earth where it helps to produce the aurora (see Chapter 7). The name given to the ejected solar hydrogen is the **solar wind.** Dur-

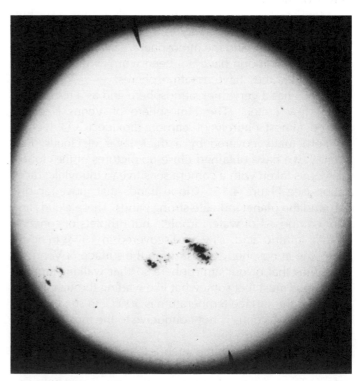

FIGURE 6.5
The sun, showing an exceptionally large sunspot group.

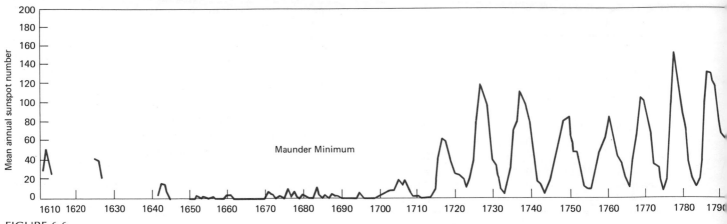

FIGURE 6.6

The sunspot cycle, including the Maunder Minimum, when almost no sunspots were observed for about 80 years.

ing a time of many sunspots and flares, the solar wind grows stronger. At these times auroras on earth grow more intense and radio reception is distorted.

Like the moon, the planet Mercury is covered by craters. Some of the craters were formed by volcanoes, but most were produced when giant chunks of rock or iron known as meteorites smashed into the surface of the planet. You can make a crater by throwing a rock into a layer of mud or soft clay.

Many craters were formed on earth in this way, but most have been worn away in the course of time by the actions of water and the atmosphere. The craters on Mercury and the moon have not been worn away, which indicates that they have no atmospheres.

Venus has a very thick atmosphere and as a result may have few craters. The atmosphere of Venus is composed almost entirely of carbon dioxide ($CO_2$). Venus is perpetually covered by a thick layer of clouds. Recently, we have obtained close-up pictures of the clouds of Venus taken with a camera sensitive to ultraviolet radiation (see Figure 4.15). Cloud bands that move rapidly around the planet indicate strong winds. These clouds are not composed of water droplets but droplets of concentrated sulfuric acid, as was discovered in 1973! In addition, the atmospheric pressure on the surface of Venus is 90 times that of our atmosphere so that walking around on Venus must feel somewhat like wading through water, whereas the surface temperature is 500°C. In short, the atmosphere of Venus is not conducive to the most basic biological function—life.

Why does Venus have such a thick $CO_2$ atmosphere? Why is there so little water vapor and virtually no oxygen at all? These are tough questions that are not yet fully answered. The earth's atmosphere has only a tiny amount of $CO_2$. The oceans hold roughly 40 times as much $CO_2$ as the atmosphere, but even that is far, far short of the amount of $CO_2$ in Venus' atmosphere.

The earth does have immense quantities of limestone (rocks formed from seashells). Limestone has the chemical formula $CaCO_3$ (calcium carbonate). When the temperature gets high the $CaCO_3$ breaks down to form CaO and $CO_2$. This is much like what happens to iron. At room temperature, iron will rust (which means that it will combine with oxygen to form iron oxide), but if the rust is heated enough the oxygen will separate from the iron again (see Figure 6.7). It is apparent that Venus is close enough to the sun and therefore warm enough to allow most of the $CaCO_3$ on its surface to break up into CaO

FIGURE 6.7

The chemical equations for rusting and for the baking of limestone.

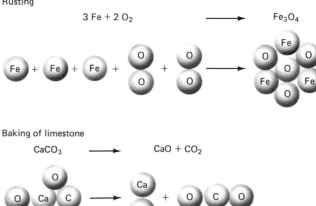

THE COMPOSITION AND EVOLUTION OF OUR ATMOSPHERE AND THE REST OF THE SOLAR SYSTEM

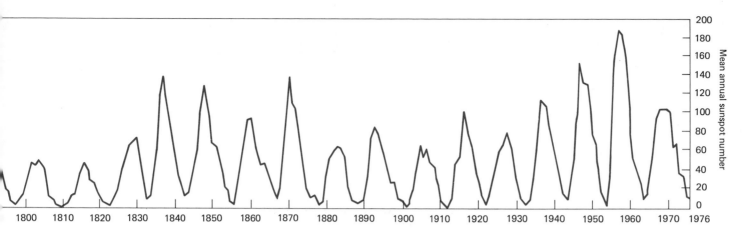

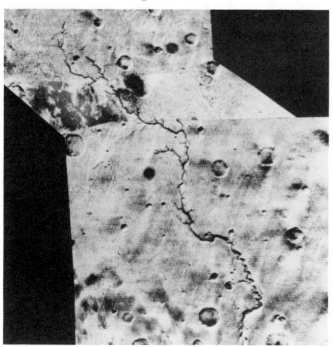

FIGURE 6.8a
Martian sand dunes indicate enormous winds in the tenuous Martian atmosphere.

FIGURE 6.8b
Giant Martian canyons that dwarf our Grand Canyon indicate that there was once flowing water on the surface of Mars.

(calcium oxide) and $CO_2$ (if indeed $CaCO_3$ ever formed). The same thing would probably happen on earth if the sun were hotter.

The atmosphere of Mars is also largely composed of $CO_2$. The surface pressure on Mars is only about 7.5 mb (about 3/4 of 1% of that on earth). At the poles Mars is covered by icecaps that are composed of frozen $CO_2$ and $H_2O$. The water stays frozen year round, but the $CO_2$ sublimes (evaporates) in the summer-half of the year and refreezes in the winter-half of the year causing air pressure changes.

Giant dust storms periodically cover large areas of the surface of Mars, and over the course of time giant sand dunes have been produced. (See Figure 6.8a.) Since the air is so thin, the winds on Mars at these times must be truly enormous in order to lift and carry the dust. When and where these dust storms occur, the color of the planet changes. Before the rocket *Mariner* reached Mars some scientists thought that this color change showed that plants were undergoing a seasonal growth. We now know that this is false, but we still don't know if there is life on Mars.

There is at least one more intriguing feature of Mars. It

has giant canyons much like the river canyons on earth. (See Figure 6.8b.) If, as is likely, these canyons were cut by flowing water, then at one time in the distant past Mars must have had a thicker atmosphere with a warmer, wetter climate that was far more hospitable to life than the present atmosphere. Where did all that water go? Some may still be chemically locked in the Martian soil but, most probably escaped to space.

Jupiter and Saturn both have very thick atmospheres (it is hard to tell just where the atmosphere ends and the planet begins). These atmospheres are composed of about 80% hydrogen and 20% helium. The clouds on Jupiter and Saturn have several layers consisting of water droplets or ice crystals but the top cloud layer consists of ammonia (see Figure 6.9). The clouds on Jupiter were first photographed by *Pioneer 11*, and the patterns indicate that Jupiter has rapidly moving jets of air, somewhat like our jet stream (see Chapter 16).

The atmosphere of Saturn also contains a significant amount of methane. Uranus and Neptune both contain hydrogen, helium, and methane. Titan, one of the moons of Jupiter, also has an appreciable atmosphere, and in the coming years we will no doubt learn a great deal more about these distant, frigid atmospheres.

FIGURE 6.9
The planet Jupiter; its cloud bands indicate the presence of extremely large winds.

# THE ORIGIN OF THE SOLAR SYSTEM[1]

Perhaps the stars form(ed) in the spiral arms of the great whirling galaxies of space much like thunderstorms form in the spiral bands of hurricanes (see Figures 6.10 and 18.13). Presumably over 5 billion years ago an ancient star died in a cataclysmic explosion known as a supernova. The debris of this star clouded up a region of space called a nebula.

We believe that the sun and the rest of our solar system all formed at about the same time—about 4.6 billion years ago when such a giant swirling cloud of gases and dust began to contract under the influence of gravity. As the gases contracted the pressure rose and as a result the temperature rose as well. The sun began to take definite shape in the center of the nebula where the density, pressure, and temperature were the greatest. As the nebula contracted, it also flattened into a giant disk because of its swirling motion. A small fraction of the material was swirling too rapidly to be drawn into the center, and selected parts of this "left over" material combined to make up what we now know to be the planets.

The "left over" material was composed of rocky and icy dust-sized particles and gases. Near the center of the nebula the ice particles evaporated because of the higher temperatures, and only in the outer cooler regions of the nebula did the icy particles remain frozen.

As the nebula contracted the particles began to be more closely packed. Since not all particles moved at exactly the same speed, some collided with others. It is apparent that only the solid particles can stick together and grow large after a collision. Some of these particles did stick and thus grew larger. As they grew larger, they naturally collided with (and sometimes stuck to) even more particles. This process is known as **accretion** or **coalescence** (see Figure 6.11).

Eventually the chunks of rock grew so large that they began to exert a considerable gravitational attraction. Then they not only collided with rocks directly in their path, but they could also attract other particles and rocks that otherwise would have been near misses. In fact, a large enough chunk of rock may be able to attract and hold gas molecules and thus start growing its own atmosphere!

The process of accretion is not only important to the growth of planets, but it has another application—raindrops, snowflakes, and hailstones also grow by accretion. It may sound very silly, but it is true that planets actually grow somewhat like raindrops. To grow to planet size

[1] The material presented in this section is hypothetical.

THE COMPOSITION AND EVOLUTION OF OUR ATMOSPHERE AND THE REST OF THE SOLAR SYSTEM

FIGURE 6.10

The spiral galaxy M51. Stars, much like our sun, may be born in the spiral arms. Notice the similarity to the structure of hurricanes.

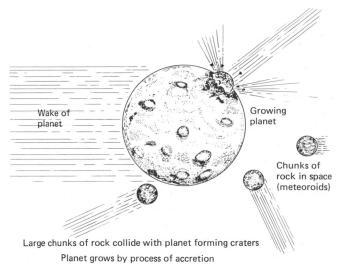

Large chunks of rock collide with planet forming craters

Planet grows by process of accretion

FIGURE 6.11

The process of accretion. Extremely large chunks of rock or planets attract even more particles and rocks because of gravity. Notice the craters.

probably takes between 10,000 and 100,000 years, while raindrops can grow in a few minutes.

If you want to see the evidence that the planets grew by accretion, simply look at the craters on Mercury or on the moon. Many of these are impressions left by colliding meteorites. The sheer number and size of these craters indicate a great accumulation of mass.

We also have a reasonable explanation for the chemical composition of the planets. The so-called four inner planets—Mercury, Venus, Earth and Mars—formed under relatively high temperature conditions; therefore, they are composed mainly of those elements that can exist in solid form at high temperatures. This may well explain why these planets have high percentages of iron, oxygen, silicon, aluminum, and other so-called high-temperature elements (see Table 6.5).

The outer planets formed under colder temperature conditions so that they have a higher percentage of low-temperature elements. Furthermore, since gas molecules move slowly at very low temperatures, they can easily be captured and retained by a planet. Thus, the outer planets have enormous quantities of hydrogen and helium.

The temperature on any of the inner planets is too high to retain the hydrogen and helium that gradually escape into space. The reason that the atmosphere of Mars is so thin is that even the heavier molecules are able to escape from its relatively weak gravitational field.

It seems that the earth lost its first atmosphere early in its life because of something that happened to the sun. The sun obtains virtually all its energy from nuclear fusion. (This is the same process that takes place in a hydrogen bomb.) During fusion deep within the sun four atoms of hydrogen combine to form one atom of helium. During this process of "nuclear burning" a tiny bit of mass is destroyed and is converted into pure energy. This energy ultimately leaves the sun in the form of electromagnetic radiation.

Fusion did not begin on the sun until it had contracted sufficiently. Then once the temperature reached a certain critical point, fusion began in the sun's center. In this early, adolescent phase of the sun's history, known as the T-Tauri stage, sunlight was perhaps 30% weaker than it is today, but the solar wind may have been a million times stronger. It was this tremendous solar wind, called the T-Tauri solar wind, which lasted perhaps a million years, that swept whatever atmosphere the young earth had acquired far away into space.

TABLE 6.5

The Highest Temperature at Which Some Elements Can Appear in Solid Form[a]

| Element | Temperature |
|---|---|
| Aluminum, oxygen | 2700°C |
| Calcium, titanium | 2460°C |
| Silicon | 2380°C |
| Iron, nickel, cobalt | 2180°C |
| Potassium, sodium | 1830°C |
| Hydrogen, carbon | 260°C |

[a] Environmental pressure of 1 mb.

# THE EVOLUTION OF LIFE AND OUR ATMOSPHERE

Although the young earth was stripped of its first atmosphere, another was soon to form. The countless collisions of rocks that took place during the accretion stage generated a tremendous amount of heat. This heat most likely caused a large fraction of the rocks of the earth to melt. Most of the iron sunk to the center, while radioactive elements such as uranium floated toward the surface regions of the earth where they concentrated and produced extra heat. Many gases were emitted from the molten earth. As the earth began to solidify on the outside, molten rock occasionally broke through the surface in the form of volcanoes (as it does today) and continued to emit gases. These accumulating gases formed the primitive atmosphere.

Volcanoes today produce significant amounts of carbon dioxide ($CO_2$), carbon monoxide ($CO$), water vapor ($H_2O$), nitrogen ($N_2$), and sulfur dioxide ($SO_2$), but not much oxygen. Although we cannot be sure that today's volcanoes emit the same gases in the same proportion as did the volcanoes early in the earth's history, we do know that the early atmosphere contained virtually no $O_2$.

Nitrogen, $N_2$, rather quickly accumulated in the atmosphere, since it does not easily combine chemically with other elements. The ocean formed on the cooling earth once the water vapor condensed. Yet oxygen was still virtually absent. We know this because oxygen readily combines with other elements such as iron. If oxygen had been present in the atmosphere in appreciable quantities, then it would have combined with the iron in the ancient rock deposits. Some of the oldest rocks that were exposed to the early atmosphere are deficient in oxygen. This can only mean that the early atmosphere of the earth had essentially no oxygen.

There is one more important piece of information that indicates that the early atmosphere had virtually no oxygen. The complex molecules that form the building blocks of all living things are easily destroyed in the presence of oxygen. Thus, oxygen would have been poisonous to the earliest forms of life on earth!

We have fossil evidence that somewhat more than 3 billion years ago one-celled creatures called blue-green algae were living in the ocean and were producing oxygen by photosynthesis. They had developed to the point where the sensitive molecules were protected from the oxygen that they were in fact producing.

The blue-green algae and other primitive forms of life that obtained their energy by photosynthesis were producing a drastic change in the atmosphere. They were manufacturing oxygen that is present in the atmosphere today. Most of the oxygen in our atmosphere has been produced by photosynthesis.

It is estimated that oxygen accumulated slowly at first. With little oxygen there is also little ozone ($O_3$) to protect animals and plants from the sun's ultraviolet rays. In the early atmosphere the lethal ultraviolet rays from the sun therefore most likely penetrated to the earth's surface. Fortunately, water acts as a shield against the deadly ultraviolet rays and thus it makes sense that life began at some depth below the ocean surface where there was ample protection.

As $O_2$ accumulated in the atmosphere, increased amounts of $O_3$ were produced and finally began to shield the land from the lethal ultraviolet radiation. In addition to this, the increased concentrations of $O_2$, both in the atmosphere and dissolved in the ocean, made it possible for more advanced forms of life to evolve and survive.

Up until about 600 million years ago we only have evidence of primitive forms of life on earth. Then, suddenly more advanced forms of life, such as trilobites, spread across the surface of the oceans (See Figure 6.12). What caused this revolution? Did it occur because the amount of oxygen had reached a critical amount, or did it occur because of some purely biological reason? We don't yet know.

FIGURE 6.12
Trilobites. These primitive creatures thrived for roughly 300 million years before becoming completely extinct.

About 450 million years ago the first plants emerged from the ocean and rapidly spread across the continents. Soon after that the first animals crawled out onto the land. Why did this happen so suddenly? Was it because until that time there had not been enough $O_3$ to shield the land from the sun's ultraviolet rays, or, again, was it because of a purely biological reason?

There are many unanswered questions on exactly how rapidly oxygen accumulated in the atmosphere, and exactly why the geological record of life on earth shows such short periods of drastic changes and tremendous advances. There is much to be learned and there is the chance that many questions will be answered within our lifetimes.

You can be reasonably sure that if we could invent a time machine to go back 400 million years, we would find that the atmosphere is very much the same as it is today. If we went back much further than that, we might not have enough oxygen to breathe.

## PROBLEMS

**6-1** Try to write out the number $10^{654}$ in long form.

**6-2** Name the two most abundant elements in rocks.

**6-3** You can view sunspots in the following manner. Take a piece of cardboard and punch a pinhole in it. Allow sunlight through and then magnify this with a pair of binoculars or a telescope. Then take a second piece of cardboard and view the image of the sun on it. (Incidentally, this is also a good way to view a solar eclipse.) What have you learned about sunspots?

**6-4** What are the most abundant gases in the atmospheres of Venus, Mars, Jupiter, and Saturn?

**6-5** Describe the process of accretion. Using this description explain why craters indicate that a planet or moon grew by collecting nearby meteorites.

**6-6** Why are there so few craters on earth?

**6-7** Did our atmosphere always have appreciable amounts of oxygen? Cite evidence.

**6-8** What is the principal process by which oxygen is produced in our atmosphere?

**6-9** Why is the atmosphere of Venus so much more massive than ours?

# LIGHT, ATOMS, AND MOLECULES: THE UPPER ATMOSPHERE AND US

At the end of the last chapter you learned that there is a protective layer of ozone above us in the atmosphere. The ozone layer actually has only a minute amount of ozone which is in constant danger of being destroyed. If the ozone were destroyed, we would soon die from exposure to the sun's ultraviolet rays.

In this chapter, we will see how ozone is produced and destroyed by the sun's rays. The discussion actually begins with a general rule discovered by Albert Einstein, which explains the interaction of light with atoms or molecules. From this general rule we learn not only about ozone but also about such phenomena as the airglow, the aurora, and ionosphere, and other layers of the atmosphere, even about human skin color.

## EINSTEIN'S LAW AND MODERN PHYSICS

Newton's laws of planetary motion (see Chapter 15) seemed to line up beautifully with a totally rational philosophy in which there was virtually no need for God. Many people during the eighteenth century believed that the universe had been created by a God who then went away and left it to operate by the laws that Newton had discovered.

The triumphs of the early years of the Industrial Revolution further increased the notion of a totally rational but mechanistic universe. People living at the time thought that it was possible to predict the behavior of the universe for all time to come, and they walked around feeling that there was no limit to what humankind could accomplish. This historical period of optimism and euphoria is known as the Age of Reason.

The seeds that destroyed the Age of Reason were planted toward the end of the eighteenth century by writers like Rousseau and by events such as the French Revolution and the Reign of Terror. Throughout the nineteenth century the prevailing intellectual mood of optimism slowly disappeared. Although industrialization did improve the lot of humanity, it also created a great deal of misery and suffering. To many people the world did not seem so beautiful or rational. Art started to grow impressionistic; music became discordant; literature more symbolic; philosophy almost irrational; and Sigmund Freud was busy at work showing that there are dark, inner recesses to the human mind that had previously remained hidden from view. However, no one would have imagined that physicists themselves would soon discover a world in which the laws of nature would prove perhaps even more irrational.

In the last few years of the nineteenth century a number of strange and surprising experimental discoveries were made. In 1896, Antoine Becquerel (1852–1908) discovered radioactivity. During the Age of Reason people had laughed at the medieval alchemists and magicians who had attempted to convert the so-called base metals (such as lead) into gold. In *Gulliver's Travels*, Jonathan Swift satirized both the scientists and alchemists of his day. In this quote we see Gulliver on a tour of the Academy.

I went into another chamber but was ready to hasten back being almost overcome with a horrible stink. My conductor pressed me forward, conjuring me in a whisper to give no offense, which would be highly resented; and therefore I durst not so much as stop my nose. The projector of this cell was the most ancient student of the Academy. His face and beard were of a pale yellow; his hands and clothes dawbed over with filth. When I was presented to him he gave me a very close embrace (a compliment I could have well excused). His employment from his first coming into the Academy, was an operation to *reduce human excrement to its original food* [emphasis added].

*Gulliver's Travels,* Part III, Chapter V

Becquerel's discovery amounted to almost the same thing.

For reasons not known, or even suspected at the time, atoms of certain elements such as uranium and radium were found to split apart (decay radioactively) and transmute themselves into atoms of different elements such as lead and helium. Perhaps the alchemists had been correct after all!

Although it is difficult to give a precise date for the beginning of certain historical time periods (such as the Age of Reason), a good case can be made that "modern" physics was born in the year 1900. A physicist named Max Planck (1858–1947) was trying to explain mathematically how rapidly an object could radiate heat away from itself (i.e., cool off). Planck found that the only way the proof would work was if he made an unheard of assumption. He was forced to assume that energy comes in little packets that could not be further subdivided. Until that time, scientists had always thought that energy could be subdivided indefinitely—but they were wrong. Planck's theory was too good to be false.

Five years after Planck's discovery came several more astounding discoveries. Albert Einstein (1879–1955) published his special theory of relativity in 1905. Although Einstein is famous for his discovery of relativity, you probably didn't know that he actually won the Nobel Prize in physics for a different discovery, which he also made in 1905.

This discovery explained the so-called *photoelectric effect*. Einstein showed that light not only behaves like waves (as all scientists of the time thought), but it can also act like a stream of tiny particles when it collides with solid objects such as atoms.

Einstein found that the energy of all electromagnetic radiation (light waves included) depends only on the wavelength. In fact, though it may sound strange, the *shorter* the wavelength of the radiation, the *more* energetic the waves.

**Rule:** Shorter waves of electromagnetic radiation have more energy than longer waves.

The energy of electromagnetic waves is often expressed in a strange unit called an *electron volt*. To give you an idea of just how small an electron volt is, here is an example. Say that you lifted a weight of 1 kilogram to a height of 1 meter. The amount of energy needed to do that is 10 joules or about

$$(6.25)(10^{19}) \text{ electron volts}$$

Einstein found that the energy ($e$) of a light wave, in electron volts, is equal to 1.25 divided by the wavelength ($L$), given in microns. Here is Einstein's equation.

**Einstein's Equation:**

$$e = \frac{1.25}{L}$$

---

Example 7.1
What is the energy of a wavelength of red light?

Information: Red light has a wavelength of approximately 0.7 microns (see Chapter 3).

Equation:

$$e = \frac{1.25}{L}$$

$$e = \frac{1.25}{0.70} = \frac{12.5}{7.0}$$

Solution
$e = 1.79$ electron volts for red light

Example 7.2
What is the energy of a wavelength of violet light?

Information: Violet light has a wavelength of approximately 0.4 microns.

Equation:

$$e = \frac{1.25}{L}$$

$$e = \frac{1.25}{0.40} = \frac{12.5}{4.0}$$

Solution

e = 3.13 electron volts for violet light

Example 7.3

Which has more energy—ultraviolet radiation or infrared radiation?

Equation:

$$e = \frac{1.25}{L}$$

Solution

Since the wavelength is in the denominator, when the wavelength is large, the fraction $1.25/L$ must be small. Therefore, the longer the wavelength, the smaller the energy.

Remember that infrared waves are far longer than ultraviolet waves (from Chapter 3). Therefore, the longer infrared waves have far less energy than the shorter ultraviolet waves!

# LIGHT, ATOMS, AND MOLECULES

You may think that Einstein's equation has little to do with the subject of meteorology. Actually, Einstein's equation is very important in meteorology, even though it has no direct bearing on day-to-day weather prediction.

The sun emits radiation of all wavelengths but, as you have read in Chapter 3, it emits visible light at maximum intensity. Roughly 9% of the sun's radiation consists of ultraviolet waves. Because ultraviolet waves are relatively short, they have a great deal of energy. They have enough energy to break apart atoms and molecules.

Our bodies are made up of atoms and molecules. Radiation with short enough wavelengths can therefore break apart the molecules of our bodies. This can lead to cancer, mutations, or even death. Enough of this short wavelength radiation from the sun reaches the top of the earth's atmosphere to kill us. Fortunately, almost all of the lethal ultraviolet radiation is absorbed by the gases in the upper atmosphere, and therefore is blocked from reaching the surface of the earth. The tiny amount of this radiation that does reach the surface causes sunburns.

Exactly how does the upper atmosphere protect us? Some of the atoms and molecules at least 15 km above the ground serve as our saviors. They absorb the very short waves of radiation so that we don't have to. The scientific problem that we must understand is how electromagnetic radiation interacts with atoms and molecules.

In 1913, Niels Bohr made an astounding discovery. He found that the electrons can use only certain specific orbits around the nucleus. These orbits have become known as shells. The amazing nature of this restriction of the permitted orbits is easy to illustrate. If you were to roll a marble around the inside of a bowl, you could make the marble take any orbit by merely adjusting the speed; therefore, you can adjust the energy of the marble. If you add a little energy to the marble, it will move a little outward and upward. If you add enough energy, the marble will spin right out of the bowl (see Figure 7.1).

There is no such freedom with an orbiting electron. You can add only certain specific amounts of energy to the electron when it is in orbit. You should notice that this sounds something like Planck's law. The electron's orbit will change by certain specific amounts. No intermediate orbits are permitted. If you do not add the precise amount of energy to the electron, the electron will not accept any energy at all and will not budge one iota from its orbit (see Figure 7.2).

This restriction no longer applies if you add enough energy to knock the electron completely free of the rest of the atom. For instance, it takes 10.2 electron volts of energy to lift an electron of the hydrogen atom from the lowest rung (called the 1s shell) to the next rung (called the 2s shell). No other amount of energy is acceptable. On the other hand, it takes 13.6 electron volts to knock the electron from the lowest rung completely out of the hydrogen atom. Any amount of energy greater than 13.6 electron volts can do this.

An important rule to remember is that electrons will always try to get to the lowest rung they can. However, in each rung or shell of the atom only a certain number of electrons is permitted at any one time. For example, only

FIGURE 7.1
A marble can circle a bowl at any height.

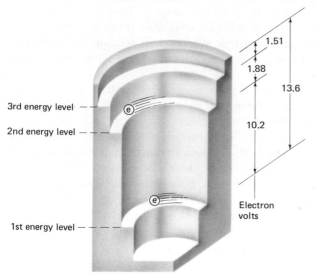

3rd energy level

2nd energy level

1st energy level

1.51
1.88
13.6
10.2

Electron volts

FIGURE 7.2
The electrons can circle the atom in certain permitted orbits only. This drawing shows some of the permitted orbits for the hydrogen atom, and the amounts of energy needed to change orbits.

two electrons are permitted in the innermost rung of any atom. Once a space is available in one of the lower rungs, an electron from one of the higher rungs will drop down to fill the vacancy.

When an electron "drops" from a higher to a lower rung, it loses the same amount of energy it gained when it was raised from the lower rung. When the electron drops, this energy is released and leaves the atom.

Do you have any idea about the nature of the energy that is released by electrons that slip to lower rungs in an atom? Do you have any ideas about the way in which energy is supplied to electrons to make them jump to higher rungs in the atom? Electrons are far too small to be moved within an atom by a hammer or a shovel or even by tweasers.

It is simple to answer both these questions. If you want to raise an electron to a higher rung or even knock the electron completely free of the atom, all you have to do is supply electromagnetic radiation with the appropriate amount of energy. On the other hand, when an electron drops to a lower rung within the atom the energy it releases (radiates) is electromagnetic radiation with the appropriate energy. By now you should all know that the energy of electromagnetic waves depends only on their wavelength.

Whenever an electron is knocked free of an atom, it is said that the atom is **ionized.** An **ion** is an atom from which one or more electrons have been removed (or added).

Example 7.4
What kind of electromagnetic radiation is emitted when an electron in a hydrogen atom drops from the second to the first energy level?

Information: It emits 10.2 electron volts.

Equation: Einstein's equation, naturally

$$e = \frac{1.25}{L}$$

Substitute:

$$10.2 = \frac{1.25}{L}$$

Rearrange:

$$L = \frac{1.25}{10.2} \quad \text{(given in microns)}$$

Solution
$L = 0.123$ microns
This corresponds to ultraviolet radiation.

**Note:** The fact that the sun puts out a great deal of radiation with a wavelength of 0.123 microns has led scientists to conclude that there is a great deal of hydrogen on the sun. Scientists have found that each atom and molecule emits its own specific wavelengths of electromagnetic radiation. It is therefore possible to tell what atoms and molecules there are on the sun, the stars, or even the atmospheres of other planets—simply by analyzing the spectrum of electromagnetic radiation coming from the star or planet. *Spectroscopy* is very important in meteorology, astronomy, physics, chemistry, and biology. In short, it is important in every branch of the sciences.

The same basic ideas that work for atoms also work for molecules. However, since molecules consist of groups of atoms, their behavior can be somewhat more complicated than the behavior of single atoms. The electrons in molecules can also change their rungs or orbits as is the case with atoms. With molecules there are several additional degrees of freedom. Molecules can also rotate or vibrate. Each molecule has only certain specific rotation rates and only certain specific vibration rates that are permitted (see Figure 7.3). When the appropriate amount of energy is supplied to the molecule, it may make the molecule either rotate faster, vibrate faster, or change the orbit

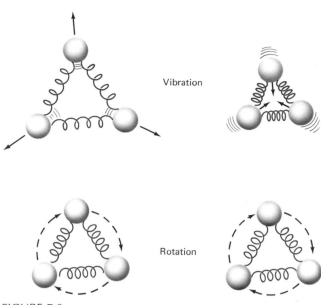

**FIGURE 7.3**
Rotation and vibration of molecules.

Vibration

Rotation

of one of the electrons in the molecule. If the wavelength is short enough, it can make the atoms in the molecule vibrate so violently that the entire molecule will split. The splitting of molecules by electromagnetic radiation is called **photodissociation.** When electrons are split from an atom or molecule by electromagnetic radiation, they are said to be **photoionized.** As I have already mentioned, if the molecules of the upper atmosphere weren't photodissociated or photoionized by the sun's short rays, then the molecules of our bodies would soon be.

## OXYGEN AND OZONE

Oxygen molecules ($O_2$) consist of two oxygen atoms. Electromagnetic waves with an energy of 5.12 electron volts are sufficient to photodissociate the oxygen molecule into its two component atoms.

Example 7.5
What is the wavelength of radiation that can photodissociate $O_2$?

Information: The energy of the waves must be at least 5.12 electron volts.

Equation:
Einstein's equation, naturally

$$e = \frac{1.25}{L}$$

*Substitute:*

$$5.12 = \frac{1.25}{L}$$

Rearrange:

$$L = \frac{1.25}{5.12} \quad \text{(given in microns)}$$

Solution
$L = 0.244$ microns (or shorter)

Oxygen molecules high in the atmosphere can therefore be split by ultraviolet radiation.

Many of the atoms and molecules that are photoionized and photodissociated during the daytime by the sun's rays recombine during the night and give off light as a result. At night a great deal of light is produced in this manner high in the atmosphere (80 km and higher) and is known as the **airglow.** Although you may think that the sky is pitch black between the stars, it isn't. On a clear night in the countryside, far away from the city lights, hold your hand up to the sky and you will see that your hand is much darker than the sky. Much of this skylight is due to the airglow and, while the airglow does not glow brilliantly, it does provide enough light to get around on a clear, moonless night.

Two of the brightest colors of the airglow are red and green. The red color is produced by oxygen atoms recombining to form $O_2$ molecules via some complex reactions and has wavelengths of 0.630 and 0.636 microns. This occurs roughly 300 km above the earth's surface. The green color is produced by oxygen atoms recombining at 90 km above the earth and has a wavelength of 0.5577 microns.

On the basis of preceding material you may have realized that oxygen is vital to our continued life on earth not only because we need it to breathe but also because oxygen, high in the atmosphere, absorbs much lethal ultraviolet radiation with wavelengths shorter than 0.244 microns. In fact, $O_2$ principally between 65 and 115 km above sea level almost completely shields us from all solar radiation of wavelengths between about 0.1 and 0.2 microns. But in the very act of protecting us, the $O_2$ molecule sacrifices itself and is photodissociated or split into two oxygen atoms ($O + O$).

The story is not over yet. The sun emits a small but significant amount of ultraviolet radiation with wavelengths shorter than 0.1 microns. These waves are highly energetic and would prove lethal if they were not largely absorbed by some of the gases in the upper atmosphere (above 100 km). Once again oxygen comes to our aid. This time (for wavelengths less than 0.1 microns) atomic oxygen, O, is our main protector. The O protects us by becoming photoionized (see Figure 7.4).

Oxygen ions, ions of other atoms, and electrons are therefore found in the atmosphere at heights greater than roughly 90 km above the earth's surface. For this reason the region is often called the **ionosphere** and the ionosphere plays an important role in radio transmissions, as you will see a little later.

In spite of all this protection, we are still vulnerable to radiation with wavelengths between 0.2 and 0.3 microns. However, neither O nor $O_2$ is able to absorb much of the sun's rays with wavelengths between 0.2 and 0.3 microns. Only *ozone*, $O_3$, is able to absorb these rays effectively. It is now time to see how ozone is produced in the atmosphere.

Imagine that the oxygen atoms are lonely and that they "want" to join with other oxygen atoms or molecules. Sometimes they will collide with another oxygen atom and recombine into a normal oxygen molecule, $O_2$. Unfortunately, there are relatively few oxygen atoms so that it is much more likely for an oxygen atom to collide with an $O_2$ than with another O. The "marriage" of an O with an $O_2$ produces $O_3$, ozone.

There is one more complication. In order for this marriage to occur, you need a "Minister" molecule ($M$). Speaking scientifically, when an O collides with an $O_2$,

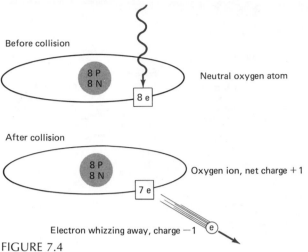

Before collision

Neutral oxygen atom

8 P
8 N

8 e

After collision

8 P
8 N

Oxygen ion, net charge $+1$

7 e

Electron whizzing away, charge $-1$   e

FIGURE 7.4
The photoionization of an oxygen atom.

there is simply too much energy for them to combine unless a third molecule is also involved in the collision. This third, "Minister" molecule serves to effect the union by removing the excess energy of the collision (see Figure 7.5).

Ozone is a very unstable, corrosive molecule that is destroyed when it reacts with the surface of the earth and various pollutants in the air near the ground. It also tends to combine with parts of your body in an undesirable manner. In other words, ozone is poisonous, and breathing it in sufficient quantities can cause death.

This might have created problems for us if we weren't living in the best of all possible worlds. Too much ozone would lead to death by poisoning; too little, to death by

FIGURE 7.5
The formation of an ozone molecule. Note the need for a "Minister" molecule—to absorb the excess energy of the collision.

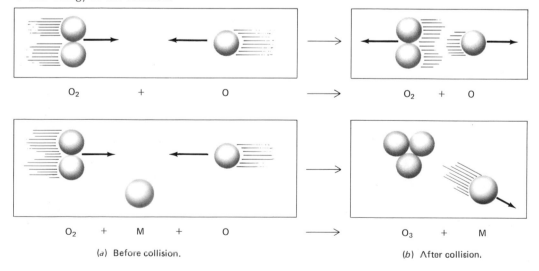

$O_2$    +    O       $\longrightarrow$      $O_2$    +    O

$O_2$    +    M    +    O       $\longrightarrow$      $O_3$    +    M

(a) Before collision.

(b) After collision.

sunburn. Fortunately, there is just the right amount of ozone and it is most concentrated far away from the ground, in a region between 15 and 50 km above the earth's surface where it can do us little harm. Very little of the ozone reaches the earth's surface and when it does it is rather quickly destroyed.

When ozone absorbs the longer ultraviolet radiation it is photodissociated into an O and an $O_2$ in the process (see Figure 7.6). Fortunately, this does little to deplete the ozone, because the O quickly recombines with any $O_2$ as soon as both collide with a "minister" molecule.

Summarizing, we find that radiation with wavelengths of

1. Less than 0.1 microns photoionize O atoms.
2. Between 0.1 and 0.2 microns photodissociate $O_2$ molecules.
3. Between 0.2 and 0.3 microns photodissociate $O_3$ molecules.

All these waves are lethal to the molecules of our bodies.

The life-protecting layers of the atmosphere are shown in Figure 7.7.

## SUNLIGHT AND SKIN COLOR

Skin color is one of the most noticeable differences among the various peoples of the world. Unfortunately, many people distort the facts about skin color. Since many of the racial differences among people have been induced by worldwide differences in climate, you might almost say that if you hate people of one race *because of their skin color* then what you really hate is a different climate!

Our bodies can synthesize vitamin D. We need vitamin D for strong bones and teeth (technically speaking, proper

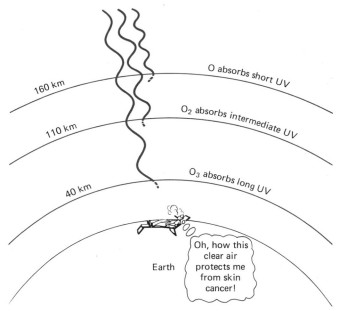

FIGURE 7.7
The life-protecting layers of the atmosphere.

calcium metabolism). Unfortunately, most foods provide little vitamin D. But the layer of cells directly under our skin will produce vitamin D when ultraviolet light from the sun penetrates the skin. We can therefore produce vitamin D when we are exposed to sunlight. It is essential to produce enough vitamin D to prevent rickets (severe bone deformity usually in the legs and spine). On the other hand, too much vitamin D will lead to brittle bones as well as calcium deposits such as kidney stones. It is therefore necessary to maintain a proper balance of vitamin D in the body and to expose the body to the right amount of sunlight.

What does all this have to do with skin color? Light skin will permit most of the sun's ultraviolet rays to penetrate to the layer of cells that produces vitamin D, whereas dark skin will block most of these rays. Light-skinned people should therefore be found where there is little ultraviolet radiation, while dark-skinned people should be found where there is too much ultraviolet radiation.

Very few short waves from the sun penetrate the atmosphere when the sun is low in the sky because of the long path through the atmosphere and because short waves are readily absorbed or scattered. Since the sun is low in the sky in the polar regions, people in the polar regions need all the sunlight they can get; thus their skin will be light. Near the equator, where the sun is high in the sky, plenty of ultraviolet radiation penetrates to ground level and people's skins must be dark to protect them from an excess production of vitamin D. In line with this, darker

FIGURE 7.6
The photoionization of an ozone molecule. Through most of the stratosphere, the oxygen atom will rapidly recombine with an $O_2$ molecule so that the net effect of the whole process is to heat the stratosphere.

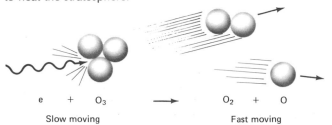

skinned people also have a lower incidence of skin cancer.

In summary, we find that human skin color generally gets lighter as you go from the equator to the poles. Of course, recent human migration patterns must be excluded, since evolutionary changes take thousands of years to adjust to external conditions.

# VERTICAL STRUCTURE OF THE ATMOSPHERE

As you go up in the atmosphere the air thins out and the pressure decreases. By the time you are approximately 5.5 km above sea level, half of the atmosphere lies beneath you. The air at 5.5 km is also considerably thinner (less dense) than the air at sea level, and no human beings live at such altitudes for extended times simply because the air is too thin to breathe. Even fires do not burn so well in the thin air, as Marco Polo observed.

If you go up another 5.5 km—to 11 km above sea level—you do not go through the other half of the atmosphere. Instead, you go through about half of the *remaining* part of the atmosphere. In other words, *for roughly each 5.5 km you ascend, the atmospheric pressure halves.* This behavior is due to the fact that the air is compressible, and the air at the bottom of the atmosphere is compressed by all the air above it. Think of air as a series of springs piled one on top of the other. Assume that each spring weighs exactly the same amount. Clearly, the bottom spring will be most crushed, while the springs near the top will be spread out (see Figure 7.8). More tech-

## FIGURE 7.8
The air pressure and density decreases with increasing height, because there is less weight above. The air thus behaves like a pile of springs or a pile of people.

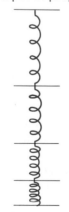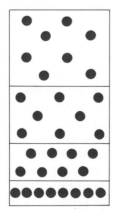

LIGHT, ATOMS, AND MOLECULES: THE UPPER ATMOSPHERE AND US

TABLE 7.1
The Average Altitudes of Standard Pressure Levels

| Pressure (millibars) | Typical Height (Meters) |
| --- | --- |
| 1013 | 0 |
| 850 | 1500 |
| 700 | 3000 |
| 500 | 5500 |
| 300 | 9000 |
| 200 | 12500 |

nically, the ideal gas equation shows the same result—namely, that volume doubles when the pressure halves (for constant temperature).

The average sea-level pressure is 1013.25 mb or roughly 1000 mb. In the lowest few thousand meters of the atmosphere it is convenient to remember that pressure drops roughly 1 millibar per 10 meters that you ascend.

Meteorologists use certain standard weather charts (called constant pressure charts) to tell them what the weather is like at various heights above the surface. Some of the standard charts are listed in Table 7.1, which also gives their approximate heights above sea level.

In the nineteenth century a number of balloonists ascended into the atmosphere. They found that the higher they went the colder it became. You must have seen pictures where mountaintops are covered with snow or ice while lowlands are bare. (See Figure 7.9.) The scientists of the nineteenth century thought that this decrease of temperature with height continued throughout the entire atmosphere.

Then around 1900, Teisserenc de Bort found that beyond about 12 km above sea level, the temperature stopped decreasing with height. He had discovered the atmospheric layer now known as the **stratosphere.** We now know that the atmosphere is composed of a number of layers, as is shown in Figure 7.10. We live in the lowest layer—the **troposphere.** The troposphere is the layer in which virtually all the clouds and "weather" occur.

When you leave the troposphere you must cross the **tropopause** in order to get into the stratosphere. Within the stratosphere the temperature may level off at first, but it soon begins to rise as you ascend. At an approximate height of 50 km the temperature reaches a maximum, and you cross the **stratopause** and enter the **mesosphere.** Within the mesosphere the temperature falls with increasing height. At a height of approximately 80 km you cross the **mesopause** and enter the **thermosphere.** Temperatures increase dramatically with height within the thermosphere, typically reaching nearly 1200°C. The thermosphere is also known as the **ionosphere.**

If you read the last paragraphs carefully, then you real-

**FIGURE 7.9**
A snow line on a mountain ridge. This illustrates the principle that temperature in the troposphere generally decreases with height. At a certain height, it becomes too cold for trees to grow; at an even greater height, it becomes too cold for snow to melt. Often, distinct tree lines and snow lines can be seen.

**FIGURE 7.10**
The layers of the atmosphere. Temperatures are shown for January and July (July values are in parentheses).

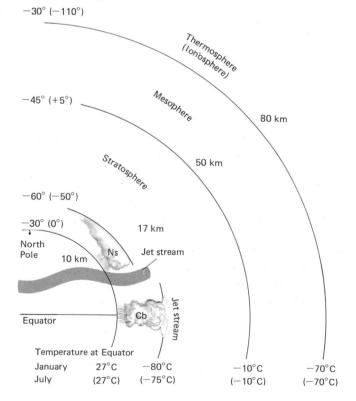

ize that the vertical temperature structure of the atmosphere is pretty complicated. However, the explanation for this behavior is not so complicated. There are three warm regions—at the ground, at 50 km, and above 80 km. These regions are warm because they have specific sources of heat. The cold regions are cold because they lack heat sources.

The source of heat for the ground is visible light from the sun. Visible sunlight largely penetrates the entire atmosphere and is then absorbed at the ground. The ground in turn acts like a heater warming the air near it. Even though the warm air rises from the ground, the upper troposphere remains cold because air cools as it rises.

The next hot region lies near the top of the ozone region where the ozone is absorbing the ultraviolet rays with wavelengths between 0.2 and 0.3 microns, and consequently this warms the atmosphere. Recall how, in the formation of ozone the collision speeds up the "minister" molecules. In the third hot region, the thermosphere, various atoms and molecules absorb the shortest ultraviolet rays and are split and heated up. The resemblance of Figures 7.7 and 7.10—which depict the atmosphere's protecting layers and the vertical temperature structure of the atmosphere—is therefore no coincidence.

It is appropriate at this point to reveal a little more about the lowest layer of the atmosphere, the troposphere. This is the atmosphere's so-called "region of change" in which virtually all the weather occurs. In fact, roughly 75% of this book is devoted to explaining and describing the events that occur in the troposphere.

A good rule to use for the troposphere is that, on the *average*, the temperature decreases 6°C for each kilometer you ascend.

---

**Example 7.6**
The temperature at sea level is 35°C. What do you expect the temperature to be on the top of a nearby 3000 m high mountain?

Information: On the average temperature drops 6°C for each kilometer you ascend.

Equation:
$$T_{top} = T_{bottom} - (6)(H)$$
Temperature ($T$) in centigrade
Height ($H$) in kilometers

Substitute:
$$T_{top} = 35 - (6)(3) = 35 - 18$$

Solution
$$T_{top} = 17°C$$

In Figure 7.10 you can see an average profile of temperature for January and July in the NH. Surprisingly the coldest part of the tropopause during either month is not near the North Pole but rather is above the equator! Seventeen kilometers above sea level at the equator, the temperature is generally −80°C!

In the polar region there is a large difference in the stratosphere between January and July. Much of this difference is due to the fact that during January no sunlight strikes the polar regions, and consequently the normal heat source at 50 km is shut off. As a result the temperature continues to fall with increasing height through most of the stratosphere in January.

The main storm belt of the atmosphere is found in the midlatitudes. This storm belt is intimately connected with the jet stream, which is discussed in Chapter 16, along with the large-scale winds. In this region the tropopause can be twisted, folded, or even split. When radioactive material is released in the stratosphere, it usually reenters the troposphere through these midlatitude breaks in the tropopause.

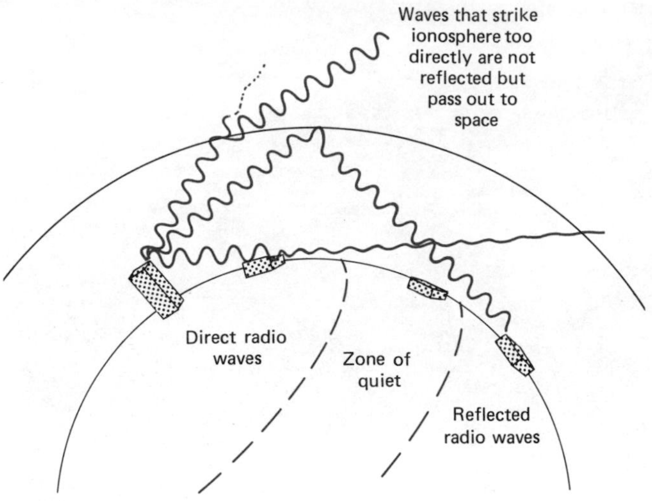

FIGURE 7.11
The transmission of radio waves. Places near the transmitter receive direct radio waves, while distant places receive radio waves reflected from the ionosphere. Often, there is some intermediate zone that cannot receive either. This is called a zone of silence.

# THE IONOSPHERE AND RADIO WAVES

When you are traveling across the country in a car, you notice that radio reception often fades out about 80 km away from the cities where the broadcasts are being transmitted from. At night a strange thing happens. Perhaps as you are driving along listening to some radio station the announcer will suddenly interrupt and say that he is speaking to you from WLS Chicago. This may seem odd if you are listening from somewhere in Colorado, which is more than 1600 km from Chicago. Cases like this happen every night.

You can thank the ionosphere for long-distance radio reception at night. Radio waves are merely one part of the electromagnetic spectrum and therefore are affected by electrical charges. Since ions are electrically charged, radio waves and ions will interact.

The ions in the ionosphere can have either of two effects on the radio waves. They can either absorb (deplete) the waves or reflect them (see Figure 7.12). In the lower part of the ionosphere, radio waves tend to be absorbed and as a result local regions of the lower atmosphere can actually heat up by some 50°C. In the higher parts of the ionosphere, radio waves tend to be reflected, and therefore the higher parts of the ionosphere act like mirrors.

For short distances on earth radio waves are received directly from the transmission station but, because the

earth is curved, this is not possible over long distances. However, it is possible that rays reflected from a layer high in the atmosphere can be transmitted over long distances, as Figures 7.11 and 7.12 indicates.

**Note:** If a radio wave hits the reflecting layer too directly, it will not be reflected but will penetrate the layer and be lost to the earth. As you can see from Figure 7.11, there is generally a so-called quiet zone in which you are too far from the transmission station to receive the direct radio waves, but too close to receive the reflected radio waves from that particular station.

The main reflection layer in the ionosphere, called the *F2* layer or the reflection layer, is found at an approximate height of 300 km above the earth's surface. This reflection layer is present both day and night. The *F2* layer is actually a better reflector during the *daytime*, despite the fact that long distance radio reception is far better at night.

The reason that the *F2* layer is a much better reflector during the daytime is simply that there are more ions during the day. Since the atoms are split into ions by the sun's ultraviolet radiation but are constantly recombining into atoms, the number of ions must decrease every night when there is no sunlight and increase again every day.

This increase of ions during the daytime is also true in the lowest part of the ionosphere, which is known as the *D* layer. The D or Depletion layer is responsible for absorbing the radio waves during the day, but at night the *D*

LIGHT, ATOMS, AND MOLECULES: THE UPPER ATMOSPHERE AND US

layer virtually disappears. Radio waves are therefore absorbed (depleted) only during the day.

We can now put the story together. Both day and night the F2 layer reflects radio waves. Unfortunately, during the day few radio waves penetrate through the Depletion layer. These few waves are then reflected by the F2 layer. Before they reach the ground they must pass through the D layer for a second time, where almost all the rest of the waves are absorbed. At night the D layer almost completely disappears. At night, therefore, radio waves easily reach the F2 layer where they are reflected back toward the ground. On the way back down there are also no obstructions so that the waves reach the ground in a well-preserved state, hundreds of kilometers from the transmission station.

## THE AURORA

The aurora is the silent fireworks of the nighttime polar sky. The aurora can take on a variety of forms and may appear like sheets or waving curtains of green, blue, or red light suspended in midair. (See Plate 10.) The words, "aurora borealis," mean the "northern dawn." The aurora was given this name because when people living in the midlatitudes see the aurora they generally must look toward the north, where the sky near the horizon may appear reddish.

The aurora can be seen on almost any clear night of the

year, *if you are in the right place.* The best time to see the aurora is around midnight. The right place is anywhere in a belt a few hundred kilometers wide that circles the magnetic north pole of the earth and is located about 2000 km from the magnetic north pole (see Figure 7.13). On rare occasions the aurora can be seen as far south as New York City or Chicago, but if you want to have a good chance to see the aurora you normally should go at least as far north as southern Canada.

Because the auroras are centered around the *magnetic* poles of the earth, it is apparent that the aurora has something to do with magnetism. When you use a compass to find direction you should know that the compass does not point directly north! The compass needle points to the magnetic north pole, and the only reason that a compass is useful is that the magnetic north pole is reasonably close to the geographical North Pole.

In the year 1600, William Gilbert discovered (or published his discovery) that the earth acts like a giant magnet. If you take a regular bar magnet and sprinkle iron filings around it, the filings will line up the same way every time.

Scientists have found it useful to draw magnetic field "lines" for the bar magnet, and it is useful to draw magnetic field lines wherever there is a magnetic field. (See Figure 7.14.) Figure 7.15 shows a drawing of the magnetic field of the earth. These field lines exhibit a very important property. Any time an electrical charge tries to move across the magnetic field lines, it will be twisted by the magnetic field.

The aurora is produced by a stream of electrons that

FIGURE 7.12
Long-distance radio reception is interrupted during the day when the depletion layer fills with ions. At night, the depletion layer disappears, so long-distance reception is restored.

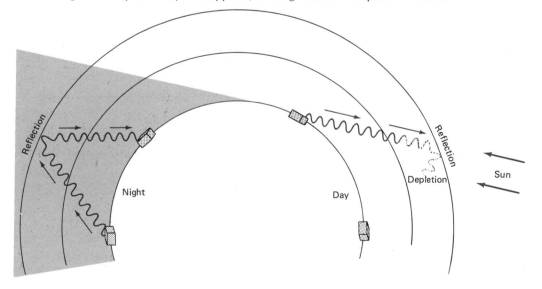

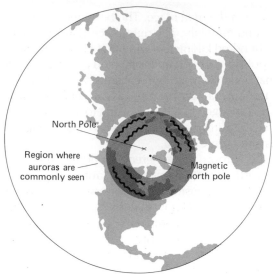

FIGURE 7.13
The region in which auroras can be seen on more than half the clear nights. Near the center of the ring auroras can be seen on almost all clear nights. Notice this ring-shaped region centers around the north magnetic pole.

have come from the sun. The sun is largely composed of hydrogen. Because of the high temperatures of the sun, most of the hydrogen is ionized and is merely a bunch of separate protons and electrons. Protons and electrons are constantly streaming out of the sun (this is the *solar wind*) and would smash directly into the earth if it weren't for the magnetic field of the earth. As these electrons and protons come into contact with the earth's magnetic field, they are twisted by the magnetic field. Some of these are trapped by the earth's magnetic field. Somewhere in the tail of the earth's magnetic field they are accelerated to tremendous velocities by a process that is as yet not fully understood.

In a mothlike manner these trapped electrons and protons (now in the atmospheric layer called the **magnetosphere**) spiral around magnetic field lines, bouncing back and forth between the north and south magnetic poles. Most of the time the particles are so high above the earth's surface that they are in an almost complete vacuum. As a result, there is little chance of a collision between a proton or electron and an air molecule. However, as the particles approach the poles, they do get close to the earth's surface where the atmosphere becomes dense. The more rapidly the electrons and protons are moving, the closer they can approach the magnetic poles and the earth's surface (see Figure 7.16).

As the electrons and protons approach the earth's surface, the chance of a collision between these particles and air molecules increases dramatically. The electrons

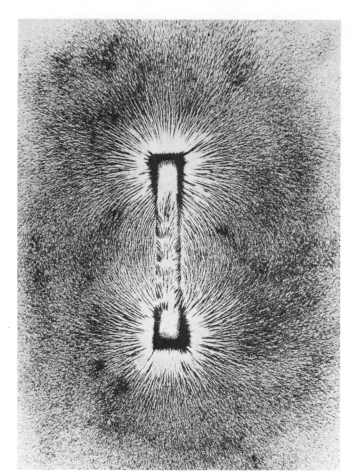

FIGURE 7.14
Iron filings line up to show magnetic field lines.

FIGURE 7.15
The magnetic field of the earth. Particles from the solar wind enter the earth's magnetic field from the side opposite the sun.

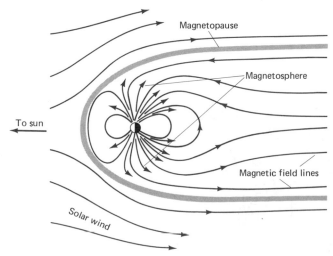

LIGHT, ATOMS, AND MOLECULES: THE UPPER ATMOSPHERE AND US

that produce the aurora are traveling at more than one-tenth (1/10) the speed of light and have energies in excess of 10,000 electron volts. This is easily enough energy to ionize atoms.

Because the collision zones occur where the magnetic field lines get close to the earth's surface and this happens around the magnetic poles, the aurora is found around the magnetic poles. Atoms are ionized by the energetic electrons, and light is emitted when these ions recombine and release the energy they absorbed during the collisions. This is the auroral light. Collisions involving protons give off much duller light and have little direct connection with the aurora.

The more energetic electrons penetrate deeper into the atmosphere, producing the green colors of the aurora at about 100 km above the earth's surface, whereas the less energetic electrons only penetrate as far as about 250 km and produce the red colors of the aurora.

Normally, the aurora can be seen only near the auroral belt. However, when sunspots and giant flares erupt on the sun, the solar wind becomes much stronger for a few days and as a result the aurora grows much brighter. These times are known as magnetic storms, and while they last the entire upper atmosphere becomes highly disturbed. Radio reception is also dramatically altered—and always for the worse. Even TV reception can become totally distorted. Then, while you sit at home cursing at a blurry television, the scientists who study the upper atmosphere are ecstatic, because the distorted radio and TV waves provide them with much important information about the nature and behavior of the upper atmosphere.

FIGURE 7.16
The aurora and its relation to the earth's magnetic field. Electrons spiral back and forth between the magnetic poles, along the magnetic field lines. The aurora only occurs when the electrons get close enough to the earth to collide with air molecules.

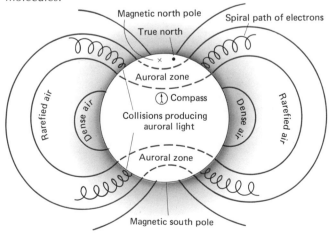

# PROBLEMS

**7-1** Why do you think X rays are so dangerous?

**7-2** Light with wavelengths of 0.56 microns and 0.66 microns are both observed in the aurora. Considering that one energy transition in oxygen produces 2.25 electron volts while one energy transition in hydrogen produces 1.90 electron volts, tell which element produced which light. To which color does each wavelength correspond?

**7-3** When electrons drop from the third to the second energy level of hydrogen atoms, what color light is emitted? Some helpful information is given in Figure 7.2.

**7-4** Why isn't ozone produced much above 50 km?

**7-5** Why aren't oxygen molecules photodissociated near sea level by the sun's ultraviolet radiation?

**7-6** Why do we need ozone even though it is poisonous?

**7-7** The fastest winds are typically found at about the 250 mb level. How high is this above sea level?

**7-8** Why does temperature generally decrease with height in the troposphere?

**7-9** Why does temperature generally increase with height in the stratosphere?

**7-10** Explain why the stratopause near the North Pole is so cold in January. (Notice that the mesopause by contrast is warmest in January!)

**7-11** Explain why the tropopause is coldest and highest over the equator.

**7-12** The average temperature in the tropics is 27°C at sea level. About how high does a mountain at the equator have to be in order to support a glacier?

**7-13** Explain why distant radio stations can sometimes be heard at night but never during the day.

**7-14** Explain the "zone of silence" encountered in radio reception.

**7-15** Widespread auroras can be forecast about three days in advance just by watching the sun. Explain.

**7-16** According to Figure 7.13, which of the following places would be best for viewing an aurora: (a) Miami, Florida; (b) Point Barrow, Alaska; (c) Madrid, Spain; (d) London, England; (e) Godthaab, Greenland; (f) Vardo, Norway; (g) Calcutta, India?

**7-17** Why should there be a relationship between the color of the aurora and the height at which it forms?

**7-18** Why do the auroras appear around the earth's magnetic poles?

**7-19** Do you think that 1987 will be a good year to observe auroras?

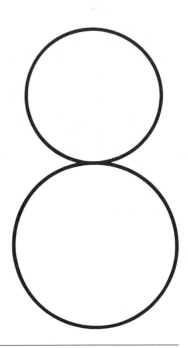

# TEMPERATURE AND HEAT

The autumn and winter of 1976–77 will long be remembered in the eastern half of the United States. From early October to mid February, temperatures averaged 5°C or more below normal. In Buffalo, New York, some snow fell on 45 consecutive days, and throughout this time the temperature remained below freezing.

The worst cold wave of the season occurred on January 17, 1977 when the frigid air poured southward into Florida. All-time low temperatures occurred in dozens of cities. In Miami, Florida, the temperature dropped to −2°C, and for the first time in recorded history snow was observed there (and in the Bahamas as well). In Cincinnati, Ohio, the temperature fell to a record-breaking −32°C, and people walked across the frozen Ohio River (see Figure 8.1). A number of people froze to death in their homes when they could no longer pay their utility bills and their power was shut off. Many factories were shut down to conserve fuel.

At the same time the western third of the United States was having abnormally warm, dry weather. This proved disastrous to agriculture and to the skiing industry. Economic losses approached $4 billion. There were even a few days when most of Alaska had warmer weather than Florida!

The record-breaking cold brought temperatures typical of Canadian winters southward to the eastern United States. In Canada temperatures of −20°C are not unusual.

FIGURE 8.1
The Ohio River frozen at Cincinnati in January 1977.

During the winter of 1976–77 the winds blew this cold Canadian air southeastward into the United States day after day for months.

Explaining the record-breaking cold weather requires two steps. The first step is to explain how the air in Canada gets so cold in the first place during the winter. Addressing such questions is one of the main purposes of this chapter. The second step is to explain why the winds are so persistent at times. We know disappointingly little about this second point, but what we do know will be discussed in Chapter 24.

This chapter is devoted to the description and explanation of the factors that determine temperatures and temperature variations on the earth and on the other planets. We will see why the earth normally has such livable temperatures, while Venus and Jupiter have temperatures far too extreme to support life as we know it.

## MEASURING TEMPERATURES

Thermometers are used to measure temperature. If you take a thermometer and place it face upward on the road on a hot summer day it is quite likely that the temperature will exceed 58°C (the world's record high *air* temperature). Under such conditions you are not measuring the air temperature—you are measuring the road temperature. Air temperature must be measured under standard conditions. The thermometer must be placed in a shaded and well-ventilated place, about 2 m above the ground.

There is still some controversy about who invented the thermometer. The thermometer makes use of the fact that some substances expand more rapidly than others when heated. Hero of Alexandria was aware of this property over 2000 years ago.

Hero designed a glass bulb filled with air (see Figure 8.2). A thin tube filled with water was connected to the bulb. When the bulb was heated the air inside it expanded, forcing the water out of the tube.

Hero published his discovery in a book that was lost for a long time. His work was rediscovered and translated into Latin in 1572, causing great excitement among the European scientists and engineers of the time. About 20 years later either Galileo, Santorio Santorre, or Cornelius Drebel experimented with and perhaps calibrated the first thermometer.

The standard thermometer is simply a glass tube containing mercury. When the temperature rises the mercury expands more rapidly than the glass and so the mercury rises in the tube. When the temperature falls below

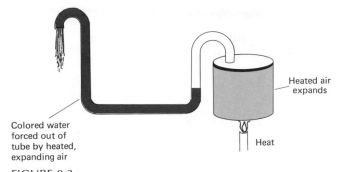

FIGURE 8.2
Hero's thermoscope. As the air in the container is heated, it expands (according to the ideal gas equation), forcing the water in the tube to rise.

−39°C, mercury freezes and alcohol must be used in its place. Most alcohols do not freeze until the temperature drops below −80°C.

The ordinary thermostat consists of two strips of metal that are soldered together (see Figure 8.3). As the temperature rises one of the metals expands more rapidly than the other, causing the strips to bend or warp. When this is designed properly the bending strips will touch an electrical switch that then turns off the heat. As the temperature falls once again the metal strips straighten out, passing another switch that will then turn the heat back on.

## TEMPERATURES ON EARTH

The average temperature of the atmosphere near sea level is roughly 15°C. The lowest temperature ever recorded on the earth's surface is −88°C, which occurred at Vostok, Antarctica, on August 24, 1960 with a wind of 15 knots! At such low temperatures spit will freeze in midair. The highest temperature ever recorded is 58°C, and it occurred in Azizia, Libya, on September 13, 1922. Remember that this was measured in the shade!

A brief picture of the temperatures of some of the world's cities is now presented to give you an idea of the representative temperatures around the world. World temperatures will be discussed in greater detail in Chapter 22.

Aristotle thought that it was impossible for human beings to approach the equator because he thought the temperature there was much too high to support life. Many people still have the impression that the equator is steaming—today we know that this impression is false. At most places within 5° latitude of the equator, tempera-

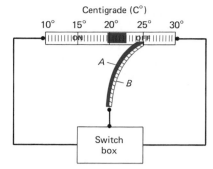

FIGURE 8.3
The thermostat. A strip consisting of two different metals is soldered together. When the temperature rises, metal *A* expands more rapidly than metal *B*, causing the strip to warp and trigger a switch.

tures are typically 27°C year round on the average. During the daytime the temperatures will rise to perhaps 32°C, while at night they will drop to about 22°C.

At this point you should recall that temperature normally decreases with height. Therefore Quito, Ecuador (elevation 2700 m), averages only 13°C, despite the fact that it lies right near the equator. In general, by the time you reach 5000 m above sea level, it is cold enough to support glaciers even on the equator!

Singapore, Indonesia, is a city with a typical equatorial climate. Located about 150 km north of the equator at the southern tip of Asia, Singapore has an annual mean temperature of 27.0°C. As you can see from Table 8.1,

January is the coldest month in Singapore with an average of 26.4°C, while May is the warmest month with an average of 27.8°C. Thus, there is very little temperature change from one month to another in Singapore, and there is certainly no such thing as winter.

In fact, the temperature changes so little in Singapore that in 30 years of records the coldest temperature they have ever experienced is 18.9°C. But what is even more surprising is that the hottest temperature there is a mere 36.1°C. Normally cool places such as London and Verkhoyansk, Siberia, have had greater all-time highs—36.8°C and 36.7°C, respectively.

In the subtropics (from about 15 to 30° latitude) many of the world's deserts are found. Thus, days are hot as a rule. However, because the air is dry and clear, this heat rapidly escapes at night and the temperature falls considerably. Sometimes, during the winter, the nights can be truly cold in the world's subtropical deserts. Diurnal (day to night) temperature ranges average about 15°C in most places and can be as high as 30°C. Try to imagine a place that is 35°C in the afternoon but only 5°C at night. This happens in the desert.

Aswan, Egypt, is a typical desert weather station. From Table 8.2 you can see that Aswan (24° N latitude) definitely has an annual temperature cycle. January weather is actually quite pleasant, though perhaps a bit cool at night. January is the winter in Aswan, but you shouldn't expect to find any snow. In July (the hottest month) the average temperature is a mere 34°C; when you take the diurnal variation into account, this means that you can expect typical afternoon temperatures over 40°C. Even

TABLE 8.1
Temperatures (°C) for Singapore (1° N)

|  | Jan | Feb | Mar | Apr | May | Jun | Jul | Aug | Sep | Oct | Nov | Dec | Year |
|---|---|---|---|---|---|---|---|---|---|---|---|---|---|
| Average | 26.4 | 27 | 28 | 28 | 28 | 28 | 28 | 27 | 27 | 27 | 27 | 27 | 27 |
| Range[a] | 7 | 7 | 7 | 7 | 7 | 7 | 7 | 7 | 7 | 7 | 7 | 7 | 1.5 |
|  |  |  |  |  | Record high | 36 |  | Record low | 19 |  |  |  |  |

[a] Range is average daily range of temperatures except under "year" where it is difference between coldest and warmest months.

TABLE 8.2
Temperatures (°C) for Aswan (24° N)

|  | Jan | Feb | Mar | Apr | May | Jun | Jul | Aug | Sep | Oct | Nov | Dec | Year |
|---|---|---|---|---|---|---|---|---|---|---|---|---|---|
| Average | 17 | 18 | 22 | 27 | 30 | 33 | 34 | 33 | 31 | 27 | 21 | 18 | 26 |
| Range | 13 | 15 | 16 | 17 | 17 | 16 | 15 | 14 | 12 | 10 | 10 | 11 | 17 |
|  |  |  |  |  | Record high | 51 |  | Record low | 2 |  |  |  |  |

though this air is quite dry, it is still rather uncomfortable during the day. Thus the world's hottest weather is not found at the equator but in the subtropical deserts during the summertime. The subtropical latitudes do not contain only deserts, but any subtropical station not right on the coastline will always have a few very hot months.

St. Louis, Missouri, has a climate that is typical of an inland city in the midlatitudes. St. Louis is located at 39° N latitude and has a summer that is almost as warm as the temperatures at the equator. The July mean temperature is 26°C. In fact, when a heat wave occurs and the temperature reaches 40°C or more for a few consecutive afternoons, St. Louis is hotter than it ever gets at the equator.

It is the winter that reminds you that St. Louis lies far from the equator. January is the coldest month in St. Louis when the average temperature is only 0°C. When cold polar winds from Canada pass through St. Louis in winter, the temperature can drop below −25°C.

London, England, lies about 1200 km closer to the North Pole than does St. Louis. Therefore, you might expect that winter in London would be colder than it is in St. Louis. However, this is not the case. Winter in London is distinctly *warmer* than it is in St. Louis! London averages 4°C during January, which is much warmer than you would expect for a city at 52½ N latitude.

London's weather is strongly influenced by the proximity of the Atlantic Ocean and the relatively warm remnants of the Gulf Stream (the North Atlantic Current). The winds blow predominantly from the Atlantic Ocean which lies little more than 200 km west of London. Without the Atlantic Ocean nearby, London would probably be at least 20°C colder during winter.

After the long winter, the land warms fairly quickly, while the ocean warms very slowly. During the summer in London the ocean actually cools the climate somewhat so that London's warmest month, July, averages only 17°C. Summer days in London are quite pleasant, whereas the nights are positively cool—so much so that if you ever intend to go to London during the summer, be sure to bring a sweater. The English have become so accustomed to the cooling influence of the ocean in summer that the occasional heat waves that produce temperatures above 30°C inevitably cause many cases of heat stroke and even death.

London's climate is a good example of a general rule.

**Rule:** The ocean has a moderating influence on climate. Its presence always serves to reduce temperature variations or extremes.

Another general characteristic of climate is that annual temperature variations increase from the equator to about 60 or 70° latitude. Subpolar locations (away from the ocean) generally have the largest annual temperature variations in the world.

Verkhoyansk, Siberia, is perhaps the world's extreme example of annual temperature variations. Located 100 km north of the Arctic Circle in the Yana River Valley, the average temperature in January is −50°C. Extremely cold air sinks into the valley where there is no sunlight to warm it up until late January. In July the temperature averages 15°C making it almost as warm as it is in London. This means that the annual temperature range in Verkhoyansk is 65°C, which is astounding. The average temperature

TABLE 8.3
Temperatures (°C) for St. Louis (39° N)

|  | Jan | Feb | Mar | Apr | May | Jun | Jul | Aug | Sep | Oct | Nov | Dec | Year |
|---|---|---|---|---|---|---|---|---|---|---|---|---|---|
| Average | 0 | 2 | 6 | 14 | 19 | 24 | 26 | 25 | 21 | 15 | 7 | 2 | 13 |
| Range | 10 | 10 | 11 | 12 | 11 | 11 | 11 | 11 | 11 | 12 | 10 | 9 | 26 |

Record high  46    Record low  −31

TABLE 8.4
Temperatures (°C) for London (51° N)

|  | Jan | Feb | Mar | Apr | May | Jun | Jul | Aug | Sep | Oct | Nov | Dec | Year |
|---|---|---|---|---|---|---|---|---|---|---|---|---|---|
| Average | 4 | 5 | 7 | 9 | 12 | 15 | 17 | 17 | 14 | 11 | 7 | 5 | 11 |
| Range | 4 | 5 | 7 | 9 | 9 | 9 | 10 | 9 | 9 | 8 | 6 | 4 | 13 |

Record high  37    Record Low  −13

TABLE 8.5
Temperatures (°C) for Verkhoyansk (68° N)

|  | Jan | Feb | Mar | Apr | May | Jun | Jul | Aug | Sep | Oct | Nov | Dec | Year |
|---|---|---|---|---|---|---|---|---|---|---|---|---|---|
| Average | −50 | −43 | −30 | −15 | 1 | 10 | 14 | 11 | 2 | −15 | −37 | −46 | −18 |
| Range | 5 | 10 | 14 | 16 | 13 | 12 | 11 | 11 | 10 | 8 | 6 | 5 | 64 |
|  |  |  |  | Record high | 37 |  | Record low | −68 |  |  |  |  |  |

drops more in Verkhoyansk during the last two to three weeks of October than it does in London during the six months from July to January!

Most inland subpolar locations have annual temperature ranges of at least 40°C. These large variations occur simply because it is cold in winter but still reasonably warm during summer. However, near the poles it remains cold even during the summer so that variations in temperature are not quite so large. At present in the earth's history the polar regions of the earth are truly in the middle of an ice age.

# RADIATION AND TEMPERATURE

Is life possible on other planets? In the past this might have been regarded as a science fiction question, but now we have reached the point where it is possible to make reasonable guesses. Life as we know it is based on the chemistry of the carbon atom and requires liquid water. Liquid water exists under a narrow range of temperatures—from 0 to 100°C under normal atmospheric pressure. Of all the planets in our solar system only the earth has an average temperature within this range (Mars is close). Why does the earth have such equable temperatures?

It is easy to give a qualitative answer to this question. If the earth were much closer to the sun, it would get more sunlight and therefore be warmer. On the other hand, if the earth were much farther from the sun than it is, the temperatures would be much colder. Thus, the earth is at just the right distance from the sun (for our purposes).

Unfortunately, a descriptive answer to such questions is not very useful. We want to know what temperatures to expect on any planet, and we want to know why the temperature averages 15°C on earth. Fortunately, it is possible to get a quantitative answer. There is a law that can be used to determine the temperature of an object so long as you know how much radiation it receives from the sun.

Planck's law is the fundamental law of radiation. It is the law you read about near the beginning of Chapter 7—the law whose mathematical derivation helped in the birth of modern physics. Planck's law is too complicated to present mathematically in this book, but it is portrayed simply in Figure 8.4. Planck's law tells us both the total rate of radiation and the wavelengths of the radiation.

Planck's law refers only to objects said to be in thermal equilibrium. This means it does not apply to objects undergoing chemical reactions such as burning. Similarly, the light from fireflies or other phosphorescent objects is also not governed by Planck's law.

Planck's law gives the maximum amount of radiation that a body can give off as a function of its temperature. An object that radiates heat at this maximum rate is called a **black body,** because such objects also absorb all the heat or light that shines on them (at the same wavelengths). Radiation rates of real substances are given in Table 8.6.

This terminology may seem somewhat confusing because both the sun and snow radiate as practically like black bodies. On the other hand, aluminum foil radiates heat at only 1 to 5% of the maximum ( or black body) rate. This is why hot food wrapped in aluminum foil stays hot for a relatively long time.

Planck's rather complicated law can be broken down into two rather simple laws. Both of these simpler laws were actually discovered before Planck's law. One part of Planck's genius was to realize and then prove that these two simpler laws are immediate consequences of the general radiation law.

The first of the two simpler laws is called **Wien's law.** Wien's law states that as the temperature of an object increases, the wavelength of the most intense radiation that it emits decreases. It can be written

**Wien's Law:**

$$L_{mi} = \frac{2900}{T}$$

where $L_{mi}$ stands for the *most intense* wavelength in microns, and $T$ is the absolute temperature.

Wien's law has two very important and interesting consequences. The first of these is very colorful. Take a piece of iron and throw it into a fire. As the iron gets hotter it begins to glow with a dull red color. As the iron con-

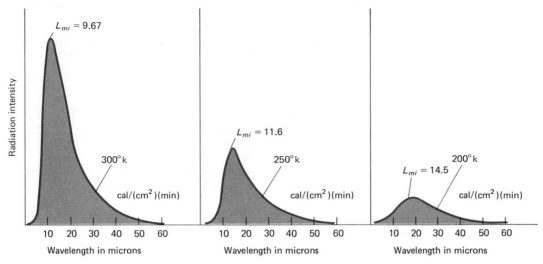

FIGURE 8.4
Planck's law of radiation for three different temperatures. The areas under the curves are proportional to the total rate at which heat is radiated (i.e., the intensity). Notice how the intensity of radiation increases rapidly as temperature increases (i.e., the Stefan-Boltzmann law), but the most intense wave length get shorter (i.e., Wien's law).

tinues to get even hotter, the color gradually changes to orange and eventually to white. At this point the iron would begin to melt but, if it could be heated further, it would glow blue!

This is true of all radiating objects. As they become hotter the color changes from red to white to blue. The explanation is rather simple. Since temperature is in the denominator of the equation for Wien's law, as the temperature increases the fraction decreases. This means that the wavelength decreases. Now you must recall that red is the color with the longest wavelength, while blue has a much shorter wavelength.

Therefore, when we look into space we can actually tell something about the temperature of the stars (and our sun) by their color—blue stars are the hottest and red stars the coolest.

TABLE 8.6
Radiation Rate (Emissivity) of Objects Compared with Black-Body Radiation Rate

| Substance | Emissivity (Percent) |
|---|---|
| Water | 92 to 96 |
| Snow | 82 to 99.5 |
| Sand | 84 to 95 |
| Forest | 90 |
| Grasslands | 90 |
| Human skin | 95 |
| Aluminum foil | 1 to 5 |

The second important consequence of Wien's law is that it leads to the well-known "greenhouse effect." The greenhouse effect will be described shortly. Basically without the greenhouse effect, temperatures on earth would be considerably colder.

When the wavelengths of the sun's radiation are determined, it is found that the most intense wavelength is $L_{mi} \cong 0.5$ microns. Now let's see how Wien's law can help us find the temperature of the sun.

Example 8.1
What is the temperature on the sun?

Information: The sun's most intense radiation has a wavelength of approximately 0.5 microns.

Equation:
Wien's law
$$L_{mi} = \frac{2900}{T}$$

or, solving for the temperature,
$$T = \frac{2900}{L_{mi}}$$

Substitute:
$$T = \frac{2900}{0.5}$$

Solution
$T = 5800°$   (absolute or Kelvin)

**Note:** This is the temperature of the photosphere that is the effective radiating surface and coldest part of the sun.

Example 8.2

The earth has an average temperature of 15°C or 288°K (Kelvin). Since the earth radiates almost like a black body all the time, why doesn't it glow at night?

Equation:
Wien's law
$$L_{mi} = \frac{2900}{T}$$

Substitute:
Use $T = 290$ for computational ease
$$L_{mi} = \frac{2900}{290}$$

Solution
$$L_{mi} = 10 \text{ microns}$$

Ten microns corresponds to infrared radiation and our eyes are not sensitive to this. Otherwise, we actually would be able to "see" the earth glowing at night.

---

Keep these two examples in mind when the greenhouse effect is discussed. The sun's radiation consists predominantly of radiation with wavelengths near 0.5 microns and is therefore called short-wave radiation, whereas the earth's radiation at 10 microns is called long-wave radiation.

The second simple law is known as the **Stefan-Boltzmann law,** after its two discoverers, Joseph Stefan (1835–1893) and Ludwig Boltzmann (1844–1906). The Stefan-Boltzmann law states that the rate at which an object radiates heat is proportional to the fourth (4th) power of the absolute temperature. In equation form

**Stefan-Boltzmann Law:**
$$T^4 = (I)(123)(10)^8$$

where $T$ is the absolute temperature and $I$ is the *intensity* of radiation in units of calories per square centimeter per minute.

The Stefan-Boltzmann law is of central importance in meteorology. It is from this law that we can determine the average temperature of a planet (or any object) just by knowing the amount of heat it receives from the sun.

The Stefan-Boltzmann law actually tells how much heat a body radiates away and not how much it receives. However, since the average temperature of the earth and the other planets does not change very much, this automatically implies that there is an almost exact balance between the incoming and outgoing radiation. Therefore, any time that the Stefan-Boltzmann law is used to tell the temperature of a planet, what is really being computed is an average *balance* temperature.

---

Example 8.3

A planet receives an average of 0.5 calories per square centimeter per minute from the sun. What do you expect the average temperature of this planet to be?

Information: Under balanced conditions the average heat leaving a planet is equal to the average radiation that it receives. Thus
$$I = 0.5$$

Equation:
The Stefan-Boltzmann law
$$T^4 = (I)(123)(10)^8$$
$$= (0.5)(123)(10)^8$$
$$= (61.5)(10)^8$$
This is $T^4$, not $T$! To find $T$ simply take the square root twice in a row. Thus,
$$\sqrt{T^4} = T^2 = \sqrt{(61.5)(10)^8} = (7.85)(10)^4$$
and
$$\sqrt{T^2} = T = \sqrt{(7.85)(10)^4} = (2.80)(10)^2$$

Solution
$$T = 280°K$$

**Note:** If you haven't guessed it by now this example applies to the *earth*. What this means is that the Stefan-Boltzmann law provides a remarkably accurate estimate of the earth's average temperature (actually 288°K) without once referring to the properties of the atmosphere!

---

The Stefan-Boltzmann law can also be used to find out the intensity of radiation if we already know the temperature of an object. For instance, since the sun is near 5800°K, we can use the law to find that the intensity of radiation on the sun is roughly 91,000 calories per square centimeter per minute. Then, multiplying by the number of square centimeters on the sun's surface, we find that the sun radiates $(56)(10)^{26}$ calories every minute. Heating at that rate would melt a one-cubic-kilometer iceberg in one-ten billionth of a second!

In order to find the balance temperature of a planet we first must find the average radiation intensity on the planet. To do this we must use the inverse square law. In Chapter 4 we used the inverse square law to find that the intensity of direct sunlight on the planet Jupiter is 0.074 calories per square centimeter per minute, whereas on earth it is 1.95 (the so-called solar constant).

However, in order to find the balance temperature of a planet we must find the *average* intensity of radiation and not the intensity of direct radiation. Direct sunlight falls on only one point of a planet at a time. Half of the planet receives no light at all (night) at any given instant, while the rest of the planet receives sunlight obliquely. Although it may seem like a very difficult task to find the average intensity of sunlight it is actually quite easy.

All planets are essentially spherical, and spheres have an area given by the formula, $A = 4\pi r^2$. This is 4 times the area of a circle that would be receiving direct sunlight. Therefore, *average* sunlight on any planet has an average intensity 0.25 that of *direct* sunlight (see Figure 8.5). This is why $I = 0.5$ applies to the earth.[1]

---

Example 8.4

What is the balance temperature on Jupiter?

Information: From the inverse square law we found that the intensity of direct sunlight on Jupiter is 0.074. This means that the average intensity is 0.25 of that, or
$I = 0.0185$

Equation:
The Stefan-Boltzmann law
$$T^4 = (I)(123)(10)^8$$
$$= (0.0185)(123)(10)^8 = (2.075)(10)^8$$

Taking square roots twice in a row we find
$$T^2 = (1.44)(10)^4$$

Solution
$T = 120°K$

---

This use of the Stefan-Boltzmann law works very well for every planet except for Venus and Jupiter. Venus is far warmer than the Stefan-Boltzmann law indicates because it has an enormous greenhouse effect and retains heat, while Jupiter produces heat internally.

By using the Stefan-Boltzmann law and the inverse square law together, we find that around each star there is a range of distances for which the temperature is neither too hot nor too cold. This range can be called the "zone of possible life" (see Figure 8.6).

Since life as we know it can only exist when water is liquid, the temperature should not be much colder than 0°C or much warmer than 100°C. Allowing some leeway (say, between −50 and 150°C) we find that only Mars, Venus, and the earth fall within the zone.

Every star has its own zone of possible life, depending

FIGURE 8.6
The "zone of possible life." Only Venus, Earth, and Mars lie within the zone. Venus has grown too hot because of its strong greenhouse effect.

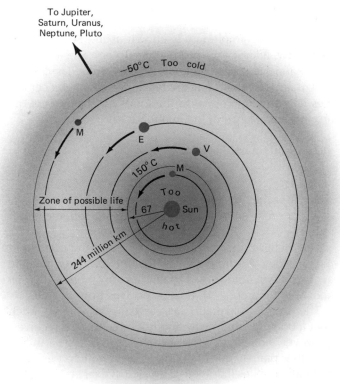

FIGURE 8.5
The average intensity of sunlight on a spherical planet is only one-fourth as intense as direct sunlight, since a direct circular ray of sunlight has to spread over the sphere, which has 4 times the area of the circle.

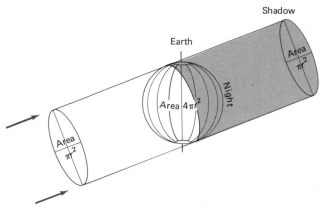

[1] For all computations we use the approximate value of 2.0 cal/ (cm²)(min) for the solar constant.

on the total radiation it emits. If the radiation of the star happens to vary considerably from time to time then the zone of possible life will also vary in size. This will mean that a particular planet may not remain in the zone of possible life over a long enough period for life to evolve at an advanced level. Thus life is most likely on planets of old, steadily burning stars.

# REFLECTION OF SUNLIGHT (ALBEDO) AND THE ICE AGES

Benjamin Franklin once performed the following experiment. After a snowfall he covered several patches of snow with differently colored cloth strips. Franklin found that the snow melted most rapidly under the black cloth and least rapidly under the white cloth. From this he correctly concluded that in a hot, sunny climate you stay cooler when you wear light-colored clothing. Europeans living in the tropics took over 100 years to discard fashion and to profit from Franklin's observation.

If you ever tried to run barefooted across a blacktop road during a summer afternoon, you quickly realized how hot the ground can get. At the same time the white line in the middle of the road is far cooler than the asphalt.

When light is reflected from any object it does not contribute anything to the heating of the object. Only that fraction of the light that is absorbed produces heat. Thus, dark objects absorb sunlight and heat up, while light-colored objects reflect much of the sunlight and remain cooler.

The **albedo** is the technical term for the reflectivity of an object. For example, an object that reflects 30% of the light that strikes it has an albedo of 30. Typical albedos of objects and planets are listed in Table 8.7.

Most soils, plants, and even water have very low albedos, while snow, ice, and clouds have high albedos. Thus, even though the sun is low in the sky in the polar lands, snow blindness has always been a problem to the hunting eskimos. Not only do the eskimos have to shield their eyes from the direct sunlight, but also from the reflected sunlight as well. Any eskimo and skier knows the value of sunglasses.

There is an even more important consequence of albedo. Since the reflected fraction of the sunlight does not contribute to heating, light-colored objects have a cooling influence. It is actually possible to change the climate of the earth by changing its albedo. For instance, in the past

TABLE 8.7
Albedos of Various Surfaces

| Substance or Planet | Albedo (Percent) | | |
|---|---|---|---|
| Water | | | |
|   With sun's elevation 90° | 3 | | |
|   With sun's elevation 30° | 7 | | |
|   With sun's elevation 10° | 24 | | |
| Snow, fresh | 75 | to | 95 |
| Ice, sea | 30 | to | 40 |
| Sand, dry | 35 | to | 45 |
| Soil, dark | 5 | to | 15 |
| Forest | 10 | to | 20 |
| Grassland | 15 | to | 20 |
| Clouds, thick | 70 | to | 90 |
| Clouds, thin | 35 | to | 50 |
| Earth | 30 | | |
| Moon | 6.7 | | |
| Mercury | 7 | | |
| Jupiter | 45 | | |
| Mars | 16 | | |
| Venus | 76 | | |

100 years because of industrialization the air has been filled with many more dust particles (called aerosols). These aerosols tend to increase the albedo of the earth and thus produce a cooling effect on earth. Some meteorologists believe that this effect is partially responsible for the global cooling that has taken place since about 1940.

Large volcanic eruptions inject dust explosively into the stratosphere where it remains suspended for months or even years. This volcanic dust also increases the earth's albedo, and there is some evidence to show that the great volcanic eruptions such as Krakatoa in 1883 and Tambora in 1815 were followed by a few years of significantly colder weather.

To some the most frightening prospect is the distinct possibility that an ice age can start itself! It is possible that due to the natural variability of weather one particular winter may be exceptionally snowy. Since snow has a much higher albedo than bare ground, more sunlight striking the earth would be reflected than normal. Thus less sunlight than normal would be absorbed on earth and temperatures would drop.

It is conceivable that this could start a chain reaction. As world temperatures begin to drop not only will it be easier for it to snow (rather than to rain), but some ocean water will start to freeze. Since ice also has a high albedo, even more sunlight will be reflected from the earth, and this in turn will produce a further cooling, and so on. This chain reaction is called a **positive feedback mechanism** (see Figure 8.7). In meteorology there are many other feedback mechanisms (both positive and negative) that can potentially affect the weather and climate.

The ice-albedo feedback mechanism constitutes an essential ingredient of most ice age theories. Calculations using the Stefan-Boltzmann law, which take the effects of albedo into account, show that if the entire earth were covered with ice the temperatures would fall so low that the ice would not even melt at the equator! According to one theory, if the ice were to spread further toward the equator than 40° latitude, the chain reaction would begin, and the ice would continue spreading right to the equator.

On the other hand, we can also look forward to the possibility of a much warmer climate. If the millions of square kilometers of polar ice were covered with a thin dark film, the albedo would decrease and more heat would be absorbed on earth. This would cause the temperatures to rise, thus melting some of the ice. This would lead to further decreases in the albedo and therefore further warming, and so on until all the ice melted. It now lies within our technological power to end the ice age in this way, but the millions of cubic kilometers of melted ice would then run into the oceans and cause sea level to rise. Most of the world's largest cities such as New York, London, Paris, Shanghai, Tokyo, Calcutta, Rio de Janeiro, Sao Paolo, Buenos Aires, and Statesboro would then be submerged.

This is the stuff that modern novels are made of. However, the issue is not a joking one—the possible consequences are enormous. In Chapter 24 this issue will be examined further. Until then, "Remember the albedo."

## THE GREENHOUSE EFFECT

If we "remember the albedo" and recalculate the balance temperature on earth (earth's albedo is 30%), the new answer is 255°K or −18°C. This result is less accurate than the simple balance temperature calculation that did not take the albedo into account (Example 8.3)! Actually, we are lucky that this new result is not accurate, since at −18°C the earth would have a perpetual ice age.

We must now answer the question, "Why does the realistic inclusion of the effect of albedo produce such an inaccurate answer for the balance temperature of the earth?" By sheer coincidence, the cooling effect of the earth's albedo is almost exactly offset by the warming due to the greenhouse effect. Actually, the greenhouse effect is slightly more important than the influence of the albedo on earth; with the greenhouse effect the earth is warmed to 15°C.

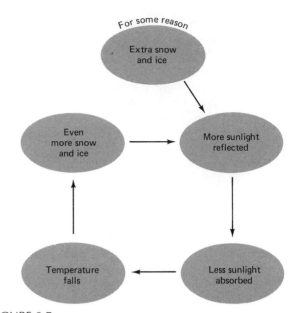

FIGURE 8.7
The albedo-climate feedback mechanism. If, for *any* reason, there is more snow and ice on earth, the feedback mechanism can begin to work like a chain reaction.

The **greenhouse effect** is the property of the atmosphere that permits sunlight in rather easily, but releases that heat back into space only with great difficulty. Because of the greenhouse effect, the atmosphere retains considerably more heat, and thus the temperature of the earth's surface is higher than otherwise (see Figure 8.8).

How does the greenhouse effect operate? This is where Wien's law is important. The short-wave radiation from the sun largely penetrates the atmosphere and is absorbed at the ground. The ground then heats up and reradiates long-wave radiation from the ground which does not penetrate the atmosphere very well.

Now, recall Einstein's law which states that each different wavelength of radiation corresponds to a different energy. Each molecule absorbs only those wavelengths with the appropriate energy. The fact is that wavelengths shorter than 0.3 or longer than 1.5 microns are readily absorbed by $CO_2$, $H_2O$, $O_3$, and cloud droplets. The intermediate range of wavelengths—from 0.3 to 1.5 microns —are hardly absorbed at all and easily penetrate to the ground. Most of the sun's radiation falls within this intermediate range of wavelengths so it easily penetrates the atmosphere. On the other hand, virtually all of the earth's radiation is longer than 1.5 microns and therefore is largely absorbed by the atmosphere.

The $CO_2$, $H_2O$, $O_3$, and cloud droplets heat up when they absorb the long-wave radiation from the ground.

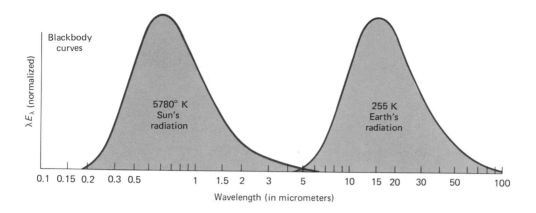

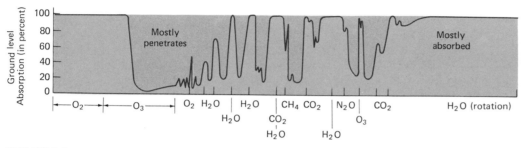

FIGURE 8.8
The sun's radiation occurs mostly at short wavelengths, which easily penetrate the atmosphere. The earth's radiation occurs at long wavelengths and is largely absorbed by the atmosphere. This produces the greenhouse effect.

These heated molecules then collide with the other air molecules, thus heating the rest of the atmosphere as well. Once these absorbing gases have heated up, they too radiate heat. Some of this radiation escapes to space, but some heads back toward the ground. *This extra radiation (in addition to the solar radiation) received at the earth's surface produces the greenhouse effect.* The greenhouse effect on earth produces a net warming of about $32°C$[2] (see Figure 8.9).

There are a number of interesting situations to which the greenhouse effect applies. Consider the case of two nearby cities that have identical temperature and humidity conditions at nightfall. The only difference is that city *A*

[2] It may now please you to read that, although the principle is correct, the term—greenhouse effect—is actually a misnomer. The main reason that greenhouses on earth are so hot is that the glass prevents the heated air from rising. Greenhouses made of rock salt (which allows infrared radiation to escape) are almost as hot as greenhouses made of glass. Furthermore, most of the heat lost by the greenhouse is lost by heat conduction (see Chapter 9) and not by radiation.

is cloudy and city *B* is clear. The question is, "Which city will be colder by morning?"

The answer is simple. Since clouds strengthen the greenhouse effect by absorbing the radiation from the ground and reradiating some of it back to the ground, the air at city *A* cools slowly overnight. At city *B*, there is a greater net loss of heat from the ground, since less heat is reradiated back to ground level. Therefore, city *B* cools more rapidly at night. In a manner of speaking, the clouds act like a blanket holding heat in.

**Rule:** Clear nights get colder than cloudy nights (all other conditions being the same).

There are even times when the temperature of the cloud base is warmer than the ground temperature, and in such cases the air temperature will actually warm slightly overnight.

Another interesting consequence of the greenhouse effect is that fog is more likely after a clear night than after a

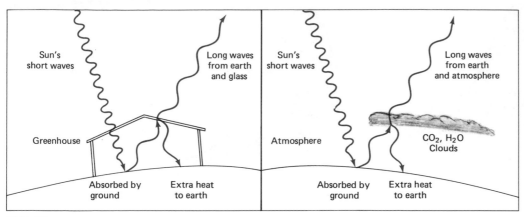

FIGURE 8.9

The greenhouse effect. The atmosphere reradiates some of the heat it received back to the earth. This represents an extra source of heat for the ground, making it warmer than it otherwise would have been.

cloudy night (see Figure 8.10). Since the air near the ground cools more on the clear night, there is a greater chance that condensation will occur. Thus fog in the morning is often a sign of a clear day to come. More will be said about this in Chapter 11.

The greenhouse effect still exists without clouds, but it is then somewhat weaker. During clear conditions the strength of the greenhouse effect varies with the amount of water vapor in the air. As a result, in the desert or in the high mountains, temperatures fall rapidly at night, and diurnal temperature variations tend to be quite large (at least 15°C).

There are also feedback mechanisms that involve the greenhouse effect. For instance, the strength of the green-

FIGURE 8.10

Clear nights get colder than cloudy nights because clouds increase the greenhouse effect. Thus, clouds and the greenhouse effect act like blankets. Numbers are merely schematic.

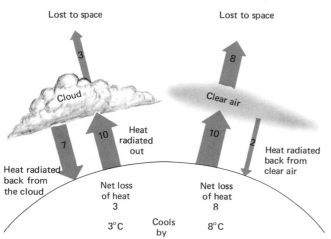

house effect depends on the amount of water vapor in the atmosphere. As the air warms, its capacity for holding water vapor increases. This potentially strengthens the greenhouse effect and could lead to further warming.

It is conceivable, though quite unlikely, that the oceans could be boiled away by such a chain reaction. It would be far more likely if the earth were as close to the sun as Venus is. Such a chain reaction is known as a **runaway greenhouse.** There is strong evidence that the greenhouse effect on Venus is an example of just such a runaway-greenhouse effect.

Scientists believe that a runaway-greenhouse effect has caused tremendous amounts of water to completely boil away from the planet Venus. The theory goes something as follows. At first Venus had an atmosphere with large quantities of water vapor, and it may even have had something of an ocean. But the enormous quantities of water vapor in the air caused the temperature to rise past a critical point. The vapor molecules then moved fast enough to escape from the gravitational field of Venus, but the high temperature had another fundamental effect on the atmosphere of Venus.

You should recall that the atmosphere is 90 times more massive than our atmosphere and is composed almost entirely of $CO_2$. Theory indicates that the 15% increase of $CO_2$ in our atmosphere over the last century has produced a small global warming of about 0.25°C. Since Venus' atmosphere contains about 250,000 times as much $CO_2$ as the earth's, you can imagine how strong the greenhouse effect is there.

Once the temperature on Venus reached a certain critical point (as you have read in Chapter 6) the rocks composed of calcium carbonate, $CaCO_3$, began to degenerate into CaO and $CO_2$. The $CO_2$ which was then released into

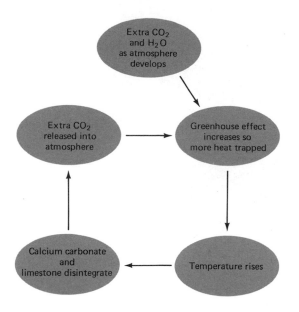

FIGURE 8.11
The runaway-greenhouse effect. Such a chain reaction is unlikely on earth because of the distance to the sun.

the atmosphere further increased the greenhouse effect, and this then caused even more $CaCO_3$ to decompose. As a result there is no longer a reasonable chance of finding life on Venus (see Figure 8.11).

Such a runaway-greenhouse effect is not expected to occur on earth, unless the sun suddenly begins to radiate more intensely. Fortunately, this is not expected to happen for about another 10 billion years.

# OCEANS, LAND, AND TEMPERATURE

The time has now come to explain why air temperature in oceanic locations varies far less from winter to summer and from day to night than it does inland.

The underlying surface has a considerable influence on the air temperature. For instance, cold arctic air streaming over the ocean is quickly warmed, whereas cool air from the ocean during summer is also warmed rapidly when it passes over the hot land. Thus, the air temperature quickly adjusts to the conditions of the underlying surface. Therefore, the question of why temperature variations are small in oceanic locations and large inland amounts to the same question: "Why is the temperature of the ocean surface so conservative compared to the land?"

On the average, 50% of the sun's rays penetrate the atmosphere and are absorbed at the earth's surface. On land the sunlight is absorbed right at the surface and therefore is concentrated; in the ocean the sunlight spreads to a considerable depth and therefore is quite diffuse. This produces an intense warming at the surface of the soil, but only a slight warming through a great depth of ocean. Evaporation of water from the ocean surface also keeps the temperature from rising much. Thus, the air in contact with the land heats up quite a bit, whereas the air over the ocean warms only slightly (see Figure 8.12).

At night (or during the winter) both land and ocean radiate heat out to space and cool. However, as the surface layer of water cools, it becomes denser and sinks. It is then replaced by somewhat warmer water from below. The heat loss is spread through the depths so that the surface temperature does not fall very much. When solid land cools it certainly cannot move. Therefore heat is replaced from below rather slowly by heat conduction. Thus the earth's surface can get quite cold, and this cools the air directly above.

Water also has an enormous heat capacity. This means that it takes far more heat to raise the temperature of water than it does to raise the temperature of rocks or soil. If, for instance, we take one gram of water and add one calorie of heat to it, the temperature will rise 1°C. If we do the same to one gram of a typical rock such as granite, the temperature will rise 5°C. Conversely, when granite and water lose the same amount of heat, the temperature of the granite will drop 5 times as much. We all know that water has a large heat capacity, simply because we know how long it takes to heat a pot full of water.

As a result of all these processes the ocean warms and cools very slowly. During the summer the ocean gradually accumulates heat and its temperature rises slowly until August or even September in the NH. During autumn the land cools rapidly, but the ocean still retains some of the heat accumulated during the summer. The ocean temperature drops slowly until late February or March, when it reaches a low. All this stored heat (or "stored cold") naturally affects the air temperature so that the warmest and coldest times of the year occur roughly a month after the solstices.

The ability of the ocean to store enormous amounts of heat is one important factor that limits large and relatively rapid changes of climate.

# THE HEAT-BALANCE SHEET

Business executives use balance sheets to keep their monetary affairs in order. Meteorologists also find it useful to keep a balance sheet for heat in the atmosphere. Over a period of time (such as a year) there is a balance of heat

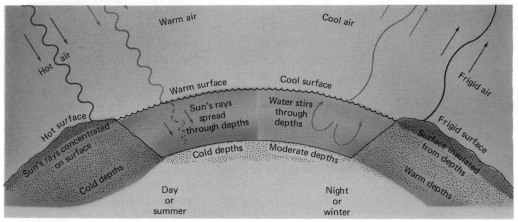

**FIGURE 8.12**
Oceanic climates have smaller temperature variations than continental climates. The underlying surface largely determines the air temperatures. Variations on the ground are extreme, because they are largely confined to the surface; whereas variations on the ocean are smaller, because they are spread through the depths.

for the earth and for the atmosphere. Whatever heat is gained will sooner or later be lost. This is verified by observations that are presented in Figure 8.13.

You can see that heat is transported by several processes. Radiation is the most important, but evaporation, conduction, and convection are also important. Millions of tons of water evaporate from the earth's surface and cool the earth, since evaporation is a cooling process. This water vapor condenses in the atmosphere and therefore heats the atmosphere. Heat is also transferred from the earth's surface into the air by contact (conduction) and is then carried aloft by convection.

Figure 8.13 gives the average balance sheet for the entire globe. The situation is not so simple when we look at individual locations on earth, because there can be large local imbalances of heat. The equatorial regions absorb more heat from the sun than they return to space, whereas the opposite is true in the polar regions. This would seem to imply that the equatorial regions are getting even warmer, while the polar regions are getting even colder.

It is the winds that rectify the situation. Cooler air blows toward the tropics, keeping them from getting too warm, and warmer air moves toward the poles, keeping them from getting too cold (see Figure 8.14).

It is indeed one of the ironies of nature that the winds which are created by the imbalances in heat (i.e., temperature differences) act precisely to eliminate those imbalances!

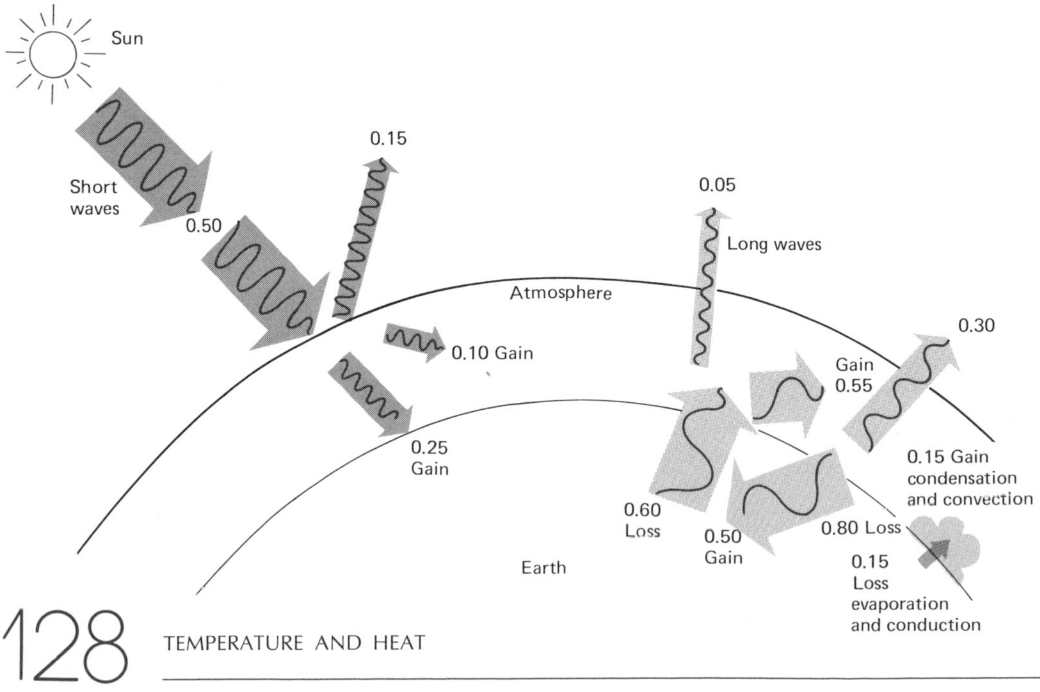

**FIGURE 8.13**
The heat-balance sheet of the earth-atmosphere system. All units are in calories per square centimeter per minute, or cal/(cm²) (min), and are approximate. The average temperature on earth does not change appreciably because there is as much heat leaving as entering.

Sun

Short waves

0.15

0.50

0.05

Long waves

Atmosphere

0.10 Gain

0.30

Gain 0.55

0.25 Gain

0.15 Gain condensation and convection

0.60 Loss

0.50 Gain

0.80 Loss

Earth

0.15 Loss evaporation and conduction

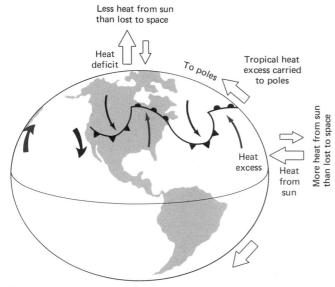

Less heat from sun than lost to space

Heat deficit

To poles

Tropical heat excess carried to poles

More heat from sun than lost to space

Heat excess

Heat from sun

FIGURE 8.14

Local heat imbalances are offset by the winds. The equatorial regions receive more heat from the sun than they radiate out to space. This excess heat is then transported by the winds and, to a lesser extent, by the ocean currents to the polar regions. There is a net transport of heat because cold air and water flows from the polar regions and is replaced by warmer air and water flowing from the equatorial regions. If the winds and currents ceased, the equator would grow warmer and the poles cooler than they now are.

## PROBLEMS

**8-1** Why are mercury thermometers worthless during winter in most of Antarctica?

**8-2** Why is London warmer than St. Louis during winter even though it is much closer to the North Pole than St. Louis?

**8-3** What is meant by saying that the sun and snow both radiate heat as *black bodies?*

**8-4** A star has a surface temperature of 15,000°K. What is its color?

**8-5** If we want to visit a planet of a very hot star, why would we be risking our lives even if the planet received the same intensity of light as the earth?

**8-6** About how hot does an object have to become to start glowing visibly?

**8-7** If the absolute temperature of a black body doubles, how much more rapidly does it radiate heat?

**8-8** What is the general relationship between the distance of a planet from the sun and its balance temperature?

**8-9** A planet receives an average of 0.66 cal/(cm²)(min) from the sun. What is its balance temperature in degrees centigrade?

**8-10** Explain why the average intensity of sunlight on a planet (neglecting atmospheric effects) is equal to one-fourth of the solar constant on that planet.

**8-11** Using the fact that the cloud-top temperature on Jupiter is warmer than the Stefan-Boltzmann law suggests (160°K vs. 120°K), show that Jupiter produces more heat internally than it gets from the sun.

**8-12** Calculate the balance temperature for Venus considering the fact that its albedo is 0.76.

**8-13** Explain why it is good to wear light-colored clothing in a hot, sunny climate.

**8-14** What do you think would be the climatic effect if you spread a sooty film over the world's polar ice caps and sea ice, thereby reducing the albedo drastically?

**8-15** If ice covered the entire globe, would sunlight be intense enough to melt the ice at the equator? Use an albedo of 50% and assume the average intensity of sunlight at the equator is 0.6 cal/(cm²)(min). Finally, add 30° to your answer to include the greenhouse effect.

**8-16** Explain the ice-albedo feedback mechanism.

**8-17** Explain and describe the greenhouse effect of our atmosphere. How is Figure 8.8 helpful in this regard?

**8-18** Explain why the greenhouse effect due to water vapor alone would probably be stronger if the earth were slightly closer to the sun.

**8-19** Explain why cloudy nights do not cool off as much as clear nights.

**8-20** Explain how temperatures may actually rise at night after clouds have rolled over, even if warmer air has not blown in.

**8-21** A good way to illustrate the fact that the ground surface heats the air above is to consider the fact that, if sunlight were used to heat the atmosphere uniformly, the temperature would rise about 0.2°C per hour during the morning hours. What is so informative about this illustration?

**8-22** Which hemisphere has a greater winter-to-summer temperature change? Explain.

**8-23** Why is the hottest time of the year about a month after the summer solstice?

**8-24** Why is the hottest time of the day around 3:00 P.M. even though the sun is highest in the sky at noon?

**8-25** Explain how it is possible for heat to be radiated from the earth's surface at a faster rate than radiation received from the sun. (Look at Figure 8-13 to verify this statement.)

**8-26** Why are the equatorial regions of earth not becoming warmer, despite the fact that they receive more heat from the sun than they radiate back to space?

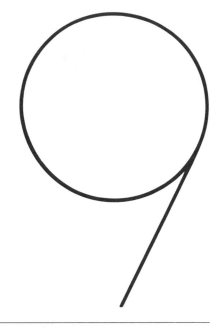

# AGRICULTURAL METEOROLOGY

Far from the madding crowd's ignoble strife
　Their sober wishes never learned to stray
Along the cool sequestered vale of life
　They kept the noiseless tenor of their way.

"Elegy Written in a Country Churchyard"

Even as Thomas Gray sat in a country churchyard writing his immortal elegy for the humble farmers of the earth, an agricultural revolution was being kindled all around him. Two of Thomas Gray's fellow countrymen began applying scientific principles to farming. Charles Townshend introduced an advanced system of crop rotation, despite jeers from neighbors. The neighbors stopped jeering when Townshend's harvest increased by 30%. Jethro Tull did even better. He improved the plow and introduced a machine that planted seeds on the ground in well-defined rows and then buried the seeds. As a result of his inventions, his harvests doubled.

Since these improvements in the 1730s, agriculture has become a highly complicated profession that makes use of many of the technological advances from all fields. Nevertheless, the farmer is still subject to the weather. The record-breaking cold wave of January 1977 in Florida killed millions of dollars worth of fruits and vegetables. During such extreme events, there is little the farmer can do to save a crop.

Fortunately, there are many times when the farmer is not the helpless victim of unfavorable weather. It is well known that in the lowest few meters of the atmosphere, there are extreme variations in weather that may be harmful but often can be modified! One example of how the weather near the ground can be modified was given in the last chapter. The temperature of the white line in the middle of a blacktop road is distinctly cooler than the pavement during sunny days. Farmers are aware of such fine points and, if the dark soil proves to be too hot for tender young plants, the farmer may protect a new crop by using lighter colored soil. Some of the things farmers do to protect their crops may sound rather silly, but they are done with good reason.

In this chapter, we will look at the variations and differences of climate that occur near the ground, and how knowledge of these variations is applied by farmers, architects, and even animals. Some specific examples of distinct microclimates—the climate of forests, cities, and even caves—are presented. Many of the phenomena presented here will illustrate principles discussed in the last few chapters.

# WIND, TEMPERATURE, AND HUMIDITY NEAR THE GROUND

On clear days, perhaps 70% of the sunlight is absorbed right at the earth's surface, and the ground becomes much warmer than the air a few meters above it. On clear nights, the ground radiates heat so efficiently that it becomes much cooler than the air a few meters above it. There is often a larger temperature difference between the ground and the air at a height of 2 m than there is between the air at 2 m and the air at 1000 m!

These differences are naturally of great concern to the farmer. Many plants begin to grow once the temperature rises above 10°C. In the spring, the farmer must be careful that tiny newborn plants are not "tricked" into growing too soon by the warm ground temperatures during the day, only to be frozen to death when the ground gets cold at night.

On sunny days, it is typical for the ground temperature to be at least 20°C higher than the temperature at the standard height of 2 m. On clear nights, the temperature at the ground is typically 2 to 3°C, and in exceptional cases up to 10°C, cooler than at 2 m. Ice often forms overnight in puddles on the ground after nights in which the "official" temperature never falls to the freezing point. Obviously, the ground temperature does fall below freezing.

A direct consequence of this information is that the diurnal temperature range is largest near the ground and gets progressively smaller with increasing distance from the ground. On the Eiffel Tower in Paris, the average daily temperature variation is 8°C at 2 m, but only 4°C at the top of the tower (300 m). Throughout most of the **free atmosphere** (higher than 1000 m above the ground) the diurnal temperature range is no more than 1 to 2°C (see Figure 9.1).

These figures make it quite clear that the ground heats the lower troposphere during the day and cools it at night. On clear nights, the ground can get so cold that some vapor in the air just above it condenses, and dew forms on the ground. If the ground continues to cool, a thicker layer of air cools sufficiently and fog forms.

Temperature is not the only quantity that changes considerably in the lowest few meters of the atmosphere. Wind speed and moisture also change rapidly in the air near ground level. On extremely windy days, crawling insects can still walk around on the ground without much trouble because the wind speed right at ground level is far slower than it is a few meters up because of friction (see Figure 9.2).

Shortly after the Green Building at the Massachusetts

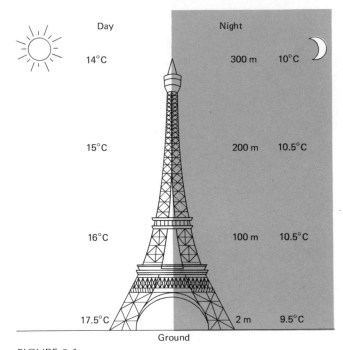

FIGURE 9.1
Average day and night temperatures at different heights on the Eiffel Tower. The temperature *variations* decrease with height. Temperature differences near the ground are more extreme on clear days and nights and are more extreme in the surrounding countryside.

Institute of Technology was completed, people noticed that it created a wind-tunnel effect. On days when the winds at the weather station were little more than 20 knots, wind gusts at the base of the Green Building reached 90 knots. (Ironically, the Green Building houses the Meteorology Department at M.I.T.) Most of us are familiar with similar idiosyncracies of the wind. Obstacles, such as buildings or hills, can drastically change the wind

FIGURE 9.2
Wind speed always drops rapidly near ground level. Even on a windy day, crawling insects can walk around.

speed and direction. The wind is therefore measured at a reasonable distance from such unusual objects, where possible, and the anemometer (used to measure wind speed) and wind vane (for direction) are usually placed 10 m above ground (or rooftop) level.

Wind speed falls to almost zero at ground level because of friction. On windy days, this means that there is a tremendous wind shear near the ground. As you may recall from Chapter 2, wind shear leads to waves in the ocean and in the atmosphere. It also produces turbulence and gustiness, which is essentially erratic waves. This turbulence and gustiness is responsible for carrying pollutants, heat, and water vapor away from the ground and for mixing the air.

In a sense, it is odd that the presence of the ground is a major factor in producing turbulence and mixing precisely because it slows down the winds. In fact, very near ground level, the wind is weakened so much that even the turbulence and mixing is suppressed. This inhibition of mixing near the ground is one of the major reasons that variables such as temperature can change so drastically in a short distance near ground level. If such large differences occurred away from ground level, the natural turbulence of the air would soon mix the air and reduce the contrasts, but near ground level the mixing is suppressed.

Mixing is roughly proportional to the wind speed near ground level. This has several important applications. Since the ground gets colder than the air on clear nights, a calm night with little mixing will tend to have a thin layer of very cold air close to ground level. On a windy night a thicker layer of air will be cooled, so the cooling near the ground will not be as severe. Thus calm nights tend to get colder than windy nights (all other conditions being the same). Now, test yourself. When will fog be more likely to form—after a windy night or after a calm night? To keep you in suspense, the answer is not given here, but in Chapter 11.

The concentration of water vapor also changes considerably near the ground. As you will see in Chapter 10, enormous quantities of water are stored beneath ground level. During the day as the earth is heated, some of this water is drawn up from the depths and evaporated. Near the ground, therefore, there is a high concentration of water vapor that decreases with height. At night, when the ground gets cold, dew may form on the ground, but the concentration of vapor is a minimum at ground level. This point typically confuses most people, yet it is quite straightforward. The air near the ground at night contains less water vapor simply because (due to condensation) the vapor becomes dew or frost if cold enough.

# PREVENTING FROST DAMAGE TO CROPS

Late in the summer as the nights get longer, the danger of frost increases in apple orchard country. Farmers dread the frost warnings issued by the National Weather Service, but they must be prepared to deal with them, since one night of frost can ruin an entire crop. Fortunately, there are several things that can be done to reduce the severity of a frost or to eliminate it completely. Many orchards are now equipped with heaters that are turned on as soon as the temperature drops to 0°C. Such heaters can warm the air a few degrees centigrade on clear nights when the winds are not too strong. The technique is reasonably successful because the coldest air is near the ground. The heated air will not rise too far and therefore it is only necessary to heat a thin layer of air. Even so, this technique is expensive; with repeated night frosts it can significantly add to the farmer's (and, therefore, the consumer's) costs.

A somewhat cheaper method for preventing frost is to use large fans. At first, this may sound like lunacy because fans are customarily used to keep people and things cool. Nevertheless, the technique works well on clear, relatively calm nights and is based on sound meteorological knowledge, which you have already acquired. On clear, calm nights the air near the ground is considerably colder than the air several meters up. Plants that grow at ground level often freeze, while there is warmer air just above them. Since warm air is light, it will not sink naturally, but must be forced down. The fan stirs the air and forces some of the warmer air down to ground level where the plants are (see Figures 9.3 and 9.4). This use of fans has saved millions of dollars of crops from destruction by frost.

There are many other methods used in frost prevention. Some of these sound as ridiculous as the use of fans, but nothing is to be laughed at if it succeeds. For example, plants have been sprayed with water on cold nights so that a layer of ice would form on the plants. As the water freezes, it releases latent heat of fusion that, oddly enough, keeps the plant itself from freezing. This technique works because for some plants frost damage does not begin exactly at 0°C, but instead begins at about −2°C. The ice coating helps to keep the plant temperature from dropping below the freezing point. Needless to say, there is a strong possibility that things can go wrong by using this technique, so extreme care and proper timing must be used.

It is also essential to choose the proper locations for orchards and vineyards. For instance, the worst thing to

FIGURE 9.3
A fan in a citrus grove in
Florida ready to go to work
preventing frost.

do is to plant an orchard in an enclosed valley. Perhaps an enclosed valley might seem a safe and protected place, but few places could be worse from the point of view of frost damage. Cold, heavy air that forms right above the ground slides down the slope and collects in the protected valley, while the "exposed" slopes have much warmer air.

An extreme example of how cold air can collect in an enclosed valley occurs in Germany near the town of Gstettner. Here there is a sinkhole that is essentially a collapsed cave or a deep natural pit in the ground. This sinkhole is about 100 m deep and is noted for the ex-

tremely cold air that accumulates there every night. On one windless night when the temperature at the top of the sinkhole was 2.3°C (36°F), it was −28.8°C at the bottom! Even the cattle, which graze in the sinkhole, quickly leave at nightfall. Dwarf pine trees, which are normally characteristic of high mountain climates, grow at the bottom, and a dog once froze to death there waiting for its master, while salvation was only 100 m away. (See Figure 9.5.)

The moral of the story is simple. Normally, temperature decreases with increasing height, but at night a relatively thin layer of cold air forms near the ground because of contact with the cold ground. This cold air will sink into

FIGURE 9.4
Fans can be used to prevent frost damage in crops! Although we normally use fans to keep cool at night, in an orchard the fan will actually modify the cold air at plant level by mixing it with warmer air from above.

Clear, calm nights

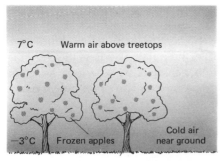

7°C  Warm air above treetops

−3°C  Frozen apples  Cold air near ground

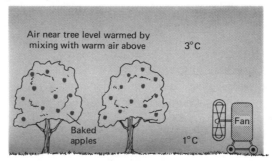

Air near tree level warmed by mixing with warm air above  3°C

Baked apples  1°C  Fan

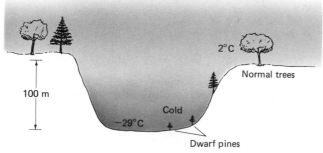

FIGURE 9.5
The Gstettner sinkhole at its best. This represents an extreme example of cold air sinking into an enclosed valley at night.

any depression in the ground and accumulate. Because of this, frost damage is more likely in such depressions or enclosed valleys, and one should hesitate before planting crops there. On nights that fog develops, pockets of fog will tend to develop first (where the air gets coldest) in such depressions.

# CONDUCTION OF HEAT THROUGH THE GROUND

Many farmers pray for some snow during the winter because they know that snow keeps the ground from freezing! The cold ground loses heat to the air by the process of conduction. This process is dramatically slowed down when the ground is snow covered. On the top of a 30 cm thick snow cover, it is possible to encounter temperatures of −20°C or lower, while under the snow at ground level the temperature is 0°C. Without the snow, the ground would be exposed to the −20°C, which might prove lethal to certain plants (see Figure 9.6). Huskies make use of the insulating properties of snow, and by instinct burrow into the snow on cold arctic winter nights.

In the air, heat can be transferred by radiation, convection (mixing) and conduction but, within the ground, conduction is the only process that transfers heat. Some substances, such as silver and gold (and most metals), transfer heat rapidly by conduction, which is why coffee cups are not made of metal. Ceramics, and most forms of rock and earth, conduct heat rather poorly and therefore are fairly good insulators. Of all substances, air is just about the worst heat conductor and just about the best insulator. Styrofoam cups are full of tiny air bubbles and, therefore, they are outstanding insulators.

The fact that the earth is a rather poor conductor of heat means that the heat of the day and the cold of the

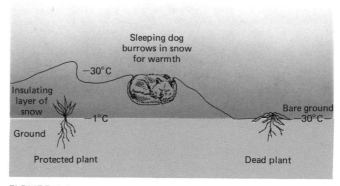

FIGURE 9.6
Snow keeps the ground warm during winter! This is true because snow acts as an insulator and protects the ground from the winter's worst cold spells. Dogs and other animals often burrow into the snow to sleep.

night penetrate very slowly into the ground. On the beach during the day the top layer of sand is very hot, but a few centimeters down the sand is cold. By evening time as the heat finally begins to penetrate to the depths, the surface is already cooling. At night the top layer of sand feels cold, while the sand a few centimeters down is noticeably warmer.

The heat penetrates so slowly that at a depth of about 40 cm, the soil is warmest shortly after midnight and coldest shortly after noon, which is the opposite of what occurs above the ground. The magnitude of temperature variations decreases with depth, so by 80 cm depth there is hardly any daily cycle in soil temperature.

The annual cycle of soil temperature penetrates deeper. At a depth of about 750 cm the soil is coldest in July and warmest in January, and this is one reason why well water is so cold in summer. Past a depth of 15 m, the annual temperature variations are barely felt; this is why caves have almost a constant temperature year round (see Figure 9.7). It is, therefore, quite reasonable that the temperature in caves is approximately equal to the average annual temperature of the air above. Actually, it may be somewhat warmer in the cave due to heat from within the earth.

Since the temperature variations are largest at the top of the soil, newborn plants are subjected to great heat stress and may die. For this reason, farmers have long known that **mulching** the soil, which means covering it with an insulating layer of straw or peat, can save many a crop. When the mulch is placed above new plants, the temper-

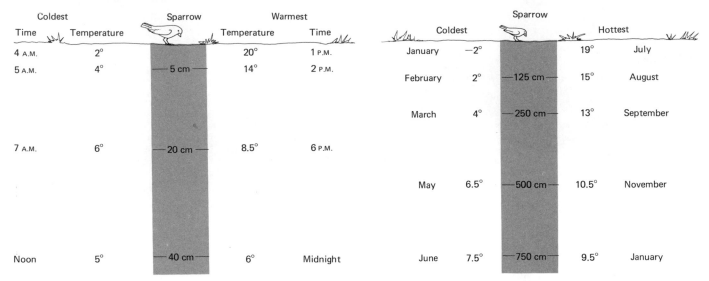

| Coldest | | Sparrow | Warmest | |
|---|---|---|---|---|
| Time | Temperature | | Temperature | Time |
| 4 A.M. | 2° | | 20° | 1 P.M. |
| 5 A.M. | 4° | 5 cm | 14° | 2 P.M. |
| 7 A.M. | 6° | 20 cm | 8.5° | 6 P.M. |
| Noon | 5° | 40 cm | 6° | Midnight |

| | Coldest | Sparrow | Hottest | |
|---|---|---|---|---|
| January | −2° | | 19° | July |
| February | 2° | 125 cm | 15° | August |
| March | 4° | 250 cm | 13° | September |
| May | 6.5° | 500 cm | 10.5° | November |
| June | 7.5° | 750 cm | 9.5° | January |

FIGURE 9.7
Daily and annual temperature variations in the soil. The deeper you go, the smaller the variations. For this reason, caves have almost constant temperatures year round. It also takes time for heat to penetrate through the ground, so that there is a depth where it will be warmest precisely when it is coldest at the surface.

ature at the top of the mulch experiences the large diurnal variations, while the plants a few centimeters below are protected from excessive heat and cold. Thomas Jefferson used a mulch of straw to cover large blocks of ice collected during the winter near his home, Monticello. Because the straw was such an effective insulator, the ice lasted through the summer, providing his guests with ice cold drinks.

The conduction of heat through the soil also affects the air temperature (see Figure 9.8). If the soil conducts heat very slowly, the heat of the day is trapped in a thin layer of earth near the surface, and the temperature rises quite a bit. At night when heat leaves the earth's surface, a poorly conducting soil will get colder, since little heat is transferred from below. With a highly conducting soil, the heat of the day can penetrate to the depths and not accumulate in the surface layers, whereas at night as the surface loses heat much of the heat is replenished from the depths. The general rule that follows from these observations is that temperature variations of the air above the surface are larger when the ground is a poor heat conductor.

After planting grass seeds in the spring, it is wise to roll or press the soil. This is done, not so much to discourage birds, but to reduce frost damage! Loose earth is full of air pockets and therefore is a poor conductor. The surface air gets at least 2°C colder on clear nights over loose soil than over tightly packed soil. Many farmers have lost a crop to frost damage because they neglected to stamp the soil after it had been loosened by plowing or weeding.

FIGURE 9.8
The conductivity of the soil affects the daily range of temperature. The larger the conductivity, the smaller the temperature range. This scene shows heat radiating from the soil at night. With the poorly conducting soil, very little heat flows from the depths to replace the heat lost at the surface, so that the temperature drops sharply. The numbers merely indicate relative units of heat flow.

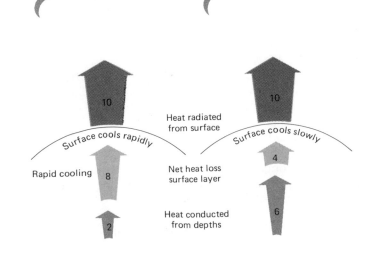

There is even an incident in which the remains of a medieval village were spotted because of heat conduction effects. Although the village had been razed and converted to farmland, the ancient foundations remained buried. The conductivity of the remains of the foundations was poorer than that of the surrounding soil; the soil temperature above the foundation was therefore a few degrees cooler every night. After one night, the soil became just cold enough so that frost formed only above the ancient foundations, and this was photographed from a plane.

The air temperatures near the North Pole during winter are far warmer (by 40°C) than near the South Pole during its winter. One of the reasons for this is that there is water under the ice near the North Pole. Enormous quantities of heat are conducted from the water through the ice and into the air. This conduction of heat certainly does not lead to tropical temperatures, but it does cause significant warming.

## TEMPERATURE, COLOR, AND MOISTURE OF THE SOIL

Since lighter color objects reflect a higher percentage of sunlight than dark objects, it is natural that they are cooler. Farmers make use of this simple fact to change soil temperature when necessary. In India, white powdered lime is sometimes added to soils to make them cooler. In one experiment, the addition of lime reduced soil temperatures by 15°C. When additional soil warmth is needed, darker soil should be added.

Some additional strange effects are produced by color of the background. Since light color surfaces are cooler, there is less evaporation from the soil, and water is conserved. This is quite important in dry climates. Another, perhaps unexpected, consequence of color is that tomatoes planted in front of white walls grow significantly faster and larger than tomatoes grown in front of black walls. The tomato *plants* grew faster in front of the warmer black walls, but their yield was lower. Similar results have been observed with peaches and grapes. Considering the enormous number of hungry people on earth, such curious results are highly significant.

Plants, such as cranberries, are typically grown in bog soils. Bog soils tend to be cold because they are wet and lose heat by evaporation. In bogs frost presents a major threat to a successful harvest. The frost danger could be considerably reduced if the evaporational cooling could be inhibited. A successful technique adds sand to the top layers of bog soils. This keeps the top layers drier and cuts

down the evaporation. As a result, when sand is added to bog soils, they become significantly warmer.

## ANIMAL AND PLANT ADAPTATION TO LOCAL CLIMATE CONDITIONS

No one should go through life unaware of the amazing behavior of *Leipoa ocellata*, the so-called "thermometer bird." This animal uses modern agricultural methods, such as mulching, to produce a nest that remains at a constant 34.5°C temperature for weeks, despite all changes in the weather. The bird usually lays about 20 eggs early in the spring and then covers them with a thick layer of leaves, which are then covered by sand. As the leaves rot, they produce heat that is trapped by the sand. As the weather grows warmer the leaves are removed, and the bird spends about two to three hours each day adjusting the thickness of sand according to the air temperature and time of day in order to keep the eggs at 34.5°C.

This represents one of a number of interesting examples of how animals either adapt to or modify local climate conditions. Many animals (and even plants) do by instinct what has taken humans centuries to learn. However, in several respects we are not so far from the animal world. Rudolf Geiger, whose book *The Climate Near the Ground* must be considered a classic in its field, has some interesting observations.

When the first warm days of spring arrive, mothers know how to take their babies' carriages to the warmest and most sheltered places, richest in radiation, without going into any theoretical calculations. In the summertime, you can come across real artists in finding the most comfortable places to lie in both on the beach and on mountain pastures. I once read that the homeless in London, wandering at night on Victoria Embankment, would get to know the temperature of every wall, and choose the outer walls of hotel kitchens as favorite places to rest against.

*The Climate Near the Ground*

There is a small lizard that lives in the Namib Desert in southwestern Africa. This lizard hunts for beetles on the sand dunes in the hours after dawn. While the side of the dune facing the sun becomes hot, the shady side remains cold. The cold-blooded lizard races to the sunny side and presses its cold body against the hot sand. As its body warms, the lizard lifts itself higher and higher off the sand until it stands on only two of its legs at a time. When the heat finally becomes unbearable, the lizard races back to the cold shady side until it gets cold, and then the process

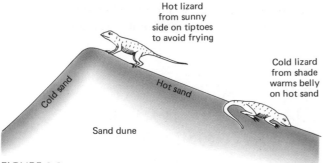

FIGURE 9.9
In the morning on sand dunes of the Namib Desert, it is too hot for the lizards on the sunny side of the dunes, but too cold on the shaded side. During its morning hunt, each lizard moves back and forth between sunny and shaded sides to keep comfortable.

starts over again (see Figure 9.9). But the lizard must work efficiently; once the sun gets high in the sky, all small creatures must bury themselves in the cool depths beneath the scorching sand at the surface.

Insects and other small climbing animals "know" that the ground is cold at night and hot during the day. Desert snails climb the scrub vegetation during the day to get away from the hot ground. There is a certain type of cricket in Denmark that spirals up the tree trunks as the night progresses just to avoid the cold air near the colder ground at night. In the forests, insects will swarm at tree-top level in the morning, since that is the first place to warm up; as the heat slowly penetrates to the forest floor, the insects move down.

The African cricket turns its body broadside to the sun in the morning to obtain maximum heating, but at noon the cricket will turn its body to get as little heat as possible. In Australia, termite nests are built on this principle and, therefore, can be used as compasses. The broad sides of these nests face east and west to get maximum heating in the cold morning, while the narrow sides face north and south to avoid the sunlight at the hottest time of day, as in Figure 9.10.

FIGURE 9.10
Termite mounds often can be used as compasses. The narrow sides of the mounds face north to south, to avoid the broiling noon sun, but the mounds face broadside to the rising sun of the cold morning.

Many people have picked wild raspberries at one time in their lives. Raspberries grow best with only moderate amounts of sunlight. Thus they are not often found out in the open and not often found deep within the forest. Berries grow best at the fringe of the forest, where there is just the right amount of sunlight. A good place to hunt for berries is along the sides of roads and along the edges of the woodcuts made along the paths of electrical power lines.

Animals can also adapt to local moisture conditions. There is a rabbit that lives in the desert near Great Salt Lake, in Utah, and never drinks water. With every breath we breathe out more water vapor than we breathe in, unless the air is extremely warm and humid. Eventually, we would dry out if we didn't drink. So would the rabbit, except that it spends most of the time hidden in its burrow where the relative humidity of the air is very high, unlike the desert air above. The rabbit, therefore, hardly loses any moisture by breathing and is somehow able to obtain enough moisture from its diet of desert plants.

Often in the mountains, a depression between two rocks provides enough shelter from the wind for a dwarf pine tree to survive, while all around it is nothing but tundra. There, each low-lying plant protects its neighbors and, like cattle in a blizzard, they manage to withstand the ferocity of nature, clumped together. Even the tallest trees are always found together in a cluster or **stand** where they can protect each other from excessive winds. They stand haughtily together, while many dwarf pines endure the passing years alone. If you ever walk the high mountains, tread lightly — some of these trees may be less than a foot high, but they deserve your respect for they may be several hundred years old. Perhaps theirs is the real glory.

# CLIMATE IN THE FOREST

When the climate permits trees to grow, a forest may form. Thereafter, the forest modifies the climate near the ground. Go into a forest on a typical summer day and you will notice several features of the weather. Within the forest, it is generally somewhat dark, cool, moist, and almost calm.

In an area where the forest is dense, less than 5% of the sunlight penetrates to the forest floor, and the light that does penetrate has a distinct greenish hue. In the tropical rain forests of the world, it is estimated that in places the vegetation is so thick in the **canopy** (upper crown layer of the trees) that only about 0.5% of the sunlight penetrates to the ground. Few good photographs of dense rain forest exist from inside, merely because it is so dark.

The result of the thick vegetation in the canopy is that the sunlight is absorbed there at treetop level, and it is there where it gets warm. Ground level is shaded and naturally remains cool. On typical sunny days, the air near the ground in the forest remains at least 5°C cooler than in the crown area or near the ground in the clearings. At night, that is blocked by the crown area and the forest is usually a little warmer than the surrounding open areas (see Figure 9.11).

Not only do the leaves provide shade, but they also slow the wind so that winds in the open are typically at least 5 times faster than they are near the forest floor. Even though the leaves may be rustling in strong winds at treetop level, all remains calm below.

The forest feels humid mainly because the air remains cool. However, the air within the forest generally contains about 10% more water vapor than the air above it. Rain is reduced in the forest with some water being trapped in the canopy and some running down the tree trunks. However, there are actually times when the forest increases the rain at the ground. When fog blows through a forest many of the fog droplets, which would not ordinarily settle on the ground, are captured by the leaves. When these droplets coalesce on the leaves, they grow large enough to fall to the ground as rain.

Often, at the edge of the forest on sunny days, a weak cool breeze blows out into the clearing. This is called the **forest breeze** and is similar to the better known sea breeze. When air over the open plains gets warm, it rises because it is light. The cooler, heavier air from the forest then moves out over the plains to replace the risen air. The forest breeze is felt only at the forest's edge, and blows at about 2 to 3 knots.

# CLIMATE IN THE CITY

Ever since 1836 when Luke Howard (of cloud fame) undertook the first study of city climates, we have known that the cities are warmer than the surrounding countryside. Howard found that London averaged about 1.5°C warmer than its surroundings, and we find that to be true today of most large cities.

The extra heat of the cities is most striking on clear, calm nights when the temperature is supposed to fall sharply. So it does all around the city, but within the city temperatures fall much less, forming the so-called **urban heat island.** On clear, calm nights central London is typically 5°C warmer than its surroundings. One extreme example of this urban heat-island effect occurred in Mon-

**FIGURE 9.11**
Little light penetrates to the forest floor. As a result, it is cooler at the forest floor than in the crown area of the trees. During the daytime, there is often a weak cool breeze called the forest breeze that blows out onto the adjoining meadows.

**FIGURE 9.12**
The urban heat-island effect at Cincinnati on the (clear) night of June 27, 1967. Notice the pockets of cold air "hugging" the ground in most of the countryside surrounding the city.

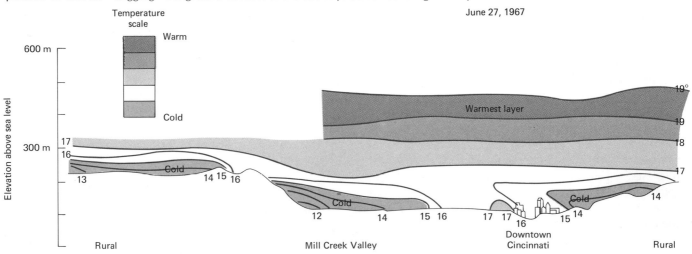

treal on January 23, 1969, when the downtown area was more than 11°C warmer than the suburbs to the northeast.

In Figure 9.12 a vertical profile (known as a **cross section**) of temperatures above Cincinnati and the surrounding countryside shows the effect of the city on the temperatures quite clearly.

The extra heat is due to a variety of factors. Sunlight, absorbed by the walls of the buildings, is reradiated into the air at night. The buildings themselves are heated on winter nights, and even air conditioning during the summer puts extra heat out onto the city streets. In fact, during December it is estimated that the artificial heat production in Manhattan is more than 5 times as large as the heat from the sun.

On windy nights the air near the ground in the countryside no longer cools as much, while the wind blowing down the city streets cools off the city more than on calm nights. When the wind speed at night exceeds 20 knots, the countryside is no longer any cooler than the city.

During the day there is also some tendency for the city to be warmer than the country, but the differences are smaller. There are even some times when the buildings act like trees in the forest, shading the ground and keeping the city cooler than the country!

The extra heat and pollution (discussed more fully in Chapter 14) of the city tend to increase precipitation. Cities tend to get roughly 10% more precipitation than the surrounding countryside, although the heat and pollution often are blown several kilometers downwind from the city before they produce precipitation. Because of the extra heat, city precipitation has a greater tendency to fall as rain rather than snow, and when snow does fall it will melt more quickly in the city.

City winds are also affected. Generally, the buildings block the flow of air, reducing the wind speed by an average 5%. However, under very light wind conditions, city winds are actually somewhat faster than in the surrounding countryside. Cities also tend to have a wind system much like a sea breeze, with warm light air rising over the city and cooler air from the suburbs flowing inward along the ground.

# PROBLEMS

9-1 Compare the temperature at ground level with the temperature at 2 m on a clear day and on a clear night in an open area. Be careful to shade the thermometer during the day.

9-2 Explain whether dew is more likely after clear nights or cloudy nights, all other conditions being equal.

9-3 Read the quote on page 338 and explain why Bill and Al both crawled back home.

9-4 How are fans used to prevent frost?

9-5 Given that the wind speed 1 km above the ground is 40 knots, when will it be windier at the ground, during the day or night? Explain.

9-6 Using the same conditions as Problem 9-5, will it be windier on a clear night or on a cloudy night?

9-7 Clear, windy days do not warm up as much as clear, relatively calm days. Why?

9-8 During the autumn why do the leaves of trees in bogs turn colors before the leaves of trees on the surrounding hillsides?

9-9 Explain why at a depth of approximately 7.5 m below the ground it is warmest during the winter and coldest during the summer.

9-10 If the continents were made of solid gold (an excellent heat conductor) instead of rock, what would happen to annual and daily temperature variations on earth?

9-11 Browse through the archaeology journals and see how many pictures you can find that illustrate the discussion at the top of page 137.

9-12 Can you find any more interesting examples of plant, animal, or human adaptation to local climate conditions? (If so, let me know—I'll put it in the next edition of this book and send you a free copy.)

9-13 Explain why snow lasts longest on north-facing slopes, and why in somewhat warm, dry climates the north slopes have more trees and vegetation than the south-facing slopes. (See page 130.)

9-14 Why are forest fires sometimes necessary for the rejuvenation of a forest?

9-15 How do you expect the city-country temperature differences to be affected by differences in evaporation rate?

9-16 If the Eiffel Tower were located in the countryside instead of the middle of Paris, how do you think the nighttime temperatures of Figure 9.1 would change?

9-17 Why do you think there tend to be more clouds over cities than over the areas surrounding cities?

9-18 Give two reasons why the sunshine in large cities is typically 15 to 20% lower than in the surrounding countryside.

9-19 Why do flowers and trees bloom earlier in the city than in the surrounding countryside?

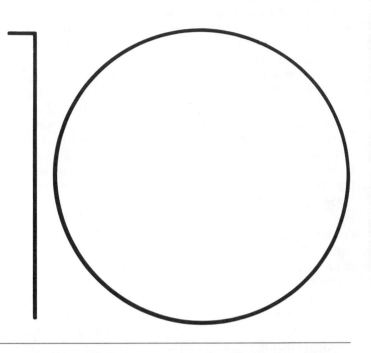

# WATER VAPOR AND THE HYDROLOGIC CYCLE

On Saturday, March 5, 1977, President Jimmy Carter conducted the first nationally broadcast telephone conversations with his fellow Americans. Forty-two people were able to speak with the President, and one of them, a 13-year-old boy from California named John Herold, asked the following question:

Since the West is having a drought and the East has too much snow, why not ship it west [in boxcars]?

President Carter answered correctly that it would be too costly. During the winter of 1976–77 the West did have a major drought, and it cost farmers several billion dollars. However, transporting water from the East would have cost far more. Most crops require the equivalent of 5 cm of rain per month. It would take 1000 huge boxcars holding 50,000 kilograms of water apiece to supply 1 square kilometer with 1 month's water! The drought in the West covered several hundred thousand square kilometers so that the total amount of water needed in the West was enormous.

The atmosphere is normally able to supply this amount of water. An average low pressure area drops at least 100 trillion kilograms of water on the earth during its brief lifetime of several days. The amounts of water involved are truly staggering when you consider that this water is hidden in the air in the form of vapor before the precipitation begins.

All of this water vapor gets into the air when it is evaporated from the ocean or the moist surface of the earth. After the rain falls the water soaks back into the earth or runs back into the sea where it can once again be evaporated. This cyclical process is known as the hydrologic cycle, and it has been understood for more than 2000 years.

All the rivers run into the sea; yet the sea is not full; unto the place from whence the rivers come, thither they return again.

Ecclesiastes, Chapter 1, 7

In this chapter we consider water vapor in the atmosphere and the consequences of evaporation and condensation. It is necessary to begin this complicated subject by methodically analyzing the vapor-holding capacity of the air. Many terms will have to be defined and explained, but once this is done the subject will no longer appear so complicated. Next, the hydrologic cycle will be analyzed, with some cases of its temporary imbalances (i.e., floods and droughts).

# WATER VAPOR
# IN THE AIR

Try to imagine two entirely different situations. The first situation is a blizzard in Greenland in which the temperature is −10°C and the relative humidity (hereafter written as RH) is 100%. The second situation is in the Sahara Desert where the temperature is 40°C and the RH is 10%. The question is—which air is holding more vapor?

Your first impression might be to say that since the air in Greenland has 100% RH it is holding much more water vapor than the Sahara air. If that is your impression, you were fooled. Pound for pound, the so-called dry Sahara air is actually holding more than 2.5 times the water vapor in the Greenland blizzard air!

The reason for this has already been given in Chapter 1. Cool air cannot hold as much vapor as warm air. Now it is time to become precise and ask: How much water vapor can air hold? To answer this and other similar questions it is necessary to define the following terms: (1) relative humidity, (2) saturation, (3) mixing ratio, (4) dew point temperature, (5) wet bulb temperature, (6) vapor pressure, and (7) vapor density. It is also necessary to understand the material on evaporation and all other phase changes, which was presented in Chapter 5.

**Relative humidity** or RH is the term that is used in the public weather forecasts to express how much vapor is in the air. The relative humidity is defined as

$$RH = \frac{\text{amount of water vapor air IS holding}}{\text{amount of water vapor air CAN hold}}$$

and it is always expressed as a percentage. The RH is nothing more than a ratio. When the RH is 100%, it means the air IS holding all the vapor it CAN at the given temperature. Therefore, by itself the RH tells nothing about how much vapor the air is holding.

There is one interesting effect related to the RH. Hair (and other objects) stretch when the RH rises. It is then possible to measure the RH with an instrument called a hair hygrometer. The hair hygrometer merely measures the length of the hair and interprets it directly as a relative humidity. In these days of longer hair styles you, too, may be able to estimate the RH if you use your head.

**Saturation** is almost self-explanatory. When the air is saturated with vapor, it is holding all the vapor that it can under normal conditions. If any extra vapor were added to the air at this point, it would rapidly condense to liquid water or ice. Saying that the air is saturated is the same thing as saying that the RH = 100%.

In Chapter 12 you will see that it is actually necessary for the air to be slightly supersaturated (i.e., the RH must be slightly greater than 100%) in order for clouds and rain to form. The degree of supersaturation is almost always very slight. To repeat an important point, under normal conditions *the RH never effectively rises above 100%.*

The **mixing ratio,** or symbolically $W$, tells how much vapor is in the air. The mixing ratio is defined as

$$W = \frac{\text{number of grams of water vapor}}{\text{number of kilograms of dry air}}$$

Thus it is the number of grams of water vapor per thousand grams (i.e., per kilogram) of dry air. The mixing ratio is always expressed as parts per thousand (%o).

Here is a question for you. If the temperature is 30°C, what is the mixing ratio, $W$? Do not strain your brain to answer this question because you simply do not have sufficient information. If you know the temperature then all you can tell is the vapor-holding *capacity* of the air. In actuality, the air may be holding anything up to its capacity.

Table 10.1 gives values of the **saturated mixing ratio,** symbolically $W_s$—the mixing ratio that the air *would* have *if* it were saturated. Table 10.1 shows that each kilogram of dry air can hold as much as 49.8 grams of vapor at 40°C, but only 1.79 grams of vapor at −10°C. This example shows that warm air can indeed hold far, far more vapor than cold air.

Here is another question for you. If the temperature is 30°C and the RH = 50%, what is the mixing ratio? This question can be answered with the information you have learned.

---

Example 10.1
If $T = 30°C$ and RH = 50%, what is the mixing ratio? How much vapor *is* in the air?

Information: Use Table 10.1. At 30°C the saturated mixing ratio, $W_s = 27.6$%o. In other words, the air at 30°C *can* hold 27.6 grams of vapor per every kilogram of dry air.

Equation:
$$RH = \frac{\text{amount of vapor air IS holding}}{\text{amount of vapor air CAN hold}} = \frac{W}{W_s}$$

Substitute:
$$RH = 50\% = \frac{W}{27.6}$$

Solution
$$W = 13.8\text{%o}$$
The actual mixing ratio is 13.8%o.

---

When the weather is cold outdoors the RH is generally quite low indoors, unless a humidifier is used. This is

WATER VAPOR AND THE HYDROLOGIC CYCLE

TABLE 10.1
Saturated Mixing Ratios of Water Vapor (Strictly True at a Pressure of 1000 mb)

| $T(T_d)(°C)^a$ | $W_s(W)$ (Grams of Vapor per Kilogram of Dry Air) | |
|---|---|---|
| 50 | 88.12 | |
| 45 | 66.33 | |
| 40 | 49.81 | |
| 35 | 37.25 | |
| 30 | 27.69 | |
| 25 | 20.44 | |
| 20 | 14.95 | |
| 15 | 10.83 | |
| 10 | 7.76 | |
| 5 | 5.50 | |
| 0 | 3.84 | |
| −5 | 2.644 | $(2.518)^b$ |
| −10 | 1.794 | (1.627) |
| −15 | 1.197 | (1.034) |
| −20 | 0.7847 | (0.6456) |
| −25 | 0.5048 | (0.3955) |
| −30 | 0.3182 | (0.2375) |
| −35 | 0.1963 | (0.1396) |
| −40 | 0.1183 | (0.0803) |
| −45 | 0.0695 | (0.0450) |
| −50 | 0.0398 | (0.0246) |

$^a$ If the first column represents the temperature, then the second column represents the *saturated* mixing ratio ($W_s$). If the first column represents the *dew point* temperature ($T_d$), then the second column represents the *actual* mixing ratio ($W$).
$^b$ Red numbers indicate value when ice is present.

rather simple to explain. No matter how high the RH outdoors during winter, the fact that it is cold means that the air cannot hold very much vapor. When this air leaks indoors it is heated. Heating air increases its capacity for holding vapor but does not add any vapor. This means that $W_s$ increases but $W$ remains constant. Thus the fraction, $W/W_s$, decreases and, since it is equal to the RH, the RH also decreases. This leads us to the following general rule.

**Rule:** As the temperature rises, the RH falls (so long as no vapor is added to the air).

Example 10.2
It is a wet winter day and the air outside is $T = 0°C$ and RH = 100%. When this air leaks into the house it is heated to 20°C. What is the RH indoors?

Information: Use Table 10.1. At $T = 0°C$, $W_s = 3.84‰$, whereas at $T = 20°C$, $W_s = 15.0‰$.

Equation:
$$RH = \frac{W}{W_s}$$

Procedure: The air outside is saturated at 0°C so that its actual mixing ratio equals its saturated mixing ratio. Thus, $W = 3.84‰$. Once the air moves indoors it is warmed to 20°C, but its actual mixing ratio, $W$, still equals 3.84‰. However, at 20°C, $W_s$ has risen to 15.0‰.

Solution
$$RH = \frac{3.84}{15.0} = 26\%$$

**Note:** If the indoor temperature were kept at 25°C, then the RH would only have been 19%. Therefore, there are two reasons for not heating too much during winter. First, excessive heating wastes expensive and valuable fuel and, second, it makes the air uncomfortably dry.

This rule explains why the RH is usually highest near dawn when the daily temperature reaches its minimum, and lowest in mid-afternoon when the temperature reaches its high point for the day.

Table 10.1 can be used to explain the popular saying, "It's too cold to snow." Technically, this is not true, but it does contain a kernel of truth. When air is extremely cold it can hold so little vapor that very little snow could form from it. At temperatures below about −50°C, the air can hold so little vapor that merely breathing will make it snow at your feet! This is commonly experienced by Antarctic explorers.

Now you are ready to prove that the air in the Sahara contains more vapor than the air in Greenland.

Example 10.3
Prove that air at 40°C with RH = 10% contains more vapor than air at −10°C with RH = 100%.

Information: Using Table 10.1 we find that the saturated mixing ratio, $W_s$, equals 49.8‰ for air at 40°C and equals only 1.79‰ for air at −10°C. (We have chosen the larger value at −10°C simply for the sake of argument.)

Equation:
$$RH = \frac{W}{W_s}$$

Substitution:

*In Greenland*
$$100\% = \frac{W}{1.79‰}$$

*In the Sahara*
$$10\% = \frac{W}{49.8‰}$$

## Solution

$$W = 1.79\%_{00} \qquad W = 4.98\%_{00}$$

Conclusion: The Sahara Desert air contains far more vapor than the Greenland air even during a blizzard. Nevertheless, the problem in the Sahara still remains—its vapor cannot be easily extracted from the air.

One complicated feature of Table 10.1 must be discussed. In the example above we had a choice of two different values for $W_s$ when the temperature was $-10°C$. This is true of all subfreezing temperatures.

The larger of the two values represents the amount of water vapor that the air can hold so long as there is *no ice* present. As soon as a single ice crystal is introduced, the vapor-holding capacity of the air is suddenly decreased and the excess vapor must sublimate onto the ice crystal, making it grow larger.

In other words, *ice crystals act somewhat like magnets for water vapor.* You will soon see that this fact is very important in the formation and growth of raindrops and snowflakes.

## Example 10.4

A tiny ice crystal is placed in a room at $-15°C$ and RH = 100%, with respect to water (i.e., when no ice is present). How much will this ice crystal grow? There are 1000 kg of air in the room.

Information: Using Table 10.1 you can see that at $-15°C$, $W_s = 1.2\%_{00}$ with respect to water, but only $1.05\%_{00}$ with respect to ice.

### Solution

Once the ice crystal is introduced, the vapor content is reduced by $0.15\%_{00}$. Since there are 1000 kg of air in the room, this means that 0.15 kg of water vapor sublimate to ice. Thus, the ice crystal grows by 0.15 kg.

On clear nights it often gets cold enough so that dew forms on the grass or on cars and other objects near the ground as in Figure 10.1. When air cools, the temperature at which dew begins to form is called the **dew point temperature,** $T_d$. Dew forms at the point of saturation. Therefore, the dew point temperature is actually a concept used to tell how much vapor *is* in the air.

**Rule:** The higher the dew point temperature, the more vapor is in the air.

## Example 10.5

A parcel of air has temperature of $30°C$ and dew point temperature, $T_d = 5°C$. A second parcel has the same temperature but a dew point temperature of $20°C$. Which parcel is holding more water vapor, and what is the RH of each parcel?

Information: Using Table 10.1 we find that the saturated mixing ratio, $W_s$, is equal to $27.7\%_{00}$ at $30°C$, $15.0\%_{00}$ at $20°C$, and $5.5\%_{00}$ at $5°C$.

### Solution

If the first parcel were cooled to $5°C$, it would reach saturation. Thus, its mixing ratio is equal to the saturated mixing ratio of air when $T = 5°C$, so it has a mixing ratio of $W = 5.5\%_{00}$. The second parcel will reach saturation when cooled to $20°C$; thus, it has a mixing ratio of $W = 15.0\%_{00}$. The air with the higher dew point temperature, $T_d$, has the higher mixing ratio—which means it contains more vapor.

Computing the RH is now a simple matter.

Equation:

$$RH = \frac{W}{W_s}$$

### Solution

For the first parcel | For the second parcel

$$RH = \frac{5.50\%_{00}}{27.6\%_{00}} = 20\% \qquad RH = \frac{15.0\%_{00}}{27.6\%_{00}} = 54\frac{1}{2}\%$$

We can formulate another general rule from this example.

**Rule:** For a given temperature, the higher the dew point, or $T_d$, the higher the relative humidity, RH.

You already know that dew and fog are more likely after clear nights, simply because the temperature drops more on clear nights. If, for instance, the temperature tends to drop by $10°C$ on clear nights but only $3°C$ on cloudy nights, and if the evening readings are $T = 25°C$ and $T_d = 19°C$, then for this example dew or fog certainly will not form on the cloudy night, but may form on the clear night (since the $T_d$ generally doesn't fall much overnight).

Dew forms on the outside of glasses that hold cold drinks or on bathroom mirrors after you have taken a hot shower. What happens is that the glass has a temperature below the dew point temperature of the air and therefore

FIGURE 10.1
Spider webs are often coated with dew by morning.

FIGURE 10.2
Frost that forms on windows often has exquisite ice-crystal patterns.

cools the air touching it enough for condensation to occur. When the temperature falls below 0°C, frost will form instead of dew. On very cold winter days or nights frost may form on the inside of house windows (or car windows). This happens when the temperature of the glass on the inside falls below the dew point (really the frost point). Sometimes the most beautiful ice crystal patterns result as in Figure 10.2.

Unfortunately, the dew point temperature is not easy to measure directly. The standard method for measuring the humidity of the air is to take a *wet* piece of gauze, wrap it around the *bulb* of the thermometer, and swing it around vigorously for a minute or so. As the water on the gauze evaporates, it cools the thermometer (since evaporation is a cooling process). The temperature reached by this process is called the **wet bulb temperature** or $T_w$ (see Figure 10.3).

The wet bulb temperature is *not* the same thing as the dew point temperature! The wet bulb temperature involves cooling by evaporation. Once evaporation takes place, vapor IS ADDED to the air, causing the original dew point temperature of the air to rise. As a result, *the wet bulb temperature is lower than the temperature, but higher than the dew point temperature,* unless the air is saturated—in which case all three are equal. Table 10.2 contains the relations between the $T$, $T_w$, $T_d$, and RH.

Here is an example of what the wet bulb temperature means. When you get out of the water after swimming, you feel cool because of evaporative cooling. How cool can your skin become? Your skin can cool down to the wet bulb temperature. When we sweat, our skin can only cool down to the wet bulb temperature. As you have read in Chapter 5, the purpose of sweating is to enable us to keep cool on hot days.

If the wet bulb temperature were higher than body temperature (37°C), then sweating would not cool us and our body temperature would rise. This has led to death on a number of occasions. It is fortunate for us that the wet bulb almost never rises to 37°C anywhere on earth. Usually when the temperature does rise above 37°C, the humidity is low enough so that $T_w$ remains well below 37°C.

I remember getting out of a swimming pool in Las Vegas, Nevada, when the temperature was 40°C and feel-

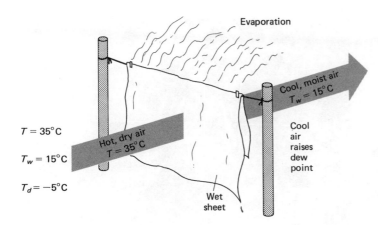

$T = 35°C$

$T_w = 15°C$

$T_d = -5°C$

Evaporation

Hot, dry air
$T = 35°C$

Wet
sheet

Cool, moist air
$T_w = 15°C$

Cool
air
raises
dew
point

**FIGURE 10.3**
Illustration of the wet bulb temperature.
When the wind blows past the sheet, the
air and sheet are cooled by evaporation.
At the same time, the added vapor raises
the dew point temperature. The final temperature reached by this process is called
the wet bulb temperature, which is always
somewhere between the temperature and
the dew point temperature. If the wet bulb
temperature is below freezing, wet clothes
that are hung out to dry can freeze even
though the temperature is above freezing.

**TABLE 10.2**
Relationships between $T$, $T_w$, $T_d$, and RH.

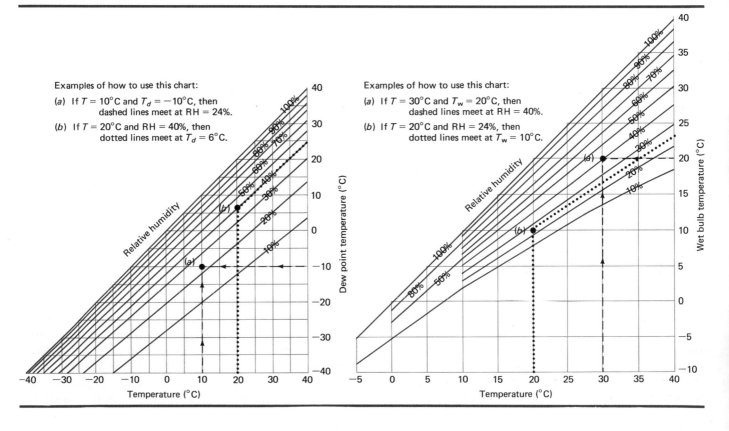

Examples of how to use this chart:

(a) If $T = 10°C$ and $T_d = -10°C$, then
dashed lines meet at RH = 24%.

(b) If $T = 20°C$ and RH = 40%, then
dotted lines meet at $T_d = 6°C$.

Examples of how to use this chart:

(a) If $T = 30°C$ and $T_w = 20°C$, then
dashed lines meet at RH = 40%.

(b) If $T = 20°C$ and RH = 24%, then
dotted lines meet at $T_w = 10°C$.

ing cold simply because the RH was so low that the wet bulb temperature was also low. On the other hand, after swimming in the Atlantic Ocean in Georgia, I rarely feel cold, because the air is quite humid. Even though the air temperature in Georgia is almost always well below 40°C, the wet bulb temperature there is usually higher than it is in Las Vegas.

When people sweat in humid heat, little cooling can result because there is little evaporation. Sweating itself doesn't cool you—it is the *evaporation* of the sweat that does. Now you can understand the saying, "Dry heat is not as bad as humid heat."

---

Example 10.6

After you get out of a pool, when will you feel colder: (1) when $T = 40$°C and RH $= 10\%$, or (2) when $T = 25$°C and RH $= 80\%$?

Information: Using Table 10.2 we see that for case (1): $T_w = 18.5$°C, while for case (2): $T_w = 22.5$°C.

Solution
The wet bulb temperature of case (1) is lower than that of case (2). Therefore, even though the air in case (1) has $T = 40$°C, so long as a wind is blowing to promote evaporation case (1) will feel colder!

---

This phenomenon is not merely a matter of feeling. Because of evaporation it is actually possible for wet objects to freeze solid on dry windy days, even when the air temperature is above freezing. Wet objects can freeze whenever the wet bulb temperature is below freezing. Thus when clothes are hung outside on lines to dry, they may well freeze on dry windy days, even when the temperature is above 0°C.

An even more important consequence of this phenomenon is that snow and hail tend to stay frozen as they fall through the *above* freezing air, so long as the wet bulb temperature remains below freezing (i.e., if the RH of the air beneath the cloud is low).

You might want to know the highest temperature at which snow can fall. The highest temperature that can still have a subfreezing wet bulb is $T = 10.6$°C. Above this temperature, snowflakes must rapidly melt. On extremely rare occasions snow has been seen falling from thunderstorms in the middle of the summer. This is possible because the downdrafts in thunderstorms are usually cool and sometimes rather dry, but also because they blow the snowflakes down to the ground very quickly (see Chapter 19).

It has been suggested that in order to cool cities on hot summer days one might evaporate huge quantities of water into the air. Unfortunately, this may not do too much to improve comfort because, even though it does lower the air temperature, it doesn't lower the wet bulb at all. In fact, once the sun heats up the newly humidified air, the wet bulb temperature will rise even higher than before, and human discomfort may actually be increased.

The **Temperature Humidity Index** or **THI** was developed to show how human discomfort from the heat depends on temperature and humidity. Basically, the THI is an average of the temperature and the wet bulb. When the THI rises above 80 virtually everyone is uncomfortable. According to Table 10.3, a day with $T = 30$°C and RH $= 60\%$ is just about as uncomfortable as a day with $T = 35$°C and RH $= 15\%$! Both have THIs of 79.

The vapor pressure and vapor density of air are two alternative ways of expressing the concentration of vapor in the air. Each of these terms has its use. Both are listed in Table 10.4. The vapor pressure is merely the part of the total air pressure that is a result of vapor molecules. The

TABLE 10.3
The Temperature-Humidity Index (THI)[a]

| Temperature (°C) | Relative Humidity | | | | | | | | | | |
|---|---|---|---|---|---|---|---|---|---|---|---|
| | 0 | 10 | 20 | 30 | 40 | 50 | 60 | 70 | 80 | 90 | 100 |
| 25 | 66 | 67 | 68 | 69 | 71 | 72 | 73 | 74 | 75 | 76 | 77 |
| 30 | 71 | 72 | 74 | 76 | 77 | 78 | 79 | 80 | 82 | 83 | 84 |
| 35 | 76 | 78 | 80 | 82 | 84 | 85 | 86 | 87 | 89 | 90 | 91 |
| 40 | 81 | 83 | 85 | 87 | 89 | 91 | 92 | 94 | 96 | 97 | 98 |
| 45 | 86 | 89 | 91 | 93 | 95 | 97 | 99 | 101 | 103 | 104 | 105 |
| 50 | 91 | 94 | 97 | 99 | 102 | 104 | 106 | 108 | 110 | 112 | 113 |

[a] When the THI reaches 75, at least 50% of all people are uncomfortably hot; at 80 or above, virtually everyone is.

TABLE 10.4
Saturated Vapor Pressures and Saturated Vapor Densities

| $T(°C)^a$ | Vapor Pressure (mb) | | Vapor Density (g/m³) | |
|---|---|---|---|---|
| 100 | 1013.25 | | | |
| 95 | 845.28 | | | |
| 90 | 701.13 | | | |
| 85 | 578.09 | | | |
| 80 | 473.67 | | | |
| 75 | 385.56 | | | |
| 70 | 311.69 | | | |
| 65 | 250.16 | | | |
| 60 | 199.26 | | 130.3 | |
| 55 | 157.46 | | 104.4 | |
| 50 | 123.40 | | 83.06 | |
| 45 | 98.86 | | 65.50 | |
| 40 | 73.78 | | 51.19 | |
| 35 | 56.24 | | 39.63 | |
| 30 | 42.43 | | 30.38 | |
| 25 | 31.67 | | 23.05 | |
| 20 | 23.37 | | 17.30 | |
| 15 | 17.04 | | 12.83 | |
| 10 | 12.27 | | 9.40 | |
| 5 | 8.719 | | 6.80 | |
| 0 | 6.108 | | 4.847 | |
| −5 | 4.215 | $(4.015)^b$ | 3.407 | $(3.246)^b$ |
| −10 | 2.863 | (2.597) | 2.358 | (2.139) |
| −15 | 1.912 | (1.652) | 1.605 | (1.381) |
| −20 | 1.254 | (1.032) | 1.074 | (0.884) |
| −25 | 0.807 | (0.632) | 0.705 | (0.552) |
| −30 | 0.509 | (0.380) | 0.453 | (0.339) |
| −35 | 0.314 | (0.223) | 0.286 | (0.203) |
| −40 | 0.189 | (0.128) | 0.176 | (0.119) |
| −45 | 0.111 | (0.0020) | 0.106 | (0.0684) |
| −50 | 0.0636 | (0.0394) | 0.0617 | (0.0430) |

$^a$ If the first column represents the temperature, then the second and third columns represent the *saturated* values of vapor pressure and density, respectively. If the first column represents the *dew point* temperature, then the second and third columns represent the *actual* values of vapor pressure and density, respectively.
$^b$ Red numbers indicate value when ice is present.

vapor density is merely the mass of vapor divided by the volume of air it occupies.

The vapor pressure is useful in determining the boiling point of water. Water boils at 100°C (212°F) at sea level *under average atmospheric pressure*. At higher altitudes, where the atmospheric pressure is lower, water boils at lower temperatures! Thus, in Yellowstone National Park, which is about 2300 m above sea level, there are pools of boiling water at 92°C. At this altitude the boiling point of water is only 92°C. Near the top of Mount Everest, the boiling point of water is little more than 70°C so that it is impossible to get a really hot cup of coffee. On the other hand, people use pressure cookers in order to boil foods at temperatures much higher than 100°C.

There is a simple rule for determining the boiling point of water.

**Rule:** Water boils at the temperature for which the saturation vapor pressure equals the surrounding air pressure.

---

Example 10.7
What is the boiling point of water in Leadville, Colorado?

Information: Leadville is the highest city in the United States and is situated at an altitude of 3094 m above sea level. At this height the typical pressure is about 700 mb. Now use Table 10.4.

Solution
Using the rule for boiling and Table 10.4 we find that the saturation vapor pressure for boiling in Leadville is 700 mb, and this corresponds to a temperature of 90°C. Therefore, the boiling point of water in Leadville is 90°C.

Example 10.8
At what height above sea level will your blood begin to boil?

Information (for this gory question): Body temperature is 37°C. The saturation vapor pressure at this temperature is approximately 62 mb (from Table 10.4).

Solution
We must determine at what altitude the atmospheric pressure equals 62 mb. Using a crude rule from Chapter 7 for the decrease of pressure with height, we know that the pressure halves every 5.5 km. Thus at sea level, $p = 1000$ mb; at 5.5 km, $p = 500$ mb; at 11 km, $p = 250$ mb; at 16.5 km, $p = 125$ mb; and at 22 km above sea level, $p = 62.5$ mb. Therefore, human blood begins to boil at heights above 22 km. This will not happen when people are protected by pressurized space suits. All this need not bother you very much because long before this altitude you would have suffocated from lack of oxygen.

---

Boiling is nothing more than rapid or explosive evaporation. As long as the air pressure exceeds the saturation vapor pressure, vapor molecules can only slowly diffuse away from the water surface. Once the saturation vapor pressure exceeds the air pressure there is nothing to "hold" the vapor molecules back, and they rush away from the water. In short, the water boils (see Figure 10.4).

The *vapor density* has two uses. Its first use is that it

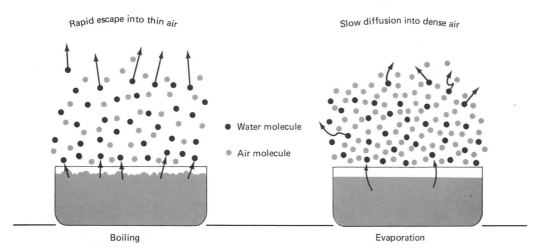

Rapid escape into thin air

Slow diffusion into dense air

● Water molecule

● Air molecule

Boiling

Evaporation

FIGURE 10.4
When the saturated vapor pressure is greater than the surrounding air pressure, vapor molecules rapidly escape from the surface of the water and the water boils. When the surrounding air pressure is greater than the saturated vapor pressure, the vapor molecules slowly diffuse away and evaporation takes place.

gives an idea of how much vapor is contained in a prescribed volume, such as a room. For instance, if you use a humidifier and wish to know how much water should be added to a room, the vapor density is useful.

---

Example 10.9
The temperature in a room is kept at 25°C and the RH is initially 0%. How many grams of water will it take to raise the RH to 100% if the room has a volume of 300 cubic meters (3 × 10 × 10)?

Information: Use Table 10.4. The saturated vapor density is 23 grams of vapor per cubic meter at 25°C.

Solution
Since the room has 300 cubic meters of air, it will take 6900 grams of water to bring the RH up to 100%.

---

The second use of the vapor density is in calculations of the rate of evaporation or condensation. Evaporation and condensation are basically diffusion processes. The vapor density at a water surface or an ice surface is always equal to the saturated vapor density at the temperature of the surface. When the vapor density of the surrounding air is less than this, vapor molecules will diffuse away from the surface and spread into the surrounding air. These molecules are then replaced from within the water or the ice. This is evaporation. Condensation will occur when the vapor density of the surrounding air is greater than the sat-

urated vapor density at the water or ice surface. You should recall from Chapter 5 that diffusion always works so that, on the average, molecules tend to drift from the more concentrated regions to the less concentrated regions.

You will notice that for temperatures below 0°C, the saturated vapor density also has two values—one with respect to water and one with respect to ice. Since the value with respect to ice is lower (as with the saturated mixing ratio), this means that when there is an ice crystal and a water drop nearby that diffusion will always take vapor from the drop and deposit it onto the crystal. Thus ice crystals grow at the expense of water droplets (see Figure 10.5).

We can now look at why warm air can hold more water vapor than cold air. You must recall two facts. First, as the temperature rises the average speed of the molecules increases. Second, at any temperature the molecules have a wide range of speeds.

A molecule must have a certain speed (or energy) in order to escape from the water or ice and become a vapor molecule. The higher the temperature, the larger the number of molecules that have this necessary speed. Thus warm air can hold more vapor (molecules) than cold air. Furthermore, no matter how cold it gets, there will always be a few molecules with sufficient energy to remain free as vapor.

Cold air can hold fewer vapor molecules when ice is present simply because the molecules need more energy (i.e., a higher speed) to escape from ice than from water.

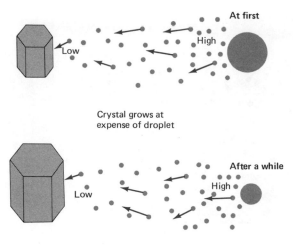

Crystal grows at
expense of droplet

Vapor density

FIGURE 10.5
Ice crystals grow at the expense of water droplets. The saturated vapor density over ice is less than over water. Therefore, vapor molecules diffuse from the droplets (where they are concentrated) to the crystals (where they are scarce). Looked at another way, it is sometimes possible for ice crystals to grow under the same conditions that cloud droplets must evaporate (see Plate 3).

This is the same thing as saying that there is a higher latent heat of vaporization for ice than for water.

## EVAPORATION AND CONDENSATION

Evaporation at the ground level goes on quietly and invisibly and helps keep the ground from getting too hot. The vapor then rises into the atmosphere where it ultimately condenses and produces clouds and precipitation. An important byproduct of condensation is the heat that is released into the atmosphere. *Every time that it rains or snows (strange as it may seem) the atmosphere is actually being heated.*

Heating of the air due to condensation can cause a chain reaction. As warm air rises clouds form and condensation occurs. This heats the warm air further and enhances its tendency to rise, which in turn increases the amount of condensation further. Such a situation is not imaginary—this is how hurricanes are produced.

Although condensation is a more dramatic process, evaporation is just as important. In fact, evaporation seems to be a more complicated process. The rate of evaporation depends on many factors such as the temperature, the RH, the wind speed, the roughness of the ground, the plant cover, the moisture content of the ground, and so on. It is possible to obtain a crude estimate of the evaporation rate by filling a pan with water and seeing how long it takes for the water to evaporate. Other, more sophisticated ways of measuring evaporation include weighing the ground and seeing how rapidly it loses weight as the water in it evaporates.

The conclusions obtained from such measurements show that the rate of evaporation increases as the wind speed increases and as the RH decreases. The most important factor, however, is the temperature. As the temperature rises the rate of evaporation increases exponentially (enormously, to say the least), simply because the vapor-holding capacity of the air increases exponentially. As long as there is any wet object nearby, the air will try to become saturated since at the surface of the object the air is saturated.

The potential rate of evaporation (from the surface of some moist object) is therefore proportional to the **vapor deficit** or the amount of vapor needed to bring the air to saturation. You can use Table 10.4 to see how the vapor deficit changes with temperature.

---

Example 10.10
Two parcels of air have 0% RH. One has $T = 0°C$ and the other has $T = 40°C$. Compare their rates of evaporation.

Information: Using Table 10.4 and the fact that the RH is 0%, the vapor deficit is equal to the saturated vapor density. This is 4.85 grams per cubic meter for the 0°C air and 51.1 grams per cubic meter for the 40°C air.

Solution
Since the vapor deficit at 40°C is more than 10 times larger than the vapor deficit at 0°C, the evaporation rate of the warmer air is also more than 10 times larger.

---

No wonder the desert is so dry! Even when it rains in the desert the water rapidly evaporates because of the enormous vapor deficits of the hot air, and soon the desert appears bone dry once again. When artificial lakes are created behind dams in the desert, enormous quantities of water are lost through evaporation.

Many of the desert areas of the world receive far more precipitation than the bulk of Antarctica; yet, because it is so cold in Antarctica, the little precipitation that does fall does not evaporate. Instead, it has accumulated over the centuries into a massive ice cap. More will be said about this in Chapter 22.

A world map of evaporation is shown in Figure 10.6. Several features of this map should be apparent. First, evaporation rates are larger over the ocean than over land. Second, they are larger near the equator where it is hot, and smaller near the poles where it is cold. Third, the largest evaporation rate in the world occurs off the eastern coast of the United States where extremely cold, dry air blows over the warm Gulf Stream waters during winter. This air is then rapidly heated and charged with vapor.

The worldwide average annual evaporation is approximately 1 m and is almost exactly equal to the worldwide average annual precipitation. This is true because whatever goes up must come down.

Not every place on earth has a balance between evaporation and precipitation. The oceans and the dry regions on earth such as the subtropics generally have less precipitation than evaporation, while the continents and humid regions such as the tropics generally have an excess of precipitation, which then must flow out in the rivers or underground.

In regions where there has been a drought or where the climate has recently become drier, lake levels can easily drop half a meter or more in a year. The Caspian Sea in Asia has been gradually shrinking over the past 10,000 years or so and may eventually evaporate completely. This is significant because the Caspian Sea is about the size of the state of California.

## WATER ON EARTH

The earth is blessed with a bountiful supply of water. There are roughly 1300 million cubic kilometers of water on earth. Slightly more than 97% of this is ocean water and therefore is salty. Perhaps one of nature's greatest ironies is that many sailors have literally died of thirst, although surrounded by water. Yet, when the sun beats down on the ocean, the water it evaporates is largely fresh and almost free of salt. Thus, the rain and snow that falls

FIGURE 10.6
The world map of evaporation. Evaporation generally decreases from equator to pole. It is surprisingly low in deserts, simply because they have so little water.

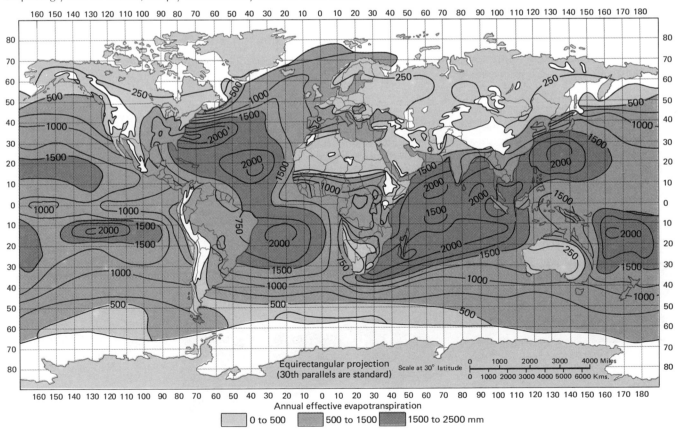

on the earth is relatively fresh. This fresh water has accumulated in glaciers, lakes, rivers, and even underground. There is slightly more than 36 million cubic kilometers of water, not including the water in the oceans.

Twenty-eight million of this is locked up in the great glaciers and ice caps of the world. At any one time the equivalent of 12,000 cubic kilometers of water is held in the atmosphere as vapor. Only about 220,000 cubic kilometers is contained in all the rivers and lakes of the earth. This still leaves 8 million cubic kilometers to be accounted for. Where could this be?

This enormous quantity of water is stored underground and is known as **groundwater.** If you have the impression that it is stored in huge subterranean caverns, then you are wrong. For the most part, this water fills the tiny pores and cracks in the soil and rocks.

Of the rain that falls to earth, somewhat less than 50% runs down the rivers into the sea. Most of the precipitation soaks into the ground temporarily but soon reevaporates. Once in the ground the water will sink until it reaches the level known as the **water table,** where the pores in the rocks and soil are filled with water. When someone digs a well, the level that the water reaches is the water table.

As you can see from Figure 10.7, in most places the water table lies well below the earth's surface. The water table does break through the surface at rivers, lakes, marshes, springs, and oases. During a dry season the water table will generally fall, then marshes and wells will dry up, and then river levels will drop. Once the rains resume, the water table rises once more.

Walden Pond, made famous by Henry Thoreau, is one example of a pond or lake that has no streams or rivers running into it. Walden Pond is fed by water running into it from underground. Walden Pond therefore provides evidence that not only are enormous quantities of water stored underground, but that this water can move about somewhat freely. The existence of oases in the middle of deserts also proves that water can travel through solid rock for hundreds of kilometers underground.

Certain rocks such as sandstone and limestone are quite *permeable*. Water travels through permeable rocks much like it does through a sponge, even if there are no large cracks in the rocks. Since large cracks do tend to develop in some rocks such as limestone, water often travels quite rapidly underground. Over the centuries, limestone will slowly dissolve in water, and great caverns can be created when groundwater flows through a limestone rock layer. However, other rocks such as granite are more impermeable and therefore block the flow of groundwater.

Groundwater tends to move "downhill" in the sense that it runs from regions where the water table is higher to where it is lower. For instance, when someone draws water out of a well the water table is lowered slightly, and water will flow through the surrounding rocks toward the well. When a river passes through a desert, the water table of the surrounding desert is usually lower so that the river loses considerable water by underground seepage. The Nile River is the prime example of a large river that passes through a major desert and loses much of its flow.

FIGURE 10.7
The water table is the level to which the water fills all pores in the rocks. The water table "breaks through" the surface at lakes, streams, marshes, and so on. When the water table slopes, the water will flow downhill through porous rocks, such as sandstone and limestone, but may be blocked by impermeable rock layers, such as granite.

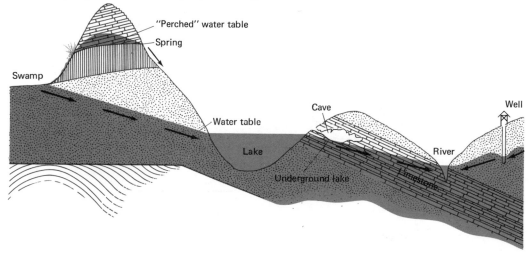

In the process, the Nile feeds many oases hundreds of kilometers away. On the other hand, when rivers pass through humid regions, they are often fed further by groundwater.

The hydrologic cycle combines all of these movements of water—above, on, and below the earth's surface. The picture of the hydrologic cycle shown in Figure 10.8 is easily worth a thousand words.

# FLOODS AND DROUGHTS

There are basically two distinct types of floods. The first type of flood occurs in coastal areas when an intense storm such as a hurricane raises sea level a meter or more and causes the sea to inundate the low-lying coastal areas. This type of flood is discussed in Chapter 18 on hurricanes.

It is the second type of flood with which we are concerned here. This type occurs when too much water pours into a river, causing the river level to rise and overflow its banks. Many times these floods can be disastrous—such was the case in the Big Thompson River valley on August 1, 1976 and in the eastern United States during hurricane Agnes in June 1972. Occasionally, however, such floods are not only predictable, they are actually beneficial. The annual flooding of the Nile has watered and fertilized the soil of Egypt every year for thousands of years. Recently, the Aswan dam has stopped these annual floods and in many respects has done more harm than good.

There are a variety of conditions that can cause a river to overflow its banks. Among these are excessive rain, excessive snowmelt, some obstruction such as an ice jam, and, finally, the collapse of a dam. The worst floods occur when several of these factors combine; for example, when torrential spring rains melt the snow and then the rising water levels cause dams to collapse.

FIGURE 10.8
The hydrologic cycle, showing where all the water on earth is, and how it moves. All amounts of storage are in cubic kilometers; all rates are in cubic kilometers per day.

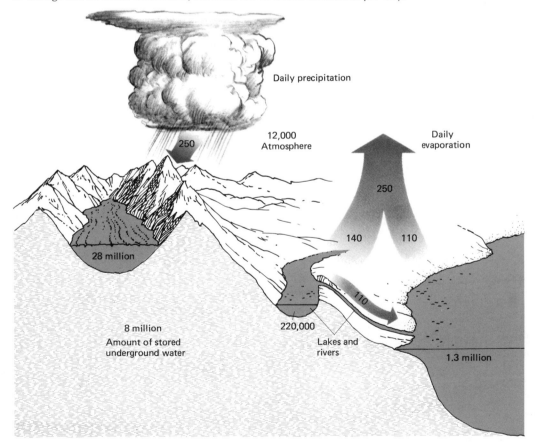

Daily precipitation

250

12,000
Atmosphere

Daily evaporation

250

140          110

28 million

8 million
Amount of stored
underground water

220,000

110

Lakes and
rivers

1.3 million

The conditions of the ground are also extremely important in determining the severity of flooding. If the ground is saturated with water from previous rains or frozen and impermeable after the long winter, flooding tends to be more serious simply because any new rain cannot soak into the ground but must flow straight overland into the rivers. Moderately dry soil can soak up quite a bit of rainwater and, thus, can prevent or reduce flooding.

Ironically, extremely dry ground can become baked and hardened like cement and therefore relatively impermeable. This makes flash floods somewhat more likely. In desert regions flash floods occur in which raging torrents of water race down preexisting, dried-out channels known as *arroyos* or *wadis*. Within a few hours all the water sinks into the ground and disappears, but not before it has drowned many inexperienced travelers (see Figure 10.9).

During the winter, in middle and high latitudes, much of the precipitation falls as snow and accumulates on the ground. In the spring this snow begins to melt, and it is at this time of year that most rivers reach their high stage for the year. Flooding is always a distinct possibility after a winter that has been especially cold and snowy. Danger of flooding is often increased in these situations when the ice flowing down the river gets jammed. This can effectively dam the flow and cause the water upstream to rise. When the ice dam finally breaks up, the water rushes downstream with increased intensity and produces even more flooding.

The spring floods usually start not with warm, sunny days but with rainy spring days. Nothing can melt the accumulated snows of the winter like a good spring rainstorm. But floods do not end once the rain stops. Often it

takes hours or even days for the rainwater to travel across the land and reach the major rivers. *Generally, the larger the river, the longer the delay between the rains and the peak of the flooding.* Flood forecasters will certainly breathe a sigh of relief when the rains stop, but they are aware that most of the trouble may still lie in the hours ahead.

On June 15, 1972, a small storm was born in the Gulf of Mexico just north of the Yucatan Peninsula. It took two days for this storm to reach hurricane intensity, and as the first hurricane of the season it was named Agnes. Never an intense hurricane, Agnes soon became the most costly storm in history in terms of property damage. On June 19, Agnes smashed into Florida and for the next seven days it plagued the eastern coast of the United States. As soon as Agnes moved overland, its peak winds dropped below hurricane strength. As a hurricane Agnes was a failure, but as a rainstorm she was a monster (see Figure 10.10).

The cruelest trick that Agnes played occurred on June 22. Just when it seemed that Agnes would blow out to sea, she made a sudden, unexpected turn inland and then stalled for four days over Pennsylvania and New York. Earlier, in Georgia, Agnes had brought rains that actually relieved a minor drought. In New York and especially in Pennsylvania there had been no drought. The ground was saturated from earlier rains, and no extra rain was needed or wanted. Despite this, in six days Agnes brought up to 50 cm of rain to parts of Pennsylvania and New York. Large areas received over 25 cm of rain, and Harrisburg, Pennsylvania, received 31.6 cm within a 24-hour period, which broke an all-time record.

The rivers responded as they had to and became raging torrents that swept away anything in their path. Estimates at the time stated that the current speeds easily exceeded 50 km/h. In Richmond, Virginia, the James River rose 11.2 m, exceeding the previous record set in May 1771 by 2 m. The Susquehanna River rose 9.9 m in Harrisburg, overflowing its banks there and then rising another 5 m to inundate the state capital (see Figure 10.11). Over $2 billion of damage was sustained in Pennsylvania alone. All in all the storm cost $3 billion.

Water levels also rose nearly 10 m in New York State. Corning and Elmira had to be largely evacuated when the raging torrent swept through them. These high water levels also broke 200-year-old records by 2 m or more. The Finger Lakes in New York State reached all-time high water levels, and yet ironically the water supply was cut off for several days in many places because of contamination due to mud and other impurities. Sewage backed up and produced rather unsanitary conditions.

After the storm had passed, over 300,000 people were left homeless. However, it was astounding that a storm of

FIGURE 10.9
A dry stream bed, known as an arroyo or wadi, can become filled quite rapidly by flash floods in the desert. Thus, it is not wise to fall asleep there. This arroyo is found west of Roswell, New Mexico.

Rainfall from hurricane Agnes
June 18–25, 1972

Precipitation
in centimeters

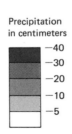

| | |
|---|---|
| | —40 |
| | —30 |
| | —20 |
| | —10 |
| | —5 |

FIGURE 10.10
Precipitation from hurricane Agnes.

this magnitude—occurring in such a highly populated region—had a death toll that was so infinitesimal. Only 120 lives were claimed—primarily the result of excellent forecasting and warning provided by the National Weather Service.

Earlier that month, on June 9, 1972, a much, much smaller flood claimed 236 lives in Rapid City, South Dakota. Rapid City lies just east of the Black Hills. Easterly winds began forcing extremely humid air up against the

Black Hills on the afternoon of June 9, and the rising air produced torrential rains. Up to 38 cm of rain fell within six hours.

The flooding began in the Black Hills, and cars, trailers, and even houses were swept downstream. Several of these large objects perched on the top of the Canyon Hill Dam which was filled to the brim. These objects then blocked the normal flow of water over the dam's floodgates so that the water level rose to more than 3 m

FIGURE 10.11
Flooding from hurricane
Agnes in Harrisburg, Pennsyl-
vania.

over the top of the dam. Eventually, this debris was forced over the dam, and the water began rushing down the canyon. Then the dam collapsed and all hell broke loose as a gigantic wave of water swept down on Rapid City.

Why did the dam collapse? It collapsed because it was an earth-filled dam, and any child who has played in the sand knows what can happen to these when the water overflows. Children build dams of sand that can contain the water so long as no water spills over the top. Once the water does overflow, it rapidly erodes the sand and then all the water pours out. On the beach this is fun, but on a larger scale it has proven many times to be monumentally disastrous. So why do engineers build earth-filled dams? They build them because the harried taxpayers living downstream refuse to pay for more expensive dams.

The world's longest river, the Nile (6695 km) flows northward from equatorial Africa. It passes through the Sahara Desert and Egypt, and waters a thin strip of land that would otherwise be absolutely unlivable. The reliability and fertility of the Nile River waters have made Egypt one of the cradles of civilization, and Egypt is often called the gift of the Nile (see Figure 10.12).

Each year, beginning in June the Nile River begins to rise and reaches its peak in September at a level some 8 to 10 m higher than at low water. The ancient Greek historian, Herodotus, commented at length on the Nile. He noted that the Egyptians had become master dam builders and irrigators to control the river's irregularities. But he wondered why the Nile crested every year in September when all other rivers flowing into the Mediterranean Sea crested in the spring.

About why the Nile behaves precisely at it does I could get no information from the priests or anyone else. What I particularly wished to know was why the water begins to rise at the summer solstice, continues to do so for a hundred days, and then falls again at the end of that period, so that it remains low throughout the winter until the summer solstice comes around again in the following year. Nobody in Egypt could give me any explanation of this, in spite of my constant attempts to find out what was the

FIGURE 10.12
This satellite picture clearly shows how the Nile River "waters" the desert.

WATER VAPOR AND THE HYDROLOGIC CYCLE

peculiar property which made the Nile behave in the opposite way to other rivers. . . .

*History of the Persian Wars.*

We know now the explanation to be simple. The source region of the Nile is mainly north of the equator where the rainy season is June, July, and August. It then takes about one month for the flood waters to reach Egypt. The Mediterranean area on the other hand receives most of its precipitation during the winter months of December, January, and February, and the rivers may peak somewhat later because of snowmelt in the spring.

Flooding has caused some rivers to change their course. Two such rivers are the Mississippi and the Hwang Ho in China. These rivers carry large amounts of sediment from the highlands that fall to the river bed as the water approaches the sea and the current slows. As the sediment accumulates on the river banks and the river bed, the level of the river is gradually raised above the level of the surrounding floodplain. Thus these rivers build natural levees. During flood times such rivers may overflow their banks and break through their own levees, carving new riverbeds. And, until the floodwaters subside, literally tens of thousands of square kilometers may be under water (see Figure 10.13).

When people arrive on the scene they build farms on the rich but vulnerable lowlands near the river. They seldom worry that the water level is higher than their land. When the river level begins to rise, people begin to act. They build up the natural levees even further. In the case of the Hwang Ho River, the levees were built almost 25 m high. When the river finally overflowed its levees in 1851, 900,000 people drowned. The Hwang Ho now pours into the ocean through a new channel, nearly 1000 km north of the old channel.

Because of the large population density of China and the fact that the land in eastern China is so flat, floods have always proven especially disastrous. When the Yangtze Kiang and Hwang Ho overflowed their banks in 1931, they flooded almost 300,000 square kilometers of land and drowned over 1 million people—which may well be the worst human disaster of all time. This flood was merely the worst of a series of ruinous floods that plagued China during the 1930s.

During this time, the midwestern United States suffered the worst drought in its history and became known as the "dustbowl." The resulting drop in agricultural productivity did nothing to spur the economy during the Great Depression. During the 1930s, summers on the Great Plains were extremely hot, so that evaporation increased enormously. The rainfall remained low throughout the decade, and the water table dropped so far that it was said

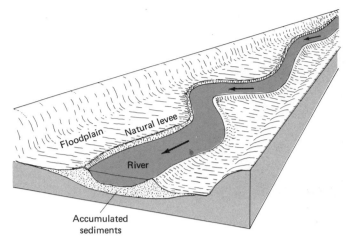

FIGURE 10.13
River levels can be higher than the surrounding flood plains. This occurs when rivers build their own natural levees by dropping sediments. This poses a distinct menace during "flood time" when a river can overflow its own banks, flooding thousands of square kilometers and perhaps even changing its course.

to take as much as a gallon of gasoline to pump up a gallon of water.

The top soil became dry and dusty, and the wind actually lifted the dust several kilometers into the air, blowing it as far as the Atlantic Ocean. Dust storms occasionally made it so dark that pedestrians walked into one another at high noon. Thousands of square kilometers of farmland had all its topsoil stripped away, and many farmers deserted what had been their homes for generations.

Except for occasional duststorms, drought is rarely a dramatic event. Instead, it is prolonged and insidious. A drought is merely a period of time in which the precipitation is much below normal. We do not speak of droughts in the desert because dry conditions are the norm there. Drought is a much feared problem in many of the world's tropical and semitropical places where rain is seasonal. India depends on the summer monsoon to bring soaking rains after many almost rainless months. When the monsoon comes late, or is weak or even absent, mass starvation inevitably follows.

From 1968 to 1973 a drought captured the attention of the entire world. Just south of the Sahara Desert there is a large semiarid region known as the Sahel. Every year, around July, a brief rainy season is part of the life in the Sahel. This season occurs when the tropical rainy belt moves north of the equator (see Chapters 16 and 22). There, as in India, some years are rainier than others. However, the Sahel is much drier than India, even in the best of years. In some years the rain belt never reaches the drier parts of the Sahel and these regions revert to desert.

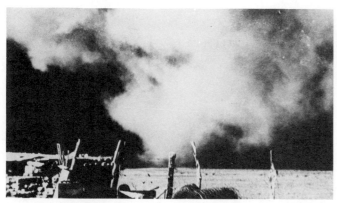

FIGURE 10.14
The leading edge of a dust storm on the Great Plains during the dust-bowl era of the depression.

FIGURE 10.15
A dust storm as viewed from space.

In the years before the drought began, the Sahel had above average rains. The human population and the animal flocks increased rapidly, and the stage was set for a major tragedy.

Then, for five consecutive years, the tropical rain belt both weakened in intensity and did not move quite so far north as usual. The precipitation fell 25 to 50% below its normal value, and the overgrazed land gave way to desert. Relief efforts, late as usual, were hampered by primitive transportation and the most up-to-date corruption. Even so, the death toll was miraculously held to 100,000 (see Figure 10.16).

Because of this drought, great shock waves were felt through the meteorological community. They asked, "Does this drought signify a change in the world's climate patterns?" "Are the years of plenty over?" We do not have the answers to such questions. Usually, lean years follow fat years in an unpredictable manner. The Sahel has experienced many such droughts over the centuries. While we are living off the fat of the land, few questions are asked. But when times get hard they "try men's souls," and we enquire as to the nature of the universe—if we are not too busy struggling to survive.

Many of the drought-prone areas of the world are areas of low rainfall in which two or three fewer rainy days a year can make the difference between survival and death. Many parts of the western United States fit into the category of being drought prone. Fortunately, however, when one area is stricken, aid is sent from other nearby areas. In other parts of the world the situation may not be so fortunate. In these places prolonged drought spells disaster; and those few who can, leave the area. Usually the drought is followed by a few relatively good years, and

people gradually move back, forgetting the hard times that went before. Within most of our lifetimes we will probably see another disastrous drought on a repopulated Sahel.

FIGURE 10.16
Drought in the Sahel.

# PROBLEMS

**10-1** Two parcels of air have the same temperature and pressure, but one has a higher RH. Which is heavier (if there is a difference)? Explain.

**10-2** If you know only the air pressure and RH, do you know how much vapor is in the air?

**10-3** The pressure is 1000 mb, the temperature is 20°C, and the RH is 25%. What is the mixing ratio ($W$) and the saturated mixing ratio ($W_s$)?

**10-4** Why is the RH generally so low indoors during the winter even if it is 100% outside?

**10-5** Now that you know from Problem 10-1 that humid air is lighter than dry air, find out which of these is lighter: (a) air with $T = 35$°C and RH = 80% or (b) $T = 37$°C and RH = 0%. To do this problem you must use the formula: virtual temperature (or $T_v$) = $T(1 + 0.61 W)$. The so-called virtual temperature is the temperature that completely dry air would have in order to be just as dense as the actual, humid air that is at temperature $T$. Careful, this formula uses the Kelvin temperature scale. (Assume the air is at 1000 mb. Then, the higher the virtual temperature, the lighter the air.)

**10-6** Why are there two values for $W_s$ in Table 10.1 when the temperature is below 0°C?

**10-7** Unless otherwise specified, RH is measured with respect to water as if no ice were present. If the RH is 100% with respect to water at $T = -10$°C, what is the RH with respect to ice? (Assume the air is at 1000 mb.)

**10-8** The temperature in a room at 1000 mb is $-20$°C and the RH is 95% with respect to water. A small ice crystal and a small water droplet are brought into the room. What will happen to each? Explain.

**10-9** Does the dew point temperature have anything to do with dew?

**10-10** The mixing ratio ($W$) is 5.50‰. What is the temperature and the dew point temperature? (Assume the air is at 1000 mb.)

**10-11** Your clothes get wet in the rain and then the wind starts to blow. Do your clothes cool to the wet bulb temperature or to the dew point temperature?

**10-12** Do beads of water condense on the outside of a glass of ice-cold water more easily on a cold, dry day or on a hot, humid day? Must the glass be cooler than the wet bulb temperature or must it be cooler than the dew point temperature?

**10-13** Describe the weather conditions if water vapor from the air condenses on your skin.

**10-14** Assuming it is windy, when will it feel colder after you get out of a pool: (a) $T = 35$°C, RH = 10% or (b) $T = 20$°C, RH = 70%?

**10-15** Light rain falls at $T = 2$°C, RH = 10%. Explain why this rain freezes on a moving car's windshield.

**10-16** The temperature is 1°C, dew point is $-20$°C, and rain begins to fall. An hour later, without any cold winds blowing into the region, the rain changes to snow. Explain this somewhat unusual situation using the given information.

**10-17** What air feels more uncomfortable: (a) $T = 35$°C, RH = 70% or (b) $T = 40$°C, RH = 60%? Compute the THI of each.

**10-18** Why does it take longer to cook spaghetti on the shores of Lake Titicaca, Peru (elevation 3812 m, pressure 635 mb), than at sea level?

**10-19** What is the approximate boiling point of water on the shores of Lake Titicaca? (Use the information in Problem 10-18.)

**10-20** At what temperature will an ice crystal grow most rapidly at the expense of a water droplet?

**10-21** How many grams per cubic meter of water can possibly be condensed out of air initially at $T = 25$°C, RH = 100%?

**10-22** A climate is adequately moist if the precipitation at least equals the evaporation. Therefore, how much more rain is needed at 30°C than at 10°C to avoid semiarid conditions?

**10-23** As you can see from Figure 10.6, in the Sahara Desert evaporation is rather low despite the high temperatures. Why?

**10-24** The evaporation rate over the subtropical oceans is somewhat high, as you can see from Figure 10.6, and greatly exceeds the precipitation there. (a) How does this relate to the fact that the surface waters of the subtropical oceans are very salty? (b) What is the annual evaporation rate around Hawaii?

**10-25** Explain, using the concept of the water table, why marshes dry up during droughts.

**10-26** What percentage of the rain that falls on the continents runs down into the sea?

**10-27** Why can flooding be so severe in the early spring in the middle and higher latitudes?

**10-28** Find some interesting records of severe droughts or floods.

**10-29** See if you can duplicate Figure 10.13 in a sandbox. Start a small stream flowing down a steep slope of sand, and let it flow on to a more level region where it may produce its own levees. Then increase the stream flow and simulate a disastrous flood.

# FOG

Fog is nothing more than a cloud that touches the ground. When the visibility is reduced below 1 km, because of tiny droplets or ice crystals, fog is reported. Thick fog occurs when the visibility is less than 300 m. In the thickest fog it may be almost impossible to see your outstretched hand and when standing even the ground may be completely obscured.

Since fog forms on the ground and often in depressions in the ground, it is apparent that it is not usually a result of rising air. Condensation occurs when air is cooled sufficiently or when enough evaporation occurs to make the air saturated. Since there are a variety of conditions that can lead to condensation, there are several types of fog.

In this chapter we will look at some foggy places and some interesting cases of fog. We will then move to a discussion of the various types of fog and their meteorological implications. To conclude the chapter, I will describe and explain a number of techniques used to disperse fog.

## THE WORLD'S FOGGY PLACES

England and especially London is known for its fog. What would Jack the Ripper, the Ancient Mariner, or Sherlock Holmes have been without fog. English literature is full of references to fog. Here is one example from the Sherlock Holmes mystery, *The Hound of the Baskervilles.*

I have said that over the great Grimpen Mire there hung a dense, white fog. It was drifting slowly in our direction and banked itself up like a wall on that side of us, low but thick and well defined. The moon shone on it, and it looked like a great shimmering ice field with the heads of the distant tors borne as rocks upon its surface. Holmes's face was turned towards it, and he muttered impatiently as he watched its sluggish drift. . . . my mind [was] paralyzed by the dreadful shape which had sprung out upon us from the shadows of the fog. A hound it was, an enormous coal-black hound, but not such a hound as mortal eyes have ever seen. Fire burst from its open mouth, its eyes glowed with a smouldering glare, its muzzle and hackles and dewlap were outlined in flickering flame. Never in the delirious dream of a disordered brain could anything more savage, more appalling, more hellish be conceived than that dark form and savage face which broke upon us out of the wall of fog.

Sir A. Conan Doyle, *The Hound of the Baskervilles*

This great, awesome hound was a dog that the villain had coated with a phosphorescent substance to make it glow ominously. When Sherlock Holmes shot the huge dog, the villain realized his game was up. He tried to escape through the Grimpen Mire, but was lost in the fog and buried alive in quicksand—a ''clear'' victory over the forces of evil!

Some of the world's thickest and most notable cases of fog have occurred in London. There the natural tendency for fog to occur has been greatly increased by the large amount of coal burned by the population for heating and industries. Coal burning produces sulfur dioxide and dust, and these facilitate condensation. Not only do they help thicken the fog, but when sulfur dioxide (a noxious gas) combines with water droplets it produces dilute sulfuric acid. The whole mixture is, needless to say, quite unhealthy, as has been known for over 300 years. In 1674 Sir Thomas Browne wrote

Mists and fog also hinder the . . . . coal smoke from descending and passing away. [So] it is conjoined with the mist and drawn in by the breath, all which may produce bad effects . . . . and produce catarrhs and coughs.

Will Durant, *The Age of Louis XIV*

The most memorable and deadly case of London fogs occurred in December 1952. On December 5, the winds died down and the fog began to form. For the next three days the fog continuously thickened, so much so that for large periods of time the visibility was only a few meters. Traffic was brought to an almost complete standstill and there were many accidents. The unknowing population heated their homes more than necessary to combat the moisture. This produced more coal dust and sulfur dioxide—making the air all the more poisonous and thick with fog (see Figure 11.1).

Still, people went to work with masks over their mouths, finding their way along the sidewalks by holding onto the walls of the buildings. Everything became covered with soot and, even with the masks, breathing was painful. To this was added the feeling of disorientation. Someone who had been in that fog recalled "One of the bad things was that you could be standing right by your house and not know it for you could hardly see your hand in front of you." All in all, 4000 deaths were attributed to this case of fog in the county of London alone. This fog contributed to the establishing of a Clean Air Act for England.

As a result of the Clean Air Act, London will never again have the pea soup fogs it is famous for. Perhaps never again will there be cases of people actually walking into canals and rivers, because they literally could not see their feet or the ground! Fog they will have, but never will it be so thick again.

There are a number of actual cases where fog has saved an army or navy. When Napoleon set out to conquer Egypt, Admiral Nelson of England followed him. Nelson unwittingly passed by the French on a foggy night. This

FIGURE 11.1
Pea soup fog in London on December 5, 1952.

event gave Napoleon time to invade Egypt, but a few weeks later Nelson backtracked and destroyed the unsuspecting French fleet.

The Americans also benefited considerably because of fog during the Revolutionary War. On August 30, 1776, an extremely rare case of thick fog during summer enabled the American army to cross the East River from Long Island to Manhattan (New York) and escape the British.

Providence further interposed in favor of the retreating army by sending a thick fog which hung over Long Island while on the New York side it was clear . . . . Had it not been for that heavenly messenger, the fog, to cover the first desertion of the lines, and the several proceedings of the Americans after daybreak, they must have sustained considerable losses. The fog resembled a thick small mist, so that you could see but a little way before you.

David M. Ludlam, *Weatherwise*, June 1975

Fog is quite common in coastal waters, especially

where there are cold ocean currents or in polar regions. Because fog is most prevalent over the ocean where there are no fixed weather stations, we are not exactly sure which is the foggiest place in the world. The Grand Banks off the coast of Newfoundland is perhaps the world's foggiest place, with 2400 hours of dense fog on over 150 days a year. Other places are not far behind. Santa Catalina off the coast of California averages 1500 hours on 158 days of heavy fog per year. A "day with fog" means that fog occurred for at least one hour in the day. Maps of fog frequency are shown in Figures 11.2 and 11.3.

Few places have more than 80 days with fog. You should immediately notice that foggy regions usually occur in coastal areas. This incidentally increases the chances for collisions at sea. Two very notable collisions have been caused by a combination of fog and carelessness at sea. The "unsinkable" *Titanic* collided with an iceberg on its maiden voyage, and quickly sunk. The *Andrea Dorea* collided with another ship in the fog and sunk off the coast of New York. At coastal airports the situation can be almost as bad. Fog in the Canary Islands led to the collisions of two 747 jumbo jets; this incident claimed the lives of over 570 people in March 1977. There are also many cases of chain-reaction accidents in fog, involving up to 200 vehicles.

At times the edge of the fog is quite abrupt, and then the foggy area is known as a **fog bank.** The behavior and appearance of the fog bank can be quite dramatic. Some-times fog literally pours over the hills from the oceanside of San Francisco, down to the landward side (see Figure 11.4). This mass of fog can take on the appearance of a gigantic avalanche or a breaking wave, which is absolutely harmless—if you don't drive too fast.

## RADIATION FOG

After a clear night there is always a chance that fog will form in a thin layer near the ground if the dew point temperature is not too low. This type of fog is known as **radiation fog,** because it forms when the ground cools by radiation at night. The ground then cools the air above it by contact, and fog is likely to form if the air is cooled down to the dew point temperature. This type of fog layer is rarely more than 100 m thick and may be called **ground fog.**

Since cooling is more dramatic on clear nights, fog is more likely after clear nights than after cloudy nights. Ironically, then, *fog in the morning is often an indication of a clear day to come.* Once the sun comes up the fog tends to "burn off" or disappear from the ground up. The fog disappears or actually evaporates when the ground and air are heated above the dew point temperature. Radiation fog also tends to form in enclosed valleys or even small depressions in the ground, since the coldest air is heavy

FIGURE 11.2
World map of fog frequency.

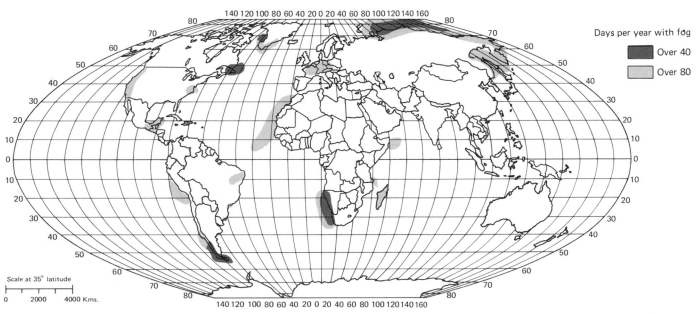

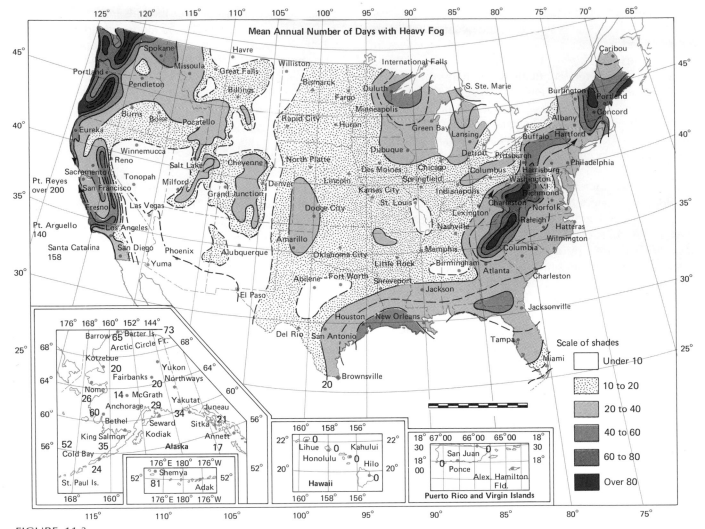

FIGURE 11.3
U.S. map of fog frequency.

and sinks and is most likely to condense. Hilltops often lie above the fog while the valleys are "socked in" (see Figure 11.5).

Here is an interesting question for you. When is radiation fog more likely—after a relatively calm night or after a relatively windy night? The answer is, after a relatively calm night. The reasoning is simple. On a relatively calm night a thin layer of air near the ground is cooled considerably. On a windy night the cold air near the ground is mixed with somewhat warmer air directly above (recall from Chapter 9 why farmers use fans to prevent frost). Therefore, calm nights usually get colder and thus fog is more likely after a calm night.

It is therefore possible to predict when radiation fog will occur. If the dew point temperature at sunset lies within 5°C of the actual temperature, and if the wind speed is slow (say less than 5 knots), there is a good chance that radiation fog will occur by dawn. The exact details of fog prediction depend strongly on the precise location; no one technique works for all weather stations.

## ADVECTION FOG

Advection is the technical word that is commonly used in meteorology. **Advection** is the *horizontal* movement of a mass of air that causes changes of properties such as temperature. (Convection is the same thing, except it refers to *vertical* movement of air.) Advection fog is then caused by the wind but, like radiation fog, it tends not to form when the wind is too strong.

FIGURE 11.4
Fog pouring through the Golden Gate Bridge in San Francisco.

Advection fog can occur any time warm air is blown over a cold surface and is the most common cause of widespread fog in the world. The fog on the Grand Banks and the fog over any cold current are advection fogs. If warm air from nearby warmer areas becomes cooled by contact when it passes over the colder surface, and if it is cooled down to the dew point temperature, fog will most likely result.

FIGURE 11.5
Fog (with varicose appearance) filling river valleys on August 2, 1977.

The reasons fog often forms near coastal areas may now become clearer. First, some coastal currents contain especially cold water because of a phenomenon known as **upwelling** (explained in Chapter 16). Second, air over the land during summer can be quite warm, and when it passes over the cold currents can easily be cooled down to the dew point temperature.

The coast of California is especially foggy during the summer-half of the year (April to October). When warm moist air from the open ocean approaches the cold coastal current, it is cooled down to the dew point. Oddly enough, this fog does not always form precisely at sea level, but usually forms a layer of low-lying stratus clouds, which make the hilly land foggy (see Figure 11.6).

It is not too difficult to explain why this type of fog caused by the cold sea prefers to form a few hundred meters up. Off the coast of California the wind stirs a layer of air 300 to 600 m thick (see Figure 11.7). Blobs of air are only partially cooled when they touch the cold ocean surface and are cooled further when they rise. They are often not cooled quite enough to form fog right at the surface. The little extra cooling caused by rising only a short distance will produce fog against a hillside. If this type of fog does occur exactly at sea level, it is a sign that it is quite thick and dense.

These fogs are often blown a few kilometers inland and keep coastal locations quite cool. During the rainless summer months they also provide a humid environment essential for the survival of the coastal Redwood trees. By limiting the rate of evaporation the fog saves the trees the job of having to pump much water through their trunks which are often over 100 m high.

Fogs similar to the California coast fogs form off the western coasts of the world's continents in subtropical latitudes. These fogs often bring some moisture to regions of the world that are otherwise among the driest in the world. This is the case in Lima, Peru, where rainfall is little over 2 cm for the year, but the earth is still green as a result of the persistent fog, known there as the *garua*. A little known song has been written about this fog (sung to the tune of ''The Rain in Spain'')

*The dew through Peru*
*Is due to the garua!*
*What could be truer?*

The suffering reader will probably recognize the author's heavy hand.

On the Grand Banks the fog rests on the ocean surface. The warm Gulf Stream current is located just southeast of the Grand Banks, while the cold Labrador current lies directly over the Grand Banks. To produce fog, the wind has only to blow from the warm Gulf Stream toward the

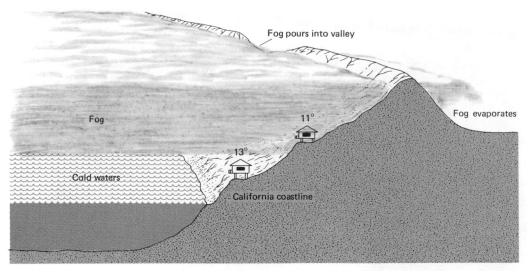

**FIGURE 11.6**
California coast fog often forms some distance above sea level. Partially cooled by contact with the water, it cools even further by rising. Conversely, as the fog pours inland and sinks, it evaporates. The evaporation is aided by contact with warm ground.

**FIGURE 11.7**
California coastal fog of May 21, 1977 as seen from space. Notice how it has crept inland in places. Closed convection cells can also be seen.

cold Labrador current (see Figure 11.8). Fogs in this region can be very dense and persistent—sometimes lasting a week or more without a break. Here the fogs form in all seasons, although there is a marked preference for summer, when fog may even occur if warm air from the land blows over the cold waters of the Grand Banks. Northeast of Japan, similar conditions exist where the cold Oyoshio current runs next to the warm Kuroshio current. This is another one of the world's foggiest regions.

In polar waters fog is especially common during the summer months, when the land heats up somewhat but the water remains at or near the freezing point. Many polar coastal locations receive 100 or more days of fog per year.

All of the cases of fog mentioned so far in this section are examples of "pure" advection fog. However, over land we often see fog that is a combination of an advection fog and a radiation fog. This is the type of fog that occurs in England when warm moist air from the Gulf Stream blows over the cold land during autumn and winter. Thus there is a natural tendency for fog over land, which is enhanced at night when the ground cools further.

In the eastern United States such advection-radiation fogs are quite common after the long autumn and winter nights when warm moist air from the Gulf of Mexico moves northward over the cold land (see Figure 11.9). Morning fog is a reasonably common sight during winter along the Gulf coast and is more likely when the ground is wet. The valleys of the Blue Ridge Mountains are especially foggy during autumn when Gulf air is still able to

FIGURE 11.8
Winds associated with some of the common fogs.

FIGURES 11.9a and 11.9b
Fog in the southeast United States on December 15, 1977. Fog is extensive at 1530 GMT, but has largely "burned off" by afternoon (2030 GMT).

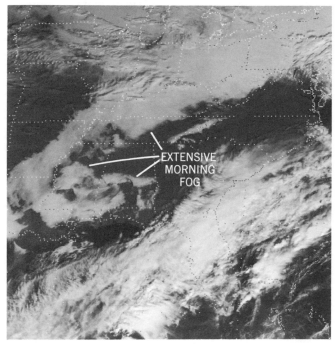 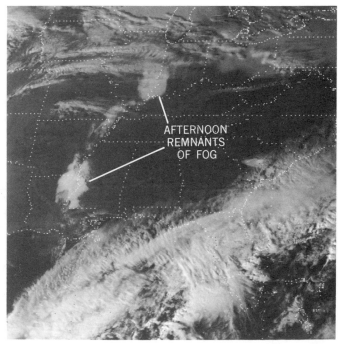

penetrate that far north with some frequency. (During winter, air from the Gulf of Mexico does not often move far north.)

## FRONTAL FOG

Before you read much further take a long, hot shower. In so doing you will (1) add to your heating bill, (2) cause all the paint on the bathroom walls to peel, and (3) essentially produce what is known as **frontal fog.** Frontal fog forms after it has been raining for hours. During that time the falling raindrops evaporate into the air beneath the clouds until the air is moistened to saturation. The same thing happens when you are taking a shower. As the water drops fly through the air they evaporate and fill the air with vapor. The warm drops fill the cooler air with more vapor than it can hold, producing fog or steam.

This type of fog occurs near fronts when rain from warmer air just above the front falls through the colder air near the ground. It occurs more often in connection with *warm* fronts (than with cold fronts) since long periods of rain precede the warm front. For this reason **frontal fog is distinguished by its coming after hours of rain; it is also often a sign that the rain is soon to end and that the weather will soon get warmer,** for at least a little while (see Figure 11.10).

## UPSLOPE FOG

An east wind on the Great Plains can bring fog. Although the Great Plains has a reputation for appearing flat, the land actually slopes up toward the west. Eastern Kansas is

less than 300 m above sea level, whereas western Kansas is almost 1200 m above sea level. Air blowing from the east may seem like it is going across level ground but it is actually rising.

Fog will form when the air near the ground is cool enough so that it will not rise and moist enough so that when it moves upslope it can cool to the dew point. As you might expect, this type of fog forms more frequently at night, so perhaps it should be called upslope-radiation fog. On the Great Plains upslope fog may occur only 5 to 10 times a year, but in the mountains upslope fog may be far more common (see Figure 11.11). Fog is reported an average of 310 days a year on Mt. Washington in New Hampshire.

I have been in the Great Smoky Mountains in North Carolina and have literally seen upslope fog form when air blew up the side of the mountain. As with cumulus clouds (recall from Chapter 2), the cloud or fog forms at the lifting condensation level. I recall the funny feeling of seeing the newly formed droplets moving up the mountain, while all the time more new drops began to form at the unchanging condensation level (see Figure 11.12). When seen at a great distance from the mountain, upslope fog may appear like a cloud (which it is).

## STEAM FOG

**Steam fog,** also known as **arctic sea-smoke,** is the same thing as steam. It is similar to the steam that you produce when you breathe out on a very cold day. Arctic sea-smoke is generally quite thin and wispy in appearance. It is also usually only a few meters thick and only forms over

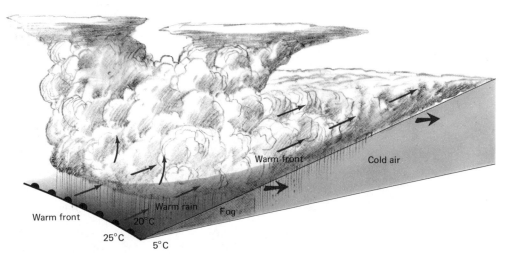

FIGURE 11.10
Frontal fog forms near the warm front when warm raindrops fall through colder air. Frontal fog often indicates the rain will soon end and the weather will get warm. Heavy arrows indicate movement of front.

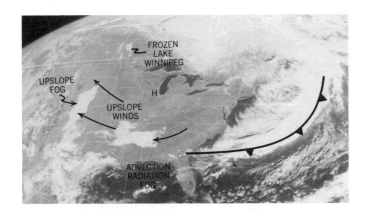

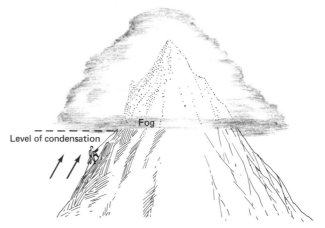

FIGURE 11.12
Upslope fog forms when air rising up a slope reaches its condensation level. The fog remains in the same place, but individual droplets can be seen moving uphill.

FIGURE 11.13
Steam fog over a stream on a cold winter morning.

FIGURE 11.11
Upslope fog on the Great Plains on April 12, 1976. This fog has a tendency to form at night. So perhaps it should be called upslope-radiation fog. Notice how area of fog has decreased from 1500 GMT to 2000 GMT.

water or a wet surface. Steam or sea-smoke forms when the air is much colder than the water directly below (see Figures 11.13 and 14.5).

You should realize from the name, arctic sea-smoke, that this type of fog most commonly occurs in the polar regions. During the long polar winter the air grows quite cold, and cracks inevitably form in the ice which covers the polar oceans. Even though the exposed water is near the freezing point, it is still far warmer than the air above.

Arctic sea-smoke occurs mainly in the autumn or winter far from the equator when cold air rushes over much warmer water. It can frequently be seen during cold snaps over lakes and rivers before they freeze over. Arctic sea-smoke was seen in Boston Harbor during the record-breaking cold wave on February 1934. A fog similar to arctic sea-smoke can occur near dawn over lakes or rivers when the air surrounding the lake or river has been cooled at night, but the water remains relatively warm. It is also seen on hot, wet pavements after a brief shower.

Steam fog is not that simple to explain, even though it is so commonly seen when cooking. Steam forms by a process involving several steps (see Figure 11.14). In the first step cold air is blown directly over a warm water surface and is simultaneously warmed and enriched in vapor. In the second step as this air (call it sample 1) whirls up away from the water surface, it mixes with some of the cold air above (call it sample 2). (Steam fog never forms exactly at the water surface but a *short* distance above it.) The resulting mixture may contain excess vapor that condenses to fog, even though neither sample 1 nor 2 contains excess vapor!

The question that must be answered is this: How can fog result from mixing two unsaturated samples of air?

---

Example 11.1

Mix 1000 grams of air at RH = 90% and $T = 10°C$, with 1000 grams of air at RH = 50% and $T = -20°C$. What is the RH of the resulting mixture?

Information: Use Table 10.1. The saturated mixing ratio $W_s$ is 7.76‰ when $T = 10°C$ and is 0.785‰ when $T = -20°C$. We also need one other piece of information. After the two equal samples are mixed, the resulting temperature will be the average of 10°C and -20°C, which is -5°C. The air at -5°C is 2.64‰.

Procedure: Compute the total amount of water vapor in the mixture by adding the vapor contents of each sample. This enables you to find the actual mixing ratio $W$ of the mixture. Then find the RH of the mixture.

Equation:

$$RH = \frac{W}{W_s}$$

Substitute:

| Air at 10°C, RH = 90% | Air at -20°C, RH = 50% |
|---|---|
| $90\% = \dfrac{W}{7.76‰}$ | $50\% = \dfrac{W}{0.785‰}$ |
| $W = 7.0‰$ | $W = 0.392‰$ |
| in 1000 g of air there are 7.0 g of vapor. | In 1000 g of air there are 0.39 g of vapor. |

The resulting mixture has a total of 7.39 grams of vapor in 2000 grams of air. Its actual mixing is therefore

$$\text{Actual mixing ratio} = \frac{7.39}{2000} \cong 3.70‰$$

Solution

Relative humidity of mixture of air at -5°C

$$RH = \frac{W}{W_s}$$

$$RH = \frac{3.70‰}{2.64‰}$$

$$RH = 140\% \qquad !!?$$

Thus, the relative humidity of the mixture is 140%! This is an impossible situation and means that some of the vapor in the mixture must condense and form fog or steam.

---

FIGURE 11.14
The formation process of arctic sea-smoke and steam.

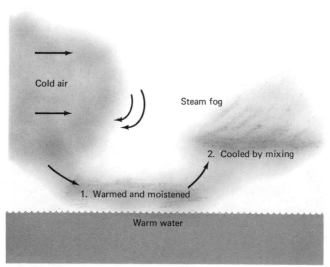

Cold air

Steam fog

2. Cooled by mixing

1. Warmed and moistened

Warm water

In the 1870s this complicated process was thought to be the cause of rain, but now we know it isn't. The process is very odd because you might think that mixing two equal amounts of air with different relative humidities should produce a mixture with an average value of relative humidities. Instead, as this example shows, when equal samples of 90% and 50% relative humidity are mixed, the air had *140%* relative humidity! This odd behavior results because, as the temperature of the air decreases, its vapor-holding capacity decreases so rapidly (exponentially) that even if the cold air had absolutely no vapor the cooling effect it has on the warm air would soon lead to saturation.

This process is also involved in the production of frontal fog, which is, of course, a form of steam.

One additional odd feature of this phenomenon is that the final temperature of the mixture actually will be slightly higher than -5°C! Why? Since the excess vapor in the mixture must condense and condensation is a warming process for the surrounding air, the mixture will be heated slightly above −5°C. I will not trouble you with the calculation, but the mixture will finally arrive at a temperature near −0.8°C!

# MAKING FOG AND CLEARING FOG

During World War II scientists worked on the problem of how to make the most effective smoke screens. In one method, some oil was burned and the resulting smoke particles obstructed visibility. The question that the scientists had to answer was how to get the maximum obscuration with using a minimum amount of fuel. They found that maximum obscuration occurred when the particles were slightly smaller than typical fog droplets!

When a given amount of water vapor condenses, the obstruction to visibility depends on the average size of the drops or crystals. Visibility is far more restricted when there are many small droplets than when there are a few large ones, as Figure 11.15 illustrates. Many small droplets scatter more light than a few large ones. Once the droplets become too small they begin to act like air molecules and do not scatter light effectively (remember Rayleigh scattering). The ideal size droplet for scattering light is about 1 micron in diameter. Fog droplets are usually

FIGURE 11.15
Small droplets obstruct visibility more than an equal *mass* of large drops.

**Severely restricted visibility**

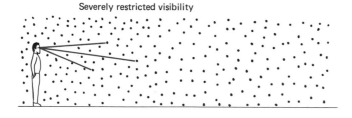

**Slightly restricted visibility**

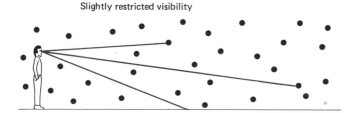

between 10 and 100 microns in diameter. The smaller they are, the more they obstruct visibility.

Now you can understand why the fogs of London were so thick. The excess water vapor had to be divided up among the countless soot particles that were suspended in the filthy London air. Therefore, each droplet remained small (about 10 microns in diameter). Inadvertently, then, the Londoners happened to do everything possible to make matters worse for themselves. While they didn't exactly manufacture their fogs, they significantly thickened them.

Since cooling the air is the principal means of producing fog, heating air should provide a straightforward means of clearing fog. The air can be heated by using heaters or even flying helicopters over the fog. The helicopters act like the fans in the orchards and mix the cold foggy air near the ground with warm dry air from above. At airports jet planes have been used to clear fog by turning on their engines and blowing the heated air down the runway. These methods work, but they are quite costly since soon after the runway clears, even the slowest wind will bring the fog drifting in from the sides.

There are more subtle ways of clearing fog or at least of increasing the visibility. If the many small fog droplets could be replaced by a few larger drops or ice crystals then visibility would improve. If the temperature is above freezing this is done by dropping a limited number of salt particles into the fog. Each salt particle attracts the water from the many tiny fog droplets that shrink to insignificance or evaporate.

When the temperature is below freezing the fog still may consist entirely of water droplets that are supercooled. This is known as supercooled fog, and it is quite common in winter over land. Supercooled fog can be cleared by dropping a limited number of ice crystals into the fog. The ice crystals then grow at the expense of the droplets, which then shrink or even evaporate (see Figure 11.16).

The few large droplets or crystals produced by these techniques are often heavy enough to fall to the ground, and thus the fog is cleared. But even if they don't fall, the visibility is improved.

There is one type of fog that cannot be cleared in this way. Any fog or cloud below about −40°C consists entirely of ice crystals and is known as ice fog. The one good feature of this type of fog is the magnificent halo displays it frequently produces. Ice fog is notorious in Fairbanks, Alaska, and in many Siberian cities where it may persist unrelentingly for days. This extremely cold air has such a small vapor-holding capacity that herds of caribou have been known to become camouflaged in the dense ice fog produced by their own breath!

FIGURE 11.16
A cloud deck cleared by seeding.

# PROBLEMS

**11-1** Find several allusions to fog in horror classics.

**11-2** Test the visibility in fog by measuring the distance of the farthest object you can see. Rarely will the visibility be less than 50 m.

**11-3** Explain the tendency in the morning for radiation fog to "burn off" from the ground upward. Take a look at the many morning and afternoon satellite pictures of fog in this chapter; note that regions of fog tend to "burn off" from the sides.

**11-4** Explain why fog in the morning is often a sign of a clear day to come.

**11-5** Explain why the last place for radiation fog to clear off is often the last place for cumulus clouds to form later that day.

**11-6** Why is radiation fog more common when the previous *afternoon* was cloudy?

**11-7** Explain the common fogs of the large central valley of California. These predominate in winter several days after humid air from the Pacific Ocean comes to rest in the valley. (This is not coastal fog.)

**11-8** Explain the meteorological conditions that prevail when fog forms in the southeast United States during autumn and winter.

**11-9** Why does fog form so frequently over cold, coastal ocean waters?

**11-10** Why are the coastal redwood trees restricted to the coastal areas?

**11-11** What kind of fog do you see wafting up from lakes or streams at night or during cold winter days.

**11-12** Why do you breathe out steam when the temperature is (a) $-5°C$ and RH $= 0\%$, but no steam when the temperature is (b) $25°C$ and RH $= 50\%$?

**11-13** All year long fog is produced on the Atlantic Grand Banks with southeast winds. Explain how it can also be produced by northwest winds during winter.

**11-14** Why is ice fog so difficult to clear?

**11-15** Why is fog after a long period of rain often a sign that the weather will soon clear up and get warmer?

**11-16** Why is an east wind favorable for fog formation on the Great Plains?

**11-17** Explain why small spherical droplets obstruct visibility more than an equal mass of larger spherical droplets.

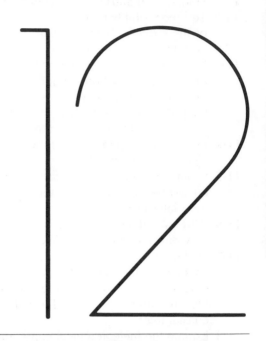

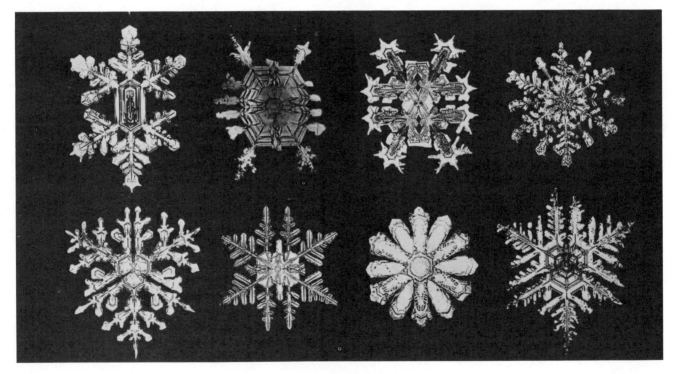

# PRECIPITATION

What makes it rain? You already know that clouds and precipitation are produced by rising air. Nevertheless, not all clouds produce precipitation. Air that rises and cools down to the dew point temperature merely marks the beginning of the rain forming process. Cloud droplets, ice crystals and, finally, raindrops must somehow be produced.

Most cloud droplets are too small to fall against the rising air that produces clouds. Then, even if they do manage to fall a short distance below the base of the cloud, they soon evaporate. Raindrops or snowflakes must grow large enough to be able to fall to the ground without evaporating completely. Therefore, the secret behind knowing what makes it rain or snow is knowing how the raindrops or snowflakes grow large enough to reach the ground.

The bulk of this chapter is devoted to describing the growth of raindrops, snowflakes, and some of the more unusual forms of precipitation. Cloud seeding, which makes use of this knowledge, is also discussed. First, however, we will begin the chapter with a presentation of extremes and means of precipitation around the world.

## SOME PRECIPITATION RECORDS

The worldwide average annual rate of precipitation is just about 100 cm. Needless to say, this precipitation is not spread evenly over the globe. The various land areas average only about 72 cm, while the oceans have an average of 112 cm.

The world's driest place may well be Arica, Chile, with an average annual rainfall of 0.05 cm! What this means is that several years may go by between brief rainshowers there. To say the least, rain is somewhat more frequent and abundant at Cherrapunji, India, and Mount Waialeale, Hawaii. For many years, Cherrapunji was thought the world's rainiest place, with about 1100 cm annually, but we know now that Mount Waialeale on the island of Kauai (Hawaii), with 1200 cm annually, is even rainier.

All such extremely high precipitation rates are restricted to quite local areas. The Hawaiian Islands are located in the trade wind belt where the wind blows from the northeast most of the year. When the air hits the Hawaiian Islands it is forced to rise over them, and rain is produced. This rain falls almost entirely on the windward (northeast) slopes of all the mountains producing tropical rain forests, while the leeward slopes a few kilometers away have very dry weather and sometimes desert vegetation (see Figure 12.1)!

Cherrapunji does not get quite as much rain as Waialeale, but this is not surprising when you consider that Cherrapunji gets just 15 cm in the four months from November through February. They make up for it during June, July, and August with a total of over 650 cm in these months alone. When you stop to think about it, that amounts to about 7.5 cm per day for 92 days. Most places seldom get more than 3 cm in a day.

At 3:23 P.M. on July 4, 1956, Unionville, Maryland, got 3.12 cm of rain in a minute! Such a high rate of rainfall cannot be maintained for very long. The most rainfall ever recorded in one hour occurred in Holt, Missouri, in June 1907, when 30 cm fell. In another record, 187 cm of rain fell in a 24-hour period at Cilaos on March 15–16, 1952, on Reunion Island in the Indian Ocean. For time periods from a week to a year, all the records are held by Cherrapunji, with 917 cm in July 1861 and 2657 cm in the one-year period, August 1, 1860 to July 31, 1861.

Now consider some snowfall records. Paradise Ranger Station located on Mount Rainier in Washington averages almost 1500 cm of snow per year. During the season of 1973–74, a record 2865 cm fell, and at times the snow was over 7.5 m deep. A 25 cm snowfall is normally sufficient to paralyze most cities for a few days, but Silver Lake, Colorado, received 193 cm within 24 hours on April 14–15, 1921.

FIGURE 12.1
Rainfall in the Island of Oahu. Rainfall is superabundant on the northeastern slopes, but often scanty a few kilometers away on the southwestern slopes. Can you see any relation to the prevailing winds.

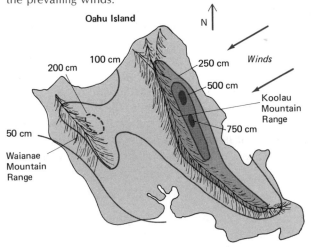

# PRECIPITATION AROUND THE WORLD

For a comparison of cities around the world, we will now look at rainfall records for the same cities whose temperatures were analyzed in Chapter 8 (see Tables 12.1-12.5).

Singapore is one of the wetter places on earth. The warm, rising, equatorial air commonly produces afternoon thunderstorms that are usually brief, but intense. (Oddly enough, in some months rain tends to fall in the early morning.) Singapore receives an average of 242 cm annually, but is rather unusual for a tropical location since it does not have any distinct rainy or dry seasons.

Most tropical locations do have distinct rainy and dry seasons. On the equator most of the year is wet, but in locations as little as 300 km from the equator there may be several months without rain. The tropical dry season usually comes during the "winter," which is called the low-sun period.

Aswan is an extreme example of the dryness of subtropical latitudes. Located in the driest section of the Sahara Desert, Aswan receives only 0.3 cm of rain per year on the average. Some years may receive one freak thunderstorm, while no rain at all may fall in many years. Do not allow the statistics of Aswan to mislead you. Whereas it is true that most places in the subtropics tend to be dry, both Waialeale and Cherrapunji happen to be located at almost the same latitude as Aswan.

St. Louis receives about 97 cm of rain per year. This is spread throughout the year, although there is a noticeable maximum from April to July. During the spring and summer in St. Louis, most of the precipitation falls from thunderstorms. During the winter-half of the year, most precipitation is the slow, steady type caused by the slowly rising air of low pressure areas.

Neither Aswan nor Singapore have ever recorded snow, but outside of the tropical and subtropical latitudes snow has fallen in most places. St. Louis gets an average of 40 cm of snow per year. The annual snowfall amounts rapidly increase as you move northward from St. Louis, doubling within 400 km. During the winter, the farther poleward you move, the greater the probability that the precipitation will fall as snow.

London receives only 61 cm of rain per year. This makes London no rainier than some places in the Sahel. London's reputation for being wet is partly due to the fact that much of the precipitation falls as drizzle from dying low pressure areas. The drizzle often falls for hours (or occasionally days) on end without accumulating very much.

Because of the warming influence of the ocean, London does not get much snow (about 25 cm per year), al-

**TABLE 12.1**
Precipitation for Singapore (cm)

|         | Jan  | Feb  | Mar  | Apr  | May  | Jun  | Jul  | Aug  | Sep  | Oct  | Nov  | Dec  | Year |
|---------|------|------|------|------|------|------|------|------|------|------|------|------|------|
| Average | 25.2 | 17.3 | 19.3 | 18.8 | 17.3 | 17.3 | 17.1 | 19.6 | 17.8 | 20.8 | 25.4 | 25.7 | 242  |

**TABLE 12.2**
Precipitation for Aswan (cm)

|         | Jan  | Feb | Mar | Apr | May | Jun | Jul | Aug | Sep | Oct | Nov | Dec | Year |
|---------|------|-----|-----|-----|-----|-----|-----|-----|-----|-----|-----|-----|------|
| Average | *a   | *   | *   | *   | *   | *   | 0.0 | 0.0 | *   | *   | *   | *   | 0.3  |

a Asterisk indicates precipitation average of less than 0.1 cm.

**TABLE 12.3**
Precipitation for St. Louis (cm)

|                  | Jan  | Feb  | Mar  | Apr  | May  | Jun  | Jul  | Aug  | Sep  | Oct  | Nov  | Dec  | Year |
|------------------|------|------|------|------|------|------|------|------|------|------|------|------|------|
| Average          | 4.7  | 5.2  | 7.7  | 10.0 | 9.8  | 11.2 | 9.4  | 7.3  | 7.3  | 7.1  | 6.3  | 5.2  | 91   |
| Average Snow     | 11.7 | 10.4 | 11.2 | 0.5  | *    | 0.0  | 0.0  | 0.0  | 0.0  | T    | 3.8  | 9.4  | 47   |

**TABLE 12.4**
Precipitation for London (cm)

|         | Jan  | Feb  | Mar  | Apr  | May  | Jun  | Jul  | Aug  | Sep  | Oct  | Nov  | Dec  | Year |
|---------|------|------|------|------|------|------|------|------|------|------|------|------|------|
| Average | 5.1  | 3.8  | 3.6  | 4.6  | 4.6  | 4.1  | 5.1  | 5.6  | 4.6  | 5.8  | 6.4  | 5.1  | 58   |

**TABLE 12.5**
Precipitation for Verkhoyansk (cm)

|         | Jan  | Feb  | Mar  | Apr  | May  | Jun  | Jul  | Aug  | Sep  | Oct  | Nov  | Dec  | Year |
|---------|------|------|------|------|------|------|------|------|------|------|------|------|------|
| Average | 0.5  | 0.5  | 0.3  | 0.5  | 0.8  | 2.3  | 2.8  | 2.5  | 1.3  | 0.8  | 0.8  | 0.5  | 13.6 |

though it did once snow even in July (1888). The highlands in western England and Wales do receive several hundred centimeters of snow, because the air at sea level is rarely much above freezing. So, only a few hundred meters above sea level, it is usually below freezing. Similarly, Paradise Ranger Station on Mount Rainier would have just about the same climate as London if it were at sea level but, since it is a little more than 1500 m above sea level, it is cold enough to snow there in winter.

Verkhoyansk receives an average of only 13.5 cm per year; this is less precipitation than most of the desert areas of the world. And yet Verkhoyansk is surrounded by dense forests, because with its cold air the little precipitation it receives hardly evaporates. Verkhoyansk gets most of its precipitation during the summer and is distinctly drier during the winter months when the frigid air can barely hold any vapor.

A seldom mentioned but important source of moisture for Verkhoyansk and other places in extremely cold climates comes from frost deposits. Verkhoyansk acts like a

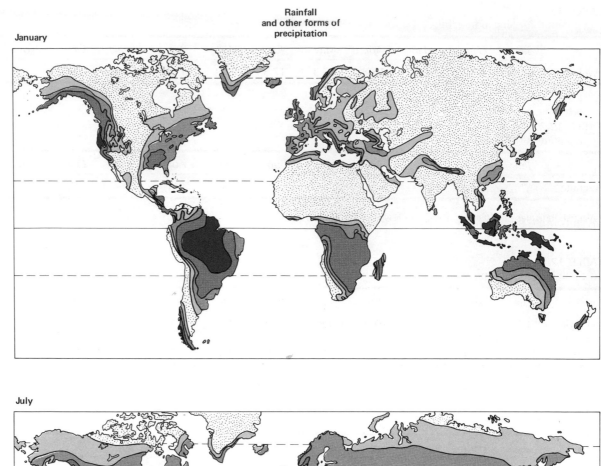

Rainfall
and other forms of
precipitation

January

July

Rainy
25.0 cm
10.0 cm
5.0 cm
2.5 cm
Dry

FIGURE 12.2
World precipitation during January and July.

giant freezer; as the air is cooled, vapor is "squeezed" out of the air and onto the ground and other objects in the form of delicate ice crystals. These ice crystals are just like the frost that forms in the freezer—they look like snow, but are fluffier. During the winter, such places get hardly any precipitation and skies are clear, but the frost accumulates on the ground like magic. Snow has been recorded in every month of the year, although it is quite rare during the summer.

World maps of precipitation for January, July, and the entire year are shown in Figures 12.2 and 12.3. Here is a list of several features you should notice immediately.

1. It is quite rainy near the equator and quite dry in much of the subtropics, 20 to 30° latitude.
2. The polar lands get very little precipitation.
3. Some of the world's rainiest areas occur along mountainous coastlines.
4. Rainy and dry areas are usually located farther north in July and further south in January, leading to rainy and dry seasons in some places.

These and other features of the world's precipitation patterns will be described in more detail in Chapters 16 and 22.

How is precipitation measured? In spite of all the advances of science and technology, rainfall and snowfall are still measured in a very basic way. Snowfall is measured by sticking a ruler through the snow in a number of places and averaging, since snow will drift as a result of wind, and since some places will have more snow than others. The liquid water content of the snow is found by simply melting the snow. When there is some snow on the ground from a prior storm, a board is placed on top of the old snow so that only the new snowfall is measured.

Rainfall is measured by allowing the rain to accumulate in a narrow bucket under a funnel and then measuring the depth of water in the bucket. Whenever possible, rainfall and snowfall should both be measured away from any obstacles such as tall buildings or trees.

# OUTLINE OF RAINDROP GROWTH PROCESS

How do raindrops grow? Since the process is complicated, I will briefly outline it first and then proceed to the details.

First, the air rises, the temperature falls, and the air loses its capacity to hold vapor. Tiny cloud *droplets* and ice crystals quickly condense around some of the dust particles in the air, growing molecule by molecule, but almost never attaining *raindrop* size.

Once the *droplets* or crystals grow large enough (this happens more easily for the ice crystals), they begin to fall appreciably. They fall on smaller *droplets* and crystals in their path, often coalescing. In this way, they quickly grow to raindrop or snowflake size and fall to earth.

You may have noticed that I have tried to distinguish between cloud *droplets* and *raindrops*. Size is the only difference. Cloud droplets typically range between 10 and 100 microns in diameter, whereas raindrops typically range from 500 to 5000 microns in diameter. Figure 12.4 should give you some idea of the sizes involved.

# NUCLEATION (BIRTH)

If the air were pure and clean with no dust at all, there would be much less rain on earth. In dirty air, condensation can begin when the RH is as low as 65%. An extremely thin layer of water condenses (or deliquesces) on the more soluble particles in the air such as salt. This tends to make humid air hazy.

Once the RH exceeds 100%, these water-coated particles begin to grow in earnest and rapidly reach cloud-droplet size. If there were no impurities or dust particles in the air, condensation would not occur until the RH exceeded 400%!

It is surface tension that makes it so difficult to start growing (nucleating) pure water droplets. Like condensation, **nucleation** is a statistical process. Molecules must collide and then stick together. Surface tension makes it very difficult for molecules to stick together when the droplets are very small. Let's see why.

Imagine for a moment that you have shrunk to the size of an insect. Now the force of surface tension will be much more important in your life and you would feel its effects as strongly as insects do. Most insects can lift several times their own weight but, if a winged insect happens to get stuck in the water, it is virtually paralyzed. Other insects, such as the water striders, have pontoonlike legs that do not float but enable them to literally walk on water (see Figure 12.5).

**Surface tension** acts to keep the surface area of a liquid to a minimum. When the water strider walks on water, its weight depresses the water surface, thus increasing the surface area of the water. Surface tension causes the water

Mean annual
rainfall
and other forms of precipitation

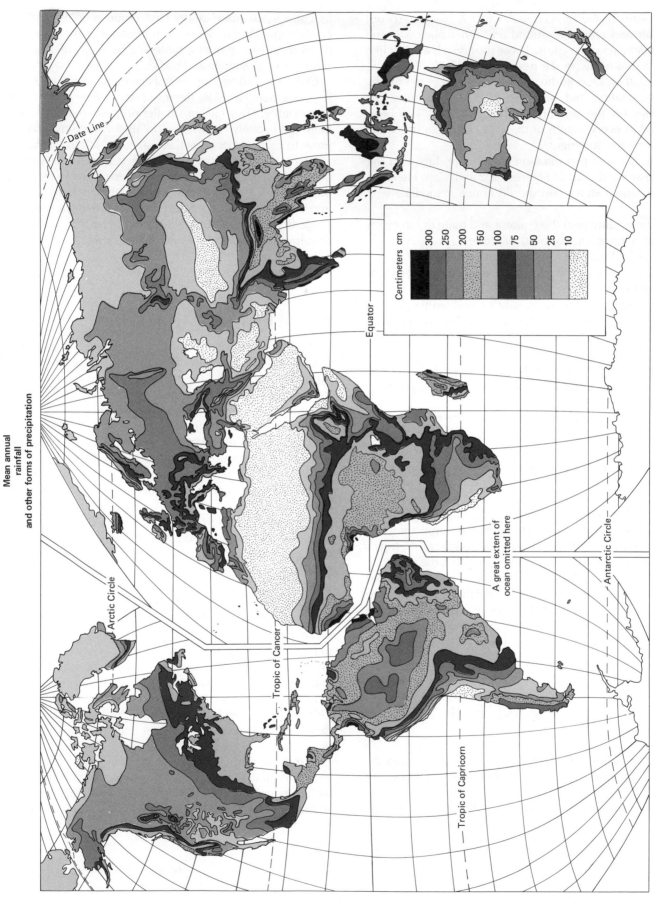

FIGURE 12.3
Annual world precipitation.

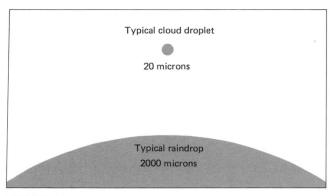

FIGURE 12.4
Comparative sizes of cloud droplets and raindrops.

FIGURE 12.5
The water strider. Notice how the water surface is depressed under the strider's legs.

surface to act like a rubber band and oppose the stretching. This keeps the bug from sinking into the water.

Now, when a water molecule enters a droplet, it increases the droplet's surface area. Since surface tension opposes an increase in area, it makes it very difficult for molecules to enter a droplet. This increase of surface area is far greater for a small droplet than for a large droplet, as Figure 12.6 illustrates. Thus the smaller the droplet, the more difficult it is for a water molecule to enter it. On the other hand, molecules can easily *leave* a small droplet (i.e., evaporate) since this decreases its area. Surface tension therefore makes it essentially impossible to start growing droplets of *pure* water.

Raindrops are held together by surface tension. In fact, surface tension tends to make raindrops and cloud droplets spherical. For any given volume, a sphere is the shape having the least surface area. But as the drops grow larger, surface tension becomes relatively less important. Larger drops begin to flatten and distort as they fall (although they never become tear shaped) and inevitably break apart before they reach 7 millimeters (mm) in diameter. There are no larger raindrops.

Salt particles and other hygroscopic particles (which dissolve in water) as well as highly soluble gases, such as $SO_2$, help droplets to overcome surface tension and to grow around a common nucleus; such particles are called **condensation nucleii.**

Concentrations of dust particles in the air (aerosols) are highly variable, but there are typically 10 billion per cubic meter over land and 1 billion per cubic meter over the ocean. Most of these particles are about 0.1 micron in diameter; but some may be as large as 10 microns in diameter.

Clouds that form over continents typically contain 500 million to 1 billion droplets per cubic meter, while clouds over the oceans contain only about 50 to 100 million droplets per cubic meter. Thus only about 5 to 10% of all aerosols aid condensation.

The growth of ice crystals is aided by other particles called **freezing nucleii.** Freezing nucleii have a crystal structure very similar to that of ice, so that they help the water molecule align properly. Clay particles (mainly kaolinite) are the most important naturally occurring freezing nucleii. Several other substances such as silver iodide are more effective, but they almost never occur naturally.

The concentration of ice crystals is far, far smaller than that of droplets and depends on the temperature. Even though large bodies of water freeze at 0°C, small droplets do not freeze until the temperature drops well below 0°C. For instance, drops of 1 mm in diameter generally do not

Number of molecules
Surface area
Area added per additional molecule

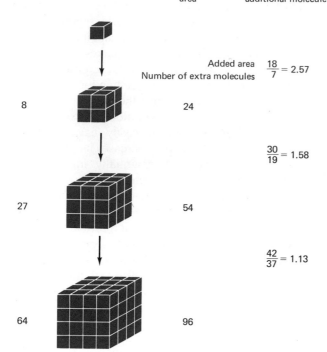

Added area
Number of extra molecules
$\frac{18}{7} = 2.57$

8                    24

$\frac{30}{19} = 1.58$

27                    54

$\frac{42}{37} = 1.13$

64                    96

FIGURE 12.6
The increase of surface area due to the addition of molecules. The larger the droplet (excuse the cubic shape), the smaller the increase of area per additional molecule.

freeze until the temperature has dropped below $-24°C$. Thus, when the temperature is $-10°C$, less than 10 ice crystals form per cubic meter of air, whereas at $-30°C$, 100,000 ice crystals form per cubic meter of air.

What is the explanation for this astonishing fact? Why is it that small droplets of water will not freeze until the temperature has fallen far below $0°C$? In ice all the molecules are lined up, while in water the molecules move around in almost random manner. When the temperature drops below $0°C$, the molecules in a water droplet move slowly enough to get frozen together, but ice will not form until the molecules line up in the correct way.

When there is a huge volume of water, it is very likely that a few of the molecules will line up the proper way and that the rest of the water will follow suit and freeze. On the other hand, when there is only a very small amount of water, it is unlikely that any molecules will line up until the temperature drops far below $0°C$. Cloud droplets generally do not freeze until the temperature cools down to about $-40°C$. When water remains unfrozen at temperatures below $0°C$, it is said to be **supercooled water.**

Even with naturally occurring freezing nucleii, it is not likely for ice crystals to grow until the temperature has fallen below about $-10°C$. Below this temperature, many of the naturally occurring freezing nucleii become active and help ice crystal formation.

Silver iodide, which does not occur naturally in the atmosphere, activates at about $-4°C$, while some organic substances activate at $-1°C$. It is this feature that makes them useful in cloud seeding, as you will see.

There is an ironic but useful aspect to this process. Aerosols are necessary to the birth process of cloud droplets and ice crystals. Many aerosols collide with the droplets and crystals and, if it then rains or snows, the aerosols are removed from the air. Thus rain or snow, which depend on the presence of *some* pollutants, acts to clear the air of pollution. Rough estimates show that a rainfall of 2.5 cm removes about 99% of all aerosols and virtually all of the soluble gases (such as $SO_2$).

# GROWTH BY CONDENSATION (DIFFUSION)

The stage of diffusion or condensation represents the childhood stage in the growth of raindrops or snowflakes. Droplets or crystals grow when the vapor density in the air nearby is greater than the vapor density just outside the surface of the droplet or crystal. Once the droplets or crystals have reached a few microns in diameter, the effects of the nucleii and of surface tension on their growth are negligible. The vapor density at the surface of the droplet or crystal is the saturated vapor density for water or ice, respectively.

What this means is that the droplet cannot grow until the air is slightly supersaturated with vapor. Therefore, in most clouds, the RH is actually over 100%. Typical values of RH in clouds are between 100.1% in slowly rising air to 100.5% in the rapidly rising air of thunderstorms. During this stage, the growing droplets warm to about 0.1°C above the surrounding air temperature because of heating due to condensation.

At first, the droplets grow rapidly by diffusion (i.e., condensation). However, since diffusion is a molecular process, it becomes less important as the droplets grow larger. Furthermore, there are a large number of droplets, and each competes for the available water supply. In a cloud that forms over land, it may take only one second for a typical droplet to grow by diffusion from size zero to one micron in diameter, but it will take 1000 seconds to reach 10 microns in diameter and 1 million seconds (about 2 weeks) to reach 100 microns in diameter. In clouds over the oceans, since there are only one-tenth as many droplets, each droplet grows 10 times as fast by diffusion, but that is still somewhat slow.

In short, diffusion alone cannot produce a rainstorm, because it produces too many small droplets but no large drops. This is why coalescence is necessary. The role of condensation is to allow the droplets or ice crystals to grow large enough so that the cloud can reach the coalescence stage. Once coalescence begins, precipitation invariably results (unless the air beneath the cloud is so dry that the rain evaporates before reaching the ground).

As you will see, droplets must grow to about 40 microns in diameter before coalescence becomes effective. In many clouds, droplets do not grow larger than about 20 microns in diameter by diffusion. Thus there is often a gap between the size that droplets reach by condensation and the size they must reach to initiate coalescence. No rain will fall unless this gap is bridged.

Because oceanic clouds have fewer droplets than clouds over land, each droplet gets a larger share of the available water supply. In many oceanic clouds, enough droplets *do* reach 40 microns in diameter, allowing coalescence to begin.

With almost all clouds over land (and many over the ocean) an extra process is needed to bridge the gap. This process is known as the ice crystal process. The idea was first suggested to be important in producing precipitation by Alfred Wegener (1880–1930) in 1911 and was more fully developed in 1933 by Tor Bergeron (1891–1977).

Bergeron had read Wegener's theory in 1917, but did not realize its significance for several years. Then in 1922, while vacationing in a forest near Oslo, Norway, Bergeron noticed that the fog never penetrated to the forest floor when the temperature was below −5°C whereas, if the temperature was above 0°C, fog always would reach the forest floor. It became apparent to Bergeron that when it was cold, the presence of ice on the trees caused the fog droplets to evaporate. However, even after this observation, it still took Bergeron over a decade to develop the idea fully. In this regard you should recall that when the temperature is below 0°C, ice crystals grow at the expense of droplets.

In a few rare cases, ice crystals can form naturally in clouds whose coldest parts are as warm as −5°C. But this does not happen in most clouds until the temperature drops below about −10°C; whereupon a small but sufficient number of freezing nucleii are activated, and a small number of ice crystals appear in the midst of the supercooled drops. The few ice crystals grow rapidly by diffusion at the expense of the droplets. In the forming of precipitation, the important point is that the ice crystals grow much larger by diffusion than the cloud droplets can. As a result, the ice crystals often grow large enough to initiate coalescence. It may seem strange but, without the subtle difference between the saturated vapor pressures of water and ice, rain or snow would be less common.

# CLOUD SEEDING

It is also this difference that people capitalize on when they "seed" clouds to produce rain. Very often, in the dry regions of the earth, there is no lack of clouds. The problem is to make the clouds produce rain. There are two reasons that clouds don't produce rain: (1) the droplets don't grow large enough to initiate coalescence and (2) the temperature is not low enough (below −10°C) to produce ice crystals naturally. However, when silver iodide crystals are dropped into a cloud, ice crystals form around them or they collide with and freeze some of the cloud droplets. These ice particles then grow large enough to initiate coalescence, even when the temperature is as high as −4°C. When this is done in a super-cooled fog, rain or snow is produced and the fog is often cleared.

All it takes is a few grams of silver iodide crystals to turn an ordinary cloud into a rain producer. This is why cloud seeding is called a trigger mechanism. The most favorable type of clouds for seeding are relatively large, growing continental cumulus clouds whose tops (coldest parts) have a temperature between −4 and −15°C. Careful studies have shown that when such clouds are seeded properly, precipitation is increased by 10 to 20% on the average. The potential economic benefits of cloud seeding to agriculture are enormous and are just beginning to be realized.

Recently, more attention is being focused on certain organic substances such as bacteria that have the potential to produce ice crystals at temperatures as high as −1°C and appear to be the "villains" causing frost damage in many crops.

Cloud seeding also occurs naturally and is thought to be important in increasing precipitation rates. Ice crystals from cirrus clouds can fall into the stratus clouds below, thus seeding them naturally (see Figure 12.7).

On occasion, other substances have been used to seed clouds and fog. Dry ice (frozen $CO_2$) crystals are so cold that when they fall through a cloud, they cool the nearby air to well below −10°C; this then activates the natural freezing nucleii. When the cloud temperature is above 0°C, salt particles have been dropped into clouds and fogs to encourage the growth of a few large drops, which then can fall to the ground.

Cloud seeding is not always successful. If too many ice crystals are produced as a result of the seeding, none will be able to grow large enough to start coalescence. Thus there is always the distinct possibility that cloud seeding, if overdone, can actually stop a cloud from producing precipitation. This form of overseeding may be one consequence of excessive air pollution. Overseeding has also been suggested as a means of preventing the growth of large, destructive hailstones.

Finally, there is the possibility that cloud seeding may prove to be too power a trigger. A cloud seeding experiment was performed the day before the Black Hills flood of June 1972. While it is likely that the flood rains would have occurred anyway, (weather on that day was almost identical to the weather that produced the Big Thompson flood, where no cloud seeding was done) no one can

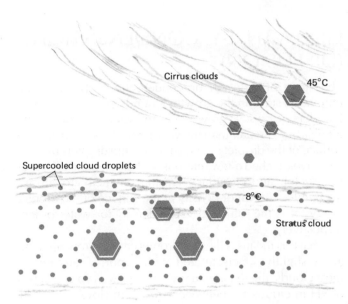

FIGURE 12.7
Ice crystals from cirrus clouds can fall into stratus clouds below, seeding them naturally. Crystals grow rapidly within the stratus cloud.

be absolutely certain. Unfortunately, it is possible to conclude that this degree of uncertainty has allowed lawyers to reap inordinate profits from cloud seeding.

## COALESCENCE (MATURITY)

Once the droplets or crystals have grown large enough to begin falling appreciably, they will collide with other small droplets and crystals in their path. It is almost as if the air were filled with ping pong balls that were coated with glue, and you jumped from the window of a skyscraper. As you fell, you would collide with the ping pong balls which would stick to you because of the glue. By the time you reached the ground, you would be coated with a thick (but I hope protective) layer of ping pong balls. This, of course, is not an experiment worth trying, but it is illustrative of the coalescence process (see Figure 12.8).

Raindrops or ice crystals often stick after colliding and, thus, grow in finite steps. The larger drops then fall faster, since they can overcome the air resistance more easily. By falling faster, they are able to catch up even more rapidly with other droplets and crystals—so that the larger they grow, the faster they grow! With condensation, the opposite is true; the larger the crystals or droplets grow by condensation, the more slowly their diameters expand.

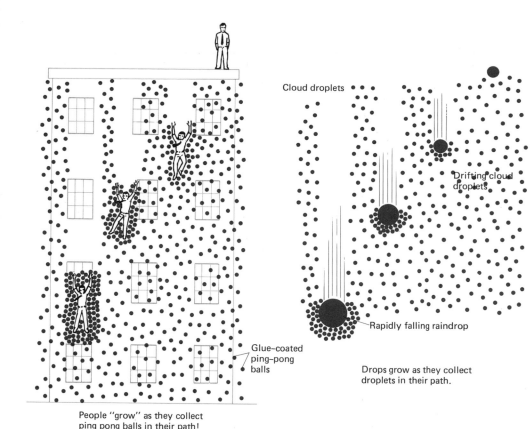

Cloud droplets

Drifting cloud droplets

Rapidly falling raindrop

Drops grow as they collect droplets in their path.

Glue-coated ping-pong balls

People "grow" as they collect ping pong balls in their path!

**FIGURE 12.8**
The growth of raindrops by coalescence occurs when drops fall and combine with cloud droplets in their paths. Here coalescence is illustrated by falling people and falling raindrops. The silly analogy which illustrates this is self-explanatory.

Condensation growth, therefore, dominates with small droplets and crystals, while coalescence (accretion) growth dominates with large droplets and crystals. Typical growth rates are shown in Table 12.6.

If the falling droplets and crystals collected all the smaller particles in their path, there would be no gap between diffusion growth and coalescence growth. The ice crystal process would still be helpful, but not as important as it is. The fact is that falling droplets and crystals do not collect all the particles in their path. Colliding droplets may bounce away from each other because of surface tension when they are small. Colliding ice crystals tend to bounce off each other when they are plane in shape and dry and cold. Another reason why the falling droplets or crystals fail to collect everything in their path is that, as they fall, they create miniature air currents that blow the small particles out of their path. This effect is similar to what happens when you try to swat a fly. As your hand approaches the fly, the pest takes advantage of the air currents and escapes on a "cushion" of air. As a result of this effect, flyswatters are built with small holes through them; this enables the air to flow through the swatter. Its effect is less beneficial to the fly.

Since raindrops do not have holes through them, the "flyswatter" effect does not work (see Figure 12.9). Therefore, the smaller the droplets, the greater the chance that the droplet will not even collide.

Because of these two effects, the **collection efficiency** is very small when the larger of the two particles is less than 40 microns in diameter. For instance, when the collecting droplet (the larger one) is 60 microns in diameter its collection efficiency is about 50%, but it is only about 10% when at 40 microns, and less than 5% at 30 microns! The collection efficiency is 100% when the drop collects all the droplets in its direct path.

When the collection efficiency and the fall rate of the droplets and crystals are both taken into account, a clear picture emerges. Coalescence is virtually nonexistent until the droplets or crystals exceed 40 microns in diameter. Once the collecting particles are larger than this, coalescence proceeds so rapidly that raindrops would often be far larger if they didn't break apart once they exceeded a few millimeters in diameter.

Snowflakes and hailstones have a smaller tendency to break apart, so that they occasionally grow larger. In fact, when you see huge snowflakes it generally means that the

## TABLE 12.6
### Gross Rates and Related Properties of Drops and Crystals

| Initial Diameter (Microns) | Terminal Fall; Speeds in Still Air (Meters per Second)[a] | Distance Drop Falls Before Completely Evaporates (Meters)[b] | Percentage Increase of Diameter in 10 Minutes[c] | | | Assumed Collection Efficiency (Percent) |
|---|---|---|---|---|---|---|
| | | | Condensation Growth of Drop (Percent) | Condensation Growth of Ice Sphere (Percent) | Coalescence Growth of Drop (Percent) | |
| 1 | 0.00003 | — | 1,900 | 13,900 | — | — |
| 2 | 0.00012 | — | 910 | 6,900 | — | — |
| 5 | 0.00075 | 0.00008 | 310 | 2,700 | — | — |
| 10 | 0.003 | 0.003 | 125 | 1,320 | — | — |
| 20 | 0.012 | 0.05 | 41 | 615 | 0.3 | 10 |
| 50 | 0.075 | 2 | 11 | 200 | 2.8 | 20 |
| 100 | 0.30 | 30 | 2 | 73 | 10.5 | 30 |
| 200 | 0.80 | 250 | 0.5 | 22 | 18 | 50 |
| 500 | 2 | 4,000 | 0.08 | 4 | 18 | 50 |
| 1,000 | 4 | 30,000 | 0.02 | 1 | 18 | 50 |
| 2,000 | 7 | Large | 0.005 | 0.25 | 15 | 50 |
| 5,000 | 10 | Large | 0.0008 | 0.04 | 9 | 50 |
| 1 cm hail | 9 | Large | — | — | 4 | 50 |
| 2 cm hail | 16 | Large | — | — | 2 | 50 |
| 5 cm hail | 33 | Large | — | — | 0.8 | 50 |
| 10 cm hail | 59 | Large | — | — | 0.35 | 50 |

[a] Terminal speeds depend on air density, so these figures are only approximate.

[b] Assuming $T = 0°C$ and RH = 90%. The distance is inversely proportional to how much the RH is below 100%. For example, if RH = 80%, all distances in this column must be divided by 2. The distances here are given for still air. Updrafts may considerably reduce these distances.

[c] Air is assumed to have $T = -10°C$, RH = 100.25% (with respect to water), and there are 20 million cloud droplets per cubic meter, each of which is 15 microns in diameter (liquid water content = 0.28 g/m³). Condensation growth of *drops* is proportional to how much the RH exceeds 100%, and it also doubles for roughly every 10°C increase of temperature. The coalescence growth is proportional to the liquid water content. Growth by coalescence occurs in discrete jumps; therefore, a small percentage of the drops will grow *much* faster by coalescence than this table indicates. *Example:* A droplet 200 microns in diameter falls 0.8 m/s in still air and would evaporate completely after a fall of 250 m below the cloud (under conditions stated above). After 10 minutes, 200 micron droplets would grow by 0.5% by condensation, but 18% by coalescence, assuming 50% of all droplets in the direct path are collected. A 200 micron spherical ice crystal would grow by 22% by diffusion during the same 10 minutes.

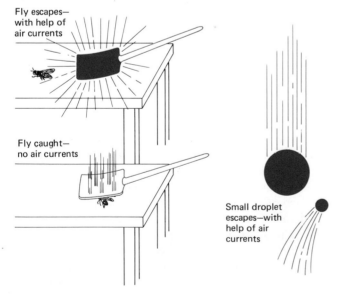

Fly escapes—with help of air currents

Fly caught—no air currents

Small droplet escapes—with help of air currents

FIGURE 12.9
The "flyswatter" effect. Since the droplets do not have the flyswatter effect, they have a low-collection efficiency when small. If flyswatters had no holes, they wouldn't catch small bugs.

snowflakes are wet (this enables them to stick together after collision), which means that the temperature is just about 0°C. Very often when you see such large snowflakes (2 to 5 cm in diameter) the snow soon changes to rain.

When do we get rain and when do we get snow from coalescence? If the ice crystals collide with water droplets and melt on the way down, the result is rain. When it is somewhat colder and the ice crystals collide mainly with other crystals, snow will result. Hail results when supercooled water droplets collide with an ice pellet and they freeze together. Sleet (frozen rain) results when rain passes through a cold layer of air and freezes.

# GROWTH OF HAILSTONES

The largest single hailstone ever accurately measured fell at Coffeyville, Kansas. It weighed 0.76 kg and was 16 cm in diameter. Hail is the "ultimate" product of coalescence. It accounts for enormous damage to crops every year and, a few hours before I wrote this paragraph, hail caused a Southern Airways jet to crash, killing more than 70 people. Hail forms only in the most violent weather—the severe thunderstorm. I will postpone the description of violent thunderstorms until Chapter 19, but will now describe the process of the growth of the hailstones.

Almost all thunderstorms have three basic characteristics:

1. The lower part of the cloud has warm air.
2. The upper part of the cloud has cold air.
3. Strong updrafts prevail through large sections of the cloud.

These characteristics are shown in Figure 12.10. Each of these characteristics plays a vital role in hail formation. The lower part of the cloud supplies the liquid water. The upper part of the cloud supplies the cooling power and freezes the water. The strong updrafts lift the droplets so that they can collide with the growing hailstones.

The water droplets get carried into the upper parts of the cloud by the strong updrafts which may reach 40 m/s. Since the air cools as it rises, the air in the upper parts of the cloud is so cool that droplets become supercooled. When the droplets collide with the ice crystals, they adhere and freeze. This forms ice pellets that then collide with other supercooled drops brought up from below. Hailstones result from this process which is called **riming.**

Riming or **icing** also occurs when airplanes fly through clouds of supercooled droplets and ice coats the plane. In the early days of aviation, before heaters were installed on the wings, this caused many crashes and almost claimed the life of Charles Lindbergh.

The ultimate size of the hailstones depends on several factors (see Figure 12.10). First, the updraft must be very strong so that large stones can be held up in the cold upper regions of the cloud. This is why large hail only occurs in the most severe thunderstorms. Second, once the hailstone begins falling it must not fall through too thick a layer of warm, moist air, or else it will melt before reaching the ground. This is one reason that hail tends to fall before the rain begins. It is also one of the reasons that thunderstorms in Florida so seldom produce hail at the ground, while thunderstorms in Colorado and Wyoming so often do. Hail is produced in the Florida clouds, but it almost always melts before it reaches the ground. In Wyoming and Colorado, the layer of warm air is not so thick, and the air around the cloud is generally so dry that the falling hailstones are kept somewhat cool by evaporation! We can remember this simple rule.

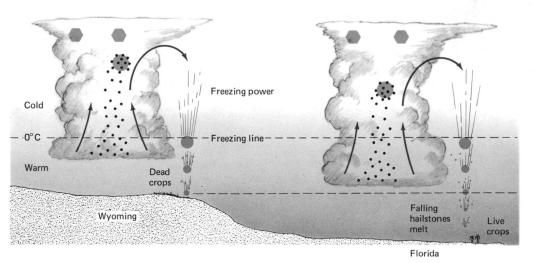

**FIGURE 12.10**
The meteorological conditions that produce hail. Wyoming thunderstorms commonly produce hail that reaches the ground, whereas in Florida thunderstorms the hail usually melts long before reaching the ground.

**Rule:** Significant melting of hailstones occurs only when the height of the wet bulb freezing level (height where $T_w = 0$) is more than 3 km above ground level.

Hail often shows a layered structure (see Figure 12.11). In some layers the ice is clear, in some layers it is opaque, whereas in others it is "spongy" or mixed with water that was trapped as the stone grew. Sometimes these layers alternate, and they can provide hints of the life story of the hailstone. For example, alternating layers often indicate that the hailstone took several round trips through the cloud as it got caught in a number of rising thermals. Thus, by studying hailstones, we can learn something of the structure of giant thunderstorms.

# SLEET AND FREEZING RAIN

**Sleet** is merely frozen raindrops. **Freezing rain** seems like ordinary rain, until it hits any solid object. It then freezes onto the object, which soon is covered by a layer of ice. Needless to say, everything becomes very slippery when freezing rain is falling, and there are many unavoidable traffic accidents. Even though people may walk carefully, they can easily slip and fall. Freezing rain and sleet both form under basically the same conditions, but sleet tends to form when it is slightly colder and therefore it freezes before it hits the ground. Both fall when a layer of air near ground level is below 0°C (so the drops can freeze), but the temperature aloft is above freezing (so rain forms instead of snow).

FIGURE 12.11

A cross section of a hailstone often shows a layered structure. A fly happened to serve as the unwilling nucleus of coalescence for this stone.

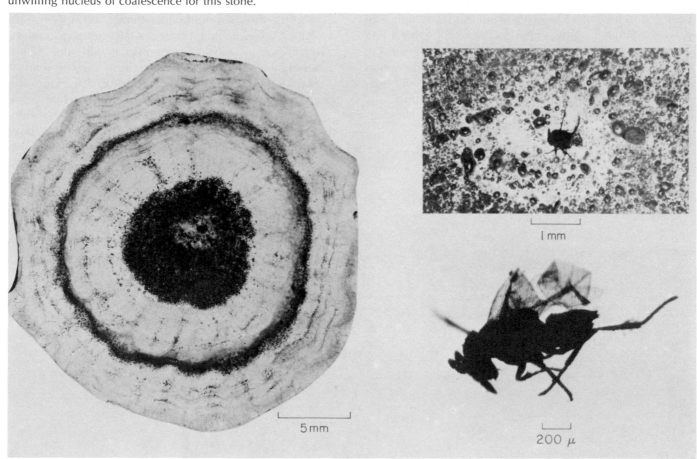

5 mm

1 mm

200 μ

Usually, the air temperature decreases as you ascend, so that the conditions accompanying sleet and freezing rain are somewhat unusual (see Figure 12.12). Under what conditions is the air aloft warmer than the air near the ground level? This is the case on clear nights, but precipitation does not fall on clear nights. It is also the case when a frontal surface is present aloft. In this case, the rising air above the frontal surface may be warm enough to produce rain, while in the cold air layer near the ground people are expecting snow. What they get, of course, is either freezing rain or sleet. When the temperature at ground level is below −3°C, sleet is more common and, when it is between −3 and 0°C, freezing rain is more common. Oddly enough, when the temperature is above 0°C, sleet may fall but freezing rain cannot since — well, you can reason this out for yourselves.

Because freezing rain and sleet occur only under such limited conditions, they are very hard to predict and seldom last long. On rare occasions, thick layers of ice have accumulated on power lines and trees and caused considerable damage (see Figure 12.13). After the weather clears, some of the most beautiful scenery imaginable can result when the trees are all covered with a glistening layer of ice. After one notable ice storm in the New York City Metropolitan area in January 1971, the wind blowing through the ice-coated trees made the forest sound like an orchestra of tinkling bells.

FIGURE 12.12
Meteorological conditions leading to freezing rain and sleet.

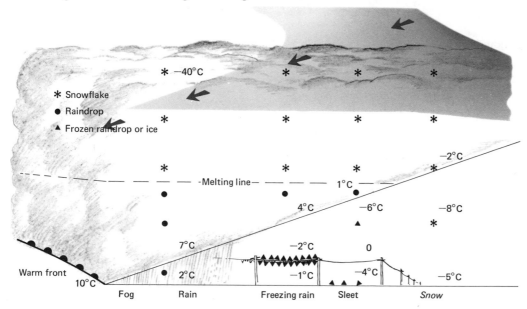

FIGURE 12.13
Freezing rain can thickly coat trees and power lines with a thick layer of glaze. This was
produced by the famous New England ice storm of November 26–29, 1921.

# PROBLEMS

**12-1** Look at Figures 12.2*a* and 12.2*b* and examine the southern part of South America. Explain why the rainfall pattern in this region implies that there are strong or persistent westerly winds south of about 30° S latitude.

**12-2** Verkhoyansk, Siberia, gets only 13.5 cm of rain a year. Why isn't it a desert?

**12-3** When does northern Australia get its rainy season? When does southwestern Australia get its?

**12-4** Name all the differences between cloud droplets and raindrops.

**12-5** Why is air near the seashore typically hazy?

**12-6** Why does air tend to become hazy when the RH rises above about 75%?

**12-7** Why will small droplets of pure water evaporate when the RH is 100%?

**12-8** Tear-shaped drops do not occur in nature, artists' preferences notwithstanding. Explain why.

**12-9** You can demonstrate the effects of surface tension with a child's bubble blower. When you blow gently the soap film surface will bend and hence expand, but as soon as you stop blowing it will straighten out again. If you have already made a bubble, hit it gently or blow on it and watch it oscillate. When the oscillations die out it will be spherical.

**12-10** Why can't raindrops grow larger than about 7 millimeters in diameter?

**12-11** How is rain responsible for removing pollutants from the air?

**12-12** Why must the RH in clouds actually be higher than 100% for rain to form?

**12-13** Explain how it can be raining when the RH at ground level is 75% or less?

**12-14** Why is the rain formation process inefficient if there are too few or too many condensation nuclei?

**12-15** Why are tropical clouds over the ocean more efficient at producing precipitation than typical clouds over land?

**12-16** Explain the rationale of cloud seeding with silver iodide.

**12-17** Can you see a possible relationship between over-seeding a large cumulonimbus cloud and hail prevention?

**12-18** Occasionally an area of cirrus clouds will form in the middle of a deck of altocumulus clouds, producing a clear ring of air in the process. Can you offer an explanation?

**12-19** Explain the analogy between the growth of raindrops and the growth of planets.

**12-20** At approximately what size does the growth of ice crystals by coalescence become dominant over condensation growth? (Use Table 12.6.)

**12-21** Explain why it is easier to dissipate fog when the temperature is below 0°C.

**12-22** Explain why drops must be larger than about 500 microns in diameter to fall out of a cumulus cloud with updrafts of 2 m/s.

**12-23** Explain why the rain falling from clouds in the desert often evaporates before it reaches the ground. (Refer to Table 12.6.)

**12-24** Explain why it would be very painful if a hailstone 10 cm in diameter hit you on the head. (Refer to Table 12.6.)

**12-25** Explain how the icing of a plane's wings is proof of the existence of supercooled water.

**12-26** Explain why icing can prove disastrous to ships at sea.

**12-27** What are the meteorological prerequisites for the formation of large hail?

**12-28** Why is sleet more likely than freezing rain when the temperature is above freezing?

**12-29** Describe the meteorological conditions that lead to the formation of sleet or freezing rain rather than snow.

13

# CHANGES IN RISING AND SINKING AIR: STATIC STABILITY

Thunderstorms and air pollution occur under opposite weather conditions. Thunderstorms occur when the atmosphere is **statically unstable.** (The term stability always implies static stability unless otherwise specified.) When the atmosphere is unstable the slightest disturbance is sufficient to cause an entire layer of air to literally turn upside down. Thunderstorms are then found in the areas in which air is rising, whereas clear weather is found in between, where the air is sinking.

On the other hand, atmospheric pollution occurs when the air layer near the ground is stable. This means that the air cannot be overturned or stirred at all. Under stable conditions whatever pollution that is put into the air is not able to disperse and therefore accumulates in the air near the ground.

Atmospheric stability is determined in a somewhat simple way. You already know that air rises when it is warm and light. To test for stability or instability, a parcel of air is lifted from its original position to new surroundings. If the lifted parcel is warmer than its new surroundings, it will continue to rise because of its lightness. This is an *unstable* situation. If, on the other hand, the lifted parcel is cooler than its new surroundings, then it will sink back to its original position. This atmosphere is stable.

Unfortunately, there is one complication. As soon as the air parcel moves up or down its temperature changes. Then, in order to determine whether the air is stable or unstable, it becomes necessary to compare the changing temperature of the parcel with the temperature of the new surroundings—so you must know *precisely* how temperature changes as air rises or sinks.

In this chapter we will begin by looking at the precise temperature and humidity changes in ascending and descending air. A chart that is used by meteorologists and that simplifies the otherwise lengthy calculations in determining temperature and humidity changes is presented. This chart, known as a **thermodynamic diagram,** is described and then used to answer several interesting quantitative questions, for example, how much rain can fall in a given storm, or how high must air be lifted to produce a cloud. This chart is also used to explain the unusual temperature and humidity conditions that sometimes accompany upslope and downslope winds such as the chinook. Finally, the conditions of atmospheric stability and instability are explained and, in using this thermodynamic diagram, we will see precisely how stable or unstable a given situation is.

## TEMPERATURE AND HUMIDITY CHANGES IN VERTICALLY MOVING AIR

You already know that as air rises its temperature decreases. The strange thing about this temperature decrease is that absolutely no heat has been removed from the air! The process is therefore called **adiabatic** (Latin for no change of heat), so long as the air remains unsaturated and no condensation occurs. In meteorology we say that rising air *cools adiabatically* and sinking air *warms adiabatically,* even though no heat change is involved.

The proper name for the adiabatic "cooling" rate for rising unsaturated air is the **dry adiabatic lapse rate.** The dry adiabatic lapse rate is almost exactly 10°C per kilometer that the air rises.

$$\text{Dry adiabatic lapse rate} = \frac{10°C \text{ (cooling)}}{\text{kilometer (air rises)}} = 10°C/km$$

(Actually the dry adiabatic lapse rate is closer to 9.8, but 10 is easier to remember and accurate enough for most purposes.)

This leads to several strange results. Imagine two parcels of air that remain unsaturated. Parcel 1 is located 10 km above sea level with a temperature of −50°C, while parcel 2 is located at sea level and has a temperature of 30°C. Which of these two parcels is warmer? At first, this may sound like a stupid question because parcel 2 is clearly much warmer.

However, if parcel 2 were lifted 10 km, it would cool adiabatically by 100°C and arrive at the new height with a temperature of −70°C. This is cooler than parcel 1! Conversely, if we had moved parcel 2 down to sea level, it would have warmed adiabatically by 100°C and reached a temperature of 50°C—again warmer than parcel 1. Thus, *when measured under equal conditions,* parcel 1 seems to be clearly warmer than parcel 2 (see Figure 13.1).

This strange behavior has led meteorologists to define a quantity called the **potential temperature.** In order to de-termine which air parcels are *potentially* warmer, all air parcels must be measured under identical conditions. This means that all air parcels must be measured under identical pressure conditions, since it is pressure changes that cause adiabatic temperature changes.

**Definition:** The potential temperature of an air parcel is the temperature that the air parcel *would* have if it were moved adiabatically to a pressure of 1000 mb.

Air in the upper levels of the troposphere (the atmosphere's lowest layer) has very low temperatures. Nevertheless, this air is potentially warm. In equatorial regions the air near the ground is warm and rises rather easily. This air rises so high and cools so much that the lowest temperatures of the troposphere are actually found near the equator about 17 km above sea level. The temperature is often below −80°C; yet, this air is potentially quite warm. If this air were brought down to sea level, its temperature would be about 90°C. Thus it is the adiabatic temperature changes that help to resolve an apparent paradox. Even though warm air rises in the troposphere, the temperature at the top of the troposphere must be colder than at the ground.

The next complication involved in rising and sinking air concerns changes in the dew point temperature. When unsaturated air rises its mixing ratio doesn't change, because the air contains the same proportion of vapor. However, the dew point temperature decreases at a rate of almost 2°C per kilometer (actually closer to 1.7°C per kilometer). The rising air expands, thus decreasing the vapor pressure. Therefore, the air must be cooled even

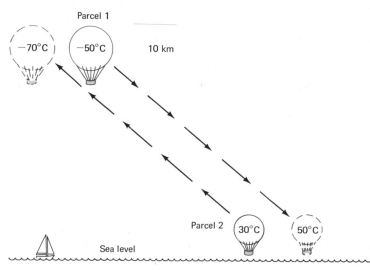

FIGURE 13.1
Which air parcel is warmer? Although the air at sea level seems much warmer, it is actually colder than the air at 10 km, *when measured under equal conditions.* Thus, the air at 10 km is *potentially* warmer.

CHANGES IN RISING AND SINKING AIR: STATIC STABILITY

further than before in order to reach the point of saturation (in other words the dew point).

In short, the

$$\text{Dew point lapse rate (for unsaturated air)} = \frac{2°C \text{ (cooling)}}{\text{kilometer (air rises)}} = 2°C/km$$

Now we have all the necessary information to tell how high air must rise for clouds to form. For unsaturated rising air, the temperature falls about 8°C per kilometer *faster* than the dew point falls. Once the temperature falls to the dew point, saturation occurs and a cloud begins to form. This means that we can make the following rule.

**Rule:** When clouds form from air that rises vertically (not at a slope) from the ground, the height of the cloud base or bottom is given by the formula

$$\text{Height of cloud base (in kilometers)} = \frac{T - T_d}{8}$$

where $T$ and $T_d$ are the temperature and dew point at the ground level.

---

Example 13.1
Cumulus clouds (which form from air rising directly from ground level in most cases) form when the temperature and dew point at ground level are 40°C and 24°C, respectively. How high is the cloud base and what is the temperature and dew point at cloud base?

Equation:

$$\text{Height of cloud base} = \frac{T - T_d}{8}$$

Substitute:

$$\text{Height of cloud base} = \frac{40 - 24}{8}$$

Solution
    2 km
Since the temperature drops 10°C per kilometer, the temperature at cloud base is 20°C cooler than at the ground or 20°C. Since the temperature and dew point are equal at this height, this is the level of saturation. Any further lifting results in condensation and cloud formation.

---

The final complication involved in rising air (but almost never sinking air) occurs once the air has become saturated. When saturated air rises it "tries" to cool by 10°C per kilometer, but the temperature drop causes some vapor to condense. Since condensation is a heating process, some heat is added to the air and some of the temperature drop is offset.

As a result of condensational heating the temperature of rising saturated air decreases somewhat more slowly than 10°C per kilometer. Meteorologists have given the names **moist adiabatic, saturated adiabatic,** or **pseudoadiabatic** to the "cooling" process of rising saturated air. Technically, the term adiabatic is a misnomer because the latent heat of condensation is added, and therefore the process is not adiabatic.

When saturated air is warm it contains a large amount of vapor that is rapidly condensed. Rising, warm, saturated air therefore cools more slowly than rising, cold, saturated air. The moist adiabatic lapse rate varies between 2°C per kilometer for extremely warm saturated air to almost 10°C per kilometer for extremely cold saturated air. Some actual values appear in Figure 13.2. If this detailed information is not available, then the best estimate to use for the moist adiabatic lapse rate is 6°C cooling per kilometer. This, of course, is only a crude approximation. Therefore,

$$\text{Moist adiabatic lapse rate} \cong \frac{6°C \text{ (cooling)}}{\text{kilometer (air rises)}} = 6°C/km$$

The moist adiabatic process generally does not apply to sinking air. It would if the water reevaporated when the air sinks, but what almost always happens is that a high percentage of the condensed water falls out of the rising air parcels as precipitation. Therefore, it is no longer present or available to be reevaporated once the air begins to sink. This leads to the conclusion that sinking air rapidly becomes unsaturated as it warms, and thereafter it warms at the dry adiabatic rate. Sinking air within thunderstorms is one important exception to this rule.

Many places in the world that are located at the base of mountain ranges will occasionally experience warm, dry downslope winds. In Germany and Switzerland such a wind is called a *föhn;* in Los Angeles, it is called a *Santa Ana;* and on the eastern slopes of the Rocky Mountains, it is called a *chinook* (from an Indian tribe in the region). When these winds begin to blow, the temperature often rises sharply, while the dew point and especially the relative humidity plummet. These unusual features can be explained by using our knowledge of the moist and dry adiabatic lapse rates.

For our present purposes it is only necessary to state that this warm, dry downslope wind is caused when air is forced to descend the mountain slopes (more will be said in Chapter 20). Often this air has been forced to ascend the other side of the mountain range a short time earlier. This air probably begins its ascent as relatively moist, cool air. Some time during the ascent it probably becomes saturated (especially if the mountain range is high), and con-

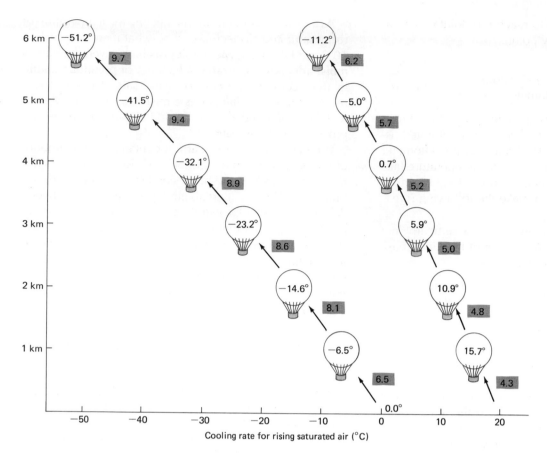

6 km — (-51.2°) ← 9.7
5 km — (-41.5°) ← 9.4
4 km — (-32.1°) ← 8.9
3 km — (-23.2°) ← 8.6
2 km — (-14.6°) ← 8.1
1 km — (-6.5°) ← 6.5

(-11.2°) ← 6.2
(-5.0°) ← 5.7
(0.7°) ← 5.2
(5.9°) ← 5.0
(10.9°) ← 4.8
(15.7°) ← 4.3

0.0°

-50   -40   -30   -20   -10   0   10   20

Cooling rate for rising saturated air (°C)

FIGURE 13.2
The moist adiabatic lapse rate. This is the rate at which rising saturated parcels of air cool when there is no mixing with the environment. The temperatures of the rising parcels are shown within the balloons, while the lapse rates are shown in the boxes. Notice that the moist adiabatic lapse rate is not constant.

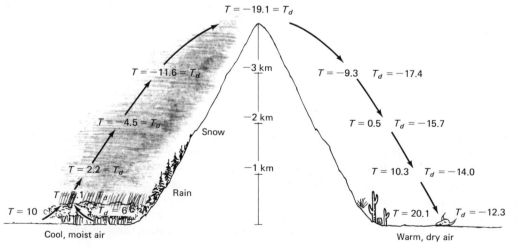

$T = -19.1 = T_d$

$T = -11.6 = T_d$

$T = -9.3 \quad T_d = -17.4$

$T = -4.5 = T_d$

$T = 0.5 \quad T_d = -15.7$

$T = 2.2 = T_d$

Snow

$T = 10.3 \quad T_d = -14.0$

Rain

$T = 6.1 = T_d$

$T = 10 \quad T_d = 6.6$

$T = 20.1 \quad T_d = -12.3$

-3 km
-2 km
-1 km

Cool, moist air

Warm, dry air

FIGURE 13.3
Cool, moist air rises over a 4-km high mountain. It cools at the dry adiabatic rate until saturation occurs; thereafter, it cools at the moist adiabatic rate. Condensation occurs and rain or snow falls. When the air descends the other side it warms at the dry adiabatic rate, reaching sea level considerably warmer and drier than it started out.

CHANGES IN RISING AND SINKING AIR: STATIC STABILITY

densation occurs. Rain or snow then falls on the wind-ward slope of the mountain range (see Figure 13.3). Thus, not only does the condensation heat the air, but it also removes much vapor from it.

Because of the released latent heat of condensation, the air arrives on the other side of the mountain warmer than before. It also arrives drier than before because of the water vapor removed by condensation and, as you can see from Figure 13.3, the final dew point is lower than before. The information for this figure has been taken from a thermodynamic diagram.

# THE THERMODYNAMIC DIAGRAM

The material in this section involves many details, so that if you read it you must read carefully.

The thermodynamic diagram is nothing more than a graph, although I must admit that it looks rather compli-cated at first. A typical thermodynamic diagram is shown in all its glory in Figure 13.4. The typical reaction of students when they see a thermodynamic diagram for the first time is one of utter confusion. Then, once the diagram is explained and the student is forced to use it (since al-most no one does this willingly) the next reaction is one of impending blindness. However, in a short while most people get used to it and become familiar with the lines on it. Then they find that a little insight can replace a lot of eyesight. You'll find that the effort is worth it. The follow-ing paragraphs are in effect an instruction manual for a thermodynamic diagram.

Since the thermodynamic diagram is nothing more than a graph, it is first necessary to define the coordinates of this graph. The vertical lines can stand for temperature, wet bulb or dew point, while the horizontal lines stand for atmospheric pressure.

**Note:** In order for you to differentiate among $T$, $T_d$, and $T_w$, you must be informed or the diagram must be carefully labeled. Frequently the temperature is denoted by a small black circle, and the dew point is denoted by an $x$.

Notice that the graph simulates the atmosphere in two ways: (1) pressure decreases as you go up and (2) pressure decreases more rapidly at first (near the ground) and then more slowly as you go up.

Now let's begin to use the thermodynamic diagram. In Figures 13.5 to 13.11 the heavy arrows either describe physical processes or show how to read the variables. In

Figure 13.5, temperature and dew point temperature are specified at a pressure of 800 mb. You should be able to read that $T = 10°C$ and $T_d = 0°C$.

The next set of lines represents the height. These are the almost horizontal dotted lines that slope slightly down-ward toward the right. The height is read from the temper-ature (and not from the wet bulb or the dew point). These height lines have been computed using two assumptions (1) that the sea-level pressure is 1013 mb and (2) that the temperature of the atmosphere decreases 6.5°C per kilome-ter (this is called the standard atmosphere). Thus, they only represent approximations. Under these conditions the air at 800 mb is found at a height of just about 1900 m or 1.9 km (see Figure 13.6).

Now, lift a parcel of this air from 1.9 to 2.9 km. On the chart, this occurs at 715 mb. During this ascent the air remains unsaturated so that it cools from 10 to 0°C. The solid, sloping black line is drawn to depict how the tem-perature of unsaturated air decreases with height, this line is one of the **dry adiabats** (see Figure 13.7). Thus, the dry adiabats (the solid sloping black lines) depict the tempera-ture changes of unsaturated air due to ascent or descent. The dry adiabats are also lines of constant potential tem-perature and are given in degrees Kelvin or absolute.

The next set of lines, the almost vertical, solid lines, depict how the dew point temperature changes when the air is unsaturated. The air parcel that was lifted from 1.9 to 2.9 km also experiences a drop in the dew point. The dew point falls from 0 to −1.7°C. This is depicted by the solid red lines (see Figure 13.8). One additional, important property of these solid red lines is that they tell the mixing ratio of the air when the dew point is known. They also tell the *saturated* mixing ratio when the temperature is known. From the chart we see that the air at 800 mb with a dew point of 0°C has a mixing ratio of 4.8‰, while the saturated mixing ratio for $T = 10°C$ is 9.6‰.

The final set of lines is the moist adiabatic lapse rate. These dashed red lines are called the moist adiabats, and they show how $T$, $T_w$, and $T_d$ decrease when saturated air rises (see Figure 13.9). Note that for *rising saturated air* all three decrease at the same rate.

Now if we continue lifting the air from 800 mb, the temperature eventually reaches the dew point. This oc-curs when both are about −2°C, and it occurs at a height of approximately 3200 m. This is the level of condensa-tion. If this air is then lifted another 1000 m, its tempera-ture, wet bulb, and dew point all fall by 6.5 to −8.5°C. (*The* **moist adiabats** *always depict how the wet bulb tem-perature changes for ascending or descending air.*)

The number on each moist adiabat gives its value of a quantity known as the **equivalent potential temperature** in degrees Kelvin. Just as the *dry* adiabats are lines of con-

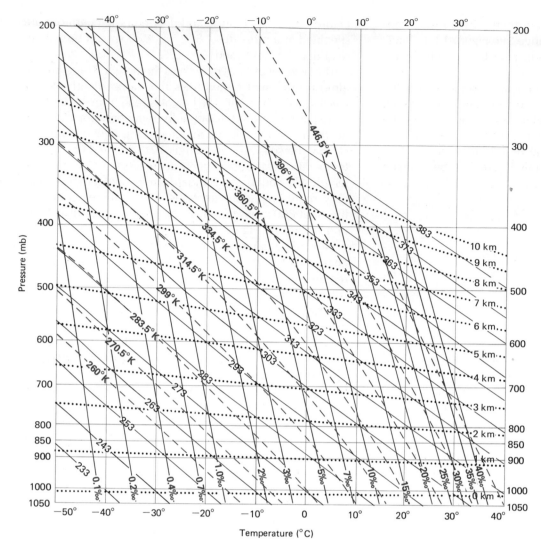

FIGURE 13.4
The thermodynamic diagram (emagram), with key to the different lines.

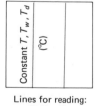

Lines for reading:
$T$, $T_w$, $T_d$

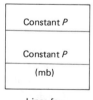

Lines for
reading pressure

Lines for
reading height

Dry Adiabats

Moist Adiabats

Mixing ratio
lines

CHANGES IN RISING AND SINKING AIR: STATIC STABILITY

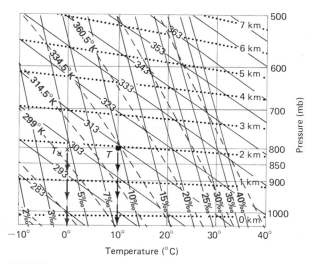

FIGURE 13.5
How to read temperature, dew point temperature, and pressure on the thermodynamic diagram. If the wet bulb temperature were plotted, you would read it the same way you read the temperature or dew point temperature.

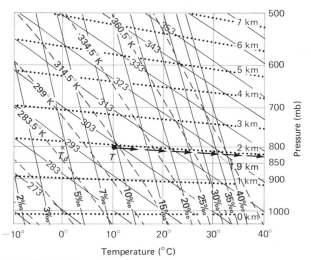

FIGURE 13.6
How to read height (dotted black lines) on the thermodynamic diagram. Height must be read from the temperature. Height is determined on this chart by using several assumptions; therefore, it will be slightly inaccurate for most cases.

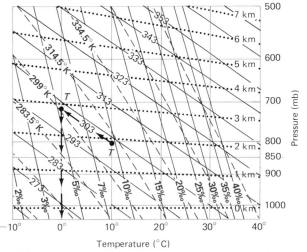

FIGURE 13.7
The dry adiabats (solid black lines) show how the temperature of rising or sinking unsaturated air parcels changes. The dry adiabats are lines of constant potential temperature.

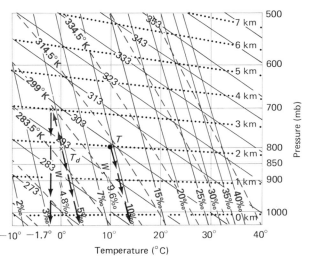

FIGURE 13.8
Lines of constant mixing ratio (solid red lines) show how the dew point temperature of rising or sinking unsaturated air parcels changes. Read the actual mixing ratio, $W$, from the dew point temperature and the saturated mixing ratio, $W_s$, from the temperature.

stant potential temperature, the *moist* adiabats are lines of constant equivalent potential temperature.

The equivalent potential temperature is the temperature the air has after all the water vapor is condensed out (this adds latent heat) and after the air is taken dry adiabatically down to 1000 mb. Because of the latent heat, the equiva-

lent potential temperature is always higher than the potential temperature (provided there is any vapor in the air).

You always determine the equivalent potential temperature from the particular moist adiabat that the wet bulb temperature lies on.

THE THERMODYNAMIC DIAGRAM 201

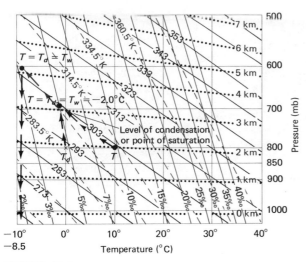

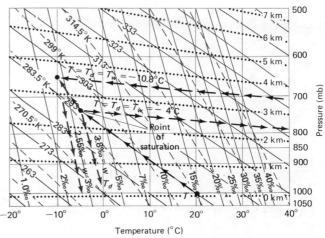

**FIGURE 13.9**
The moist adiabats (dashed red lines) show how the temperature, wet bulb, and dew point all decreases for rising, saturated air parcels. The moist adiabats are lines of constant equivalent potential temperature. Air rises along the moist adiabats as soon as the temperature and dew point meet. This point is called the level of condensation or the point of saturation.

**FIGURE 13.10**
An air parcel with $T = 20°C$ and $T_d = 0°C$ is lifted from 1000 mb. It cools at the dry adiabatic rate until the point of saturation at 2.5 km. After that it cools more slowly at the moist adiabatic rate. Notice that once the air rises above the point of saturation, its mixing ratio decreases. The decrease of mixing ratio tells you how much vapor condensed.

---

**Example 13.2**
Air at 1000 mb has $T = 20°C$ and $T_d = 0°C$. Lift this air until it becomes saturated.

  **A.** At what height does saturation occur?
  **B.** At what pressure does saturation occur?
  **C.** What is the temperature and dew point at this height?
  **D.** What is the mixing ratio at this height?

Procedure: Use the thermodynamic diagram (see Figure 13.10).

Solution
  **A.** 2.5 km  **B.** 740 mb  **C.** −4°C  **d.** 3.8‰

**Example 13.3**
Now that this air has become saturated, lift it another 1000 m.

**A.** What is the new tempertaure, dew point, and wet bulb?
**B.** What is the new mixing ratio?

Information:
  Again, use the thermodynamic diagram.

Solution
  **A.** −10.8°C  **B.** 2.55‰

There are a number of other important uses of the thermodynamic diagram. Several of these uses are listed below.

  **1.** If any two of the temperature, wet bulb, or dew point are known, then you can use the thermodynamic diagram to find the other value. When the air is saturated all three ($T$, $T_w$, $T_d$) are equal. When the air is unsaturated temperature goes along the dry adiabats, wet bulb goes along the moist adiabats, and dew point goes along the mixing ratio lines. For any sample of air, all three of these lines must meet at a point.

---

**Example 13.4**
Air at a pressure of 800 mb has $T = 10°C$ and $T_d = 0°C$. Find the wet bulb temperature.

Instructions: Using the thermodynamic diagram, as shown in Figure 13.11, go up the dry adiabat from 10°C and go up the mixing ratio line from 0°C. Then from the point of saturation, come down the moist adiabat to the original level of 800 mb. This is the wet bulb temperature.

Solution
Thus, $T_w = 5.3°C$.

---

  **2.** With the thermodynamic diagram, you can also compute the relative humidity of the air. Since the relative humidity or RH is given by the formula

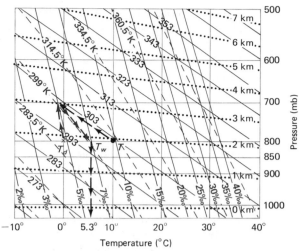

FIGURE 13.11

How to find the wet bulb when you know the temperature and the dew point. First, temperature goes up dry adiabat, and dew point goes up mixing ratio line until they meet. Then, come down the moist adiabat to original level to find wet bulb. You should be able to generalize this technique to find any one of the $T$, $T_d$, or $T_w$—so long as you know the other two. This is a computation procedure and does not represent lifting or sinking of air.

$$RH = \frac{\text{actual mixing ratio}}{\text{saturated mixing ratio}} = \frac{W}{W_s}$$

finding the RH is no problem. From the dew point temperature you can read the value of the *actual* mixing ratio. In the same way, using the temperature, you can read the value of the *saturated* mixing ratio.

---

Example 13.5

Find the RH of the air at 800 mb, with $T = 10°C$ and $T_d = 0°C$.

Information: Using the thermodynamic diagram we find the actual mixing ratio, $W = 4.8‰$, and the saturated mixing ratio, $W_s = 9.6‰$. This has already been done in Figure 13.8.

Equation:
$$RH = \frac{\text{actual mixing ratio}}{\text{saturated mixing ratio}}$$

Solution
$$\frac{4.8}{9.6} = 50\%$$

---

**3.** Another important use of the thermodynamic diagram is that it can provide an estimate of the amount of precipitation that will fall from a cloud. In Example 13.3

the air was lifted from 1000 mb to a height of 3.5 km. At first it had a mixing ratio of 3.8‰, but after condensation had occurred the mixing ratio fell to 2.55‰. Thus, for every kilogram of air, 1.25 grams of vapor condensed.

This means that every kilogram of rising air produced 1.25 grams of water or ice. Much of this may reach the ground as rain (the rest reevaporates). Considering that every thunderstorm involves literally billions of kilograms of rising air, this means that billions of grams of rain are produced.

## ATMOSPHERIC STABILITY

People are called unstable if any little disturbance is sufficient to alter their behavior completely. Similarly, the air is said to be unstable when it accelerates away from its initial position after being displaced vertically. The psychological stability of individuals is measured by comparing them to the society or environment. In meteorology, the stability of a parcel or blob of air (the individual) is measured by comparing it to the surrounding air (the environment). When a parcel of *lifted* air is warmer than its surroundings, it will continue to rise because of its lightness or buoyancy. This represents an unstable situation, because once disturbed the air will never return to its original position.

**Warning**

Before discussing instability, try to answer the following question. Air at sea level has $T = 15°C$ and $T_d = 4°C$. What is the air temperature and dew point at 1 km above sea level?

If you obtained an answer to this question, you are wrong! You were not given sufficient information to answer that question. The actual temperature structure of the atmosphere represents a *situation* and this *situation* constantly changes. On the other hand, you could have answered this question: If a parcel of this air at sea level were *lifted* to 1 km, then what would its temperature and dew point have been? Lifting air represents a *process* and, for the above information, the *process* is always the same.

With these distinctions in mind (individual vs. environment and situation vs. process) you are now ready to understand stability. Imagine the following situation. The temperature is 15°C at ground level and 10°C 1 km above ground level, and there is no vapor in the air to produce condensation. Is this a stable or an unstable situation?

We proceed as follows. Take a parcel of air at ground level and lift it 1 km. During its ascent the parcel remains unsaturated and by 1 km its temperature drops dry adiabatically to 5°C. At this height the parcel is surrounded by air that has $T = 10$°C. Thus the parcel is colder and heavier than the environment. Therefore, the parcel will sink back to ground level—which means that this situation is stable.

Take three other situations. All three will have the same ground conditions as before, but now imagine that the air at 1 km has (A) $T = 12$°C, (B) $T = 8$°C, and (C) $T = 1$°C. Which of these cases represents an unstable situation?

In all three cases the lifted parcel cools dry adiabatically to 5°C at 1 km. Since the parcel temperature is colder than the environment temperatures of cases A and B, these are both stable. In case C the parcel temperature is warmer than the environment temperature, so that the lifted parcel will continue to rise freely. Case C is therefore unstable (see Figure 13.12).

This leads us to a general rule. Notice that in case A, the environment temperature decreases 3°C per kilometer. In other words, the *atmospheric* lapse rate in case A is 3°C per kilometer. Similarly, you can see that in case B the atmospheric lapse rate is 7°C per kilometer and that in case C it is 14°C per kilometer. By generalizing we find the following rule.

**Rule:** *Unsaturated* air is unstable when the atmospheric lapse rate is greater than the *dry* adiabatic lapse rate and stable when the atmospheric lapse rate is less than the *dry* adiabatic lapse rate.

In the same manner we could show that when the air is saturated the stability criterion becomes

**Rule:** *Saturated* air is unstable when the atmospheric lapse rate is greater than the *moist* adiabatic lapse rate and stable when the atmospheric lapse rate is less than the *moist* adiabatic lapse rate.

When these two rules are combined we find (1) that both saturated and unsaturated air are stable if the atmospheric lapse rate is less than the *moist* adiabatic lapse rate (this is called absolutely stable); (2) that both saturated and unsaturated air are unstable when the atmospheric lapse rate is greater than the *dry* adiabatic lapse rate (this is called absolutely unstable); and (3) that if the atmospheric lapse rate lies between the dry and moist adiabatic lapse rates, then the air will be unstable if it is saturated and stable if it is unsaturated (this is called *conditionally unstable*). Now, read this paragraph once again, slowly.

On sunny days the air near the ground is always absolutely unstable. Air at the ground is often 20°C warmer than it is one *meter* up. This is equivalent to a lapse rate of 20,000°C per *kilometer*, and it should be obvious that such large lapse rates cannot be sustained throughout the entire atmosphere. Air with such a large lapse rate is so unstable that it almost instantly mixes up. Therefore, it can only be maintained near where the sun's rays are constantly heating the ground.

FIGURE 13.12
Air in the rising balloon always cools at the same rate (i.e., by the adiabatic *process*). On the other hand, the actual *situations* vary from day to day. A situation is stable if air in the rising balloon is colder than the surrounding air, and unstable if air in the rising balloon is warmer than the surrounding air.

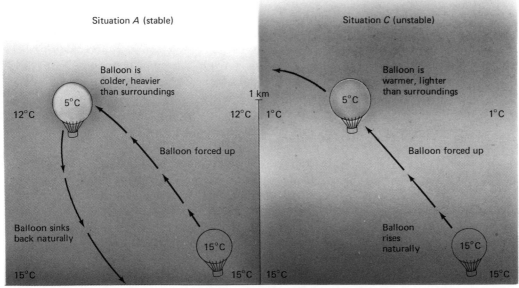

CHANGES IN RISING AND SINKING AIR: STATIC STABILITY

At night the ground temperature may be 5°C colder than the air 1 m up. Under this condition, when the air gets warmer with height, the atmospheric lapse rate is negative. This situation is called an **inversion.** Negative lapse rates are very stable situations. That is why the meteorologist talks about inversions when the air is highly polluted. It is also why farmers must use fans to force the very stable air to stir so that the warmer, light air above can be brought down to plant level.

Absolutely unstable conditions are rare except right near the ground, even in the most violent thunderstorms and tornadoes. Once the air reaches saturation all that is needed for thunderstorms is that the atmospheric lapse rate should be slightly greater than the moist adiabatic lapse rate.

Air that is conditionally unstable must be watched with caution. So long as the air remains unsaturated, it will be stable. However, if this air is forced to rise, it will become unstable as soon as it reaches the saturation point, and then thunderstorms may spring up spontaneously.

## ATMOSPHERIC SOUNDINGS AND STABILITY

Every 12 hours at literally hundreds of weather stations around the world, helium-filled balloons are released into the atmosphere. Each balloon carries a package of instruments, known as a radiosonde, for the purpose of measuring the temperature, pressure, humidity, and winds (see Figure 13.13).

Over the oceans and over other, less civilized places where there are no radiosondes released, such information is obtained remotely from satellites. The soundings obtained by the radiosonde or satellite can be plotted on a thermodynamic diagram. When this is done, the stability characteristics and other meteorological properties of the atmosphere can easily be interpreted.

When the temperature of the sounding is plotted on the thermodynamic diagram, the atmospheric lapse rate is an automatic byproduct. An example of a simple sounding is shown in Figure 13.14. Its stability must be tested layer by layer.

In the lowest layer of this sounding the temperature falls by 11°C in 900 m. The atmospheric lapse rate in this layer is therefore 11/0.90 = 12.2°C per kilometer. Since this is greater than the dry adiabatic lapse rate, the air layer from the ground to 900 m is absolutely unstable.

In the next layer the temperature actually increases 2°C in a kilometer. This constitutes an inversion, and the atmospheric lapse rate in this layer is −2°C per kilometer.

FIGURE 13.13
The radiosonde.

Since this is much less than the moist adiabatic lapse rate, this layer is absolutely stable.

In a similar manner the atmospheric lapse rate in the third layer (from 800 to 700 mb) is found to be 8/1.05 = 7.6°C per kilometer. At this height the moist adiabatic lapse rate is roughly 6.4°C per kilometer. Since the atmospheric lapse rate lies between the moist and dry adiabatic lapse rates, this layer is conditionally unstable.

The top layer has a lapse rate of 4°C/1.2 = 3.3°C per kilometer. Since this is less than the moist adiabatic lapse rate, this layer is absolutely stable.

Now that you have read these few preceding paragraphs you may be pleased to find out that all this work was unnecessary. This is so because it is possible to deter-

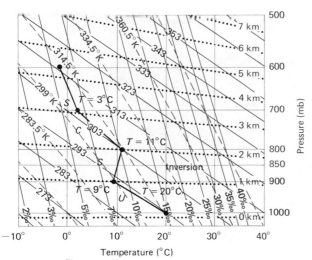

FIGURE 13.14
A sounding composed of four layers. *U* means the lapse rate is absolutely unstable; *C* means conditionally stable (or unstable); and *S* means absolutely stable. Notice how the stability characteristics can be determined from the slope of the sounding line (see Figure 13.15).

mine the stability characteristics of each layer *visually* from the thermodynamic diagram!

For example, when the line depicting the sounding slopes upward more sharply to the left than the dry adiabats, the layer is absolutely unstable. All such lines are marked *U*. When the sounding line slopes less to the left than the moist adiabats, the layer is absolutely stable. All such lines are marked *S* in Figure 13.15. When the sounding line actually slopes up to the right, then this represents an inversion—a very stable situation. Finally, when

FIGURE 13.15
How to determine the stability of soundings from their slopes on the thermodynamic diagram. Any sounding line that tilts upward more to the left than the dry adiabat is absolutely unstable; any sounding line that tilts upward more to the right than the moist adiabat is absolutely stable; while any sounding line with an intermediate slope is conditionally unstable.

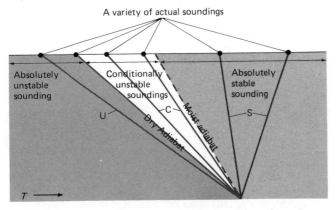

the slope of the sounding line is in between the slopes of the moist and dry adiabats, the sounding in that layer is conditionally unstable, and all such lines in Figure 13.15 are marked *C*.

## SOME ACTUAL SOUNDINGS AND THE ASSOCIATED WEATHER

It always amazes me to find out how much can be understood about different weather situations simply from studying the soundings. In this section several actual soundings are analyzed and a wealth of meteorological information is revealed.

Point Arguello is located along the California coast about 200 km WNW of Los Angeles. This is one of the regions where fog is common, especially during the summer months, because of the cold ocean current (see Chapter 11, "Fog"). During the summer months precipitation is extremely unusual along the California coast. These features of the weather can be inferred from the sounding taken at Point Arguello on June 30, 1961 at 6:00 P.M. Pacific Daylight Time (see Figure 13.16).

The sounding shows a very thin layer of cool and relatively humid air near the ground ($T = 17°C$, $T_d = 9.5°C$, RH $= 60\%$ at 1000 mb). Since the dew point is well below the temperature, no fog is present. (Fog did form by the following morning.) This layer of cool and moist air is topped (meteorologists typically say capped) by a very strong inversion in which the temperature increases 8°C in only 160 m. This means that anyone living on the top of a hill will be in warm air, while anyone living near the bottom of the same hill will be in air about 8°C cooler.

This inversion is so common along the California coast that it is often called the California coastal inversion. Fog is always restricted to the cold air layer near ground level and never penetrates the inversion, although its top is often right at the inversion level.

The warm air above the inversion is conditionally unstable, but it is so dry (RH between 10 and 20%) that the instability is not realized. Above such inversions the air is generally crystal clear. Although the air below the inversion is humid, it is cool and therefore too heavy to rise.

You may wish to know why the air above the inversion is so very dry, even though the ocean lies less than a kilometer below. The answer is that the air above the inversion is sinking. Sinking air warms dry adiabatically at 10°C per kilometer, whereas the dew point increases far more slowly (at a mere 2°C per kilometer). This results in

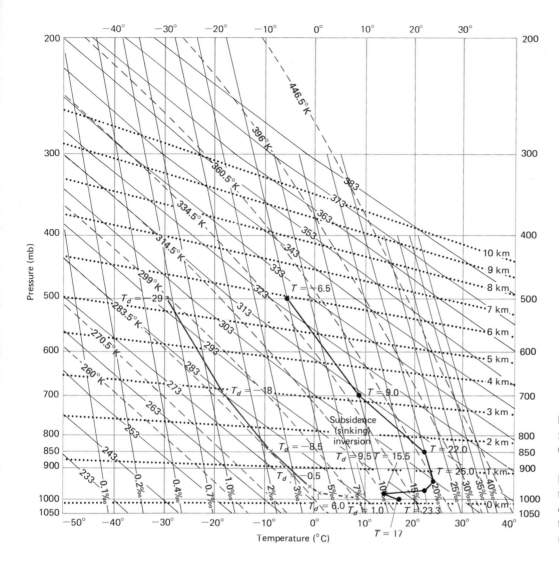

FIGURE 13.16
Sounding at Point Arguello, California, June 30, 1961, 1800 PDT. Cool, humid air beneath the inversion is topped by warm, extremely dry air. Dryness of air above the inversion indicates sinking motion.

air whose temperature is quite high but whose dew point is quite low.

Whenever you see a sounding that contains very dry air above an inversion, it indicates that the air has been sinking. This feature of soundings is common in high-pressure areas and also throughout the dry subtropics where it is sometimes called the Trade Wind Inversion. It will even occur in the hottest deserts of the world where, in fact, it is rather common. This inversion restricts rising motion to a thin layer near the ground so that thunderstorms are unusual, but pollutants are easily trapped. Now, you guess if coastal areas of California such as the Los Angeles Basin have a pollution problem.

The first half of January 1961 had been quite warm in New York City. On January 15 the thirty-day outlook published by the weather bureau was printed in the newspapers and it called for a continuation of the above-average temperatures. Then, suddenly on January 17 the temperature dropped well below freezing and remained there for the next 16 days. Two blizzards occurred during this period and caused great joy among the schoolchildren because school was closed for a total of 5 days. The first of these two blizzards occurred on January 20 and it was the smaller of the two because it dumped only 26 cm of snow on the city. (The second storm occurred on February 3–4 and dumped 44 cm.) The sounding for the January 20 storm, taken at 7:00 A.M. Eastern Standard Time, is shown in Figure 13.17.

You should notice the difference between this sounding and the California sounding immediately. The New York City sounding has very humid air from the ground right up to 500 mb (and beyond). The air throughout the sounding is quite cold so that it is not surprising that snow was falling at the time. This sounding also contains an inversion,

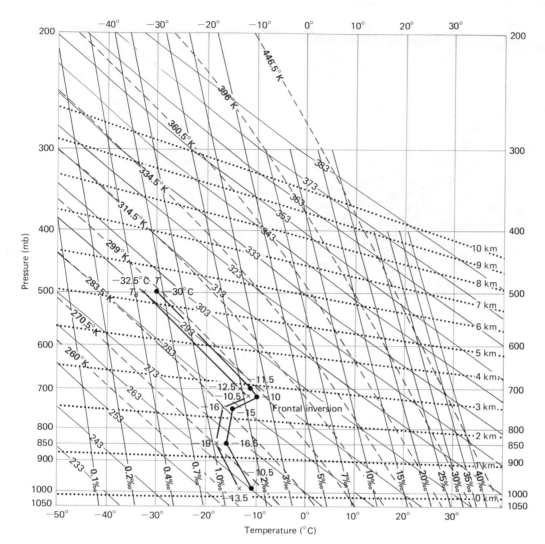

FIGURE 13.17
Sounding at New York City, January 20, 1961, 0700 EST. The air is very humid above and below the inversion. This is a frontal inversion and the air above the inversion (or front) is rising and producing snow.

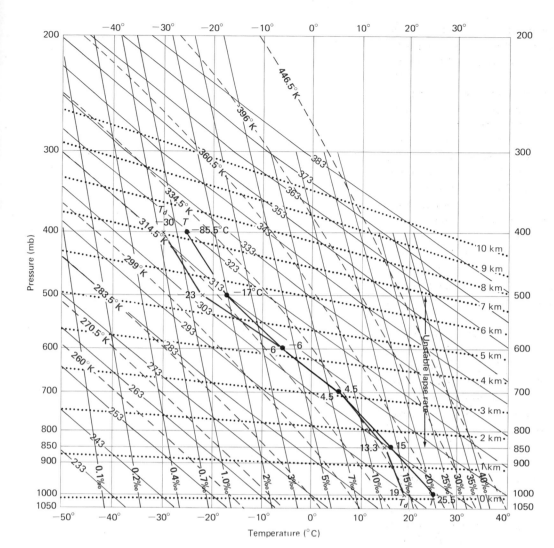

FIGURE 13.18
Sounding at Shreveport, Louisiana, May 1, 1961, 1900 CST. The air is saturated and quite unstable through a thick layer. No wonder there was severe weather all around.

but it is a far different type of inversion from that of the California sounding. The air above this inversion has certainly not been warmed (or dried) by sinking. What then is the cause of this inversion?

This is an example of a **frontal inversion.** Warm air came in contact with a mass of cold, heavy air and formed a front. The warm air was forced to rise over the cold air and soon saturation occurred. (You will notice that the air is not quite saturated with respect to water but, since there was ice or snow present in the atmosphere at the time, you can rest assured that the air at the time was saturated with respect to ice.) As it rose it cooled moist adiabatically but still remained warmer than the air below the front.

Frontal inversions are common features in the cold sector of low pressure areas, and the air above these inversions is generally quite humid since it is usually rising. (This is not true in the section of the storm behind the cold front.)

Notice that the air above the inversion is just barely stable, because the atmospheric lapse rate is slightly less than the moist adiabatic lapse rate. If you allow for the fact that the temperature is not completely uniform, then it is apparent that some localized areas will have slightly unstable lapse rates and a few cumulonimbus clouds may form. Many low pressure areas do have such isolated cumulonimbus clouds embedded in the broader stratus cloud cover.

On May 1, 1961 severe thunderstorms were recorded in Shreveport, Louisiana, and tornadoes were reported in the nearby countryside. It should come as no surprise to you that the atmosphere was unstable at this time. The atmospheric sounding for 7:00 P.M. Central Standard Time is shown in Figure 13.18.

This is one of the most unstable soundings that I have ever seen! It contains a deep layer of moist air with an atmospheric lapse rate that is considerably greater than the moist adiabatic lapse rate. Such situations almost never last very long, because as soon as the lapse rate becomes unstable the warm air near the ground rises and the cold air aloft sinks. The entire atmosphere literally overturns, and the atmosphere is thereby stabilized. Very unstable soundings such as the Shreveport sounding are produced by larger scale wind patterns and are described in detail in Chapter 19.

# PROBLEMS

**13-1** Which air parcel is potentially warmer: (a) $T = 10°C$ at 3 km or (b) $T = -6°C$ at 5 km?

**13-2** A kilogram of dry air at 6 km and 0°C is brought adiabatically to sea level and mixed with another kilogram of air 0°C. Half the mixture is then kept at sea level, while the other half is returned adiabatically to 6 km. What is the temperature of each half?

**13-3** On a day with cumulus clouds the $T = 25°C$ and $T_d = 13°C$ at ground level. How high is the cloud base?

**13-4** Explain why air is often drier and warmer when it returns to its original level on the other side of a mountain range than when it first passes over a mountain range.

**13-5** Explain why rising air cools more slowly when it is saturated.

**13-6** Explain why rising saturated air cools more slowly when it is warm.

**13-7** Using the thermodynamic diagram (Figure 13.4) for air originally at $T = 30°C$, $T_d = 15°C$, and $P = 1000$ mb, find the following: (a) the mixing ratio, $W$; (b) the saturated mixing ratio, $W_s$; (c) the RH; (d) the wet bulb temperature, $T_w$; (e) the potential temperature; and (f) the equivalent potential temperature. Now lift this air to $P = 400$ mb and find: (g) at what pressure saturation occurs; (h) at what approximate height saturation occurs; (i) $T$, $T_w$, $T_d$, and RH at 400 mb; (j) the mixing ratio at 400 mb; (k) the potential temperature at 400 mb; and (l) the equivalent potential temperature at 400 mb.

**13-8** Explain why in Problem 13-7 the potential temperature increased, but the equivalent potential temperature remained the same.

**13-9** Explain why in Problem 13-7 the mixing ratio decreased. What happened to the water vapor?

**13-10** Explain why it feels cool when you blow on your hand if you have constricted your mouth so that the air cannot escape easily, but why it feels warm when you breathe out on your hand with your mouth wide open.

**13-11** Explain the concept of atmospheric (static) instability. Be sure to include in your explanation the effects of adiabatic temperature changes and of condensation.

**13-12** Plot the following sounding on a thermodynamic diagram. Label the stability characteristics of each layer by $C$, $S$, or $U$, as in Figure 13.14. Where, if anywhere, are there inversions?

| $P$(mb) | $T$(°C) | $P$(mb) | $T$(°C) | $P$(mb) | $T$(°C) |
|---------|---------|---------|---------|---------|---------|
| 1000 | 10 | 700 | $-9$ | 400 | $-30$ |
| 900 | 2 | 600 | $-11$ | 300 | $-58$ |
| 800 | 3 | 500 | $-22$ | 200 | $-55$ |

**13-13** When the lapse rate is stable a strong wind can actually produce warming at the ground. Explain why this happens.

**13-14** Explain the relationship between the degree of atmospheric stability and the incidence of thunderstorms and pollution.

**13-15** How can you tell that the air above the inversion in Figure 13.16 is sinking and clear, but that the air above the inversion in Figure 13.17 is probably rising and cloudy?

**13-16** To what temperature must the air at the ground in Figure 13.18 be heated (without changing $T_d$) to allow it to rise freely above 500 mb?

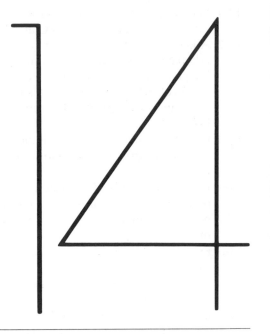

# AIR POLLUTION

The problem of air pollution was largely overlooked in the United States until the 1960s. However, within the past few years it has become in some respects an over-rated problem. Mere hypotheses about air pollution quickly become headlines while major discoveries in other areas of meteorology go unnoticed by the public. This is unfortunate, because not only is this type of activity dishonest, it also needlessly polarizes people's viewpoints. In such times it is not easy to have a balanced perspective of the impact of air pollution, as the wit of Russell Baker shows us.

The situation is bad. The crisis has become acute. Time is running out. This may be our last opportunity. What must we do when the Weather Bureau says the air quality is unacceptable?

Avoid breathing. . . .

Is there hope in scientist's recent discovery of rings around Uranus?

It is too soon to tell. Some scientists believe the Uranian rings may provide a rich new source of ionospheric patching material to plug holes around the earth caused by aerosol sprays. Most students of science, however, believe science will discover that the rings cause cancer in mice.

"Sunday Observer," *New York Times Magazine,* April 24, 1977

Scare tactics can be used on us—sometimes for a good purpose. For example, the Air Quality Act of 1967 was partly the result of frightening the public into an awareness of the harmful effects of unrestrained pollution of the air.

Over the past decade the scare tactics have become far more subtle and sophisticated than before. It is no longer fashionable to speak about the filth of the air caused by automobiles and industry. Instead, discussions usually turn to some obscure substance such as the chlorofluorocarbons (freons) in aerosol sprays.

The freons are virtually indestructable in the troposphere and eventually rise into the stratosphere, where they decompose (by photodissociation), releasing chlorine which then destroys the protective layer of ozone. Since this process allows more ultraviolet light to reach the earth's surface, more people will get skin cancer, and so on.

There is an increasing list of substances that can destroy atmospheric ozone and that we will probably be hearing much more about over the next few years.

This can place us in a serious dilemma. When do we ban the use of a chemical? The decision to ban the freons was not that difficult, but some of the potentially harmful chemicals may be of great importance to our industrialized society.

This chapter begins with a brief account of the history of air pollution and some of its notable outbreaks. This is followed by an account of how pollution is cycled into, through, and (we hope) out of the atmosphere. The chapter concludes with an account of the actual pollutants in the air—their chemical reactions and their effects on plants, animals, and humans.

## HISTORY AND CASES OF AIR POLLUTION

On May 8, 1902, Mt. Pelée on the island of Martinique erupted violently, sending a gigantic fireball of burning, poisonous gases and ash down the slope and into the city of St. Pierre. Everyone there was killed except for one badly burned prisoner who was saved by the thick dungeon walls. The death toll exceeded 30,000 people, people who had foolishly ignored the warnings that the volcano had given for several days prior to the eruption.

The eruption of Mt. Pelée is an example of air pollution. Although we tend to think of air pollution as being artificial, there are many *natural* sources of pollutants. Some of these, such as volcanoes, forest fires, and dust storms, are quite dramatic, but many natural (e.g., biological) processes quietly and continuously pollute the air. Most major pollution outbreaks are artificially created simply because the sources are quite localized.

Before the Industrial Revolution cases of air pollution were not often documented, but we can be assured that they occurred. When the Romans occupied England, coal was burned. Since coal produces quite a bit of soot and sulfur dioxide ($SO_2$), there must have been days almost 2000 years ago when the air in the larger towns was badly polluted. This may also have been the case in ancient China. During the Dark Ages, coal largely fell out of use in Europe, but sometime after A.D. 1100 it began to be used again. Marco Polo informs us that coal was widely used in China. As the forests were cut down and wood became scarce, coal became a popular substitute.

In 1237, Queen Eleanor had to leave Nottingham Castle because of the choking air produced by the burning coal in the town below. One is tempted to think that Robin Hood fled from Nottingham to Sherwood Forest merely to escape the ravages of the fetid air. In 1273 and again in 1301, laws were passed forbidding coal burning in London. Violators of these laws abounded so that in 1306 a man was reportedly executed for burning a highly impure form of coal (sea coal). One can see that air pollution in London has had a long and impressive history.

By 1661, John Evelyn, who was commissioned to conduct a study of London's air pollution, would write

The immoderate use of . . . coal . . . exposes London to one of the foulest inconveniences and reproaches. . . . While these (smokestacks) are belching (smoke from) their sooty jaws, . . . London resembles rather the face of Mt. Etna. . . . or the suburbs of hell than an assembly of rational creatures. . . . The weary traveller, at many miles distance sooner smells than sees the city to which he repairs. This acrimonious soot . . . carries away multitudes by languishing and deep consumptions, as the bills of mortality do weekly inform us.

From W. Durant and A. Durant, "Age of Reason," *The Story of Civilization*

In spite of these harsh words and a long succession of extremely thick smogs that were recorded over the next 300 years, virtually no action was taken to restrict the polluting of the air until the great killer smog of December 5–9, 1952. I have already described this five-day episode in Chapter 11. As a result of this pollution outbreak, a clean air law was finally passed for London in 1956—just in time to save many lives because another pollution episode hit London in December 1956. From the meteorological point of view the potential danger of this situation was even worse than it was for the 1952 case but, fortunately, some controls on coal burning had been instituted. In the 1956 episode 800 people were estimated to have died as a direct result of the pollution. Without any controls it might have been as high as 8000! With good controls it might have been 8.

Tremendous progress has been recorded in the fight against air pollution throughout England and especially in London—London now receives 50% more sunlight during the winter months than it did before 1956; the mortality rate from air pollution has dropped to nearly zero; and finally, wildlife is benefiting.

The peppered moth (*Biston betularia*) had always been a lightly colored and spotted moth up till 1845 when a black-colored mutant was first observed and captured in the industrial city of Manchester. Over the next 100 years the black form (*carbonaria*) gradually became dominant, and the spotted form virtually disappeared.

What could have caused such a remarkable change? After a few years of research, scientists discovered that these moths depended on camouflage for protection. In the pollution-free days the normal form was almost invisible when it rested on the tree trunks. However, as industrialization and coal burning spread, the tree trunks gradually became blackened from soot. The spotted form was easily visible when it rested on these blackened trunks, and the birds who ate these moths had a field day (see Figure 14.1). For once their ancient menu was clearly written! Then, by a random mutation a blackened form of the moth appeared, and it thrived because it was so well camouflaged on the blackened tree trunks.

The point of this story is that the spotted form has once

FIGURE 14.1
Two forms of the peppered moth. The dark form is a mutation that became prevalent because of air pollution; this form is now dying out again because of the Clean Air Acts.

again become the dominant form in the past few years throughout England, and the blackened form is dying out. This is proof that the air pollution controls in England must be working. I wonder how many people will mourn the death of *carbonaria* which may be headed for extinction.

In the United States there has never been an air pollution incident that matched the London killer fogs but, to say the least, air pollution has been quite costly to Americans. Legislation against air pollution has come as the result of continued high levels of pollution rather than as a result of any single incident. This does not imply that no outbreaks have occurred in the United States, for they have.

One infamous incident did claim seventeen lives in the small industrial town of Donora, Pennsylvania. Donora is situated in the Monongahela River valley about 50 km south of Pittsburgh. The chief culprits in the air pollution disaster were a steel mill, a zinc plant, and a sulfuric acid plant. These all released sulfur dioxide and soot into the air. Under normal conditions the air there was merely smelly because winds dispersed the pollutants and kept them from reaching too high a concentration.

On October 26, 1948, the winds died down over Donora and an inversion formed. (The same meteorological conditions led to the killer smog of London in 1952). Pollutants began to accumulate since they were literally trapped in the river valley around Donora. For five days the inversion kept the pollutants from rising while the extremely light winds could not blow them away. By Saturday, October 30, the situation became critical and people began dying. On Sunday, mercy came from the heavens in the form of rain and wind, and the population was saved from further injury.

Since Donora had a population of only 15,000 it actually had a higher death rate than London did in 1952. At the same rate, more than 8000 people would have died in London. Although Donora made the national news, nothing was done to improve the situation. The mills and factories continued to burn away as they had done before and even during those few days.

Although New York and Los Angeles never experienced such killer smogs, these cities are known for their continued high levels of pollution. The eyes, noses, throats, and lungs of the people living in these metropolitan areas tend to be constantly inflamed or irritated. And, of course, there were times when the pollution was far worse than normal. The city of Los Angeles now posts smog alerts during times of bad pollution, but for years the only precaution was to urge motorists not to drive their cars too much.

New York City has had some scares due to air pollution when the mortality rates rose by about twenty people per day. From November 28 to December 4, 1962, a temperature inversion accompanied by very light winds allowed pollution levels to build up alarmingly. During the Thanksgiving weekend of 1966, similar conditions prevailed. I can clearly recall that weekend because I drove from Boston to New York City. A thick haze filled the air all the way, even in sections which normally have rather clean air. Visibility in places was well below 1 km. Within a few months of that weekend clean air standards began to be enforced, and by 1970 all new automobiles had to be equipped with antipollution devices.

Polluted skies are common in many other populated and industrial centers of the world. Residents of cities such as Tokyo, São Paulo, Milan, Venice, and Mexico City have to put up with air that is, to say the least, somewhat befouled a good portion of the time. Even so, there is evidence that we have not yet destroyed our planet's atmosphere. What is this redeeming evidence?

# THE POLLUTION CYCLE

I have seen days in New York City when the air was so thickly polluted that it made you feel as if you were living underground. Then a cold front would pass through bringing crystal clear Canadian air.

Unfortunately, the air around New York City never stays clean for very long. As soon as the wind dies down pollutants begin to accumulate. Crisp, clean air can become quite hazy.in six hours or even less.

It is miraculous that there is still so much clean air left on the earth, since there is no reservoir of untarnished air anywhere on earth. Luckily, the pollution that is put into the air by a variety of sources gets mixed around and finally removed from the atmosphere by several natural processes. Thus, pollutants are actually cycled into and out of the atmosphere.

We can begin our discussion of the pollution cycle with the first phase—that is, with sources of pollution. Most of these sources are found at or near ground level. As a result, pollution is most concentrated near the ground and becomes less dense with increasing altitude (see Figure 14.2).

A fact that may astound most city dwellers is that the natural sources of pollution far outweigh the artificial sources! Although the artificial sources may be more concentrated locally, the natural sources of pollution tend to be global (see Table 14.1). The natural sources dominate both the gaseous pollutants and the solid or liquid pollutants (called aerosols).

The most abundant gaseous pollutant is carbon dioxide, $CO_2$. Carbon dioxide is put into the air largely by respiration of plants and animals, both on the land and in the ocean. An increasing amount of $CO_2$ is now being put into the atmosphere as a result of the burning of coal, gas, and oil—the so called **fossil fuels**—as well as wood. Since

## TABLE 14.1
Amounts of Pollutants Produced by Human Activities in the United States (in Millions of Metric Tons)

| Year | Sulfur Oxides | Carbon Monoxide | Aerosols[a] | Hydrocarbons | Nitrogen Oxides |
|---|---|---|---|---|---|
| 1940 | 21 | 66 | 41 | 15 | 6 |
| 1950 | 23 | 75 | 30 | 20 | 8 |
| 1960 | 21 | 90 | 27 | 24 | 10 |
| 1968 | 28 | 101 | 25 | 27 | 17 |
| 1969 | 29 | 104 | 32 | 26 | 19 |
| 1970 | 30 | 92 | 24 | 24 | 20 |
| 1971 | 30 | 91 | 24 | 24 | 20 |
| 1972 | 30 | 98 | 18 | 25 | 23 |

[a] In 1968 worldwide aerosol production by humans amounted to an estimated 408 million metric tons, whereas an estimated 2150 million metric tons is produced each year by natural phenomena.

the 1890s the concentration of $CO_2$ in the atmosphere has increased nearly 15%, prompting concern over its potential climatic impact (recall the discussion of the greenhouse effect in Chapter 8). Enormous quantities of carbon monoxide (CO) are also a by-product of burning and occasionally have been concentrated enough along urban highways to cause headaches.

Plants naturally emit a variety of gaseous hydrocarbons (chemicals containing both hydrogen and carbon)—in fact, this is what we smell as we enter a forest. These hydrocarbons often produce a bluish haze over dense forests and are responsible for the naming of the Great Smoky Mountains. Automobiles also emit hydrocarbons, but these don't smell quite as nice.

Methane, $CH_4$, also known as natural gas or marsh gas, and large quantities of sulfur dioxide, $SO_2$, also form from rotting plants. $SO_2$ starts out as hydrogen sulfide but quickly reacts once in the air. Sulfur dioxide is also produced directly by the burning of coal.

Solid particles in the air, or aerosols, also have a variety of sources. These aerosols also come in many sizes. First, there is the compact size, from about 0.001 up to 0.2 microns in diameter. These are called **Aitken nuclei**, and they result largely from burning. Their natural sources include volcanoes and forest fires.

Aitken nuclei sometimes produce odd optical effects such as blue moons and suns and incredibly red sunsets. They have also been known to cause "dark days." The larger particles or **giant nuclei** are often swept up by the wind in dry or desert areas. Some of these aerosols are the clay particles that serve as freezing nuclei.

An important natural source of aerosols in China is

FIGURE 14.2
The concentration of pollutants drop off sharply with altitude.

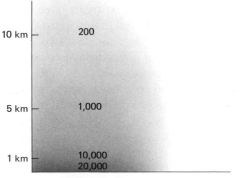

Millions of aerosols per cubic meter

loess, which is swept up from the wastes of the Gobi desert. During the winter season when the wind blows more or less continuously from the Gobi toward China, the air is filled with dust, and many Chinese wear gauze face masks.

The bursting of air bubbles on the sea surface is a surprisingly important source for all sizes of aerosols. Whenever a bubble bursts, tiny droplets called jet drops are shot as high as 15 cm into the air (see Figure 14.3). This process is similar to the fizz that forms above a glass of any carbonated drink. Some of these jet drops evaporate and leave behind a salt residue in the form of a particle often 2 or more microns in diameter, which can then be swept up into the atmosphere by the winds. Smaller salt nuclei are also produced from the evaporated residue of the many tiny droplets that also form when the water film on the top of the air bubbles bursts.

Why is this process so significant? First, it provides the atmosphere with condensation nuclei so that it can rain. Second, it also provides a source of positive electrical charges to the atmosphere and therefore may ultimately be of some importance in the formation of lightning.

The size distribution of the aerosols is shown in Table 14.2. From this you can see that highly polluted urban air has many more small aerosols. You can also see that even clean air over the continents has many more aerosols than oceanic air, and the continental aerosols tend to be somewhat smaller. You may recall that the smaller aerosal concentration over the oceans facilitates the formation of rain over the oceans without the help of the ice crystals.

The second phase of the pollution cycle is the dispersal and transport of the pollutants through the atmosphere. This turns out to be an enormously difficult scientific problem (much like evaporation). When scientists try to compute how pollution spreads through the atmosphere from its sources, they find it easiest to first treat **point sources**—such as smokestacks or volcanoes (see Figure 14.4). Next, **area sources**—such as forests or cities—are treated as areas consisting of a continuum of point sources.

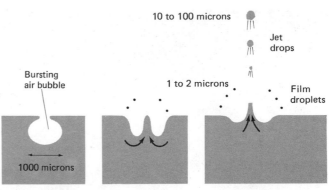

FIGURE 14.3
Jet drops and film droplets produced by bursting air bubbles at the ocean surface. When these drops evaporate, salt nuclei remain in the air.

In order to understand how pollutants spread, it is easiest to begin by seeing how they move from a single point source, such as a smokestack, and then to generalize by showing how pollution spreads out from larger areas.

When pollutants are emitted from a smokestack, they may initially be warmer than the surrounding air; as a result, they will rise a short distance until the newly polluted air reaches equilibrium with the surrounding air. The polluted air is then carried downstream by the winds, stretching out into a long plume. Near the smokestack or source the plume is rather narrow, and the pollutants here will be quite concentrated. The air in the plume gradually mixes with the surrounding air and spreads out. In the process the pollutants gradually disperse, and the plume eventually becomes invisible.

The stability of the air has a great effect on the precise shape of the plume, as you can see in Figure 14.5. In very stable air, any parcel of plume air can neither rise nor sink freely; in this case a flattened, almost horizontal plume results. When the air is unstable, the air parcels in the plume rise and sink freely; the plume then assumes a wavy shape.

On the most polluted days the air near ground level is extremely stable. Often a temperature inversion is present

TABLE 14.2
Representative Concentration of Aerosols[a] (millions of particles per cubic meter.)

| | Aitken Nuclei— Diameter up to 0.4 microns | Large Nuclei— Diameter from 0.4 to 2 microns | Giant Nuclei— Diameters Larger Than 2 microns |
|---|---|---|---|
| Large city | 100,000 | 200 | 2 |
| Rural over land | 10,000 | 100 | 1 |
| Ocean | 1,000 | 20 | 1 |

[a] All such figures are highly variable, depending on wind and weather conditions.

FIGURE 14.4
Plume from a volcano erupting in the Galápagos islands. The plume continuously widens. Although the plume eventually *appears* to become narrower, this is because the particle concentration has decreased so much so that the fringes of the plume disappear.

FIGURE 14.5
Plumes in stable and unstable air. The lower plume is confined to a cold stable layer of air near the ground while the upper plane is in unstable air above (where the wind direction happens to be different). Notice also the arctic sea smoke just above the water. This photo was taken on a frigid February morning in Boston harbor.

and this acts like a lid, keeping the pollutants near ground level. This is commonly the case in the Los Angeles area.

Dispersal of pollutants from a plume is known as **turbulent diffusion,** because it is similar in some ways to the process of diffusion (discussed in Chapter 5). Unfortunately, it is far more complicated than diffusion, because the air motions are turbulent (which means highly erratic). With simple diffusion, pollutants would slowly and evenly spread from a more concentrated region to a less concentrated region. With turbulent diffusion, puffs of polluted air alternate with puffs of clean air, and the plume spreads out in a highly irregular and unpredictable manner.

Even though it is virtually impossible to predict the exact shape of a plume at any one instant, it is not too difficult to predict the average shape of the plume and the average concentration of the pollutants. Notice that the plume in Figure 14.6 has a uniformly spreading boundary, despite the highly variable smoke concentration within which tends to average out over time. Thus a composite or "average" plume looks much smoother and can be described by simple diffusion.

When working with the "average plume" we arrive at several general conclusions. First, the plume spreads out like a cone until it hits the ground or an inversion. Second, the concentration of pollutants is at a maximum in the center of the cone at the source (i.e., the smokestack) and the concentration decreases both away from the smokestack and toward the perimeter of the cone. Third, the cone spreads out at a wider angle when the wind is weaker or the atmosphere is more unstable. Finally, when the wind is light not only does the cone spread out over a wider angle, but the pollutants are more concentrated (see Figure 14.7).

Once the plume hits the ground or an inversion, the pollutants can no longer spread out quite so freely and begin to accumulate. If you then add the output from other plumes in the area, you soon have a full-fledged area source—which can be a real mess.

It often pays to make smokestacks tall for several reasons. First, when the smokestack is tall the plume usually has diffused somewhat by the time it hits the ground, so that no extremely high levels of pollutants are encountered. Second, if an inversion is low enough, a tall smokestack may emit its pollution in the unstable air above the inversion so that the pollutants can disperse aloft.

There is at least one other complicating factor. Believe it or not, there are actually times when large particles (several microns in diameter) are emitted from smoke-

FIGURE 14.6
The spreading of a plume. Even though the concentration of smoke has irregularities at any one moment, the *average* behavior is to diffuse or spread gradually as is suggested by the uniform spreading of the plume outline.

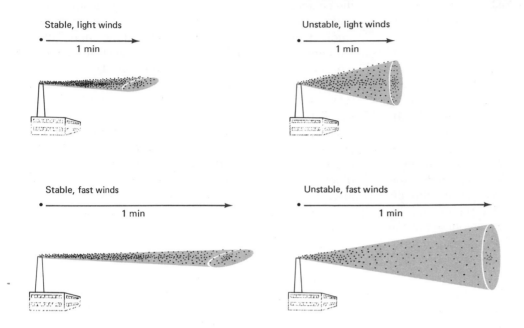

Concentration of pollutants

Stable, light winds
|← 1 min →|

Unstable, light winds
|← 1 min →|

Stable, fast winds
|← 1 min →|

Unstable, fast winds
|← 1 min →|

FIGURE 14.7
Each plume was produced in the same time and each contains the same total amount of pollution. Pollution is therefore most concentrated in stable, almost calm conditions, because the plume is shortest and narrowest.

stacks or chimneys. Since these large particles can fall at an appreciable rate through the atmosphere, they do not form a nice conical plume; instead, they tend to have trajectories like cannonballs. And all nearby objects such as houses and people might just get covered by a layer of soot (see Figure 14.8).

This brings us to the subject of how pollutants leave the air—the third and final phase of the pollution cycle. At the beginning of this section I mentioned that some of the clearest days we get are very cold. There is something ironic in this because it implies that almost identical weather patterns are responsible for some of the cleanest air and also some of the most polluted air.

FIGURE 14.8
Solid particles quickly fall out of the plume and settle on all objects nearby. Fortunately most pollution controls strictly limit the number of solid particles that are emitted.

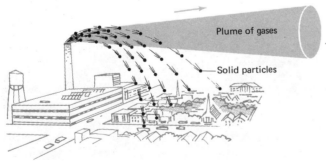

Plume of gases

Solid particles

What type of weather produces the worst pollution? Ideal pollution weather is the almost calm, clear weather found near the center of slow moving high pressure areas. There, during the late fall and early winter, when the sun is at its low point for the year and the days are shortest, the air layer near the ground cools and marked inversions result. The combination of light winds and a temperature inversion allows pollutants to accumulate, but absolutely prevents them from dispersing aloft.

Most of the extreme pollution outbreaks have occurred between the months of October and January. The weather maps for the London killer smog of 1952 and the New York Thanksgiving smog of 1966 are shown in Figure 14.9. In both cases the worst pollution occurred near the center of slowly drifting highs, under light winds. The soundings taken during these incidents are shown in Figure 14.10, and you should easily find the culprits—the temperature inversions in the lower troposphere which trapped the pollutants.

Oddly enough, the clearest, crispest air we breathe in the midlatitudes comes from the polar wastelands. This air actually grows somewhat cleaner right in the middle of slowly drifting highs with light winds and inevitably with a huge temperature inversion in the lower troposphere. The main difference between this polar air and our industrial, cosmopolitan air is that the polar lands have almost no sources of pollution.

This leads us to the question: "How is the air cleaned?" The most important cleaning process has already been mentioned in the chapter on the formation of raindrops. When rain or snow falls the raindrops and snowflakes "sweep" up many of the larger particles in their path by coalescence.

This sweeping or scavenging mechanism does not work so well for extremely small aerosols because of the "flyswatter" effect; the very small aerosols tend to blow around the falling drops. Nevertheless, another process does remove them from the atmosphere during times of precipitation. This process is called **Brownian motion.**

You should recall that air pressure is produced by the average force of the molecular impacts. Now, we are far too large to feel the impact of individual molecules, but very small aerosols can be pushed around by individual molecules! Very small aerosols (diameters less than 0.1 micron) have been observed under the microscope and they constantly move about in jumps. These jumps are known as Brownian motion and are caused by the collisions of the aerosols with individual molecules.

Believe it or not, it was by observing the Brownian motion of small aerosols that scientists were finally convinced that gases and liquids are composed of rapidly moving molecules. Another interesting historical note is that Albert Einstein was the scientist who developed the mathematical theory of Brownian motion.

How does Brownian motion help clean the air? The answer is quite simple. Since individual molecules can push around the smaller aerosols, it is apparent that many of these are pushed into cloud droplets and ice crystals and a smaller percentage into falling raindrops and snowflakes, despite the aerosols tendency to drift around them. Brownian motion also causes these smaller aerosols to collide, and often they will stick or coagulate as a result. Aerosols less than 0.02 microns in diameter grow rapidly by this coagulation process.

This leads us to the final cleaning process. Precipitation accounts for 80 to 90% of the aerosols removed from the air. The remaining fraction simply falls out of the atmosphere. But because of air resistance, small particles fall somewhat slowly and therefore only tend to settle out when the wind is calm for long periods (see Figure 14.11). The smallest aerosols would hardly ever settle out of the atmosphere if they didn't coagulate.

Now you can see why the cold, crisp air from the polar wastelands is so pure. In addition to the normal cleaning mechanism due to precipitation, the air must come to rest for several days before the ground can truly cool it. During this time of very light winds the larger aerosols are able to slowly settle onto the snow covered ground, and the air is purified further. The longer the air remains resting on the cold polar ground, the *colder and cleaner* it becomes.

During the 1952 pollution episode, many tons of soot settled on the ground, buildings, and people of London. This process would soon have significantly cleaned the air if only the people had abstained from burning coal at the enormous rate they did throughout this time.

This reasoning is fine for the aerosols, but how are the gaseous pollutants removed? In order to remove gaseous pollutants from the air the gases must adhere to solids or liquids. Any gas that is soluble in water is therefore quite quickly removed from the atmosphere by the rain (it is said to have a short residence time in the atmosphere). Unfortunately, this can have some bad side effects. Both sulfur trioxide ($SO_3$) and carbon dioxide ($CO_2$) easily dissolve in water to form sulfuric acid ($H_2SO_4$) and carbonic acid ($H_2CO_3$), respectively. Written as chemical equations they appear as

$$SO_3 + H_2O \rightarrow H_2SO_4$$
$$CO_2 + H_2O \rightarrow H_2CO_3$$

What this means is that the rainwater downwind from industrial locations has become rather acidic. This is especially due to the $H_2SO_4$. Thus, it eats into buildings (especially those constructed of limestone) and is harmful to plants and animals. The acid rain falling in the Adirondack Mountains of New York State and in the highlands of Scandinavia dissolves aluminum from the rocks which then runs into the lakes and rivers. Together with the high acidity this is decimating the fish populations.

Carbon dioxide is removed from the atmosphere by a variety of other processes. One of these is photosynthesis, whereby plants take in $CO_2$ and give off oxygen, $O_2$. Carbon dioxide also dissolves in the ocean where it is used by shell-building animals to form calcium carbonate. (Recall the runaway-greenhouse effect on the planet Venus.)

Carbon monoxide is also highly soluble and therefore is easily removed from the atmosphere. Unfortunately for us, carbon monoxide is also soluble in our blood where it chemically combines with hemoglobin, rendering it incapable of carrying oxygen through the bloodstream.

# SOME EFFECTS OF POLLUTANTS AND THEIR REACTIONS

Hop aboard an aerosol and take an imaginary voyage into someone's lung. Fortunately, most aerosols are stopped before they reach the lungs. This means that we have to choose the proper aerosol to make the voyage. Large aerosols (diameter more than 10 microns) are often caught on hairs in the nose or else smash into the walls of the respiratory tract at every bend. Since the walls of the nasal cavity and trachea are lined with mucus, these larger aerosols get stuck and do not penetrate into the lung. Extremely small aerosols also do not penetrate because they easily get knocked into the walls of the respiratory tract by impacts from individual molecules (Brownian motion).

The ideal aerosol for penetrating deep into the lung has a diameter of approximately 1 micron (see Figure 14.12). Aerosols of this size do not easily collide with the walls of the respiratory tract, since they are carried with the airstream around every bend (because of the "flyswatter effect"). Many of these particles are carried out of the lung just as they entered and therefore no damage is done. Unfortunately, too many of these particles *do* get embedded deep within the lung, since the passages become progressively smaller. Some of these aerosols even become embedded within the alveoli (the smallest, deepest terminus of the lungs).

It is when the particles reach the bronchial tree and especially the alveoli that they have a damaging effect. Cigarette smoke contains aerosols that cause cancer of the bronchi. A variety of other aerosols such as asbestos fibers, nickel, chromium, and soot are also carcinogenic.

Bronchitis, the inflammation of any part of the bronchial tree, is often caused by aerosols. Chronic bronchitis often results from the destruction of the lung's protective mechanisms so that scar tissue tends to form. This blocks the normal flow of air through the lungs and may even impair the delicate alveolar walls that are needed to diffuse $O_2$ into and $CO_2$ out of the bloodstream.

In the advanced stages of alveolar decay, the victim suffers from emphysema, and this places a tremendous strain on the heart and renders the system extremely susceptible to infections in the lungs. Emphysema claims 20,000 lives or more annually, and many of these can be attributed to air pollution levels.

The small aerosols that work their way deep into the lung cause extra damage, because many of these aerosols carry sulfur dioxide. As these $SO_2$-bearing aerosols move down the respiratory tract, they react with the moist air and form sulfuric acid. When they then reach the lower part of the respiratory tract and lodge against the bronchial walls or alveoli, they disintegrate the cells at the point of impact.

This brings up an interesting point. $SO_2$ is very soluble in water, so that *in gaseous form* it completely dissolves in the upper part of the respiratory tract where it merely causes throat irritation. Aerosols without any $SO_2$ may lodge in the lungs without causing too much damage and eventually may be eliminated from the lung by the body's

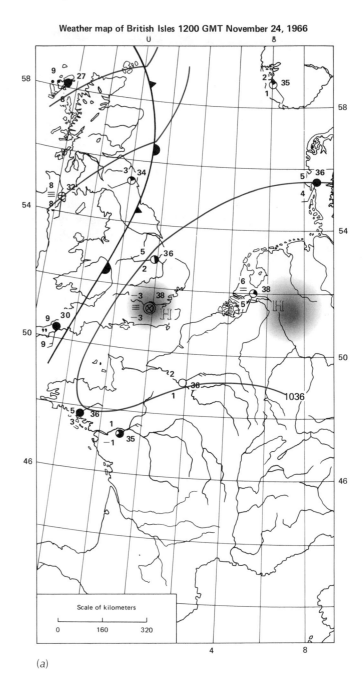

Weather map of British Isles 1200 GMT November 24, 1966

Scale of kilometers

0    160    320

(a)

natural defense mechanisms. When they act together, $SO_2$ and aerosols are far more dangerous since the aerosol gets the $SO_2$ deep into the lungs where it is damaging.

Thus two relatively harmless effects can combine to produce an extremely dangerous process. This is known as **synergy**. There are many other examples of synergy that are related to air pollution. One of these is the combination of smoking and inhaling asbestos fibers. Asbestos workers who also smoke suffer lung cancer rates that are 10 times higher than their nonsmoking co-workers.

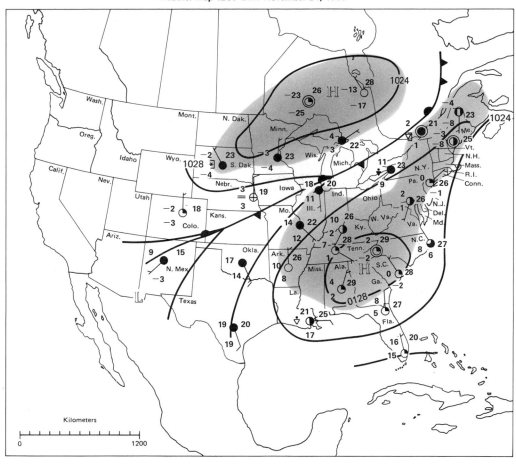

**Weather map 1200 GMT November 24, 1966**

(b)

**FIGURE 14.9**
Surface weather maps for (b) Thanksgiving Day pollution outbreak, November 24, 1966, and (a) killer smog of London, December 7, 1952, 1230 GMT. Areas shaded in red had wind speeds of 5 knots or less. Notice both London and New York are near the center of highs. The thick fog helped to reinforce itself by reflecting the sunlight, keeping London cool.

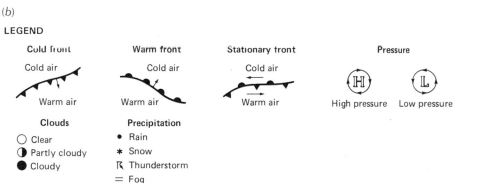

**LEGEND**

| Cold front | Warm front | Stationary front | Pressure |
|---|---|---|---|
| Cold air | Cold air | Cold air | |
| Warm air | Warm air | Warm air | High pressure  Low pressure |

**Clouds**
○ Clear
◐ Partly cloudy
● Cloudy

**Precipitation**
• Rain
\* Snow
℞ Thunderstorm
= Fog

Now, we return to sulfur dioxide damage. There are two basic types of air pollution—sulfurous pollution and photochemical smog. Sulfurous pollution is caused by coal burning and by the smelting of many metal ores. Coal consists mainly of carbon but usually contains a fairly high percentage of impurities such as sulfur. When coal is burned, $SO_2$ is released with tons of soot particles (and of course, $CO_2$). Many valuable metal ores consist of sulfide minerals (minerals containing sulfur) such as lead sulfide and copper sulfide. Smelting consists of roasting these minerals until the other elements, such as the sulfur, separates from the metals. Unfortunately, the sulfur then combines with the oxygen in the air to form $SO_2$, which would easily escape into the air without good pollution controls.

Once sulfur dioxide is in the air it can undergo a variety of chemical reactions. The most common reaction in humid air occurs within tiny droplets. Sulfur dioxide causes water to condense when the relative humidity reaches 70% (thus it is a good condensation nucleus).

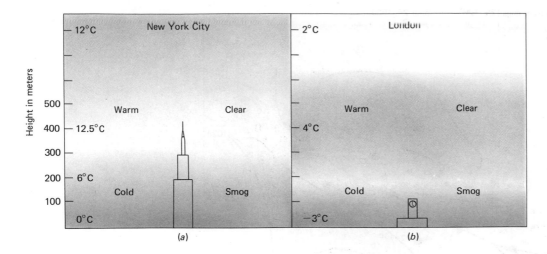

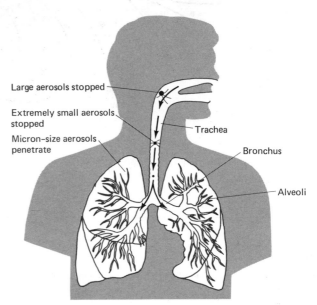

FIGURE 14.10
Soundings for (a) New York City at 1200 GMT, and (b) London at 1230 GMT, December 7, 1952. Notice the strong inversions in each case.

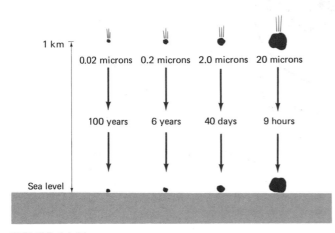

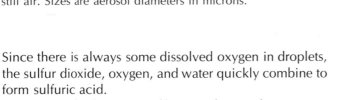

FIGURE 14.11
The time it takes for different sized aerosols to fall 1 km in still air. Sizes are aerosol diameters in microns.

FIGURE 14.12
The human respiratory tract, showing penetration of different sized aerosols. The aerosols about 1 micron in diameter can penetrate deepest into the lungs and so can potentially cause the greatest damage.

Since there is always some dissolved oxygen in droplets, the sulfur dioxide, oxygen, and water quickly combine to form sulfuric acid.

Anyone who has seen sulfuric acid poured onto sugar or plants or human skin is surely aware of its corrosive strength. I will never forget seeing how a small amount of sulfuric acid changed a few tablespoons of sugar into a mass of steaming water and black carbon. In a similar manner the acid can disintegrate the tissues in our lungs and the leaves of plants. Vegetation has sometimes been virtually eliminated for several kilometers around smelters.

The second type of pollution is the so-called photochemical smog. This is quite different from sulfurous smog in a number of respects. Photochemical smog is composed of nitrogen oxides, hydrocarbons, some ozone, and some lead. Photochemical smog results from the burning of oil and gasoline, and this is the type of smog that Los Angeles has made famous. It is not so deadly as sulfurous smog, but it is not really pleasant either.

The predominant characteristic of photochemical smog is that it reduces visibility and gives the air a brownish color. It also hurts the eyes for reasons not yet completely understood. When you first breathe this type of polluted air, you can detect a slight unpleasant odor, but soon

FIGURE 14.13
Views of Denver, Colorado on clear and polluted days.

most people become desensitized and then only the visual effect remains. Cities such as Los Angeles and Denver that were once renowned for their clean air attracted so many people (who drive automobiles) that the mountains that lie nearby are seldom visible.

Automobiles release a variety of hydrocarbons into the atmosphere. The hot engine exhaust also causes oxygen and nitrogen to combine chemically to form nitrogen oxides. Then the hydrocarbons, nitrogen oxides, and oxygen act together with sunlight to produce a number of compounds including ozone. Ozone concentrations around cities always rise markedly during sunny days. These pollutants are probably damaging to humans and certainly disastrous to agriculture, as has been demonstrated in the Los Angeles area.

Air pollution controls on automobiles have proven to be a somewhat mixed blessing. By recycling the exhaust, most of the hydrocarbons are burned and destroyed. Unfortunately, this raises the burning temperature of the engine which creates more nitrogen oxides. Even so, we have gained tremendously by demanding tough controls. The air in most major American cities and European cities is not worse than it was around 1970 despite greatly increased fuel consumption. Even in some of the developing countries, strict pollution controls are helping to keep pollution from becoming overwhelming.

# PROBLEMS

**14-1** Why do pollution levels on sunny afternoons decrease near the ground, but increase at higher levels?

**14-2** Coastal structures and plants typically suffer from salt damage. How does the salt get there?

**14-3** Why are general pollution levels highest when winds are light and there is a sharp inversion near ground level?

**14-4** Explain why the pollution at ground level from a chimney may be *momentarily* larger under unstable conditions.

**14-5** Why do marble and copper statues suffer in much of northwestern Europe and in the northeastern United States?

**14-6** Explain why very small and very large aerosols rarely penetrate the lungs.

**14-7** Explain what is meant by synergy. Do you know of any other examples (besides the one mentioned in this chapter) that relate to air pollution?

**14-8** Explain why ozone concentrations in the city rise during the day.

**14-9** Explain why ozone levels are often higher in the suburbs than in the urban centers.

**14-10** You fly out of a large polluted city. As you rise the air suddenly becomes much clearer. Explain why this is so.

**14-11** Explain why the sun sometimes disappears before it should set on cloudless polluted days.

15

# THE WINDS

On Sunday, August 8, 1976, I climbed Long's Peak in Colorado with some friends. It was a perfect day for climbing. The air was cool and the cloudless sky was deep blue and crystal clear. At the base of the mountain there was hardly any wind, but as we climbed the wind speed gradually increased. This is characteristic of mountains so there was no cause for alarm. But I *was* somewhat frightened when we had to cross over a notch known as the Keyhole; the wind tends to get funneled right through this passage. I would estimate that when we crawled through the Keyhole the wind speed was at least 50 knots. If I had stood up there, I might have been blown off the mountain.

Experienced mountaineers might sneer at this story and tell of winds well over 100 knots that were accompanied by subzero temperatures and blinding snow, and some of them would be telling the truth. Other people could tell of high winds they experienced in hurricanes or tornadoes or storms at sea. Pilots of commercial jets typically fly in a region of the atmosphere known as the jet stream where the winds exceed 100 knots. Even so, a wind of 50 knots is strong enough to knock over trees and, needless to say, that can be quite impressive.

In this chapter I will explain and analyze the various forces that produce the winds. The chapter begins with some descriptive information about the winds and wind records. After that, you will see how power is obtained from the winds by using windmills and sailboats. Then, the bulk of the chapter will be used to analyze the forces that affect the winds. This will lay the groundwork for the material in the next few chapters.

## WIND RECORDS

The wind is simply the horizontal motion of air. When, for example, the wind speed is 60 knots that means that if the air kept on moving at that rate it would move 60 nautical miles in an hour. Near the ground the wind is measured by an instrument called an anemometer which is shown in Figure 15.1. When the wind blows, the cups of the anemometer rotate because the outside of the cups have less air resistance than the inside. The faster the wind blows, the faster the cups rotate. In effect, the anemometer is designed to work like a small electric generator (i.e., a windmill); the faster the cups rotate, the larger the electric current. As you have already seen, the wind speeds *above* the ground are determined every 12 hours when the radiosondes are released and soundings are taken.

Because the wind speed is irregular and gusty, it is necessary to distinguish between the momentary gusts and the steady part of the wind. Wind records are therefore classified either as the average over a 5-minute period, a 1-minute period, or the shortest time it takes the air to move a mile. Anything of shorter duration is considered a

FIGURE 15.1
An anemometer.

on September 27, 1955. The anemometer there recorded a wind gust of 150 knots before it was blown away. Since the wind speed at the time was increasing, it is estimated that the peak wind exceeded 170 knots.

All other places that have recorded faster winds have some topographic feature that can greatly enhance the speed. The trouble with determining record winds is that most anemometers have a tendency to be blown away, as at Chetumal, once the wind speeds exceed 100 knots. To measure such high wind speeds it is necessary to install specially built anemometers, which are rather expensive. Then, even after these have been installed it is quite likely that they will be bombarded by flying debris that is inevitably picked up by extreme winds. During the severe arctic storm of March 8–9, 1972, which produced peak wind gusts of 180 knots at Thule, Greenland, Knud Rasmussen made these observations.

This storm was the worst that I have seen. During the height of the storm the sides, and for the first time ever, the roof were constantly pelted with rocks and chunks of ice. All of us here became very worried when three windows scattered throughout the complex were smashed by rocks and ice. I estimate the wind reached 140 knots up here. [This location was slightly more sheltered than the location of the anemometer that recorded the 180 knots.]

*Weatherwise,* 1972

Wind speeds in a few large tornadoes have been calculated at just over 175 knots. These calculations were made from movie pictures by computing the speed of debris flying around the tornado. Other means of estimating wind speed may be somewhat less reliable. Serious research has been performed to determine what wind speeds are necessary to drive straw into wood planks, since this has been known to happen in tornadoes. Unfortunately, the estimates are highly variable and depend on the condition and type of wood and straw.

Tornadoes have also been known to strip the feathers off chickens. This aroused the scientific curiosity of various people in the nineteenth century, and in 1842 Professor Elias Loomis conducted an experiment to determine what wind speed was necessary for stripping the feathers. Loomis loaded a dead but fully feathered chicken into a cannon and then fired. The chicken sailed through the air at 296 knots. Not only were the feathers stripped off, the entire chicken was ripped to shreds. This experiment led Professor Loomis to conclude correctly that somewhat smaller velocities could still strip chickens of their feathers while leaving them alive.

More recently biologists have learned that when chickens are under stress their follicles relax, thus making the removal of feathers somewhat easier than under nor-

gust. Therefore, a weather report which states that the wind is blowing at 50 knots and is gusting to 75 means that the average is 50 knots but at certain instants the wind may be as high as 75 knots. It is usually these gusts that knock people over on windy days.

In Chapter 5 you learned that air pressure is proportional to the square of the velocity of the molecules. Similarly, the pressure of the wind against your body is roughly proportional to the square of the wind speed. This means that if the wind speed doubles, the wind feels *four* times as strong.

At 10 knots whitecaps form on the water and leaves move constantly in the trees. At 20 knots spray begins to form on the water and entire branches sway. At 30 knots walking becomes difficult and umbrellas are useless. At 50 knots the wind can knock you down, and at 100 knots it can pick you up!

The fastest wind ever recorded on the surface of the earth occurred on Mount Washington in New Hampshire on April 12, 1934, when a wind gust reached 201 knots. However, since that time the anemometer has been moved away from the eastern rim of the mountain to a slightly more sheltered location, and thus it is doubtful if we will see any further records at Mount Washington.

The fastest wind ever recorded on level ground occurred in Hurricane Janet at Chetumal Airport in Mexico

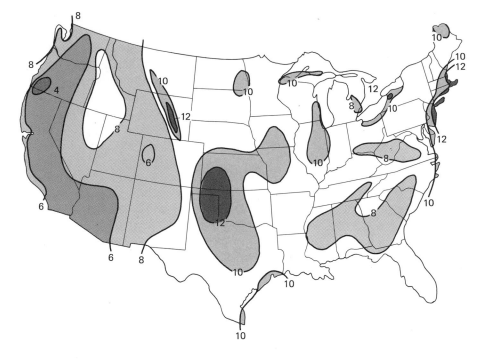

FIGURE 15-2a
Mean wind speed in the
United States in February. Con-
tours are given in knots.

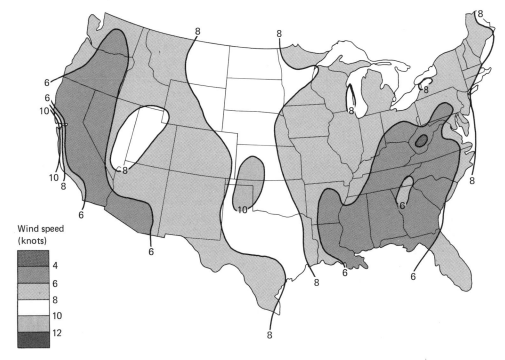

FIGURE 15-2b
Mean wind speed in the
United States in August. Con-
tours are given in knots.

Wind speed
(knots)

| | |
|---|---|
| | 4 |
| | 6 |
| | 8 |
| | 10 |
| | 12 |

mal conditions. From all this evidence we can see that much more research still needs to be performed on this vital subject.

On the average the wind is far less dramatic. Most places on the surface of the earth have wind speeds that average between 5 and 15 knots (see Figure 15.2). On

the average, Cape Dennison on the coast of Antarctica is the windiest place on earth. The average annual wind speed there is 38 knots. On Mount Washington (the run-ner up) the wind averages a mere 30 knots year round and peaks at 40 knots during January.

All the winds on the surface of the earth are strongly affected by friction. Friction significantly slows the winds so that over land it is unusual to have winds above 30 knots in most locations. Since there is less friction over water, winds over the ocean tend to be somewhat stronger but, even so, winds above 40 knots are not common at sea.

A few kilometers above the earth's surface the story is somewhat different. Here the winds aloft are quite commonly strong, especially in the middle and high latitudes during winter when speeds commonly exceed 100 knots. The strongest winds are usually found near the top of the troposphere in a meandering belt known as the jet stream that circles the globe from west to east. Wind speeds over 250 knots are occasionally observed in the jet stream (see Figure 15.3).

## POWER FROM THE WINDS

When a human being works at an average speed he or she generates about 75 watts of power. This is only about one-tenth of a horsepower. Therefore, mankind took a giant step forward when the first animal was hitched to a plow. This dramatic event went unrecorded perhaps because writing had not yet been invented. In fact, it is quite possible that the use of animal power provided our ancestors with the leisure time needed to invent writing.

The next giant step in the use of power was most likely

FIGURE 15.3
The jet stream. The jet stream varies in position and strength from day to day. Notice the wavy flow pattern.

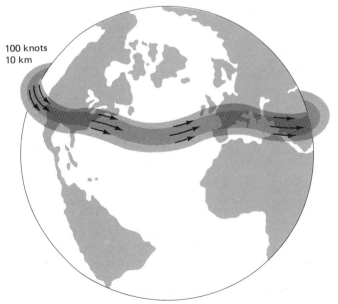

100 knots
10 km

the sailboat. Sails have been used on boats at least since 6000 B.C. These early sailboats were not designed quite so well, because galley slaves were used to provide a more reliable source of power. Similarly, thousands of years later in the early days of steamships, sails were used to provide power during the all too frequent emergencies.

After the sailboat, the next significant advance in the use of power probably took place in Egypt around 200 B.C. At that time many of the world's greatest scholars were assembled in Alexandria. You will recall that many of the ancient astronomers lived and worked here. It was here too that the steam engine and probably the water mill were invented.

Water mills were well known to the ancient world. The Romans actually built a large grain-grinding installation consisting of a series of water mills, but further construction of such labor-saving devices was stopped by the emperor, Trajan. He was afraid that such devices would produce widespread unemployment in a land already plagued by that problem. Thus, the Industrial Revolution may have been delayed 1700 years.

However, much of what was the Roman Empire was not suitable for water mills because of the climate around the Mediterranean Sea. Distinct rainy and dry seasons each year cause the river levels to change drastically from season to season. Therefore, it is necessary to divert the water from rivers in order to run the mills, and this is a very expensive process. The weather is somewhat more reliable in northern Europe, but with the advent of the Dark Ages water mills went out of use for almost 1000 years.

There is also not much use for water mills on the dry, windy plains of Persia (Iran). Perhaps for this reason it was here that the first windmills were reported. The earliest known report that mentions windmills was written by traveling merchants about the middle of the seventh century. However, there is no telling when they first were invented. These early Persian windmills were quite unlike typical Dutch windmills, since the former had vertical axes (like most anemometers). They were used to grind grain and are still used in the more remote areas of Iran.

Windmills were first mentioned in Europe in A.D. 1105. After 1200, they rapidly spread through Europe and by 1908 there were over 33,000 of them in Denmark alone. The windmills were used to grind grain and pump water uphill so that it was the windmills that reclaimed Holland from the sea. The widespread use of windmills in Spain around 1600 is alluded to in Cervantes' novel, *Don Quixote*.

As they were thus discoursing, they discovered some thirty or forty windmills that are in the plain; and as soon as the Knight had spied them, Fortune, cry'd he, directs our Affairs better than we ourselves could have wished: look yonder, Friend Sancho, there are at least thirty outrageous Giants whom I intend to encounter. . . . What Giants, quoth Sancho Panza? Those whom thou see'st yonder, answered Don Quixote, with their long extended arms. . . .

Cervantes, *Don Quixote*

As you probably well know, Don Quixote then attacked the windmills. The battle ended when one of the windmill's long swinging arms swept both Don Quixote and his horse, Rosinante, far into the air, thus giving an indirect proof that windmills generate more than 1 horsepower.

Many of the Dutch-style windmills had arms 20 m long and generated 30 kilowatts. By 1850, windmills produced an amount of energy equivalent to the burning of about one billion kilograms of coal annually in the United States alone. Then, as other sources of energy such as the burning of coal became cheaper and more reliable, windmills were gradually phased out around the world.

During the early twentieth century there was a minor resurgence in the use of windmills of a different type. These are the fanlike, multiblade windmills still seen on the Great Plains today. Before cheap electricity became available in the 1930s, these small windmills were used to pump needed water for agriculture and for the railroads. One nice thing about these small mills is that they can operate for decades without requiring much maintenance.

Now that fuel is becoming more expensive and harder to find, interest in windmills has been rekindled. Large experimental windmills which look like giant airplane propellers are now working in several European nations and in the United States. The noble forerunner of these mills was built shortly before World War II by Smith and Putnam. This huge, two-propeller windmill on top of Grandpa's Knob near Rutland, Vermont, delivered a peak power of 1250 kilowatts and operated until March 26, 1945 when one of the giant 20 m blades was torn off during a windy night and sailed 325 m away.

Harold Perry, the foreman, was aloft standing on the side of the house away from the control panel and separated from it by the 24-inch rotating main shaft. A shock threw him to his knees against the wall. He started for the controls but was again thrown to his knees. He tried again and again was thrown down. Finally, collecting himself, he dove over the rotating shaft, reached the control and brought the unit to a full stop in about 10 seconds.

Wind Energy Hearing before the Subcommittee on Energy of the Committee on Science and Astronautics, U.S. House of Representatives, May 21, 1974, #49

FIGURE 15.4
A Dutch windmill.

Wind power remains attractive, but it has several drawbacks. It can only be used in relatively windy areas, because wind power is proportional to the *cube* of the wind speed (i.e., $v^3$). And even in these areas the wind is intermittent so that expensive power storage systems would be needed. At its best, wind power may be able to provide us with about 5% of our total power needs, but its beauty is that it is totally nonpolluting (but noisy).

The energy crisis is forcing people to think once again of using sails. Modern sailboats are designed so well that they can sail almost directly against the wind! Julius Caesar was astounded when he saw the ancient Gallic sailboats doing this (Roman ships, it seems, did not have this ability).

The earliest sailboats were probably flat bottomed (i.e., they had no keels). A keel is a long, finlike plank that runs the length of the boat's bottom. The keel gives the boat stability and prevents the boat from slipping sideways in the water. Thus a boat with a keel is forced to go either forward or backward as if it were on tracks (see Figure 15.5). Since the earliest sailboats had no keel, they slipped sideways too easily and therefore were forced downwind.

POWER FROM THE WINDS 233

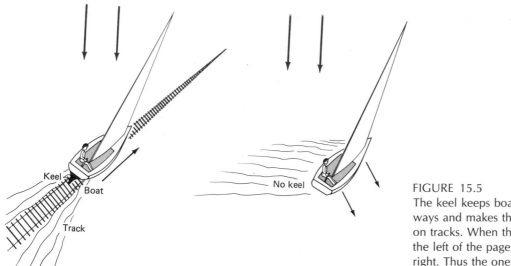

FIGURE 15.5
The keel keeps boats from slipping sideways and makes them act as if they were on tracks. When the sail turns the wind to the left of the page, boats are forced to the right. Thus the one with the keel actually heads into the wind!

It is not that hard to understand how sailboats with keels can sail almost directly into the wind. In Figure 15.5 the wind comes from the top of the page; after it hits the sail the air is deflected to the left of the page. As you will soon see this means that the boat is forced to the right of the page, to compensate for the air deflection. The boat with its keel must go either forward or backward; only by going forward and thus heading into the wind can it have a *component* of motion to the right. It is understandable if you still find this strange. After all, it took thousands of years for people to realize that it could be done.

## NEWTON'S LAWS OF MOTION

You have already read about Newton's discoveries in optics back in Chapter 3. These were dwarfed by his discoveries in mechanics. Johannes Kepler (1571–1630) had discovered three mathematical relationships concerning the motion of the planets, but he was not able to explain them. Newton was!

In fact, not only did Newton explain Kepler's laws, but in the process of doing so he discovered the basic laws that govern the motion of all objects. Newton showed that the very same laws that govern the motion of planets also govern the motion of objects such as apples. This was not a simple job, since along the way he also had to invent the calculus. Thus, Newton revolutionized mathematics as well.

Newton's laws provided a new system of thought from which many aspects of the behavior of the universe could be explained. The laws seemed to imply that if you only knew the motion of all objects at some instant in time that you could, in theory, predict all subsequent motion (and therefore happenings) in the universe. Many philosophers interpreted these laws as the tools that would provide human beings for the first time with the ability (in theory) to predict all future events in the universe. Newton's discoveries therefore had much to do with the optimistic outlook on life during the Age of Reason.

We will now turn to Newton's three laws of motion. You must read the following few pages very carefully if you have never seen Newton's laws before. In fact, you will probably have to read these pages several times. (You will note that I have mentioned the third law before the second law because the second law is more difficult to understand.)

**Newton's First Law:** If no force is exerted on an object then that object will continue to move at the same **velocity** (the same speed *and* direction) that it was moving before.

This law is fundamentally different from the earlier, *mistaken* notion that the natural state of things is be absolutely still. In reality, any object offers resistance when someone tries to change either its speed or its direction of motion. This resistance to change is called **inertia.** In everyday life the term inertia has a different meaning. It is used to describe lazy or stubborn people who find it hard to get started on anything and who tend to remain motionless or *inert*. In science, inertia means resistance to change of motion; this happens to be characteristic of all objects. There is *no* natural tendency to become or remain motionless.

**Newton's Third Law:** Whenever an object exerts a force on a second object, the second object always exerts a force back on the first that is equal in magnitude but in the opposite direction.

This is the law that explains why rocket ships are forced forward when their fuel is forced out of the back of the rocket. It also explains why cannons backfire when the cannonballs are shot out. And it explains why the sailboat in Figure 15.5 moved to the right when it deflected the wind to the left. In Newton's own words, "To every action there is always opposed an equal reaction (see Figure 15.6).

Now for the difficult one.

**Newton's Second Law:** The force on an object equals the mass of the object times the acceleration which is produced.

This is written in the famous formula

$$F = ma$$

where *F* is the force, *m* is the mass, and *a* is the **acceleration.**

In order to understand this law you must first know what the three terms mean. **Force** is almost self-explanatory. It includes such things as pushes, pulls, stretches (of a spring), and friction. These forces involve contact; there are also noncontact forces such as gravitational attraction, electrical attraction, and repulsion.

In these times everyone has a fairly good (though perhaps incomplete) notion of what *acceleration* is. For instance, it is with good reason that the gas pedal in a car is called the accelerator, because it makes the car speed up. Before a plane takes off it must accelerate along the ground until it reaches a certain critical speed. When either the plane or the car accelerate you are pushed back in your seat. This feeling is due to the inertia of your body and shows that you must be *forced* to accelerate.

The more rapidly you are accelerated the more you get pushed back in your seat. Now, imagine that you are riding on a rocket ship that is blasting off from earth. The acceleration is so large that you are pinned to your seat, and even the skin on your face gets pushed back. (At takeoff you certainly would not win any beauty contests.)

There is one complication. Acceleration can be tricky. Tests performed on accelerating astronauts are conducted in a giant centrifuge that spins at a *constant* rate! This might surprise you and you might ask, "How can anything be accelerating if its speed remains constant?"

The answer is actually not very difficult. Acceleration involves not merely change of speed—it also involves change of direction. Objects moving in a circle or in any curved path are changing their direction and therefore are accelerating.

When you stop to think about it you will find that it makes sense. When a car makes a turn you get a similar feeling to the feeling you get when a car is speeding up. The only difference is that when the car speeds up you are forced back, but when it makes a turn in one direction you are forced toward the opposite side of the car.

Any time an object changes its direction of motion it is accelerating even if the speed remains constant. The name for this type of acceleration is **centripetal** acceleration. This is what you feel when a car or plane makes a turn; it is also what you feel when you get on a merry-go-round or any of the other more rapidly spinning amusement park rides (especially those designed to make you sick; see Figure 15.7).

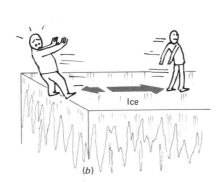

FIGURE 15.6
Newton's third law. To every action there is an equal and opposite reaction.

FIGURE 15.7
The day the merry-go-round went haywire. People feel thrust outward because of centripetal acceleration inward.

**Definition:** Acceleration is the rate of change of velocity with respect to time.

Thus, in equation form, this can be written

$$a = \frac{\text{change of velocity}}{\text{time interval}}$$

Now we must also define velocity.

**Definition:** Velocity is the rate of change of location (position) with respect to time.

Thus, in equation form, this can be written

$$v = \frac{\text{change of position}}{\text{time interval}}$$

Remember that with velocity not only must the speed be specified, but the direction must also be specified.

---

Example 15.1
An Olympic runner does the 100-meter dash in 10 seconds flat. What is the runner's *speed*?

Equation:

$$\text{speed} = s = \frac{\text{distance covered}}{\text{time interval}}$$

Solution
$$\frac{100 \text{ meters}}{10 \text{ seconds}} = 10 \text{ meters per second} = 10 \text{ m/s}$$

Notice that this is called the speed rather than the velocity. In order to be called a velocity the direction must also

be specified. For instance, 10 m/s *toward the north* is an example of a velocity.

---

Now here are some simple examples of acceleration.

---

Example 15.2
A blob of cold, heavy air pours down the slopes from the icecaps of Greenland. Starting from an initial velocity of 0, it reaches the bottom 60 s later with a velocity of 30 m/s. What is its acceleration?

Formula:
$$a = \frac{\text{change of velocity}}{\text{time interval}}$$
$$= \frac{(30 - 0) \text{ m/s}}{60 \text{ s}}$$

Solution
$$\frac{30 \text{ m/s}}{60 \text{ s}} = \frac{1}{2} \text{ m/s}^2 \text{ down the slope}$$
or 1/2 meter *per second per second* down the slope.

**Note:** Such winds are common in Greenland and Antarctica where they are known as katabatic winds. Katabatic winds gradually pick up speed (i.e., accelerate) as they descend the slopes. In certain places (as at Cape Dennison) they will occasionally exceed 50 m/s (or 100 knots)!

---

The most common example of acceleration is the acceleration that is due to gravity. Any object that is dropped falls faster and faster with time. Near sea level the acceleration of gravity is virtually constant and is the same for all objects—no matter how large or small, no matter how light or heavy.

The acceleration of gravity is almost exactly 10 meters per second per second or 10 m/s² downward. It is referred to so often that it is given its very own letter, g. Therefore, g = 10 m/s²; this means that after each second any free falling object not affected by air resistance falls 10 m/s faster that it did the second before. (Actually, g = 9.81 m/s² is more accurate.)

---

Example 15.3
King Kong falls from the top of the World Trade Center. (We are tired of dropping balls from the Leaning Tower of Pisa.) Describe Kong's progress in terms of velocity and acceleration.

Procedure: It is useful to construct a table of Kong's downfall. At the very instant that Kong begins to fall, his velocity is 0 (t = 0), but one second later (t = 1) he has accelerated to a downward velocity of 10 m/s.

Therefore, during the first second his average falling speed has been the average of 0 and 10 m/s, or 5 m/s. Therefore, in the first second, Kong fell 5 m. Thus,

| $t$ | $v$ | Distance Covered During Preceding Second | Total Distance Covered from the Start |
|---|---|---|---|
| 0 | 0 | 0 | 0 |
| 1 | 10 | 5 | 5 |

Now, by time $t = 2s$, Kong has accelerated another 10 m/s to 20 m/s. During this second his average velocity was 15 m/s (the average of 10 and 20 m/s), so that he fell another 15 m for a total distance of 20 m. You can see that Kong is beginning to get places (see Figure 15.8).

| | | | |
|---|---|---|---|
| 2 | 20 | 15 | 20 |

Here is the rest of Kong's descent.

| | | | |
|---|---|---|---|
| 3 | 30 | 25 | 45 |
| 4 | 40 | 35 | 80 |
| 5 | 50 | 45 | 125 |
| 6 | 60 | 55 | 180 |
| 7 | 70 | 65 | 245 |
| 8 | 80 | 75 | 320 |
| 9 | 90 | 85 | 405 |

Solution

Since the World Trade Center is 411 m high (1350 feet), it is all too obvious that King Kong will hit the ground a little more than 9 s after he began falling. Since he will be going more than 90 m/s (175 knots), he will surely be crushed by the impact. (In this problem the effect of air resistance has been neglected.)

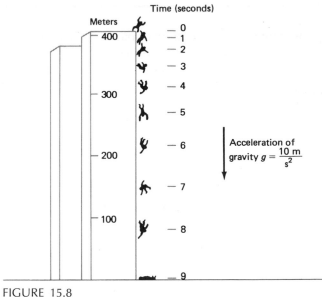

FIGURE 15.8
King Kong accelerates to his doom. In this figure and in Example 15.3 the effects of air resistance have been neglected.

The next two examples concern centripetal acceleration. This is important in meteorology because in many situations the air moves in curved paths such as in high and low pressure areas, hurricanes, tornadoes, and meanders in the jet stream. In fact, even when objects appear to be going straight on earth, they are actually changing their direction because of (1) the curvature of the earth and (2) the rotation of the earth. You will soon see how this is related to the Coriolis force.

Centripetal acceleration ($a_c$) is given by the simple formula

$$a_c = \frac{v^2}{r}$$

where $v$ is the velocity and $r$ is the radius of curvature. The simplest case is when the air moves in a circle so that $r$ is the radius of the circle, as in Figure 15.9.

FIGURE 15.9
Centripetal acceleration for a ball moving in a circle is directed inward to the center of the circle. It is given by the formula $a_c = v^2/r$.

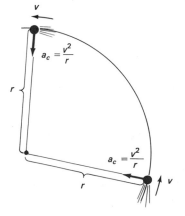

## Example 15.4

An astronaut in a giant centrifuge spins at 30 m/s around a circle 9 m in radius. What is his acceleration? Compare that to the acceleration of gravity.

Equation:

$$a_c = \frac{v^2}{r}$$

Solution

$$\frac{(30 \text{ m/s})^2}{9 \text{ m}} = \frac{900 \text{ m}^2/\text{s}^2}{9 \text{ m}} = 100 \text{ m/s}^2$$

This is 10 times greater than the acceleration of gravity which means that the man is subjected to 10 times the force of his own weight.

## Example 15.5

What is the typical centripetal acceleration in a tornado and in a low pressure area?

Information: For each case we need a typical velocity and a typical radius. The typical values for a tornado are $v = 50$ m/s and $r = 200$ meters, while the typical values for a low pressure area are $v = 10$ m/s and $r = 500,000$ m.

Equation:

$$a_c = \frac{v^2}{r}$$

Solution

For the Tornado

$$a_c = \frac{50^2}{200}$$
$$= 12.5 \text{ m/s}^2$$

This is larger than the acceleration of gravity. No wonder tornadoes can lift houses.

For the Low

$$a_c = \frac{10^2}{500,000}$$
$$= 0.0002 \text{ m/s}^2$$

This is miniscule compared to the acceleration of gravity. Even so, it is very important in weather prediction.

---

Now it is time to define mass. **Mass** is the amount of material of an object. Technically, it is the amount of inertia an object has. For instance, a large ship that is drifting very slowly can crunch into a dock and make it sway simply because the ship has such a large mass.

You must be careful. Mass is not the same thing as weight. Objects in space may *weigh* almost nothing, but they still have the same *mass* as when they were on the earth. Yes, even in outer space where objects are weight-less, they still exhibit the same resistance to any changes in motion.

The weight of an object is equal to the force of gravity on that object. Therefore, *weight is a force* and forces are governed by Newton's second law. Thus, weight or $w$ is represented by the formula

$$w = mg$$

where $m$ is mass and $g$ is gravity.

Mass is given in units of grams or kilograms, while technically weight is given in units of newtons. At sea level, a mass of 1 kilogram weighs 10 newtons. However, on the moon this same kilogram would weigh only 1.8 newtons, whereas on the sun it would weigh 145.5 newtons.

# THE FORCES THAT MOVE AIR

There are several different forces that move the air. Like velocity, each force not only has a magnitude, it also has a direction.

Weight is a force always acting downward (toward the center of the earth). The air is also forced from regions of high pressure to regions of low pressure by the **pressure-gradient force.** As soon as the air starts moving, it is forced to *its* right in the NH (and to its left in the SH) by the rotation of the earth. This is called the **Coriolis force.** Moving air is also slowed down by the force of **friction.** At any time all of these forces may be acting on the air.

**Force 1: Weight**

This always acts downward and by itself would produce a downward acceleration equal to $g$ (10 m/s²).

**Force 2: Pressure-Gradient Force**

When you shake up a bottle of soda, beer, or champagne, the pressure builds up within the bottle. As soon as you open the bottle, the contents are forced out into the open air where the pressure is lower that it is inside the bottle (see Figure 15.10). Similarly, if you are in a crowded subway car, you will be forced out of the car when the doors open, unless it happens to be more crowded on the station platform. In that case, you will be shoved even farther into the subway car.

Just because an object is under great pressure does not mean that there is any force on the object. In order for there to be a force, there must be a pressure *difference;* only then will objects be forced from high pressure toward low pressure.

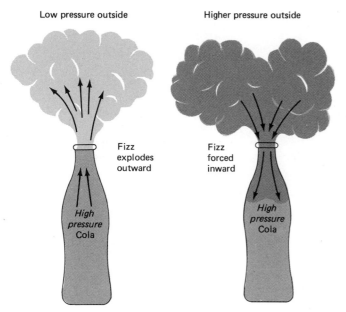

**Low pressure outside**

**Higher pressure outside**

Fizz
explodes
outward

Fizz
forced
inward

*High pressure* Cola

*High pressure* Cola

FIGURE 15.10
Air (as well as any liquid or gases) is forced from regions of higher pressure to regions of lower pressure.

There is one additional factor to be considered with the pressure-gradient force. An object will accelerate more rapidly down a steep hill than down a hill with a small slope (small gradient)—no matter what the height of the hills are. The same thing is true in the case of the pressure-gradient force. The larger the pressure difference from one place to another (i.e., the larger the pressure gradient), the larger the acceleration it will produce (see Figure 15.11).

The pressure gradient is portrayed by isobars. Between any two isobars there is a fixed pressure difference. Therefore, *the closer the isobars, the larger the pressure gradient* (see Figure 15.12). Drawing isobars is also useful because (1) the wind usually blows almost parallel to the isobars, and (2) the closer the isobars, the faster the wind.

The acceleration produced by the pressure-gradient force is given by the formula

$$a_{PG} = (1/\rho)(PG) = (1/\rho)\frac{\text{pressure difference}}{\text{distance}}$$

This formula will work as is when density ($\rho$) is expressed in kilograms per cubic meter, pressure is expressed in newtons per square meter, and distance is expressed in meters. Since we normally express pressure in millibars we must be careful to convert to newtons per square meter before substituting it into the equation.

FIGURE 15.11
The larger the gradient (i.e., the steeper the slope), the larger the acceleration. Successive positions of the ball are shown on each slope at uniform time intervals. Notice also that the larger the gradient, the closer the contour lines on the hill. The units are relative.

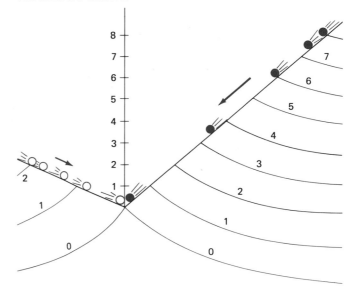

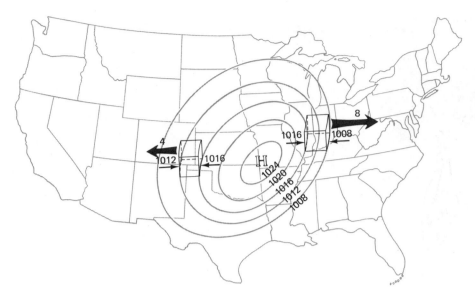

FIGURE 15.12
The pressure-gradient force. Two equal-sized boxes are both forced toward lower pressure. Since there is a larger pressure difference, and hence a larger pressure gradient on the box on the right, it is subjected to a larger pressure-gradient force. This diagram illustrates the fact that the closer the isobars on a weather map, the larger the pressure gradient.

## Example 15.6

A pressure (difference) of 10 mb is equal to (a difference of) how many newtons per square meter? What is a simpler name for newtons per square meter?

### Solution

A newton per square meter is called a **pascal.** Simply multiply the number of millibars by 100 to get the number of pascals. Therefore, 10 mb corresponds to 1000 pascals

## Example 15.7

In a hurricane the pressure difference between two cities is 10 mb. The two cities are 100 km apart. Using a density of $\rho = 1.2$ kilograms per cubic meter (kg/m³) what is the acceleration that the pressure gradient would produce?

Formula:

$$a_{PG} = (1/\rho)\frac{\text{pressure difference (in pascals)}}{\text{distance (in meters)}}$$

A pressure difference of 10 mb equals 1000 pascals. A distance of 100 km equals 100,000 m.

### Solution

$$a_{PG} = (1/1.2)\frac{1000}{100,000} = (0.83)(10)^{-2}$$

This is much smaller than the acceleration of gravity.

## Force 3: Coriolis Force

*The sun came up upon the left*
*Out of the sea came he*

*And he shone bright and on the right*
*Went down into the sea.*

S. T. Coleridge, *The Rime of the Ancient Mariner*

Now, we know that the earth goes around the sun, but it certainly *seems to us* as if the sun is moving. In the NH the sun seems to circle the sky from *left to right,* but that is merely an optical illusion caused by the fact that we on earth are spinning from *right to left* (i.e., counterclockwise). The direction of spinning is the opposite in the SH (see Figure 15.13).

Newton's first law states that all objects will move straight *in an absolute sense* when no forces act on them. Then, since the earth rotates toward the left in the NH, all objects that move straight in an absolute sense will curve (i.e., accelerate) to their right *with respect to the earth.*

According to Newton's second law any acceleration

FIGURE 15.13
The rotation of the earth. As seen from space the NH spins counterclockwise, and the SH clockwise.

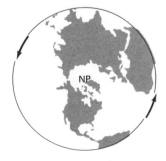

must have been produced by a force. We then *say* that moving objects on earth are *forced* to curve or accelerate to their right as a result of the Coriolis force. In an absolute sense, the object is not accelerating at all; so, there is no real force. Thus the Coriolis force is merely a convenient artifice. There are several interesting examples of the Coriolis force.

During World War I the Germans invented a long-range cannon with which they bombarded Paris. They soon found that all of their shots landed to the right of where they had aimed. In reality, the cannonballs had moved straight, but during the time they were in the air the earth turned enough to cause a miss (see Figure 15.14)! Similarly, when rockets are launched the Coriolis force must be taken into account.

There is also a tendency for the railroad track on the right to wear out somewhat faster in the NH, because the Coriolis force impels the moving train to tilt ever so slightly toward its right.

It is difficult for most people to envision the Coriolis force. The best way to do this is to take a ride on a merry-go-round. Most merry-go-rounds spin to the left or counterclockwise when you look down at them from above. When you get on be sure to take something with you that you can throw off. Once aboard begin throwing. Everything that you throw will be moving in straight lines, but *to you* they will all *seem* to curve to the right. Notice that in Figure 15.15 the man must turn his head farther and farther to the right as the cannonball moves.

Many people will still be unbelievers. They will say, "But when I throw things on earth they do not seem to curve at all." To that I answer, "Yes, you are right because objects that you throw land within 10 seconds, while it takes the earth an entire day to turn once. In 10 seconds the earth turns only a miniscule fraction of a rotation toward the left, so that the object you throw will only turn a miniscule fraction of a rotation to the right. But if you could throw an object that would continue to move for several hours, you would indeed see it begin to turn."

Thus the Coriolis force will only be effective for winds covering considerable distances. It certainly has no significant effect on the wind you create when you blow out the candles on your birthday cake!

There are several more technical aspects of the Coriolis force. The Coriolis force is proportional to the speed of the moving object. Thus, it cannot start or stop objects. It also can never change their speed, because it acts to the right and neither forward nor backward. The only effect of the Coriolis force is to make all moving objects change their direction of motion.

Because the earth is round and not flat like a merry-go-round, the Coriolis force on earth varies with latitude.

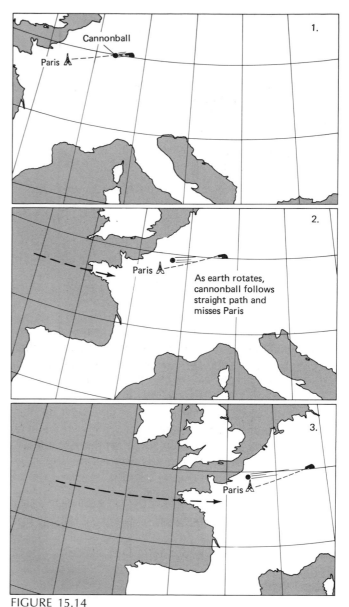

FIGURE 15.14
Long-range cannonballs fired directly at Paris missed because of the rotation of the earth. The cannonballs traveled straight, but the earth had turned.

This is probably its most complicated feature. The Coriolis force is largest at the poles and zero at the equator. Mathematically, it is proportional to the sine of the latitude!

It is not too difficult to see that the Coriolis force depends on latitude. To do this you should envision yourself taking a ride on a Ferris wheel in the (uncomfortable) position shown in Figure 15.16. When you then throw something off the Ferris wheel, it will not appear to curve to the right or left and, thus there is no Coriolis force. The

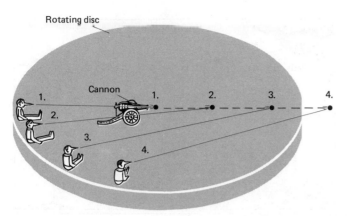

FIGURE 15.15
The Coriolis force can be illustrated by a person on a merry-go-round. Even though the cannonball moves in a straight line, as time goes on the person must turn his head farther and farther to the right to see it. Therefore, the person on the merry-go-round thinks the ball is curving.

difference between the Ferris wheel and the merry-go-round is that the merry-go-round rotates about an axis that is vertical, whereas the Ferris wheel rotates about an axis that is horizontal (see Figure 15.17). Now let's apply this to the situation on earth.

The North and South Poles rotate like the merry-go-round, while the equator rotates like the Ferris wheel. Therefore, there is no Coriolis force on the equator.[1] At any other place on earth there is an intermediate situa-

[1] Technically this statement should read that there is no horizontal Coriolis force on the equator. There is a vertical Coriolis force but it is negligible compared to the other forces in the vertical direction (such as gravity).

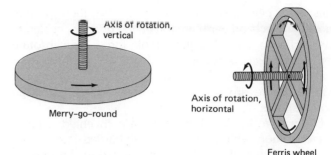

FIGURE 15.17
The merry-go-round has a vertical axis of rotation, whereas the Ferris wheel has a horizontal axis of rotation.

tion. The best way to illustrate this is to show people sitting at various latitudes on earth (see Figure 15.18). Imagine everyone sitting facing the north. Then watch how their points of view change as the world turns. The viewpoint of the person on the equator never changes (this person is always watching the North Star, Polaris), whereas the viewpoint of the person at the North Pole changes most rapidly. The closer you get to the North Pole, the more rapidly the observer turns to the left. In the *SH* the closer you get to the South Pole the more rapidly an observer turns to the *right*.

The acceleration produced by the Coriolis force is called the Coriolis acceleration. It is represented by the formula

$$C = (1.46)(10)^{-4}\sin(\text{latitude})(v)$$

The constant, $(1.46)(10)^{-4}$, is determined by the rotation rate of the earth.

FIGURE 15.16
There is no Coriolis force on a Ferris wheel. When an object is thrown, it does not appear to curve to the right or left. Notice that at successive time periods the person still looks straight ahead (along the axis of rotation) to see the ball.

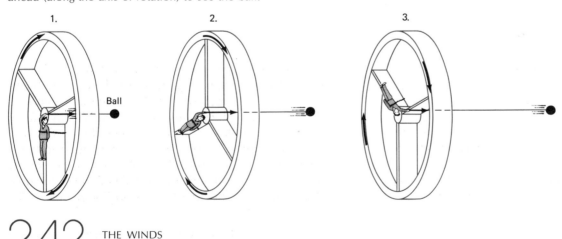

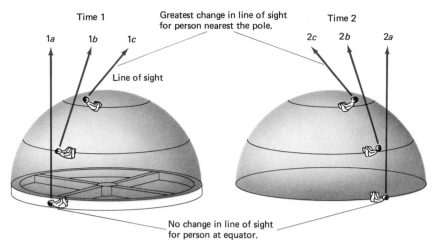

Time 1

1a 1b 1c

Greatest change in line of sight for person nearest the pole.

Time 2

2c 2b 2a

Line of sight

No change in line of sight for person at equator.

Arrows depict absolute direction in space toward which the viewers are looking. From Time 1 to Time 2, the viewpoint of viewer *a* at the equator does not change.

Viewpoint of viewer *c* changes most. Thus the Coriolis force is larger at higher latitudes.

FIGURE 15.18

The Coriolis force depends on the latitude. At the equator, which spins like a Ferris wheel, the observer's viewpoint never changes; so there is no Coriolis force. The closer you get to the poles, the more rapidly your viewpoint changes as the world turns; thus the larger the Coriolis force.

**Note:** In all the examples that follow we will assume that the latitude is 43° because at that latitude the acceleration due to the Coriolis force is always

$$C = (10)^{-4}(v)$$

This makes all the calculations much easier.

### Example 15.8

A railroad train traveling through Milwaukee (latitude 43°) is moving at 30 m/s. Each car in the train has a mass of 10,000 kilograms. What is the Coriolis force on the train?

Equation: Use Newton's second law with the Coriolis acceleration.

$$F = ma; \quad a = C = (10)^{-4}(v)$$
$$= (10)^{-4}(30)$$
$$a = (30)(10)^{-4} \text{ m/s}^2$$

Solution

$$F = (10,000)(30)(10)^{-4} \text{ newtons}$$
$$= 30 \text{ newtons}$$

This force is directed to the right, so that the track on the right is under slightly more pressure.

### Example 15.9

What is the typical acceleration due to the Coriolis force in a tornado and in a low pressure area at Syracuse, New York (latitude 43°)?

Information: Refer back to Example 15.5 so that for the tornado, $v = 50$ m/s and, for the low, $v = 10$ m/s.

Equation:
$$a = C = (10)^{-4}(v)$$
$$C = (10)^{-4}(v)$$

Solution

| For the Tornado | For the Low |
|---|---|
| $C = (10)^{-4}(50)$ | $C = (10)^{-4}(10)$ |
| $= (5)(10)^{-3} \text{ m/s}^2$ | $= (10)^{-3} \text{ m/s}^2$ |
| This is much *smaller* than the *centripetal acceleration* in a tornado. | This is much *larger* than the *centripetal acceleration* in a low. |

This example shows at least two very important facts. First, the Coriolis acceleration is always far, far smaller than the acceleration of gravity—no wonder you don't feel it. Second, the Coriolis acceleration is insignificant when compared to the centripetal acceleration in the tornado. This is because the tornado is a small-scale motion. On the other hand, the Coriolis acceleration is quite important in the low pressure area because the airflow around the low is a large-scale motion. Thus, when you flush your toilet bowl the spinning motion is barely affected by the Coriolis force.

**Force 4: Friction**

Friction almost always acts to slow down moving objects. Friction is much more complicated for water and air than it is for solid objects, so that it is impossible to give any simple formula for the acceleration that friction produces.

In fact, friction does not always slow down air, and it often speeds up water. Believe it or not, many of the world's surface ocean currents are dragged or driven by the friction of the winds at the surface.

Friction acts to slow the winds near any solid boundary, such as the ground. In this case, friction always acts in the opposite direction from the wind. It also tends to reduce all velocity *differences* throughout the fluid and tries to make the entire fluid move at the same velocity.

Friction is quite important for small-scale motions, such as the small whirlwinds which whip around building corners. It is less important for large-scale motions, at least 1 km above ground level, such as the winds of the jet stream, and for most purposes can be neglected. However, even for the large-scale motions, such as the winds around highs and lows, the effects of friction must always be included in the lowest kilometer of the atmosphere.

# COMBINATIONS OF THE FORCES

When several forces act on an object, they have to be added up to get a total force. When adding the forces their directions must be taken into account. This often proves troublesome to beginning students unless the subject is limited to the simplest cases. Since this is not a physics text, only the simplest examples will be presented.

Example 15.10

Two men pull on a lump of gold. One pulls to the right with a force of 700 newtons, while the other pulls to the left with a force of 500 newtons. What is the net force on the gold?

Solution

Both force and acceleration involve direction as well as magnitude. Since both men are pulling in the opposite direction, 500 newtons of the force to the right are canceled by the force to the left, and only 200 newtons to the right are left over. Therefore, the net force on the gold is 200 newtons to the right, and the gold will accelerate to the right (see Figure 15.19).

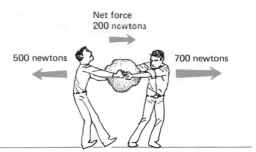

FIGURE 15.19
A force like acceleration also involves direction. The two forces on the lumps of gold partly cancel, since they act in opposite directions.

With this simple example in mind you are ready to analyze combinations of the forces.

**The Hydrostatic Approximation**

Why isn't the sky falling? After all, the air does weigh something. What holds up the atmosphere? The weight of the air constantly tries to make the air accelerate downward. Since the air is not falling, there must be some other force opposing the force of gravity. (Either that or Newton's second law is wrong.)

Newton's law is correct. The force that opposes the force of gravity is simply the vertical component of the pressure-gradient force. After all, pressure decreases quite rapidly with increasing height. This means that there is an upward pressure-gradient force. Most of the time this upward pressure-gradient force almost exactly balances the downward force of gravity. When the upward pressure-gradient force exactly balances the downward force of gravity, the air is said to be **hydrostatic.**

It is because the atmosphere is approximately hydrostatic that the pressure is essentially equal to the weight of air above you (divided by the area, of course). At the level of individual molecules, the impacts between the molecules would send many off into space if it weren't for gravity, which tends to pack them as close to the earth as possible. On planets with smaller gravity (such as Mercury) and satellites (such as the moon) where gravity is weaker, the air molecules that once were there gradually escaped into outer space. Thus the smaller planets and satellites have no atmospheres.

The hydrostatic approximation is more than 99% accurate except in violent thunderstorms, tornadoes, and extreme katabatic winds. In all these cases, there is appreciable vertical acceleration of the air, but these cases are extremely rare.

When the atmosphere is in hydrostatic balance, the upward acceleration due to the vertical pressure gradient cancels the downward acceleration due to gravity. Thus there is no net vertical acceleration, and the formula we have

$$g = (1/\rho)\frac{\text{pressure difference}}{\text{height difference}}$$

---

Example 15.11

How high do you have to go to get to 950 mb when the sea level pressure is 1000 mb at the equator and at the pole? At the equator the air is warm, so its density is only 1.1 kilograms per cubic meter (kg/m³); whereas the air at the pole is much colder and denser, and its density is 1.3 kg/m³.

Equation:
The hydrostatic approximation
$$g = (1/\rho)\frac{\text{pressure difference}}{\text{height difference}}$$

Solution

At the Equator

$$10 = (1/1.1)\frac{5000 \text{ pascals}}{\text{height difference}}$$

At the Pole

$$10 = (1/1.3)\frac{5000 \text{ pascals}}{\text{height difference}}$$

Rearranging to solve for the height difference, we find
Height difference = 455 m  Height difference = 384 m

---

Even without doing this example you can see that the denser the air the smaller the height difference between two different pressure levels. Since the air is dense when it is cold, *the colder the air the smaller the height difference between two different pressure levels.*

The hydrostatic approximation has several very important applications. Since this approximation relates height changes to pressure changes, airplane pilots are able to make altimeters out of barometers! They must only be careful to correct for unusual pressure and temperature conditions at the ground. Failure to do so has caused innumerable crashes.

The hydrostatic approximation can also help us to understand how temperature differences produce the winds. Consider the following simple situation, which is also depicted in Figure 15.20.

As the sun rises over a small island in the middle of the ocean, the ground begins to heat up faster than the water surface, so that the air above the island also begins to heat up. Imagine that before the sun rose there was no wind and there were no *horizontal* pressure differences to accelerate the air. But as soon as the air over the island is heated, it expands slightly and becomes less dense. Although the pressure at sea level remains essentially constant, this is not true of the pressure at higher levels. The hydrostatic approximation tells us that the pressure drops more slowly with height in the warmer, less dense air over the island.

Thus, at higher levels there is a high pressure area over the island, and the air at higher levels begins to accelerate

FIGURE 15.20
Temperature differences create pressure differences, which then produce the winds. The four frames present an explanation of the sea breeze, using the hydrostatic approximation. The units used are millibars.

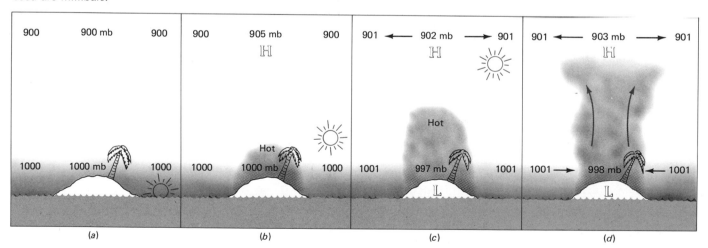

away from the island. As soon as this happens there is less air above the island than over the sea; therefore, the sea level pressure over the island drops, and air near sea level begins to accelerate from the sea toward the land. This is called the sea breeze!

Basically, I have just explained why hot air rises in terms of Newton's laws. This explanation can be used in many diverse situations, for example, to explain why air rises near the equator; it can even be used to help explain how hurricanes develop. In very general terms, however, you have now finally read the technical explanation of how temperature differences produce the winds.

### The Geostrophic Wind Approximation

One of the strangest things about the large-scale winds on earth is that they do *not* flow directly from high to low pressure. Instead, the wind flows almost parallel to the isobars, which is the same thing as saying that the wind flows with low pressure on one side and high pressure on the other side. This strange behavior of the winds was described by Christopher Buys-Ballot (1817–1890).

**Buys-Ballot's Law:** If you stand with your back to the wind in the NH, then low pressure is on your left (see Figure 15.21).

The geostrophic wind is an idealization that gives you a good approximation of the actual large-scale winds that are more than 1 km above the ground. The **geostrophic wind** is the wind that results when the pressure-gradient force and the Coriolis force balance each other. It flows in a straight line on the earth, with low pressure to its left in the NH (to its right in the SH), and it is a steady wind.

A good question to ask is, "How does the wind manage to *avoid* flowing directly toward low pressure?" Imagine the following situation. There is uniformly low pressure in the north and uniformly high pressure in the south, and at first there is absolutely no wind. However, since there is a pressure gradient, there will be a pressure-gradient force that will accelerate the air toward the north (i.e., toward lower pressure).

No sooner does the wind begin to blow than the Coriolis force starts to act, twisting the wind to its right, which in this case is toward the east (see Figure 15.22). The pressure-gradient force continues to speed up the wind; but this only increases the Coriolis force, which then makes the wind curve even further toward the right. Finally, a balance is established, so long as the isobar pattern remains the same and there is no friction.

This balanced (and ideal) wind is the geostrophic wind. In this case, it will be moving toward the east, but it always moves with low pressure on its left in the NH and on its right in the SH. You might ask why the wind doesn't continue turning even further to the right. If it did, then the pressure gradient would begin slowing it down. This would weaken the Coriolis force, since the Coriolis force is proportional to the wind speed, and therefore the wind would turn back somewhat to its left.

The geostrophic wind is just an approximation because of several reasons. Near the ground there is a significant amount of friction, so that the geostrophic wind approximation is never very accurate there. But even above the ground, the winds are almost never steady; rather than

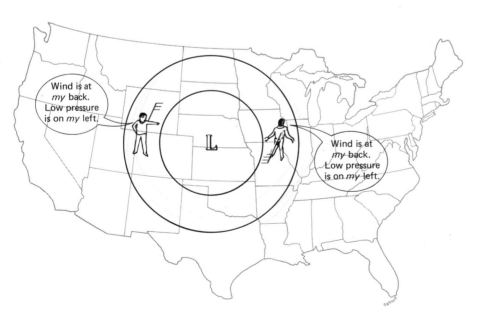

FIGURE 15.21
Buys-Ballot's law. Stand with your back to the wind, and low pressure will be to your left (in NH). Note positions of wind arrows.

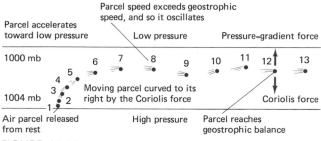

FIGURE 15.22
The geostrophic wind. An air parcel released from rest will accelerate toward low pressure. As it accelerates, the Coriolis force becomes progressively more important. The geostrophic wind (12, 13, etc.) occurs when the pressure-gradient force exactly balances the Coriolis force.

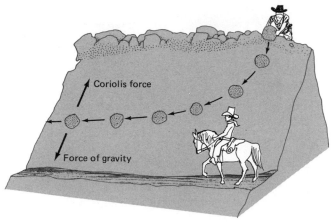

FIGURE 15.23
The analogy of the geostrophic wind for a moving rock on a hillside. The rock eventually will move "sidehill." For this to work, the hill would have to be *many* kilometers high.

flowing in straight lines, which is rare, they tend to flow in curved paths. Even so, the geostrophic wind approximation is reasonably accurate for the large-scale winds, especially away from the ground.

Here is a question for you. Since the pressure-gradient force and the Coriolis force balance and, therefore, cancel each other, there is absolutely no net force on the wind. Why then doesn't the wind stop?

The answer is actually quite simple. The natural state of things is *not* to remain still. It is to *continue* moving in the same direction at the same speed so long as no net forces are acting on it.

There is one strange feature about the geostrophic wind that astounds most people when they first learn about it. As an analogy for the geostrophic wind, let's imagine a ball or rock that will *not* roll downhill, but instead rolls with the downhill side on its left (in the NH)!

Imagine John Wayne riding on a mountain path with a nasty villain above. The villain tries to kill our hero, John Wayne, by rolling a huge boulder down the mountain. But the villain has forgotten about the Coriolis force. This makes the rock curve to its right, and John Wayne is saved.

Ridiculous, you say. Anyone knows that rocks roll straight downhill and don't curve to their right. But you must remember that the effect of the Coriolis force is only noticeable when objects move freely for several hours; in reality a rock takes a minute or two at most to roll down a hill.

However, if the mountain were much, much higher, then the rock *would* turn to its right. Our rock analogy for the geostrophic wind is illustrated in Figure 15.23; notice that the rock is rolling "sidehill" with the downhill side on its left!

Silly as this situation may seem, it actually does happen in the ocean. Sea level at Bermuda is actually more than a

meter higher than it is along the east coast of the United States. Why doesn't the water flow downhill toward the United States? If it began to flow downhill, the Coriolis force would turn it toward its right, and it never would reach the east coast. The actual current flows parallel to the coast, with the low water level on its left. You probably know the name of this ocean current—it is the Gulf Stream. The Gulf Stream is an example of a geostrophic current (see Figure 15.24).

There is a very simple formula for the geostrophic wind. Since the pressure-gradient force balances (or equals) the Coriolis force for the geostrophic approximation, we have

$$C = a_{PG}$$

or

$$(1.46)(10)^{-4}\sin(\text{latitude})(v) = (1/\rho)(PG)$$

Since we often want to find the wind, we can formulate an equation called the geostrophic wind equation. Thus we have

$$v_g = \frac{(6.85)(PG)(10)^3}{(\rho)\sin(\text{latitude})}$$

At 43° latitude

$$v_g = \frac{(10)^4(PG)}{\rho}$$

where $v_g$ stands for the geostrophic wind.

Example 15.12
Flint and Port Huron, Michigan, are both at 43° latitude and are 100 km apart. They have a pressure difference of

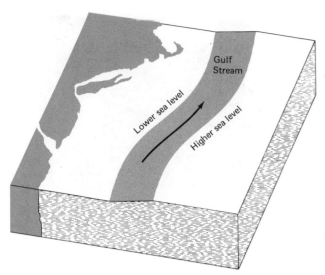

FIGURE 15.24
The Gulf Stream is a geostrophic current. Sea level (i.e., pressure) is lower on its left! Thus sea level is actually higher at Bermuda than along the entire eastern seaboard of the United States.

10 mb during an intense low. The density of the air is $\rho = 1.2$ kg/m³. What is the geostrophic wind?

Equation: The geostrophic wind equation.

$$v_g = \frac{(10)^4\,(PG)}{\rho}$$

Solution

$$v_g = \frac{(1000/100,000)(10)^4}{1.2}$$
$$v_g = 83.3 \text{ m/s}$$

**Note:** The geostrophic wind equation always overestimates the actual wind speed in a low pressure area, as will be explained in the next section. Furthermore, the geostrophic wind equation is usually an overestimate, because it does not include the slowing influence of friction. Usually, friction will slow the wind by 25% or more.

### The Gradient Wind Approximation

Why does air flow counterclockwise around lows and clockwise around highs in the NH?

Imagine that we have a circular low pressure area. The pressure-gradient force then accelerates the air toward low pressure. As the air accelerates toward the center of the low, the Coriolis force twists the air parcels to their right; Figure 15.25 shows that this automatically implies a counterclockwise flow. To show why the air flows clock-

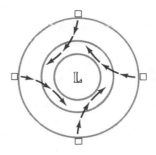

FIGURE 15.25
As air parcels are released from rest, they accelerate into the low, and the Coriolis force twists them to their right. This produces counterclockwise flow.

wise around highs, all you have to do is start the air moving away from the high, and then let the Coriolis force deflect the winds.

Now as soon as the flow is curved, the geostrophic assumption is no longer valid because there is centripetal acceleration. The **gradient wind** approximation takes into account the centripetal acceleration of curved flow patterns.

For most large-scale motions the Coriolis acceleration is much larger than the centripetal acceleration. What this means is that the pressure-gradient force almost balances the Coriolis force in large-scale flows such as highs and lows. The small residual produces the centripetal acceleration.

When the Coriolis force is *larger* than the pressure-gradient force, the flow curves to the right or clockwise (in the NH). This corresponds to a *high* pressure area.

On the other hand, when the Coriolis force is *less* than the pressure-gradient force, the air curves in a counterclockwise sense (in the NH); this corresponds to a low pressure area.

What does this mean? Recall that the wind speed and the Coriolis force are proportional to each other. If a high pressure area and a low pressure area both have the same pressure gradient, the high will have the larger Coriolis force and therefore the faster winds.

This effect may be subtle, but it is extremely important. High in the troposphere (at the jet-stream level) the wind patterns often resemble those of Figure 15.26. In the region near the low (L) the winds are slower than in the region around the high (H), *assuming the pressure gradient to be the same everywhere.*[2] Therefore, downstream from the low the air is spreading out (technically, this is called **divergence,** and some *other* air is needed to replace the diverging air. Since this process is most promi-

[2] One look at the actual wind patterns at 500 mb (see, e.g., Figure 17.28 or Figure 17.33) will convince you that this is an idealization designed merely to help point out an important tendency of the airflow.

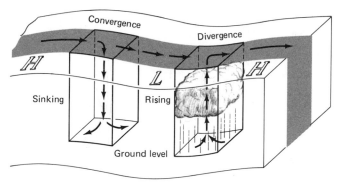

FIGURE 15.26
The vertical wind patterns induced by divergence and convergence at jet-stream levels. Where do you think the low and high pressure areas will tend to form at ground level?

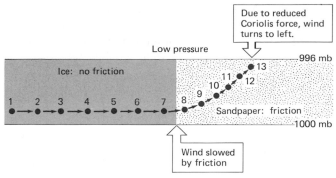

FIGURE 15.27
As soon as friction is introduced, the wind slows. This reduces the Coriolis force so that the wind then angles in toward low pressure.

nent at the *top* of the troposphere (because the winds there are strongest), the *other* air must come up from below.

This reasoning implies that downstream from upper-tropospheric lows, there is a tendency for air to rise. Therefore, downstream from the upper-tropospheric lows, the weather is expected to be cloudy with precipitation. Downstream from upper-tropospheric highs, the weather should be mostly clear because there tends to be **convergence,** which ''squeezes'' the extra air downward. You will see the importance of Figure 15.26 in Chapter 17.

## Frictional Flow

Near the ground friction is important. Friction has two major effects on the winds. First, it slows the winds. This is just what anyone would expect. Second, friction also causes the wind to turn somewhat to the left (in the NH) of where it otherwise would have gone. This effect is somewhat strange and is due to the rotation of the earth.

Imagine the wind flowing parallel to straight isobars without feeling any friction. Suddenly, Mother Nature remembers that friction cannot be neglected. Instantly, the wind begins to slow down (see Figure 15.27). As soon as it does begin to slow down, the Coriolis force (which is proportional to the wind speed) also decreases. This destroys the geostrophic balance, and the wind accelerates (i.e., turns) toward its left, into low pressure.

Finally, a new balance is established between the pressure-gradient force, the Coriolis force, and the force of friction. Because of friction, the real wind near the ground is typically 25 to 75% slower than if there were no friction, and it also blows obliquely across the isobars, heading into low pressure at an angle between 15 and 50°

(depending on latitude and on how strong the friction is). Nearer the equator, where the Coriolis force is weaker, the wind tends to cross the isobars at a larger angle. Over the ocean, where there is less friction, the wind typically slows by blowing across the isobars at an angle between 15 and 30°; while over the land, where there is more friction, the angle is usually between 25 and 50°.

**Note:** One immediate consequence of the cross-isobar flow is that we must modify Buys-Ballot's law. It should actually read: Stand with your back to the wind (in the NH). Next, turn about 30° to your right. Then low pressure will be to your left.

In Chapter 1 I stated that air spirals counterclockwise and *inward* toward lows (clockwise in the SH). Now, after more than half the book, you are finally beginning to read the explanation. Friction plays an important role, since it slows the winds and causes them to deviate toward low pressure. Because of friction, air cannot simply spiral around highs and lows, but must cross the isobars toward low pressure. Thus in both hemispheres air spirals in toward lows and outward from highs (see Figure 15.28).

FIGURE 15.28
Winds spiral near the ground into lows and out of highs largely because of friction.

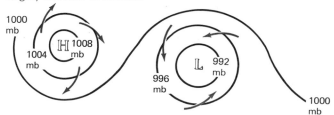

You can actually see something similar to this happen in a teacup. Stir loose tea so that it spins rapidly. The tea then piles against the walls of the teacup, and the level of the tea becomes low in the center so that the pressure is also low there. When you stop stirring the tea it will gradually stop spinning. At the bottom of the cup, where there is more friction, the tea rapidly stops spinning; then the pressure-gradient force makes the tea drift toward the center. You can see that this is so because the tea leaves on the bottom will pile up *in the center!*

As you rise above the ground the influence of friction becomes progressively smaller, until it is virtually insignificant above a height of 1 km. Not only does the wind become stronger with height, but it is less deflected to the left with height. Thus, *the wind spirals around to the right as you rise through the lowest kilometer of the atmosphere* (or the friction layer). This is called the **Ekman spiral** after the Norwegian scientist, Vagn Walfrid Ekman, who first explained it mathematically (see Figure 15.29).

Actually, Ekman had explained the Ekman spiral for the ocean and not for the atmosphere. In 1893, Fridtjof Nansen (1861–1930) led a scientific expedition onto the ice of the Arctic Ocean. For the next three years, blown about by the winds, Nansen and his crew drifted around on the polar sea ice, which covers most of the Arctic Ocean. During his expedition Nansen noticed a very strange phenomenon. The ice was not blown straight downwind; instead, it drifted off to the right of the winds, at angles between 20 and 40°. He realized that this was due to the Coriolis force, which began to twist the ice floes to the right of the wind after the wind had started the floes moving (see Figure 15.30).

This is another natural phenomenon that may seem to be merely a curio to you at this point, but it is actually quite important; its practical significance becomes evident when **upwelling** is discussed in Chapters 16 and 22.

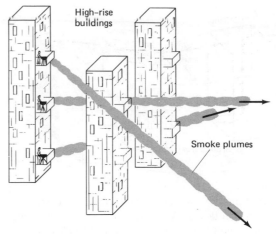

FIGURE 15.29
The Ekman spiral. The wind veers further to the right and strengthens (note longer smoke trails) with increasing height because friction diminishes.

FIGURE 15.30
Ice floe drifts at an angle to the right of the wind because of the Coriolis force.

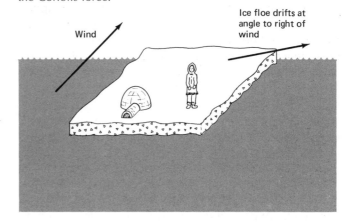

# PROBLEMS

**15-1** Explain how a keel enables a sailboat to sail into the wind.

**15-2** Why are pollution levels often high during the summer even though there is considerable solar heating?

**15-3** What is meant by a wind gust?

**15-4** Give an example of Newton's third law of motion.

**15-5** Explain how it is possible for an object moving at a constant *speed* to be accelerating.

**15-6** A ball is dropped from a tower. Starting with an original velocity of 0, in 5 seconds it is falling at 20 m/s. What is its average acceleration?

**15-7** The acceleration of gravity on the planet Mercury is 3.4 m/s². Recompute how long it would take King Kong to hit ground after falling off the top of the World Trade Center if it were on Mercury. How fast would he be going just before he hit? (Follow the procedure of Example 15.3.)

**15-8** Evaluate and compare the typical centripetal acceleration and the Coriolis acceleration for (a) a hurricane using $r = 50$ km, $v = 50$ m/s and (b) a toilet bowl using $r = 0.1$ m, $v = 0.5$ m/s.

**15-9** You travel through space visiting many planets. What will vary from place to place — your weight or your mass? Explain.

**15-10** A pressure gradient of 1 mb/km is equal to how many pascals per meter? Keep this useful conversion in mind for future problems.

**15-11** What is the approximate pressure gradient around Pittsburgh in Figure 17.3 in units of pascals per meter? Assuming air density to be 1.2 kg/m³, what is the acceleration due to the pressure gradient there? (Use the formula for 43° latitude.)

**15-12** While sitting on a merry-go-round explain the Coriolis force to a friend.

**15-13** Explain why the Coriolis force acts toward the *left* of your direction of motion in the SH.

**15-14** Why isn't the Coriolis force apparent to you when you throw a ball?

**15-15** Why is there no horizontal component of the Coriolis force on the equator?

**15-16** Explain why high pressure areas form at the surface as a result of prolonged, intense cooling in a region.

**15-17** Explain why low pressure areas tend to form at the surface in the desert during the summer.

**15-18** Give Buys-Ballot's law for the SH.

**15-19** Explain the strange feature of the geostrophic wind (i.e., its direction).

**15-20** Using the formula valid at 43° latitude, find the geostrophic wind speed for Pittsburgh in Problem 15-11. Compare this with the actual wind speed.

**15-21** In later chapters you will be using constant pressure charts (i.e., 500 mb charts). On these charts, by definition, there is no pressure gradient; then, the geostrophic wind is determined from the height gradient (i.e., slope). The geostrophic wind formula is

$$V_g = 10^5 \left( \frac{\text{change in height}}{\text{change in distance}} \right)$$

at 43° latitude. (a) What is the height gradient in units of meters per meter in Figure 17.33 over Pittsburgh? (b) Compute the geostrophic wind speed there and compare it with the actual wind speed. (c) Show how this formula is a combination of the hydrostatic formula and the geostrophic wind formula given in terms of the pressure gradient.

**15-22** For a given pressure gradient the geostrophic wind is slower at higher latitudes. Is this statement true or false? Justify your answer.

**15-23** What two factors make the geostrophic wind formula overestimate the wind speed in a low?

**15-24** Explain why the geostrophic wind formula is more accurate over the ocean than over land, and why it is almost worthless in the mountains.

**15-25** Explain how wind direction depends on stability.

**15-26** The fastest wind gusts tend to blow at an angle to the right of the average wind in the NH. Why?

**15-27** Explain why polar sea ice blown by the wind in the SH drifts off at some angle to the left of the wind.

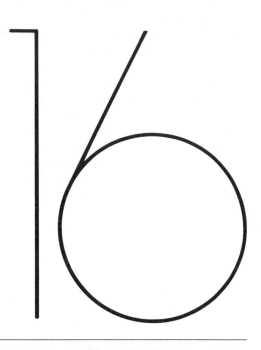

# GLOBAL SCALE WINDS

When Christopher Columbus set sail on August 3, 1492, he was armed with a knowledge of the winds. He had previously sailed the Atlantic extensively and had even reached the Azores Islands. The Azores are located due west of Spain at about a 37° latitude and almost one-third of the way from Spain to America. Even the discovery of the Azores had been a monumental sailing feat. Several mariners before Columbus had tried to sail westward from the Azores, but never got very far because of the persistent westerly winds that kept blowing them back east.

Columbus was a bit smarter than these mariners. He realized that the westerly winds around the Azores would only be good for returning to Spain, and he would have to look elsewhere for *easterly* winds that could blow him across the Atlantic.

Columbus knew where to find easterly winds. For more than half a century, the Portuguese had used the northeasterly trade winds to sail down the coast of Africa. Columbus knew that these trade winds were located in subtropical latitudes, south of Spain. Therefore, when he left Spain he immediately sailed south to the Canary Islands and, after a brief affair with the Queen of the Canaries, used the trade winds to drive his ships quickly across the Atlantic.

On the return voyage, he first sailed northward and then used the midlatitude westerly winds to blow him back to Spain, stopping at the Azores to emphasize his point. He also followed a similar route on his three subsequent voyages (see Figure 16.1).

Mariners have used their knowledge of the large-scale winds for centuries. Airplane pilots and balloonists now use their knowledge of the winds in the upper troposphere in the same manner. A knowledge of the winds is also useful to people in many other walks of life. In the Midwest, farmers have learned to plow their fields at right angles to the prevailing winds to minimize erosion. Since the world's ocean currents flow mostly parallel to the general winds, a knowledge of the winds is still useful to mariners and to fishermen as well. Finally, as you will see, a knowledge of the general winds is almost equivalent to a knowledge of the world's climates!

This chapter begins with a description and explanation of the large-scale surface winds. You will also learn most of the large-scale ocean currents. You will see how climate information can be inferred from the winds. Then we will discuss the large-scale winds of the upper troposphere (including the jet stream). This chapter concludes by showing how the atmosphere acts like a giant engine.

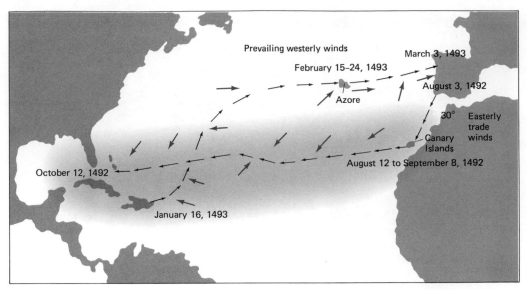

FIGURE 16.1
The winds for Columbus' first voyage.

# THE LARGE-SCALE SURFACE WINDS

The tropics are dominated by the trade winds. In 1735, George Hadley (1685–1768) presented the first reasonable explanation for the trade winds. He realized that the air near the equator is warmed more than the air farther poleward, so that the air near the equator is lighter and rises. Air near the ground then moves toward the equator to fill the "void" created by the rising air. The air returning to the equator is then deflected toward the west by the rotation of the earth (i.e., the Coriolis force), and the result is the trade winds.

The region where the trade winds converge and air rises is known as the **intertropical convergence zone** *(ITCZ)*, and it can often be seen from satellites as a band of cloudiness more or less parallel to the equator. After rising this air spreads out from the ITCZ and sinks on either side of it, causing clear weather. Thus a circulation cell exists, which is now called the **Hadley cell** (see Figure 16.2).

In general, Hadley's picture works quite well for the tropical and subtropical latitudes, and his short paper was a remarkable piece of work. Immanuel Kant and John Dalton each independently proposed the same scheme, unaware that Hadley had done so over 50 years earlier.

Nevertheless, there are several problems with Hadley's explanation. During the summer months the air in the Sahara desert is definitely warmer than the surroundings throughout the lower half of the troposphere. In spite of this, the air still sinks! This is a situation that must be forced—much the same way that water in rapids is temporarily forced uphill over rocks and boulders.

There is no simple explanation for this situation, but a related fact about the Sahara may begin to resolve the problem. Unlike the bulk of the tropics, the Sahara actually loses more heat to space than it receives from the sun. Excess heat from the nearby tropics then pours into the Sahara. Thus the Sahara acts as a heat-escape valve for the tropics, keeping them from growing too hot.

FIGURE 16.2
An idealized picture of the trade winds and the ITCZ.

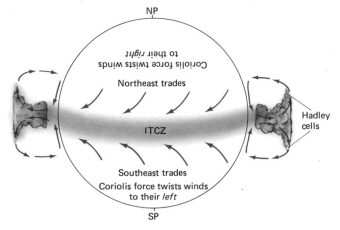

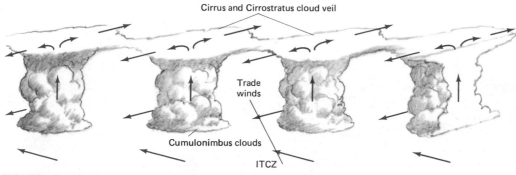

Cirrus and Cirrostratus cloud veil

Trade winds

Cumulonimbus clouds

ITCZ

FIGURE 16.3

The ITCZ is a zone in which thunderstorms tend to form. Its continuous appearance is usually due to the cirrus and cirrostratus clouds that tend to spread out and form a complete veil.

If the air in the Sahara suddenly began to rise, clouds would form. Then, because of the increased greenhouse effect, heat could no longer escape so rapidly from the Sahara, which would then no longer serve as a heat-escape valve. This might then warm the nearby tropics so much that they would become far warmer than the cloudy Sahara, forcing the air in the Sahara to sink once again.

The next problem with Hadley's explanation is that the ITCZ is not a zone of continuously rising air. Instead, it is a zone in which thunderstorm development is favored. Almost all the rising air of the tropics ascends through these thunderstorms in the ITCZ, which have been named "hot towers" because of the enormous amounts of latent heat of condensation and of sensible heat they pump into the upper troposphere. The air from these "hot towers" then spreads poleward and sinks over broad regions in the subtropics.

The occasionally continuous appearance of the ITCZ on satellite pictures is a result of the cirrus and cirrostratus clouds that spread out from the tops of the individual thunderstorms and overlap to form a complete cloud cover (see Figure 16.3).

Over the oceans the ITCZ often appears as a relatively narrow, well-defined cloud band. Over land it is often a broad, poorly defined zone of increased thunderstorm activity and is occasionally impossible to identify since it is so diffuse. Therefore, in the future discussion of tropical rains, the term ITCZ is used somewhat loosely.

Over the ocean the ITCZ is very sensitive to the temperature of the sea surface. Since thunderstorms are more likely over a warm surface, the ITCZ tends to form where the water is warmest. Generally, this occurs slightly north of the equator in the Pacific and Atlantic Oceans.

Occasionally, two ITCZs will form. This occurs most frequently in the eastern Pacific Ocean near the coast of

Columbia and shortly after the equinoxes. The clear area between the two ITCZs is largely due to the band of cool waters right on the equator that suppresses thunderstorm formation (see Figures 16.4, 16.5 and 16.6).

**Seasonal Changes in the Winds: The Indian Monsoon**
Columbus was not the first sailor to use a knowledge of the large-scale winds. At least 2000 years earlier the Phoenicians (who may have sailed around Africa) were also aware of the trade winds. The Arabs also learned to use the large-scale winds to sail around the Indian Ocean.

The situation in the Indian Ocean is somewhat different than in the Atlantic. In the Indian Ocean half the year the winds blow from the northeast (like the trades); it was at this time that all ships would sail westward to Africa or Arabia. During the summer months the winds blow from the southwest, and at this time all ships would sail for India or the Far East. This famous reversing wind system is known as the Indian monsoon (see Figure 16.7).

The monsoon is similar to a giant sea breeze. During July when the sun is overhead north of the equator, India and all of South Asia have the warmest air. Thus the air rises and this is where the ITCZ is found.

To the south, over the Indian Ocean, the winds follow a long path. They start in the SH looking like the normal trade winds, but once they cross the equator they veer toward the northeast (i.e., they become southwest winds). They curve once they cross the equator because the Coriolis force starts deflecting them to the right in the NH. This "recurving" happens to any trade winds that cross the equator.

During the months around January when the sun is overhead in the SH, the warmest air is south of the equator; thus here is where the ITCZ will be found. This situation produces the normal northeast trades in the north Indian Ocean.

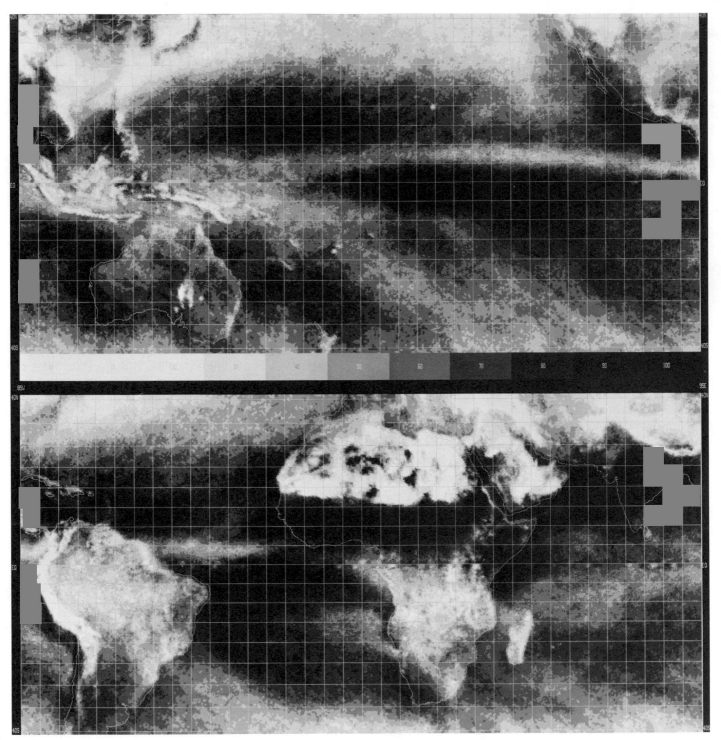

FIGURE 16.4
The average cloudiness during January 1967 to 1970. Find the ITCZ. The bright areas in this figure and Figure 16.5 over the Sahara and Arabian deserts represent sand, not clouds. Note also (1) extensive cloudiness over Africa south of the equator, (2) cloudiness off western coast of Africa centered about 15°S latitude (due to cold waters) (3) lack of clouds over India.

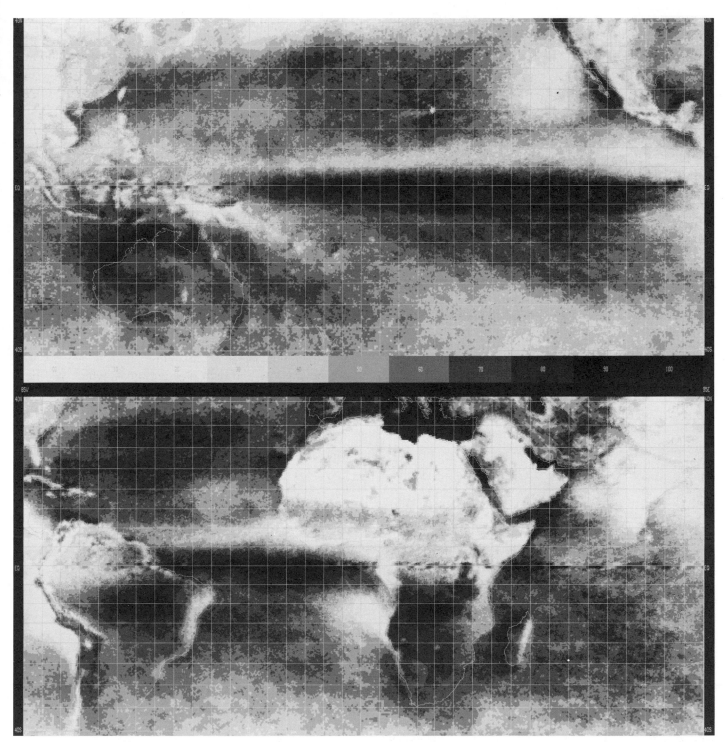

FIGURE 16.5
The average cloudiness during July 1967 to 1970. Notice how
the ITCZ is broader and moves more over land than over the
oceans from January to July. Note also that it is now mostly
clear over Africa South of equator but quite cloudy over India.

FIGURE 16.6
Two ITCZs over the eastern Pacific Ocean on April 7, 1976. The clear gap is caused by cold water right on the equator.

The ITCZ of the Indian Ocean therefore moves over the course of the year (see Figure 16.8). It reaches its northernmost position in July and its southernmost position in January. It is as if the ITCZ follows the sun!

This behavior works for almost all large-scale wind systems and can be expressed as a general rule.

**Rule:** Large-scale wind patterns (i.e., wind systems) "follow" the sun, moving north after January and moving south after July.

As you will see, this rule is very important for understanding seasonal changes in the climate.

### The Global Picture of the Winds

The westerly winds that Columbus used to return to Spain prevail in the middle latitudes of each hemisphere—from about 30 to 60° latitude. Then from 60° to the poles, easterly winds dominate (see Figure 16.9).

One of the major features of the winds outside the tropics is their high degree of variability. This is largely due to the moving low and high pressure areas, which will be discussed at length in Chapter 17. In most places the wind hardly ever comes from the same direction for more than two days in a row. Therefore, it is only in an average sense that the midlatitude winds are westerlies and the polar winds are easterlies.

FIGURE 16.7
The reversing winds of the Indian monsoon. Notice how the winds curve *toward* the east (i.e., become westerly) after they cross the equator.

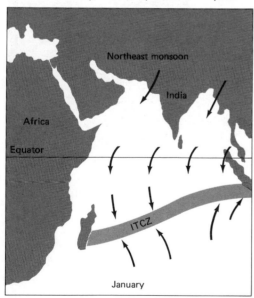
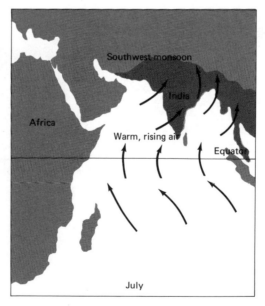

The *average* winds and pressure patterns on the earth's surface are shown in Figures 16.10 to 16.12. Most people take one glance at these charts and then say to themselves "I give up." It might surprise you to find out how small an effort is needed to memorize the entire pattern! There is after all some logic to the large-scale winds.

There are two features about the pressure patterns that stand out. Large, high pressure areas are centered over each subtropical ocean at about 30° latitude, whereas low pressure areas are centered over each subpolar ocean at about 60° latitude. Now, all you must know is how air flows around highs and lows in each hemisphere and you know the winds of the world!

One general observation is that *at any particular latitude* the pressure tends to be higher where the temperature is lower than average. In the subtropics the pressure is therefore higher over the ocean, because it is cooler than the land. This is especially true during the summer when the land gets very hot. Thus the subtropical highs over the ocean are stronger during the summer season. In similar fashion, the subpolar oceans are much warmer than the land during the winter season, and this is when the subpolar lows over the ocean are more intense.

The other seasonal changes in the wind patterns occur because the wind patterns follow the sun. Thus the ITCZ, the subtropical highs, and the subpolar lows are farther north in July than they are in January.

The large, subtropical oceanic highs are notorious to sailors. Winds in the center of highs are usually very light or even calm, and many vessels have been becalmed when they have drifted into the center of these highs. The latitudes around 30° have been called the **"Horse Latitudes"** by the Spanish who often had to throw their horses into the sea when their ships were becalmed so long that the water supply began to run out.

Another region of frequent calms occurs at the ITCZ or between two ITCZs when there are two. This region is called the **doldrums,** and the wind is often light or calm because neither trade wind dominates.

The doldrums have been immortalized in the poem *Rime of the Ancient Mariner* by Samuel Taylor Coleridge.

*The fair breeze blew, the white foam flew,*
*The furrow followed free:*       (The trades)
*We were the first that ever burst*
*Into that silent sea*

*Down dropt the breeze, the sails dropped down,*
*Twas sad as sad could be;*       (The doldrums)
*And we did speak only to break*
*The silence of the sea!*

*All in a hot and copper sky,*

*The bloody Sun, at noon,*       (The equator)
*Right up above the mast did stand,*
*No bigger than the Moon.*

*Day after day, day after day,*
*We stuck, nor breath nor motion:*
*As idle as a painted ship*
*Upon a painted ocean.*

This beautiful poem contains an amazing amount of information about the winds and the weather, in addition to its eerie mystical message. Indeed, the poem tells a tale of redemption through love of the beauties and wonders of nature.

## SURFACE OCEAN CURRENTS

Since the surface ocean currents largely parallel the average winds, it is a simple matter to learn both together. There are only a few places in the world where the ocean currents are tricky (from our point of view). The currents in the north Indian Ocean change completely from January to July, but that is because the winds do as well. The real tricky current is the equatorial countercurrent that is slightly north of the equator and that runs from west to east.

It is also easy to tell which ocean currents are cold and which are warm. Currents flowing away from the poles are cold, whereas currents flowing away from the equator are warm. The world's ocean currents and temperatures are shown in Figures 16.13 and 16.14.

Now you can see why northwestern Europe has such warm winters—their air is warmed by the warm waters of the Gulf Stream. Similarly, Newfoundland's cool summers are a result of the cold Labrador Current. In short, the ocean plays such an important role in determining the climate of a region that any meteorologist must know the ocean currents.

There is another interesting feature about ocean currents that has a tremendous influence on the climate of many coastal regions such as California and Peru. This feature is known as **upwelling.**

I was at Atlantic Beach, New York, during the record-breaking heat wave on the July 4 weekend of 1966. On July 4, a strong west wind suppressed the usual cooling sea breeze. Temperatures at the beach rose to 37°C.

But the wind had the opposite effect on the ocean waters. It blew the warm surface waters away from the shore. The surface waters were then replaced by much cooler water always present below. The cool water *upwelled* all along the shoreline, and the water temperature fell to 14°C.

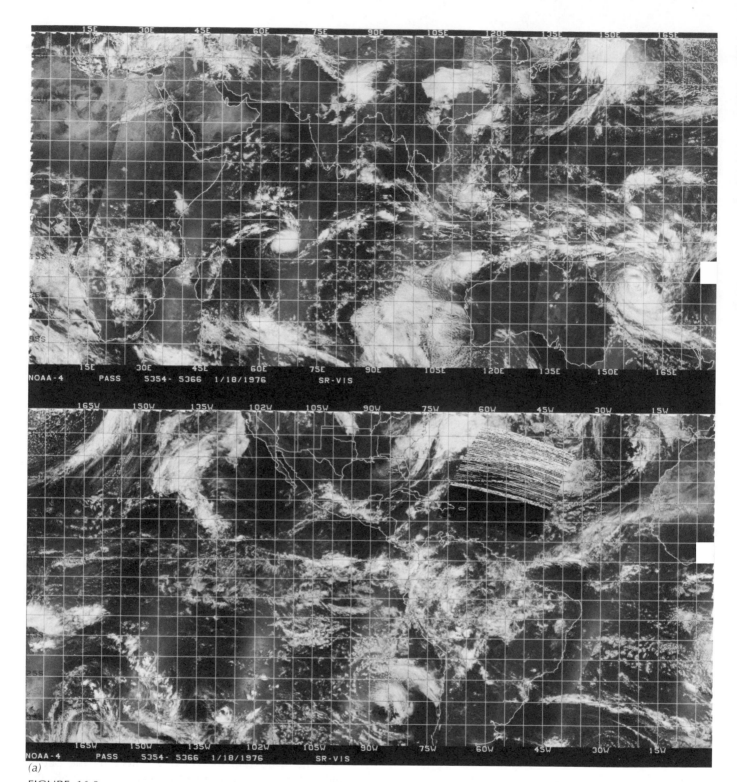

(a)

FIGURE 16.8
Location of the ITCZ over the Indian Ocean on (a) January 18,
1976 and (b) July 15, 1976. Notice that it is difficult in certain
places to identify the ITCZ. Compare with Figures 16.4 and
16.5.

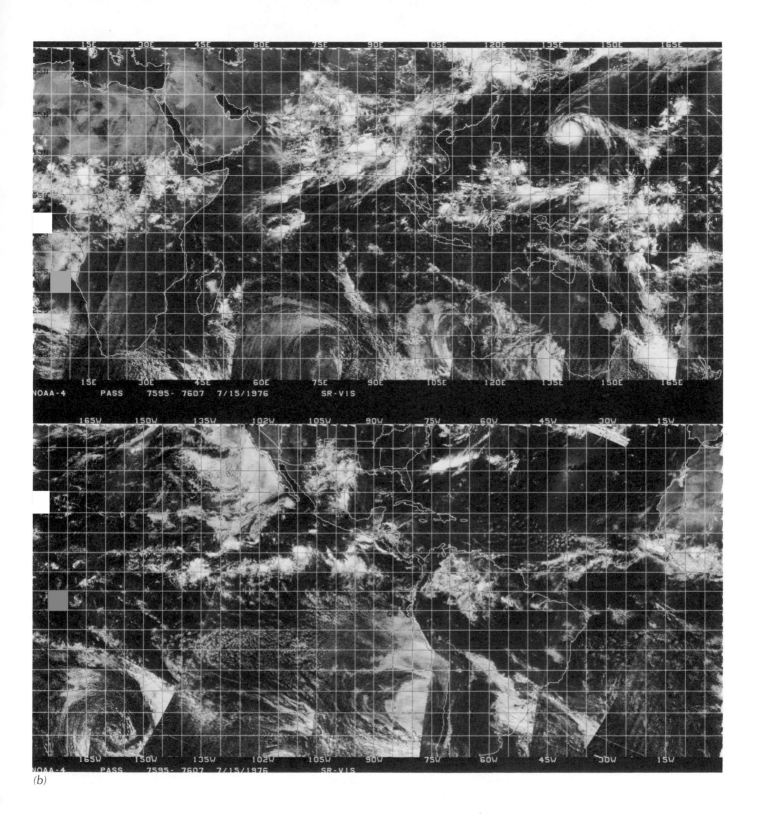

NOAA-4    PASS    7595- 7607   7/15/1976         SR-VIS

NOAA-4    PASS    7595- 7607   7/15/1976         SR-VIS

(b)

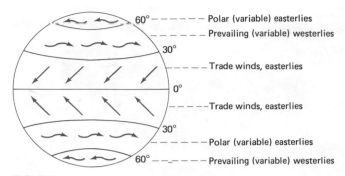

**FIGURE 16.9**

The principal wind zones on earth. Each hemisphere is an up-side-down mirror image of the other.

This meant that it was too hot to stay out of the water, but too cold to swim in it. Thousands of people lined up in knee-deep water all along the miles of beach.

Upwelling is only an occasional event along New York's shoreline, but in certain coastal waters it is commonplace. Upwelling occurs most of the year along all the subtropical west coasts of the continents and also along the Horn of Africa during the summer.

You should recall from the last chapter that when the wind blows the surface currents are dragged along; but once they start moving the Coriolis force twists them to the right in the NH. The deeper layers of water, dragged by the surface waters, are then twisted even farther to the right by the Coriolis force. This results in the Ekman spiral. The net effect of the Ekman spiral is to move the upper 200 m of the ocean to the right of the wind in the NH and to the left of the wind in the SH.

When you look at the global winds along the sub-tropical west coasts, you can see that upwelling is a fact of life there. The surface waters are blown away from the shore and are invariably replaced by colder water from below, as in Figure 16.15. This cold water cools the surface layer of air, stabilizing it and suppressing rising motion and precipitation. It also commonly produces fog—the fog of San Francisco and the *garua* of Peru.

## LARGE-SCALE WINDS AND THE CLIMATE

Many features of the world's climates can be inferred from the winds. For example, when a place has winds from the poles, it will be colder than most other places at the same latitude. Similarly, if a place receives its winds from the nearby ocean, it will have smaller temperature

**FIGURE 16.10**

The large-scale wind and pressure patterns of the world in January.

**January pressure and winds**

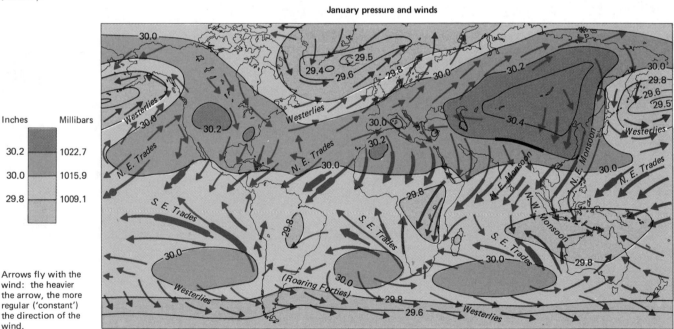

Inches    Millibars

| 30.2 | 1022.7 |
| 30.0 | 1015.9 |
| 29.8 | 1009.1 |

Arrows fly with the wind: the heavier the arrow, the more regular ('constant') the direction of the wind.

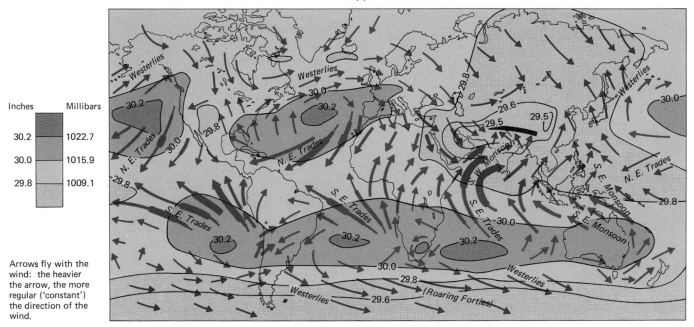

Inches | Millibars
--- | ---
30.2 | 1022.7
30.0 | 1015.9
29.8 | 1009.1

Arrows fly with the wind: the heavier the arrow, the more regular ('constant') the direction of the wind.

FIGURE 16.11
The large-scale wind and pressure patterns of the world in July.

FIGURE 16.12
Names for some of the wind and pressure systems. Shading denotes area of frequent calms.

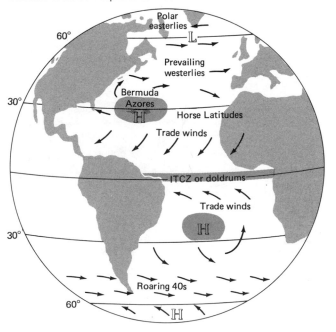

variations than places farther inland. The few examples chosen in this section will show how knowledge of the winds can also be used to infer the precipitation characteristics of a region. (See also Figures 12.2 and 12.3.)

Whenever the air rises the weather will be wet, and whenever the air sinks the weather will be dry. Any time the *surface* winds converge, they are forced to rise. You already know that the ITCZ is an area of rising air and wet weather. And indeed, the equatorial regions on earth are quite rainy for the ITCZ is never too far away.

The seasonal rainfall patterns around the equator are rather easy to remember. Simply keep in mind the rule: The wind patterns follow the sun (see page 258). Therefore, in July when the sun has "moved" north of the equator, it is the rainy season in the tropics north of the equator but the dry season south of the equator. In January the rainy belt (i.e., the ITCZ) is centered south of the equator; at this time it is the rainy season in this region. Since the ITCZ crosses the equator twice a year (March and September), there is a tendency for places on the equator to have two especially rainy seasons.

The parts of the subtropics that are just outside the reach of the ITCZ remain dry all year. They remain in the sinking branch of the Hadley cell (and also near the subtropical highs) and form many of the world's deserts such as the Sahara, the Arabian, the Australian, the Namib, the Atacama, and the Mohave.

LARGE-SCALE WINDS AND THE CLIMATE 263

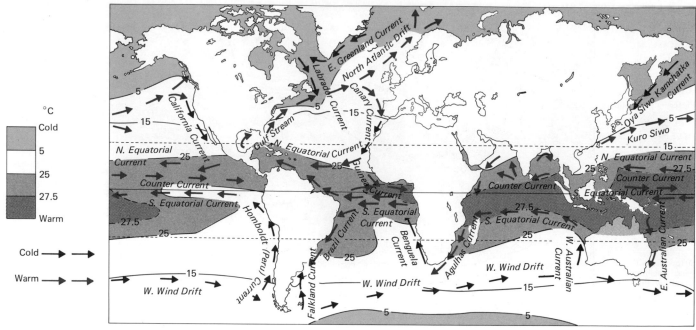

FIGURE 16.13
World ocean currents and temperatures in February.


FIGURE 16.14
World ocean currents and temperatures in August.

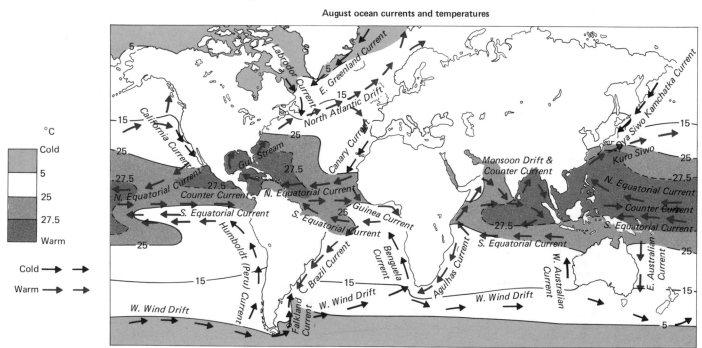

Look once again at the world rainfall maps to see the rainfall along the western side of South America. Between 5 and 30° S latitude there is far more precipitation on the eastern side of the Andes Mountains. South of 30° S latitude the pattern quickly reverses and there is much more precipitation on the western side of the Andes. This difference is easily explained in terms of the winds.

Between 5 and 30° S latitude there are easterly winds that blow upslope on the eastern slopes but downslope on the western slopes of the Andes. This makes the eastern slopes wet and the western slopes dry. Poleward of 30° S latitude the prevailing westerlies take over and the pattern reverses. These features show up clearly on the world maps of precipitation.

What about the seasonality of the precipitation? Look again at the western coast of South America (in Figures 12.2a and 12.2b). The rain belt there is farther north in July, because the westerly wind belt is also located farther north then. Thus places around 30° S latitude on the coast of Chile have their rainy season around July (their winter) and their dry season around January.

FIGURE 16.15
Upwelling as illustrated along the coast of Chile and Peru. The warm surface waters are forced to the left of the wind and away from the coast. They are then replaced by colder waters from the depths.

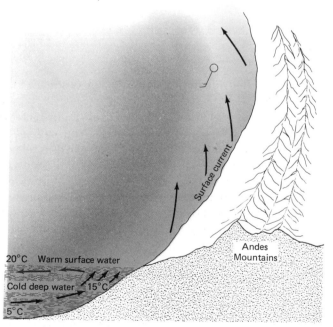

## WINDS ALOFT

In the years after World War I a few scientists had suspected the existence of a core of extremely strong westerly winds high in the midlatitude troposphere. This strong wind belt is now known as the **jet stream.** It was first discovered and documented by pilots during World War II. One German reconnaissance pilot flying westward at 17 km altitude above the Mediterranean Sea encountered a headwind in excess of 175 knots that held him in place until he ran out of fuel and had to ditch into the sea.

American pilots also ran into the jet stream during World War II when attacking Japan. Several bombing missions had to be aborted because the planes were not able to move westward against the strong jet stream winds. The one thing that the pilots liked about the jet stream was that it enabled them to get back home *real fast* once their missions were over.

Because of the jet stream, planes flying from America to Europe take about an hour less than when flying westward. Throughout most of his famous solo flight, Charles Lindbergh imitated the gliding nature of birds and flew so low that he was almost touching the water. Therefore, the jet stream winds of the upper troposphere were of no use to him. Had Lindbergh maintained an altitude of about 4 km, he would have arrived in Paris hours before he did.

But Lindbergh didn't know about the jet stream. Perhaps he should have expected something. *The Spirit of St. Louis* was manufactured in California, and Lindbergh had to fly it across America before taking off on his transatlantic flight. When he flew over the Rocky Mountains from California to St. Louis, he did it in record time because he flew at a high altitude. Without knowing it, the jet stream had helped him across America.

Figure 16.16 represents the average west to east wind at all heights in the troposphere during January and July. Because the regions of strongest winds shift around from day to day, this average picture makes the core of the jet stream seem wider but slower than it is on any particular day. In fact, there are often two distinct jet streams—the subtropical jet located about 30° latitude and the far more variable polar-front jet stream.

Several features of these winds are characteristic and easy to remember. First, the winds usually blow west to east through the bulk of the atmosphere. Since the earth spins from west to east, this implies that the atmosphere as a whole actually spins faster than the earth! 

Second, the winds are always faster during the winter hemisphere. This is understandable because winds are

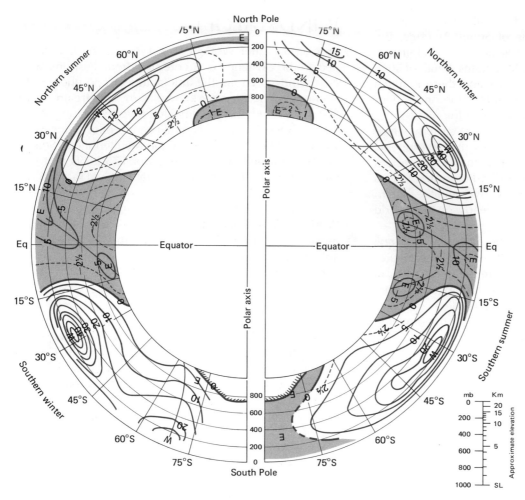

FIGURE 16.16
Mean west-to-east winds in the troposphere. Units are meters per second. Regions of easterly winds are shaded. Notice that westerly winds dominate.

produced by temperature differences, and the temperature differences are stronger during the winter.

Third, the maximum wind regions are located farther north in July than in January. This is another example of the rule that tells us that wind systems follow the sun.

The winds aloft on any particular day are far more irregular than the average picture. Figure 16.17 presents the winds at the 500 mb level (the jet stream often is found between the 200 and 300 mb level). Notice that the winds in midlatitudes move eastward in a wavy pattern. These upper-tropospheric waves in the westerlies are known as **Rossby waves** after Carl Gustav Rossby (1898–1957) who first discovered and described them.

Some of the properties of these waves were already described in the last chapter—that is, there is a tendency for rising motion to occur downstream from the troughs of

these waves and sinking motion to occur upstream from the troughs.

These waves drift slowly eastward at about half the speed of the 500 mb winds themselves. You will see in the next chapter that each trough in these waves is related to a low pressure area at the surface and each ridge to a high pressure area at the surface. Therefore, the movement and development of these waves is intrinsically tied to the movement and development of the surface weather systems. Understanding the waves in the westerlies is essential in weather forecasting.

Occasionally, the jet stream can be seen on satellite pictures because of the cirrus clouds it produces. South of the jet-stream axis there is a tendency for rising motion and clouds, while north of the axis it is clear. In such conditions the edge of the cirrus clouds correspond to the axis

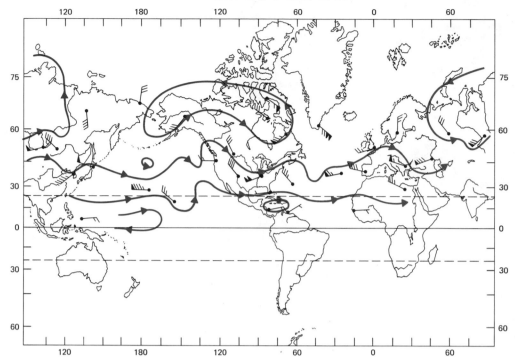

FIGURE 16.17
Winds at the 500 mb level in
the NH on January 29, 1970.

of the jet stream. In Figure 16.18 you can see a wave in the jet stream outlined by cirrus clouds.

## ISOBARIC SURFACES AND THE WINDS

Why are westerly winds found in the bulk of the atmosphere? To explain this and many other features of the winds it is convenient to use a concept called **isobaric surfaces,** which are surfaces of constant pressure. Whenever an isobaric surface touches the ground it forms an **isobar.** Isobaric surfaces are helpful because they make high pressure areas look like hills and low pressure areas look like depressions.

Imagine that a high pressure area is centered over St. Louis. The pressure at St. Louis is 1024 mb. The surrounding cities of Chicago, Nashville, Little Rock, and Omaha all have pressures of 1020 mb. However, since pressure decreases with height, it is also 1020 mb about 40 m above ground level at St. Louis.

If we connect all points where the pressure is 1020 mb, we get more than just an isobar. Not only must we con-

nect all points along the ground, but even all points in the air where the pressure is 1020 mb. This produces an isobaric surface that looks like a hill centered over St. Louis. Similarly, the 1016 mb isobaric surface will form another hill above the 1020 mb surface, and so on (see Figure 16.19). It should now be simple for you to show that isobaric surfaces make low pressure areas look like depressions.

Now, the geostrophic and gradient wind approximations imply that low pressure is found to the left of the wind (see Figure 16.20). So in terms of isobaric surfaces, we can formulate two rules.

**Rule:** The geostrophic and gradient winds blow with the downslope side of the isobaric surfaces to their left (in the NH).

**Rule:** The steeper the slope of the isobaric surfaces, the faster the geostrophic and gradient winds.

You can see that this represents a close analogy to the situation involving John Wayne and the villain, mentioned in the last chapter, on page 247.

Now, let's see how the changes that occur in the winds

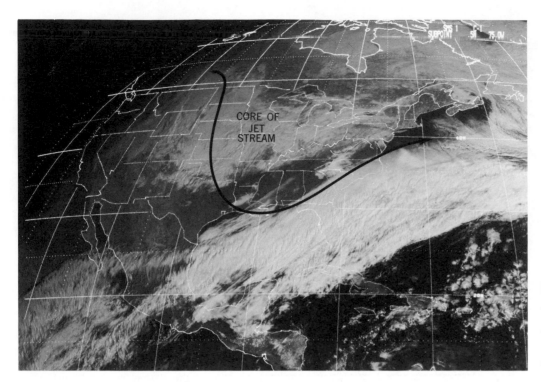

FIGURE 16.18
Jet-stream cirrus clouds on
February 20, 1978. Notice the
wave in the jet stream outlined
by the northern edge of the
cirrus clouds.

FIGURE 16.19
A 1020 mb isobaric surface. Wherever this touches the
ground, it forms a 1020 mb isobar (dashed line).

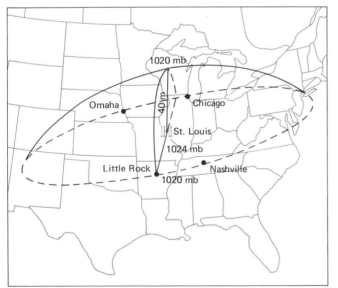

FIGURE 16.20
On an isobaric surface the downslope side lies to the left of
the wind. The wind speed is proportional to the slope of the
isobaric surface.

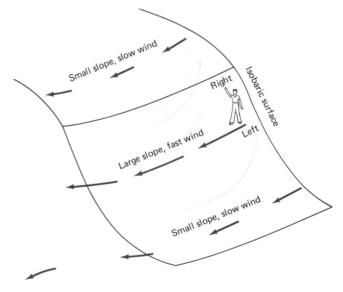

as you go up in the troposphere are indicated by changes in the isobaric surfaces.

There is always a specified pressure difference between two isobaric surfaces (see Figure 16.21). When we used the hydrostatic approximation in Example 15.11, we found that *the colder the air, the smaller the height difference between two different pressure levels.* In terms of isobaric surfaces, we can formulate another rule.

**Rule:** The colder the air, the smaller the height between two isobaric surfaces.

Imagine that we have two low pressure areas, one colder than its surroundings and the other warmer. The isobaric surfaces for the low with the cold core slope more and more steeply with increasing height, producing very fast winds aloft. Thus *cold lows intensify with height.*

On the other hand, the isobaric surfaces above *warm-core lows* level off with increasing height up to a certain point, beyond which they take the shape of hills (see Figure 16.22). This means that *warm-core lows weaken with height and eventually turn into highs aloft!*

This is precisely what happens above mature hurricanes. Although hurricanes are extremely intense lows at the surface, they always weaken with height and turn into highs by the top of the troposphere.

By using isobaric surfaces you should now be able to show that warm-core *highs* strengthen with height, while cold-core highs weaken with height, eventually becoming lows aloft. In Chapter 17 we will return to the concept of isobaric surfaces to show the close relationship between the surface pressure patterns and the Rossby waves of the upper troposphere.

## Why Westerly Winds Predominate

Now you should recall that westerly winds predominate through the troposphere, generally getting stronger with

FIGURE 16.21
The colder and denser the air, the smaller the height between two isobaric surfaces.

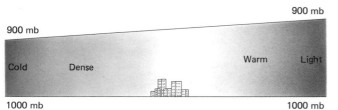

height. This is explained simply using isobaric surfaces.

In most basic terms, *the predominance of westerly winds is due to the fact that the air is colder at the poles than at the equator.* Imagine that the 1000 mb isobaric surface is horizontal from equator to pole. Therefore, because the air is colder and denser at the poles, the 900 mb isobaric surface will be lower over the pole and thus slope downward from equator to pole. Each succeeding isobaric surface will slope more steeply than the one below it, as you can see from Figure 16.23.

The implication for the wind should be immediate. When the isobaric surfaces slope downward toward the north (in the NH), (south in the SH) the geostrophic wind comes from the west. Since the isobaric surfaces slope more and more steeply as we rise through the troposphere, the speed of the west wind increases with height.

The technical name for the change of the geostrophic wind with height, which is the result of the temperature pattern, is the **thermal wind.** *The thermal wind always blows with cold air on its left,* in the NH (and right in the SH) as you can see in Figure 16.24.

Since most of the temperature difference is confined to the midlatitudes, this is where the isobaric surfaces will generally have their greatest slopes. Therefore, it is in the midlatitudes near the top of the troposphere that the west winds are strongest. There are times when most of the temperature differences are confined to a narrow frontal zone; then, the strong westerlies are also confined to a narrow zone called the jet stream!

Looking back to Figure 16.16, you should notice that in each summer hemisphere the *easterlies* between the equator and about 15° latitude strengthen with height. This, too, can be simply explained in terms of the temperature and the isobaric surfaces.

During the summer season the hottest air is not found at the equator but is found in the summer hemisphere. Therefore, there is a zone between the equator and about 20° latitude where the temperature increases toward the poles. This causes the isobaric surfaces to slope downward to the *south* (in the NH), and the easterlies result.

The strongest summer easterlies of the upper troposphere are found in two places—just north of the Gulf of Guinea over the western bulge of Africa and south of India and the continent of Asia. These easterlies are due to the somewhat large temperature difference between the hot land to the north and cooler waters to the south, and they are of great interest because they help produce enormous squall-line thunderstorms (see Chapter 19). The resulting weather disturbances that are produced over Africa by these easterlies then move westward over the Atlantic Ocean and occasionally spawn hurricanes.

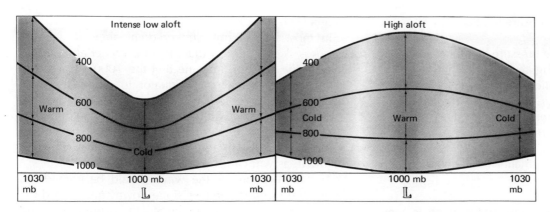

FIGURE 16.22
Cross-sectional view of isobaric surfaces shows how
cold lows intensify with
height, but warm lows
weaken with height and may
eventually become highs
aloft.

# THE ATMOSPHERE AS AN ENGINE

An engine is a device that produces motion (kinetic energy) from other forms of energy such as heat. The first steam engine was invented by Hero (Heron) in about 200 B.C., but it was used only as a toy.

In Hero's engine water was boiled in a closed kettle. We know that when water boils it expands to about 1000 times its original volume, producing a tremendous pressure whenever it is confined. Hero allowed the steam to escape through two bent pipes. By Newton's third law we know that when steam rushes out one way, it will force the pipe in the opposite direction. When both pipes are bent in the same sense the pipes and kettle will begin to spin, as you can see in Figure 16.25. In this process, heat

energy has been converted to kinetic energy. The loco-motive was invented when someone had the brains to put a steam engine on wheels and tracks!

The atmosphere is also an engine, because it uses some of the heat of the sun to produce the winds (kinetic energy). The atmosphere is also responsible for carrying water vapor aloft; when the rain falls and then rushes downstream back to the sea, its energy can be used to produce hydroelectric power.

Kinetic energy is produced at a truly enormous rate in the atmosphere. The rate at which energy is produced is called power, and the power of the winds is about $(2.5)(10)^{15}$ watts. Since the worldwide consumption of

FIGURE 16.23
Isobaric surfaces show why westerly winds strengthen with altitude. Colder air to north makes isobaric surfaces slope downward to the north. Thus the geostrophic wind is from the west. The higher isobaric surfaces slope more steeply, so the west wind becomes stronger with height. Therefore, the predominance of westerly winds is due to the temperature decrease from equator to pole.

FIGURE 16.24
The thermal wind is the change of the geostrophic wind with height. The thermal wind always blows with colder air to its left.

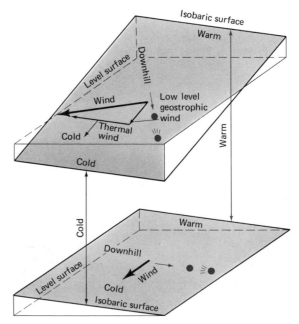

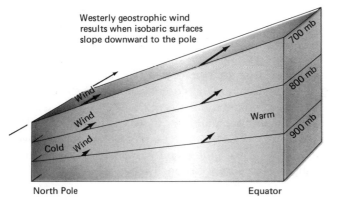

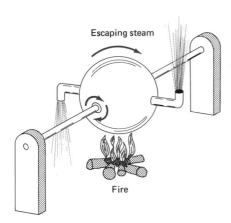

FIGURE 16.25
Hero's steam engine.

power by humans is only $(10)^{13}$ watts, the winds could conceivably provide us with all our energy needs for centuries to come. Unfortunately, this would require covering the earth with windmills several kilometers high; at present, this is not considered to be practical.

What is the efficiency of the atmosphere as an engine? Engines are rated in terms of their efficiency, which is defined as the percentage of energy that they can convert into useful kinetic energy. Most engines have an efficiency between 25 and 50%. The efficiency of the atmosphere is much lower than this, because it converts only $(2.5)(10)^{15}$ of the $(1.8)(10)^{17}$ watts of solar energy that strike the earth. Thus, its efficiency is only $(2.5)(10)^{15}/(1.8)(10)^{17} = 1.5\%$.

A more efficient atmospheric engine would not be desirable. It would mean stronger winds on earth—a potentially dangerous situation—and might also mean larger pole-to-equator temperature differences, making a larger part of the earth unlivable.

## PROBLEMS

**16-1** How did Columbus use his knowledge of the winds during his voyages?

**16-2** What gives the ITCZ its more or less continuous apearance over the oceans?

**16-3** What time of year would Sinbad sail from the Persian Gulf to Africa, and when would he return?

**16-4** Explain why the westerlies are much stronger and more persistent in the SH.

**16-5** Explain why the trade winds curve and become westerlies if they cross the equator.

**16-6** Find an ocean current that moves almost opposite to the general winds above it.

**16-7** Why do the ocean currents in the northwest Indian Ocean reverse between January and July?

**16-8** Explain why upwelling is such a prominent feature on all west coasts of continents in subtropical latitudes.

**16-9** Since it is the rotation of the earth (i.e., the Coriolis force) that is responsible for breaking up a direct wind circulation (as is depicted in Figure 1.4), explain why the Hadley cells are restricted to tropical latitudes.

**16-10** The western side of the south island of New Zealand is much rainier than the eastern side. (a) Explain why this is so. (b) Explain why this is not so for the northern island.

**16-11** The rainy belt on the western coast of South America in Chile extends farther north during July. Why?

**16-12** The eastern side of the island of Madagascar (Malagasy Republic) is rainier than the western side. Why?

**16-13** Estimate how much faster the Los Angeles to New York flight should be than the return flight, considering the plane flies at 500 knots, the distance is 2500 nautical miles, and the wind speed is blowing at 100 knots straight from Los Angeles to New York.

**16-14** The *Double Eagle II* was the first manned balloon to cross the Atlantic Ocean. (a) Which way did it go and why? (b) Why did it go in August even though the winds are weakest then?

**16-15** What is the relationship between the thermal wind and the geostrophic wind?

**16-16** Explain why the pressure systems of Problems 15-16 and 15-17 are generally shallow.

**16-17** Stand with your back to the wind. Then turn about 30° to your right (clockwise). (a) If high- or middle-level clouds blow from your left to your right, what temperature change does this imply for the bulk of the troposphere above you? (b) Why was it necessary to turn about 30° to your right?

**16-18** What would be the corresponding rule in the SH?

**16-19** Judging from the temperature profile in Figure 7.10, tell whether westerlies or easterlies will prevail at the stratopause (a) in January, and (b) in July. Explain.

**16-20** Explain why westerly winds predominate in the troposphere, and why they are strongest during winter.

**16-21** Explain why in many locations in the NH the winds at ground level are strongest in March even though the winds aloft are somewhat stronger in January.

# EXTRATROPICAL HIGHS AND LOWS

In the short story, "To Build a Fire," Jack London describes—with remarkable meteorological insight and accuracy—how a brash, young prospector freezes to death in the wild Yukon. This prospector had ignored warnings and set out alone for his stake during the coldest weather of the winter.

Day had broken cold and gray, exceedingly cold and gray. . . . It was nine o'clock. There was no sun nor hint of sun, though there was not a cloud in the sky. It was a clear day, and yet there seemed an intangible pall over the face of things, a subtle gloom that made the day dark, and that was due to the absence of sun. This fact did not worry the man. He was used to the lack of sun. It had been days since he had seen the sun. . . .

As he turned to go on, he spat speculatively. There was a sharp explosive crackle that startled him. He spat again. And again in the air, before it could fall to the snow, the spittle crackled. He knew that at fifty degrees below zero spittle crackled on the ground, but this spittle had crackled in the air. Undoubtedly it was colder than fifty below—how much colder he did not know.

Traveling over the frozen Yukon River where the going was easier, the man had to watch for thin ice in those places where underground springs keep some water flowing.

And then it happened. At a place where there were no signs, where the soft, unbroken snow seemed to advertise solidity beneath, the man broke through. It was not deep. He wet himself halfway to the knees before he floundered out to the firm crust.

The man then had to build a fire because at such low temperatures it is imperative to stay dry. He worked carefully on his fire, knowing that no mistake is tolerable at such low temperatures. But he *had* made a mistake.

He should not have built the fire under the spruce tree. He should have built it in the open. But it had been easier to pull the twigs from the brush and drop them directly on the fire. Now the tree under which he had done this carried a weight of snow on its boughs. *No wind had blown for weeks* and each bough was fully freighted. Each time he had pulled a twig he had communicated a slight agitation to the tree—an imperceptible agitation so far as he was concerned, but an agitation sufficient to bring about the disaster. High up in the tree one bough capsized its load of snow. This fell on the boughs beneath, capsizing them. This process continued spreading out and involving the whole tree. It grew like an avalanche, and it descended without warning upon the man and the fire, and the fire was blotted out! Where it had burned was a mantle of fresh and disordered snow.

The man was shocked. It was as though he had heard his own sentence of death.

Jack London, "To Build a Fire"

This story has been quoted here because it is so rich in meteorological insight. It accurately describes the conditions under which frigid air masses, known as continental polar air masses, form.

During the winter in the far north the sun is either absent or so low in the sky that it is virtually ineffective in heating the air. And, yet, it is always clear during the coldest winter weather, for this allows heat to escape to space. The longer the air remains in the far north the colder it grows, so that it tends to be coldest near the end of an extended period of light winds. And, as you have just seen, Jack London carefully noted all these conditions in "To Build a Fire."

Eventually this frigid air does begin to move and, when it pours out of northern Canada or Siberia and moves southward, it brings the coldest weather of the season to the midlatitudes and subtropics.

Crossing the leading edge of this cold air you encounter a sharp change in temperature and enter an area of tropical air. The boundary lines between the polar and tropical air masses are the fronts.

In this chapter you will read about the structure, development, and weather associated with air masses, fronts, and *extratropical* (outside the tropics) cyclones and anticyclones. A **cyclone** is a region of winds that spin counterclockwise in the NH (clockwise in the SH) and is almost always equivalent to a low pressure area. Throughout this chapter and in other places in the book, cyclone is used interchangeably with low, and anticyclone is used interchangeably with high.

This is one of the most important chapters in the book from a "weather" point of view. You will also find that many of the physical principles dealt with earlier in the book are referred to frequently.

## AIR MASSES

At the end of Chapter 8 you learned that the equatorial regions of earth receive more heat from the sun than they radiate back out to space, whereas the opposite is true in the polar regions. These imbalances are then rectified by the winds that bring warm air poleward and cold air toward the equator.

If we could stop the winds, the equator would grow warmer and the polar regions would grow colder. To a minor degree, this is precisely what happens when an air mass forms. It is possible to think of the winds and weather in the following, oversimplified manner.

When the temperature differences are small, the winds calm down. Then the air in the polar regions begins to cool and the air in the tropical regions begins to warm. Thus distinct air masses form. This process continues until the temperature differences are large enough to begin producing winds once again. The air begins moving, forming fronts, highs, and lows. The air gets stirred so that the temperature contrasts are reduced. Then the winds

drop down again. Therefore, in the crudest sense, the atmosphere undergoes something of a cycle.

An air mass is a large region of air that has somewhat uniform temperature and humidity. Cold air masses originate in the polar regions and are called *polar* (P), while warm air masses originate in the tropics and so are called *tropical* (T). Dry air masses originate over the continents and so are called *continental* (c), while moist air masses originate over the oceans and so are called *maritime* (m).

There are only four combinations that result — cP (continental polar), cT, mP, and mT. It is also possible to distinguish arctic (A) and equatorial (E) air masses, but at this basic level there is no need to do so.

Air masses tend to form under two distinct conditions. First, since air masses are simply large areas in which there are only small contrasts of temperature and humidity, they usually form over surfaces that have uniform properties. Thus, for example, maritime tropical (mT) air tends to form over the tropical and subtropical oceans where the water temperature of millions of square kilometers typically lies between 20 and 30°C.

Second, air masses will usually form in high pressure areas (see Figure 17.1). In high pressure areas the air near ground level spirals outward (i.e., diverges or stretches).

FIGURE 17.1
Contrasts are reduced by diverging air of highs. Thus air masses tend to form in highs. Conversely fronts tend to form in lows.

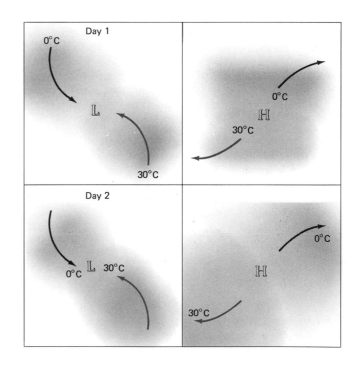

This spreading or stretching causes any contrasts to diminish in much the same way that the lettering on balloons grows less distinct when the balloons are blown up. Fronts, which are basically the opposite of air masses, therefore tend to form *between* high pressure areas where the air converges, so that contrasts are increased.

The formation of continental polar (cP) air masses has already been briefly described several times already. The process is now repeated in somewhat greater detail. Consider air that has newly arrived in northern Canada during the winter. This air is already cool, but is not always very dry. A cP air mass will begin to form if the wind settles down long enough so that the air can remain over the cold ground. While the snow reflects the feeble winter sunlight, it radiates what little heat it has rapidly into space. The air directly above the ground then cools by a combination of contact and radiation. Because the air cools mainly near ground level, an enormous inversion inevitably forms. Temperatures at ground level may be as much as 30 or 40°C cooler than the temperature 2 or 3 km above. This is one situation in which it is warmer on the mountain slopes than on the nearby lowlands, as you can see in Figure 17.2. For example, Verkhoyansk averages 21°C colder than the surrounding hilltops in January.

As the temperature falls, the air becomes denser and quite stable. Aloft, air pours into the region raising the surface pressure. All these factors ensure clear skies, with no precipitation. How then does the air lose its vapor and dry out? The answer is simple—air dries much like air in your freezer. Ice crystals sublimate on the ground or even in midair and then fall through clear skies to the ground

(along with aerosols). This frost then accumulates on the ground and on the trees to depths that may exceed 10 cm.

It is not possible to give any one representative temperature in a cP air mass. In its source regions during the winter, cP air may be −30 or −40°C or colder at ground level, but once this air moves equatorward it modifies considerably. By the time it reaches the Gulf of Mexico, cP air is rarely much below 0°C. One way to identify cP air is to compare it to the average conditions in a place. When cP air invades a place, temperatures are typically 5 or 10°C below normal, and the RH is lowered.

## MODIFICATION OF AIR MASSES

As cP air moves equatorward, it passes over warmer ground. This heats the surface layers of the air, making the lapse rate in the lowest kilometer or two of the atmosphere unstable. The weather then becomes quite gusty and the visibility is often exceptional, because the air is clean at the outset; however, now any pollutants injected into it can be dispersed aloft.

Clouds are generally suppressed, since the air is usually quite dry and only unstable throughout a rather small depth near the ground. What clouds do form are usually stratocumulus or small cumulus that may occasionally provide a few brief showers but hardly ever anything serious.

The cP air is also modified by the winds it helped to produce. Whenever the air is forced to blow over a mountain range, turbulence results and the air gets stirred. This brings the warmer air aloft down to ground level in much the same manner as fans do in orchards overnight. Such turbulent stirring can warm the air at ground level by 10°C or more.

The cP air rapidly loses its characteristics when it passes over the ocean. Satellite pictures now show quite dramatically just how rapidly this modification occurs. The lowest layers of air acquire both heat and moisture from the ocean surface. This heated, moistened air then begins to rise, forming long bands of stratocumulus clouds that line up more or less parallel to the winds, as you have seen in Figure 2.37.

This process represents the death of cP air. Depending on whether the air heads out over the cool, subpolar ocean waters or the warm, subtropical waters, either mP or mT air will form. The two areas of the world that have the greatest rate of heating of air and evaporation at the surface are the warm Gulf Stream waters off the coast of the United States and the warm Kuroshio waters off the coast

FIGURE 17.2
Typical vertical temperature and mixing ratio profiles of a cP air mass. In such conditions it is typically warmer on mountain slopes than in the lowlands!

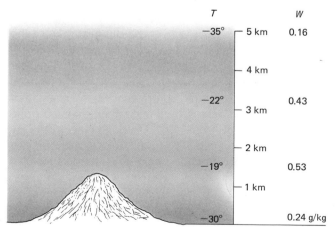

| T | | W |
|---|---|---|
| −35° | 5 km | 0.16 |
| | 4 km | |
| −22° | 3 km | 0.43 |
| | 2 km | |
| −19° | 1 km | 0.53 |
| −30° | | 0.24 g/kg |

of Japan. The cP air which commonly flows from the northwest over these waters literally sucks up heat and moisture and soon turns into mT air.

Maritime tropical (mT) air masses are the most widespread on earth. The mT air consists of warm, moist air with a reasonably steep lapse rate, although it may contain an inversion known as the **trade wind inversion.** The trade wind inversion is produced by the sinking air of the Hadley cell and is found throughout much of the tropics and subtropics.

The mT air is found throughout the tropics and also over the subtropical oceans. Cumulus and cumulonimbus clouds are relatively common in mT air, and this is the air mass in which hurricanes form.

When mT air moves poleward it passes over cooler ground or water and begins to modify. This process has already been described in Chapter 11. The poleward moving mT air gets cooled by the surface below. During the fall and winter, it is frequently cooled to the dew point so that fog forms. Since the air is cooled from below an inversion forms, rendering the air stable and trapping any pollutants. Therefore, mT air brings the warmest winter weather to the midlatitudes, but it often brings the most polluted weather as well.

Maritime polar (mP) air consists of a deep layer of cool, moist air with a steep lapse rate. The sea level temperature of this air is generally several degrees above freezing, and it is found over the subpolar oceans and the western coasts of continents above 45° latitude.

This air often brings dreary, drizzly, damp weather to places such as Seattle and London, especially during the winter half of the year. The one good thing about mP air is that it never gets too cold, since the ocean surface below keeps the air from falling far below freezing. Thus London and Seattle tend to get rain rather than snow during the winter. On the other hand, the nearby highlands tend to get snow, because the temperature inevitably decreases enough so that subfreezing temperatures are quite common more than 1 km above sea level.

Continental tropical (cT) air forms over the subtropical deserts of the world. It is the hottest of the air masses at the surface and typically produces daytime temperatures above 40°C during the summer season. The lapse rate of this air is quite steep in the lowest kilometer or two, but the trade wind inversion forms directly above this and is so strong that it largely suppresses vertical motion and clouds. Therefore, although the desert is hot, the air is not hot enough to rise very far.

When cT air leaves its source regions in the deserts, it rapidly picks up water vapor and changes to mT air. Of all the air masses, cT wanders least from its source regions before losing its identity.

## FRONTS

One of the most startling features of midlatitude weather is the suddenness of its changes. Many of these sudden weather changes occur when fronts pass by. A front is merely the narrow boundary zone between two different air masses. It is a zone *across* which the wind, temperature, humidity, and weather can change rapidly. Fronts can remain stationary for a time but, generally, they move over the face of the earth. The sense of motion often determines the name of the front—for instance, if the cold air mass is advancing, the front is a cold front.

The weather changes across the front vary in intensity. An example of a quite intense front is presented in Figure 17.3. This front occurred on November 19, 1969. The first feature to notice is that the isobars are all V-shaped. The line connecting all the vertices or kinks in the V-shaped isobars is called the trough line, because it occurs in the low pressure valley with higher pressures on either side.

On the western side of this trough line virtually all the winds are coming from the northwest, so that the air is cold. On the eastern side of the trough line the winds are blowing from the south and southeast, so that the air is warm. When you look at the temperature of the stations in Figure 17.3 the temperature contrast between the air on the western side of the trough line is dramatically colder than the air on the eastern side.

Perhaps you have already realized that the trough line is identical to the front. It always seems strange to the beginner in meteorology that fronts are often found on the weather map by drawing isobars and locating the low pressure troughs. Then the winds near the trough line are consulted, and the front is the line that, for example, separates regions of north winds from regions of south winds.

Sometimes, though not in this case, the temperature contrasts across a front are indistinct, so that fronts cannot generally be located by temperature contrasts alone.

You can also see that precipitation is occurring all along the front. Since air converges in the vicinity of the fronts, some air is forced to rise and precipitation frequently results.

Even though the cold front of November 19 advanced very slowly eastward, the contrasts were so sharp that wherever the front passed by the weather changed quite suddenly. When this front passed through New York City the temperature dropped 10°C in a little over an hour. The temperature drops (shown in Figure 17.4) for a number of cities describe this front as it moved across the country.

Generally, fronts do not have such sharp contrasts. Nevertheless, fronts are among the miracles of the atmosphere, often bringing noticeable changes in the weather in a very short time. One of the most beautiful changes

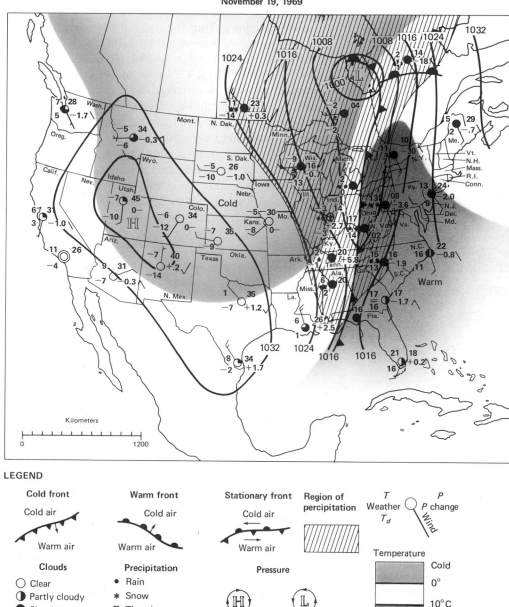

**Surface weather map**
**1200 GMT**
**November 19, 1969**

LEGEND

**Cold front**

Cold air

Warm air

**Warm front**

Cold air

Warm air

**Stationary front**

Cold air

Warm air

**Region of precipitation**

$T$ Weather $\bigcirc$ $P$ change
$T_d$ Wind

**Clouds**

$\bigcirc$ Clear

◐ Partly cloudy

● Cloudy

**Precipitation**

• Rain

✳ Snow

☇ Thunderstorm

= Fog

**Pressure**

Ⓗ High pressure

Ⓛ Low pressure

**Temperature**

Cold

0°

10°C

Warm

Kilometers

0          1200

FIGURE 17.3
Surface weather map for 1200 GMT, November 19, 1969.
This map shows an exceptionally intense cold front. Dark
shading indicates temperatures below 0°C; red shading indicates temperatures above 10°C. On all surface weather
maps, the interval between isobars is 8 mb, and regions of
precipitation are denoted by hatching. The shading used to
denote warm (red) and cold (dark) air masses does *not* use the
same numerical standard on each map; so check each map
individually for its specific code.

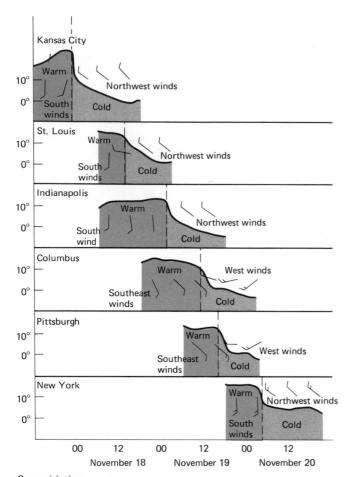

FIGURE 17.4

Progress of the cold front from west to east across the United States. Notice the sharp temperature drops as the front passes (dashed line) each city and the changes in the wind direction. Time (GMT) proceeds from left to right.

brought about by fronts is the change from hazy, polluted (and in the summer, hot) mT weather to the crisp, cool, clear weather of a cP air mass. Often this change occurs overnight, so to speak, and leaves everyone feeling bright and cheerful for a day or so.

Fronts can also be pictured with a sideways view or a cross section. The cross section is quite useful in depicting the characteristic clouds and weather. It also shows that the interface between the two air masses tends to slope. Warm air will often rise or "slide" over the cold air and produce clouds and precipitation. A cross section of the front that occurred on November 19, 1969 is shown in Figure 17.5.

The fronts do not slope very steeply. (The vertical scale of Figure 17.5 is enormously exaggerated.) Generally, the

cold front has a steeper slope than the warm front, especially near the ground. The cold frontal surface usually rises more than 1 km for each 100 km that you move away from the front. Thus the slope of the cold frontal surface is almost always greater than 1/100, while the slope of the warm frontal surface is generally less than 1/100. However, in the lowest kilometer of the atmosphere the cold front slopes far more steeply and may even appear as a vertical wall, as might be inferred from a look at Figure 10.14, because the advancing cold air at ground level is retarded somewhat by friction.

The steep slope of advancing cold fronts often forces the warm air to rise rapidly and abruptly. Conversely, the warm air near warm fronts is able to slide upward slowly and gradually because of the more gentle slope of warm fronts. This is part of the reason that precipitation at cold fronts is often intense but brief, whereas the precipitation associated with warm fronts tends to be less intense but long lasting. Thunderstorms often form along cold fronts,

FIGURE 17.5

Cross-sectional analysis for cold front of 1200 GMT, November 19, 1969. This shows the side view of the front. The front's position occurs where lines of constant potential temperature are close together. The slope of this front is about 3/100 (vertical scale is greatly exaggerated).

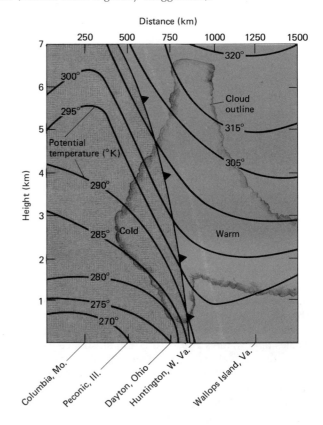

while steadier precipitation from nimbostratus clouds tends to form near warm fronts.

The fact that frontal surfaces slope is rather strange—indeed, it is almost as strange as if the water level in your bathtub tilted permanently. (Recall that the Gulf Stream is tilted.) Normally, you would expect the colder air to spread directly beneath the warmer, lighter air, but this does not happen.

The frontal slope is due to the Coriolis force. When the two air masses have differing winds (as they almost always do), the resulting Coriolis force on each air mass is different as well. In Figure 17.6 you can see a frontal slope develop, because each moving air layer in the channel is forced to *its* right (in the NH). The larger the wind difference and the smaller the temperature difference between the two air masses, the steeper the front.

You should also notice that the winds at a front always have a counterclockwise (i.e., cyclonic) sense of rotation. This is one more important reason (in addition to converging air) why fronts are located in low pressure troughs. There is no consistent way for the wind at a front to have a clockwise or anticyclonic sense of rotation, unless the colder, heavier air layer were to lie on top of the warmer, lighter layer. This is obviously an absurd situation.

Fronts tends to form in certain favored locations, which are shown in Figure 17.7. These fronts form a somewhat incomplete ring that encircles each pole and is known as

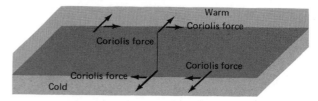

Unbalanced initial state

Final balanced state

FIGURE 17.6
Frontal surfaces slope because of the Coriolis force, and the winds at the front always have a cyclonic sense of rotation. The dashed lines are now winds, but are only used to indicate the sense of rotation of the winds.

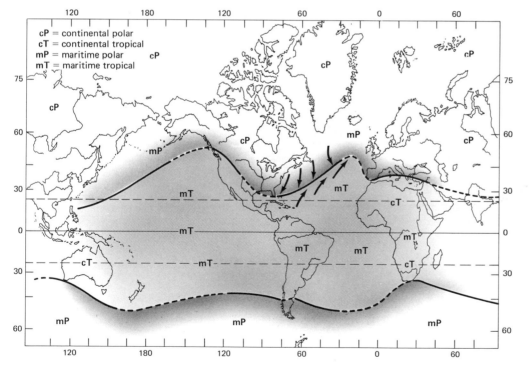

FIGURE 17.7
The average position of the polar front in January, with location of typical air masses. When the front is dashed, its position is not well-defined. The winds shown here act to intensify the front.

the **polar fronts.** The polar front is both more intense and closer to the equator during the winter. Since storms (lows) tend to develop on the polar front, you can guess that they will be more intense and located closer to the equator during the winter.

Take a closer look at the section of the polar front in Figure 17.7 that forms along the east coast of the United States during winter. The front tends to develop between the warm, moist mT air flowing from the south around the Bermuda high and the cold, dry cP air flowing from the north around a Canadian high. This temperature contrast is further increased by the difference between the cold land and the warm ocean surfaces during winter. The wind pattern in Figure 17.7 is the ideal wind pattern for the development and intensification of fronts (i.e., **frontogenesis**).

Fronts seldom appear as straight lines on the weather map. The frontal line often has a wavelike appearance with distinct kinks. At each kink in the frontal line there is a low pressure center. Often the front begins to look like a series of water waves. These waves are known as frontal waves or cyclone waves, with a typical wavelength of several thousand kilometers (see Figure 17.8).

A series of such cyclone waves is known as a cyclone family. The entire cyclone family consists of a few waves and usually moves parallel to itself, from west to east. Whenever a cyclone family passes by a place, good and bad weather will alternate every day or two for perhaps a week or so.

# THE STRUCTURE AND LIFE CYCLE OF CYCLONES AND ANTICYCLONES

Since much of the weather in the middle and high latitudes is governed by the passing highs and lows, it is essential to present a rather detailed picture of them.

The ideal picture of a developing cyclone (i.e., low) starts with a barely noticeable irregularity on the polar front. When conditions are right, the irregularity will begin to amplify into a well-defined cyclone wave. The developing cyclone wave not only looks like a developing ocean wave, it also has several common properties.

The ocean wave builds up to its maximum height just as it begins to break. After the wave breaks the water becomes turbulent and mixes up, but soon it dissipates. The evolution of a cyclone wave is closely analogous to this and is depicted in Figure 17.9. A cyclone center develops at the kink in the front. To the east of the low, southerly winds push warm air northward, making that section of the polar front a warm front. On the western

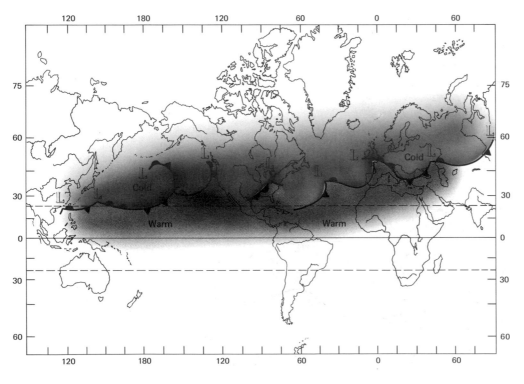

FIGURE 17.8
Cyclone waves from January 29, 1970. Low pressure areas are found at the crest of each wave.

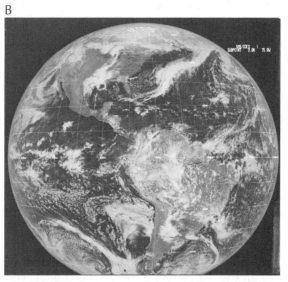

FIGURE 17.9a
A cyclone wave grows on the polar front. Notice that the low (L) moves to the northeast.

FIGURE 17.9b
Satellite photos of the growth of a cyclone wave on the polar front from March 23–25, 1976.

B

A

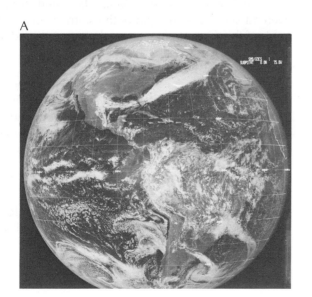

C

side of the low center, north winds push cold air far southward, making that section of the polar front a cold front. Thus the relative location of the cold and warm fronts is an automatic consequence of the counterclockwise circulation around lows.

The cold front almost always travels faster than the warm front and eventually catches up to it. When this happens the warm air is forced up away from the ground, and the cyclone is said to be occluded. The occlusion of cyclone waves is similar to the breaking of ocean waves. Cyclones generally reach their greatest intensity when they first become occluded; they then begin to weaken gradually over the next few days as the air in the storm "mixes up," and the temperature contrasts are destroyed. This deprives the storm of its energy source, and so it dies.

A cross-sectional diagram provides a simple way of depicting the weather associated with a developing cyclone wave. In Figure 17.10 the four cross sections show the weather at 12-hour intervals. Here are some points that highlight several of the more important features in this figure.

1. The weather moves from west to east. For instance, the weather that hits Kansas City gets to St. Louis about 12 hours later.
2. From time 0 to time 12 h the weather at St. Louis gets much colder, whereas the weather at Columbus gets much warmer. Thus, a cold front has passed through St. Louis, while a warm front has passed through Columbus.
3. The cold front moves faster than the warm front, so that the warm sector becomes progressively smaller. By the 36 h, the cold front has caught up to the warm front, and all the warm air has been forced aloft. The new front that is simply a combination of a warm front and a cold front is called an occluded front.
4. The cloud sequence at Philadelphia is typical of approaching cyclones (described in Chapter 2). These clouds change from cirrus to cirrostratus (with a halo) to altostratus, after which rain begins.
5. As the precipitation approaches Philadelphia, the pressure falls. A falling barometer is generally a good indicator of an approaching low pressure area and precipitation. Notice the falling barometer at Kansas City from time 12 h to time 36 h.
6. The weather *sequence* at Pittsburgh is typical of situations through which low pressure areas pass. At time 0 h, it is cool and cloudy and rain has just begun. Twelve hours later, the rain is still going strong. At time 24 h, fog has just ended and it has just cleared and grown

warmer because the warm front has passed. The warm, hazy weather only lasts a short time at Pittsburgh, since the cold front is not far behind the warm front. By time 36 h, the cold front and a thunderstorm have passed, and the weather has become clear, crisp, and cold.
7. The weather sequence tends to repeat itself. For instance, there are strong indications that the next low pressure area is approaching Kansas City from the west at time 36 h. Not only is the barometer falling in the last two frames of the figure but the cloud sequence points to precipitation and an approaching warm front.

The weather of the cyclone wave can also be presented in a standard weather map form (see Figure 17.11). In discussing the weather associated with this system, it is convenient to divide the storm into three regions. Region 1 is located to the north and east of the storm center and contains cold, cloudy air with more or less continuous precipitation. Region 2 is located to the south of the low center and is also known as the warm sector; it contains warm hazy air. Region 3 is located to the west of the storm center and contains mostly clear, crisp, and cold air.

**The Weather of Region 1**
Since storms approach from the west, you invariably encounter the weather of Region 1 first. This region, typically located to the north and east of the storm center, may on occasion extend several hundred kilometers to the north*west* of the storm center as well. This is the region of the storm containing the dreariest weather, with more or less continuous precipitation and completely overcast skies. This section of the storm is located to the north of the warm front, so that the air at ground level is cool or cold. It is also humid once the precipitation begins.

The wind in region 1 typically blows from the east or northeast. After the typical cloud sequence of cirrus, cirrostratus, and altostratus, rain or snow begins. The precipitation is usually not too intense, but it is rather prolonged since it is produced by the gradually rising warm air above the frontal surface. The precipitation typically lasts between 12 and 24 hours but, if you are really lucky, it can last several days.

One of the important prediction problems for Region 1 is to be able to distinguish between rain and snow situations. Many people ask, "What are the conditions that make snow more likely than rain?"

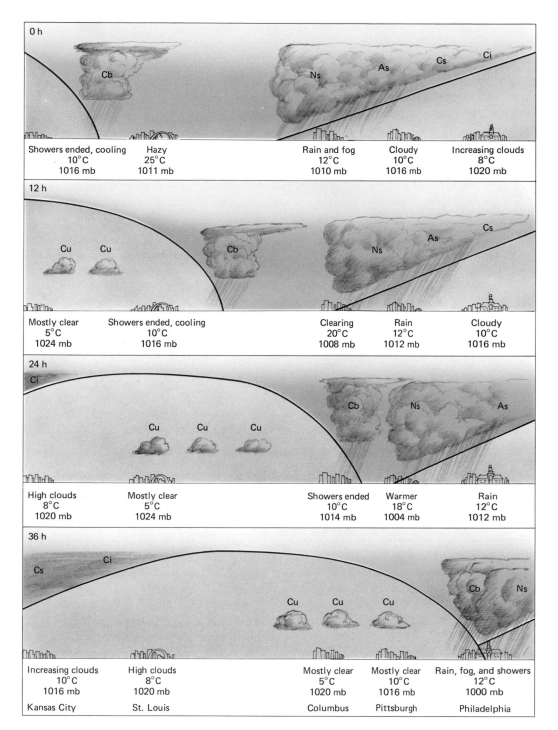

| | | | | | |
|---|---|---|---|---|---|
| **0 h** | | | | | |
| Showers ended, cooling 10°C 1016 mb | Hazy 25°C 1011 mb | | Rain and fog 12°C 1010 mb | Cloudy 10°C 1016 mb | Increasing clouds 8°C 1020 mb |
| **12 h** | | | | | |
| Mostly clear 5°C 1024 mb | Showers ended, cooling 10°C 1016 mb | | Clearing 20°C 1008 mb | Rain 12°C 1012 mb | Cloudy 10°C 1016 mb |
| **24 h** | | | | | |
| High clouds 8°C 1020 mb | Mostly clear 5°C 1024 mb | | Showers ended 10°C 1014 mb | Warmer 18°C 1004 mb | Rain 12°C 1012 mb |
| **36 h** | | | | | |
| Increasing clouds 10°C 1016 mb | High clouds 8°C 1020 mb | | Mostly clear 5°C 1020 mb | Mostly clear 10°C 1016 mb | Rain, fog, and showers 12°C 1000 mb |
| Kansas City | St. Louis | | Columbus | Pittsburgh | Philadelphia |

**FIGURE 17.10**

Cross section showing progress of fronts and weather. An occluded front forms when the cold front "catches up" to the warm front.

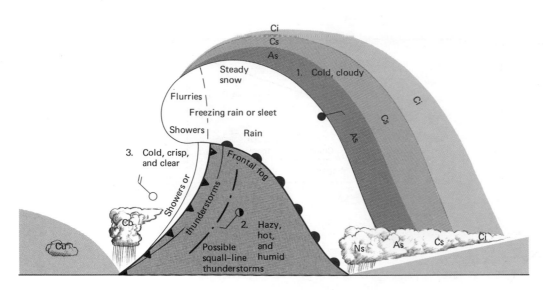

FIGURE 17.11
Weather associated with a model low pressure area.

In order for snow to fall and stick, the air temperature must be near or below freezing, both at ground level *and* above the warm front. Because the air above the frontal surface comes from the warm sector, it can only be sufficiently cold if it has risen a considerable height. Near the warm front the warm air has not yet risen far and, as a result, rain is usually produced in the clouds above the frontal surface. Then, if the air beneath the front happens to be below freezing, the most that a snow lover can hope for is freezing rain or sleet. The further you get from the warm front, the more likely the warm air above has been lifted and therefore cooled sufficiently to produce snow (as you have seen in Figure 12.12).

A simple, but general, forecasting guide for snow can be helpful at this point — avoid predicting snow unless the storm center passes at least 100 km to the south of your location. One notable example of this was the record-breaking snowstorm of February 8–10, 1973 in the southeast United States (see in Figure 17.12).

Like many meteorologists, I have always loved large snowstorms. I can remember one particular snowstorm when I was a child in which the snow finally stopped and the sky cleared for a short while. Then, suddenly, it became cloudy and began snowing heavily again. My heart skipped several beats because I *thought* that there might be much more snow still to come; perhaps I might *never* have to go back to school! But, alas, this was only a brief snowshower that added no more than a dusting to the total snow accumulation. I know now that there was little reason to hope for much more accumulation since large snowstorms typically end with a series of diminishing showers, which are soon followed by clear weather.

These diminishing showers and flurries are located to the northwest of the low pressure center. They are almost always an afterthought of sorts, although in mountainous areas or regions such as the Great Lakes they can be significant. Otherwise, they are a sign that the weather will clear soon. In any case, it will remain cold, for when such weather occurs the warm sector lies far south of the area.

Fog often forms in the cold air of Region 1, just north of the warm front. You may recall that frontal fog is produced when the warm raindrops falling from the warm air a short distance above enters the relatively thin layer of cold air near the ground. Frontal fog often occurs after many hours of steady precipitation and may have a depressing effect on people, but it should not. Since frontal fog occurs near the warm front, it is often a sign that the rain will soon end and that it will clear up and become much warmer.

Although frontal fog forms in the cold, stable air near the ground, the warm air above the frontal surface may well be quite unstable. Thus thunderstorms are possible at such times and, occasionally, they might be very severe. Fortunately, such weather combinations are unusual, for when they do occur they can be disastrous.

### The Weather of Region 2 — The Warm Sector

Now, consider the weather in Region 2 — the warm sector. We enter the warm sector right after frontal fog is encountered and the warm front passes. In the warm sector the weather is warm and generally humid for the season, with winds that tend to blow from the south or southwest. The worst heat waves of the summer season occur in the warm sector. Because of the high humidity, the air tends to be hazy. Pollution, and radiation-advection fog are quite common when the sun is low during the late fall and early winter.

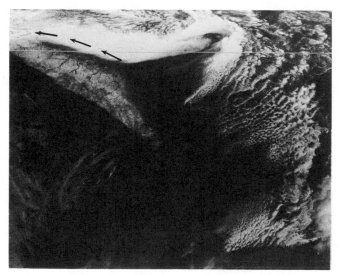

FIGURE 17.12
The record-breaking snowstorm of February 8–10, 1973 in the southeast United States. Notice how the snow (light band laced with dark river channels) lies to the north of the storm track.

The warm sector does not usually have widespread, steady precipitation, although occasionally this can happen. Generally, if there is any precipitation, it takes the form of thunderstorms, which will develop when the air is sufficiently unstable. In fact, some of the most violent thunderstorms, known as squall-line thunderstorms, form in the warm sector along a line several hundred kilometers in advance of the cold front. This line can also extend a short distance poleward of the warm front. Intense thunderstorms can also break out in a line along the cold front. The frontal and squall-line thunderstorms are dealt with in much greater depth in Chapter 19. For the moment, it is sufficient to note that these are the thunderstorms that are occasionally associated with tornadoes and large hail.

Sometimes the warm sector contains cT air rather than the normal mT air. Warm sectors containing cT air are most commonly found during the spring and fall seasons in the African and Asian countries bordering the Mediterranean Sea. At such times, unbearably hot and sometimes incredibly dusty air from the Sahara and Arabian Deserts gets drawn northward toward the lows which form over the Mediterranean Sea. The hot, dry, dusty winds that result are known as the Scirocco or Khamsin and are described in Chapter 20.

As the cold front passes, there may be a thunderstorm if the air is sufficiently unstable; or there may be brief showers; or, at times, if the air is stable and dry enough, it clears up instantly. In any case soon after the cold front

passes there is usually a refreshing and welcome change in the weather. Now we enter Region 3.

**The Weather of Region 3**
If the storm center passes by to your south, then you never experience the weather of the warm sector. Instead, after steady precipitation, you can look forward to showers and flurries that will diminish within a few hours and then bring the mostly clear weather of Region 3.

In Region 3, the wind typically blows from the west or northwest, and the weather often turns noticeably cooler and drier. During the winter the temperature may drop far below freezing, except along the west coasts of the continents, but at least the air is clean and the skies are blue.

When clouds form in the cP air of this region, they tend to be stratocumulus or cumulus clouds that are somewhat suppressed. Generally, these clouds do not produce any precipitation but, sometimes, a brief shower or flurry is possible.

However, if you happen to live in the mountains especially near the West Coast or in a place such as Watertown, New York, something different often happens in Region 3. Instead of clearing up, the weather may get far worse than it was before! Aware of the peculiarity of their weather, residents of Watertown and many other places around the Great Lakes use the expression, "During winter the weather clears up stormy." Figure 17.13 shows quite clearly the effect of the Great Lakes on the winter snowfall.

Watertown is located to the east of Lake Ontario in the foothills of the Adirondack Mountains. The cold west winds that bring clear weather to most other places must first pass over Lake Ontario before reaching Watertown. When the air passes over the lake, it picks up vapor and enough heat to make it unstable. Then, as it reaches Watertown and rises over the foothills of the Adirondacks, it dumps as snow the vapor it had picked up. Enormous snowfalls have been produced under these conditions.

On January 28, 1977, 30 cm of snow fell on Watertown with the passage of a cold front. Then, after it would have cleared up in other places, Watertown received an additional 120 cm of snow. Even so, it was far worse some distance to the southeast of Watertown where over 250 cm of snow fell and piled into drifts as high as 9 m.

Such "lake effect" snowstorms are not new to the area (see Figure 17.14). One of the most notorious (although by no means the largest) of these occurred on February 4–5, 1972 when a "lake effect" snowstorm dumped 142 cm on the town of Oswego, New York, trapping over 100 people who had been attending the Annual Eastern Snow Conference!

FIGURE 17.14
The effect of the Great Lakes snowfall on annual snowfall. Snowfall is greater to the south and east of the Lakes.

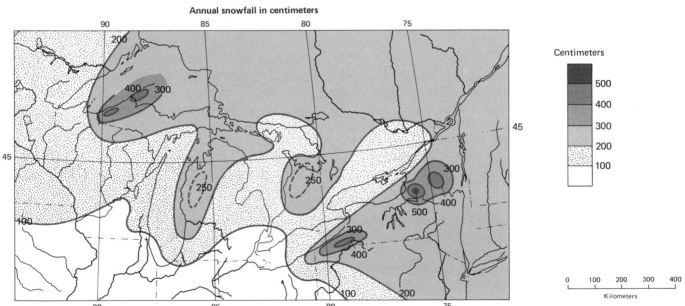

**Annual snowfall in centimeters**

Centimeters

| | |
|---|---|
| | 500 |
| | 400 |
| | 300 |
| | 200 |
| | 100 |

0   100   200   300   400

Kilometers

## The Typical Weather Sequences

The sequence of the four cross sections of Figure 17.10 suggests that there is a typical weather sequence for passing lows. There are actually two distinct weather sequences—one for when the low pressure center passes to your south, and one when it passes to your north. Since weather systems move from west to east in the midlatitudes, you can simulate the associated weather changes by moving Figure 17.11 of the ideal cyclone from left to right (west to east), while keeping your head and eyes *motionless*.

A. When the low center passes by to the north, the typical weather sequence is:

1. Cirrus, cirrostratus (with a halo), and altostratus clouds, in that order.
2. Twelve to 24 hours of steady rain (snow is unlikely except perhaps briefly at the beginning).
3. A brief period of fog as the rain tapers off.
4. Clearing and turning warmer, hazy, and humid.
5. Possible thundershowers.
6. Turning sharply colder, clear, and crisp.

B. When the low center passes by to the South, the typical weather sequence is:

1. Cirrus, cirrostratus (with a halo), and altostratus clouds, in that order.
2. Twelve to 24 hours of rain or *snow*, possibly mixed with sleet or freezing rain.
3. Precipitation tapering off as showers or flurries for a few hours.
4. Clearing, turning crisp, and remaining cold or possibly getting even colder. When the storm passes to the south it never gets warm.

The ideal picture of the cyclone wave—its weather and its weather sequences—will rarely match any storm perfectly. There is so much variability in the weather that it is not surprising that the weather is so difficult to forecast precisely. Nevertheless, the ideal picture is useful simply because it contains the principal features of most cyclone waves.

Low pressure areas tend to form in certain preferred regions. They can form anywhere along preexisting fronts and more commonly form near the standard position of the polar front. Lows also frequently form on the eastern slopes of mountain ranges (the lee side for westerly winds) for reasons which will soon be explained. Figure 17.15 shows the preferred regions for cyclone formation and the

paths or storm "tracks" taken by American lows. Once again, you should be forewarned that no two storms are ever exactly alike and never take the same "track." Even so, knowledge of where lows tend to form is quite useful to the meteorologist. For example, meteorologists in Denver, Colorado, know that lows there can develop rather suddenly; occasionally, this will produce easterly (and therefore upslope) winds, which quickly transform a clear day to a rainy or snowy one. Therefore, meteorologists in Denver are always ready for the first signs of a developing low.

# RELATIONSHIPS BETWEEN WEATHER PATTERNS AT THE SURFACE AND ALOFT

The conception of the extratropical highs and lows with their air masses and fronts, which you have just read about, was formulated in the years immediately following World War I. In 1918, the noted Norwegian meteorologist, Vilhelm Bjerknes (1862–1951) was asked by his government to establish an institute for studying the weather. A dense network of weather stations supplied detailed information, and this almost immediately led to many fundamental discoveries about the weather.

In 1919, Vilhelm's son, Jakob Bjerknes (1897–1975) announced the discovery of air masses and fronts. In the next decade, virtually all of the information contained in the last few pages was discovered by Jakob and his colleagues, Halvor Solberg (b. 1895) and Tor Bergeron (whom you have already met in Chapter 12) under the guidance of the elder Bjerknes.

It is much to their credit that they produced such a detailed picture of lows and highs in the days before satellites and radiosondes. However, since their information was largely confined to the surface weather, it is no wonder they were not aware of the important processes occurring high in the troposphere. As a result, the frontal picture of the lows and highs is somewhat oversimplified, idealized, and incomplete.

Rossby's discovery of the waves in the jet stream in the late 1930s (actually waves in the westerlies, because the jet stream per se had not yet been discovered) redirected the attention of meteorologists. They found that the high and low pressure areas at the surface are intimately related to the wind and temperature patterns aloft, with highs corresponding to ridges aloft and lows to troughs

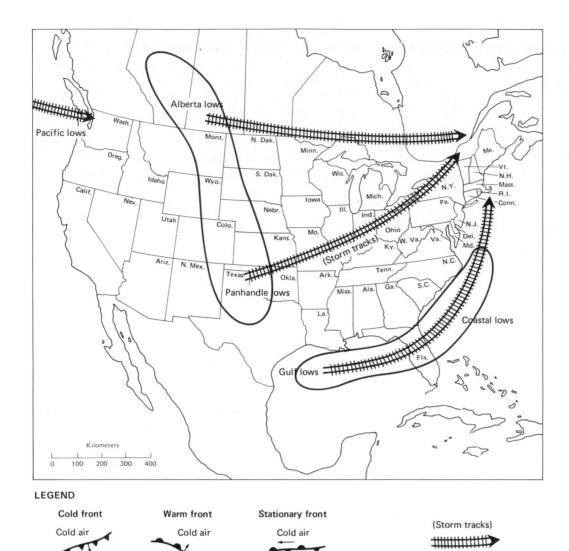

**Cold front**

Cold air

Warm air

**Warm front**

Cold air

Warm air

**Stationary front**

Cold air

Warm air

(Storm tracks)

**Clouds**

○ Clear
◐ Partly cloudy
● Cloudy

**Precipitation**

• Rain
✱ Snow
⅃ Thunderstorm
= Fog

**Pressure**

Ⓗ High pressure   Ⓛ Low pressure

FIGURE 17.15
Typical regions of origin for American lows and their storm "tracks."

aloft. (Ridges are regions where there is a clockwise curvature of the winds and contours; troughs, a counterclockwise curvature.)

They also found that the *surface weather systems move almost parallel to the direction of the upper-tropospheric winds directly above.* Finally, they found that the development of lows at the surface (technically called **cyclogenesis**) is often a result of processes taking place in the middle and upper troposphere.

It is interesting to note that J. Bjerknes, who discovered fronts, also played an important part in explaining how the processes in the upper troposphere contributed to the development of cyclones. In 1944, he and Jorgen Holmboe worked out the first modern picture of cyclogenesis. Since their arguments were rather simplistic, a more complete mathematical theory was called for. This was provided independently by Jule Charney (b. 1917) in 1947 and E. T. Eady (1915–1966) in 1949. Both theories were quite similar and describe a process known as **baroclinic instability.** Some of their ideas are discussed in a highly simplified form later in this chapter. The importance of baroclinic instability was demonstrated by the success of the early weather forecasts that were made by computer.

Satellite pictures now demonstrate quite clearly the importance of the upper troposphere in storm development. For example, consider occluded fronts. The early ideas maintained that an occlusion occurs when the cold front catches up to the warm front. In fact, the spiral cloud pattern which characterizes occluded lows (see Figure 17.17) often forms mainly as a result of the winds of the midtroposphere. This is demonstrated schematically in Figure 17.16, which shows how the mid- and upper-tropospheric winds can distort the cloud patterns into a spiral shape.

Furthermore, occluded fronts will often form when the jet stream brings a "tongue" of warm air over a layer of cold air near the ground, as is shown in Figure 17.18. The technical name for this process is *warm-air advection*

FIGURE 17.16
Schematic diagram showing how winds in upper troposphere distort clouds into the typical spiral pattern of the occluded stage.

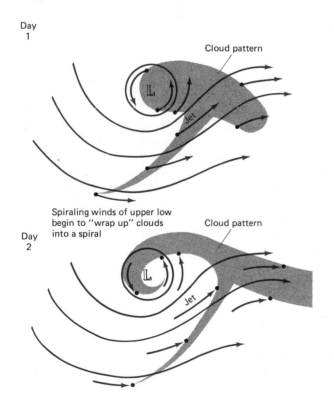

Day 1

Cloud pattern

Day 2

Spiraling winds of upper low begin to "wrap up" clouds into a spiral

Cloud pattern

(recall advection from the discussion on fog in Chapter 11). The cross section of Figure 17.18 shows that warm-air advection has produced the same temperature structure as that of the occluded front in Figure 17.10, without any cold front catching up to any warm front.

**The Use of Isobaric Surfaces**
By using isobaric surfaces, it is simple to show how the ridges and troughs in the upper troposphere correspond to highs and lows at the surface. The 1000 mb isobaric surface for a high and low is shown in Figure 17.19. Notice how the winds are related to the slope of the 1000 mb surface that sometimes slopes downward to the north, but in other places slopes downward to the south.

Now, use the fact that the air is colder to the north. This causes the isobaric surfaces at greater heights (e.g., at 500 mb) to slope downward to the north in most places. Then the winds at upper-tropospheric levels do not move in circular paths, but in a wavelike manner from west to east. Furthermore, the trough of each wave corresponds to a low at the surface, whereas the crest corresponds to a surface high.

When you look at most actual weather situations, you find that the upper-tropospheric troughs and ridges do not lie directly above the lows and highs at the surface. Instead, the *upper troughs and ridges are generally located farther west than the corresponding surface systems.* This behavior is due to the fact that the *air tends to be colder on the western side of lows* than it is on the eastern side, *and warmer on the western side of highs* than it is on the eastern side. Since the isobaric surfaces are closer together when it is colder, you can see from Figure 17.20 that the lowest point of each succeeding isobaric surface slopes toward the cold air, while the highest point slopes toward the warm air. Thus, *with cold air on the western side of lows, the lows slope to the west with increasing height* and, *with warm air on the western side of highs, the highs slope to the west with increasing height.*

**Steering**
According to Figure 17.21, there are southwest winds directly above the surface low and northwest winds above the surface high. This has an immediate application to weather forecasting.

Now, you may recall that the surface pressure systems move parallel to the upper-tropospheric winds directly above them. Thus, Figure 17.21 shows why low pressure areas generally move from southwest to northeast, and

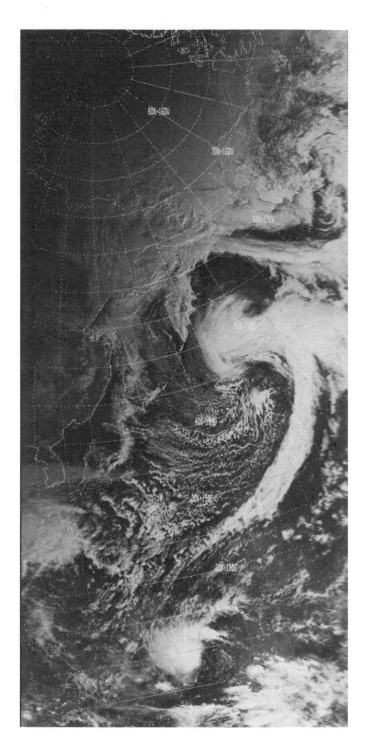

FIGURE 17.17
Satellite pictures showing extratropical cyclone of March 13–14, 1974 over the North Pacific as it evolves into the occluded stage. This spiral cloud pattern is typical of occluded cyclones.

why high pressure areas (except for the subtropical highs) usually move from northwest to southeast. Confirmation of this fact is provided on the chart of preferred storm tracks (Figure 17.15).

Meteorologists have made use of this knowledge and say that the surface systems are *"steered"* by the upper-tropospheric winds. They have developed a very simple forecasting rule known as the "steering rule."

Day 1

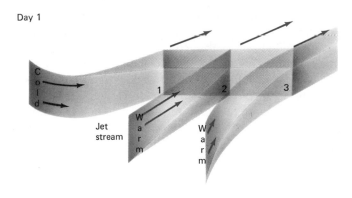

Day 2

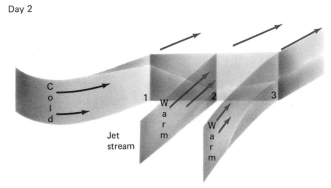

FIGURE 17.18
Occluded fronts often form because of the winds aloft. The jet stream is bringing warm air most rapidly over Point 2; therefore, an occluded front is forming there.

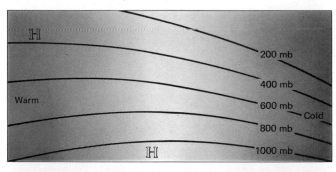

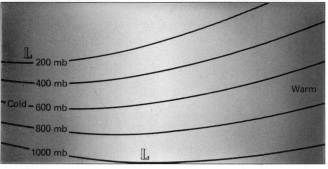

FIGURE 17.20
Cross-sectional view of isobaric surfaces showing why lows slope toward cold air and highs toward warm air with increasing altitude.

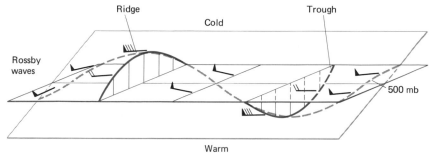

FIGURE 17.19
Comparison of 1000 mb (lower) and 500 mb (upper) isobaric surfaces. Circular highs and lows at the surface change to a corrugated surface with Rossby waves at 500 mb, when air is uniformly colder in the north.

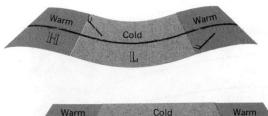

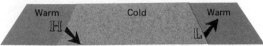

FIGURE 17.21
When cold air lies to the west of lows and to the east of highs, systems slope to the west with increasing height. This leads to southwest winds *above* lows and northwest winds *above* highs. Thus, lows typically move from southwest to northeast and highs from the northwest to southeast. Compare this to Figure 15.27.

**Steering Rule:** Surface pressure systems in the midlatitudes tend to move in the direction of the winds directly above at the 500 mb level and at a little more than half the speed of the 500 mb winds.

The steering rule is very easy to apply, but extremely complicated to explain. It works better for lows than it does for highs. It also works better for smaller, less intense systems. And it works rather well for hurricanes.

**Note:** Here is a practical outdoor weather forecasting technique for which you don't need any weather maps. First, stand with your back to the wind and then turn about 30° to your right. Low pressure is then on your left. Now look at any high or middle clouds. If they move *from* your left, then a low is approaching and it will probably rain or snow within a day. If they move *to* your left the weather should become clearer, although showers are possible.

You should realize that this simple forecasting technique represents an application of two principles—Buys-Ballot's law and the steering rule. Have patience when you use the rule since it may take a full minute or more to accurately determine the motion of the high clouds.

# CYCLOGENESIS

Two distinct processes are responsible for cyclogenesis (cyclone development). The first process works on the eastern slopes of mountain ranges and is simple to explain. It is analogous to the so-called ice-skater effect. Ice skaters always spin faster when they bring their arms as close as possible to their bodies.

Low pressure areas often weaken or even lose their identity when they cross mountain ranges. This is understandable in terms of the ice-skater effect, if you consider the low to be a column of air. When the column passes over a mountain range it squashes down and spreads out. This makes it spin more slowly (see Figure 17.22). Then, as the air column descends the eastern slope of the mountain range, the column stretches vertically and becomes thin again. This makes it spin faster once more.

**Note:** This rule does not work for high pressure areas. Because of the rotation of the earth, even still air on earth has an inherent cyclonic rotation that must be added to the rotation of the winds in order to produce the ice-skater effect. Generally this inherent cyclonic rotation is dominant so that high pressure areas invariably appear *more* intense when they cross the mountains!

One easy way to predict the appearance of low pressure areas on the eastern side of mountains is to watch lows as they approach the mountains from the west. Then, merely calculate how long it would take the low to reach the other side of the mountain range. Do not worry if the low disappears while in the mountains. More often than not, it will reappear on the eastern side of the mountains at the expected time, although probably somewhat south of where you would expect; this is depicted in Figure 17.23.

Cyclones that develop by the ice-skater effect alone would soon die because they lack an energy source. The

FIGURE 17.22
The ice-skater effect makes lows weaken or disappear while in the mountains. Tall, thin columns spin more rapidly than short, wide ones.

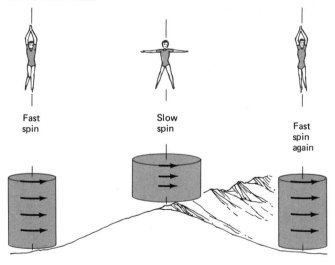

Fast spin      Slow spin      Fast spin again

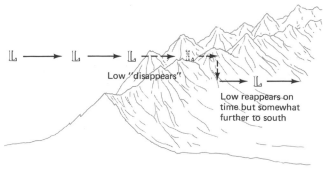

FIGURE 17.23
Lows often lose their identity in the mountains, but reform at the expected time on the other side, though somewhat equatorward of their original path.

energy source of cyclones comes from the contrast between warm and cold air masses, with cold air sinking and warm air rising. The second type of cyclogenesis involves the contrasting air masses as well, and once it begins it produces something of a chain reaction (*baroclinic instability*). The entire development process is shown schematically in Figure 17.24.

In order for the chain reaction to begin, several conditions must be satisfied simultaneously. At the surface, there should be a weak low with some temperature contrast (there may be a front). A 500 mb trough should lie some distance west of the surface low, and strong jet-stream winds should be moving directly over the surface low.

The presence of the jet stream increases the normal tendency for rising motion downstream from the upper troughs (recall from Chapter 15). Often, weak lows at the surface will remain rather inactive until the time that the jet stream passes overhead.

The rising motion above the surface low stretches the air column vertically, intensifying the winds by the ice-

FIGURE 17.24
The ideal wind and temperature patterns for cyclogenesis. Stage 1, beginning; Stage 2, well-formed but still rapidly developing; Stage 3, occluded and fully developed. After this point, storm slowly disintegrates.

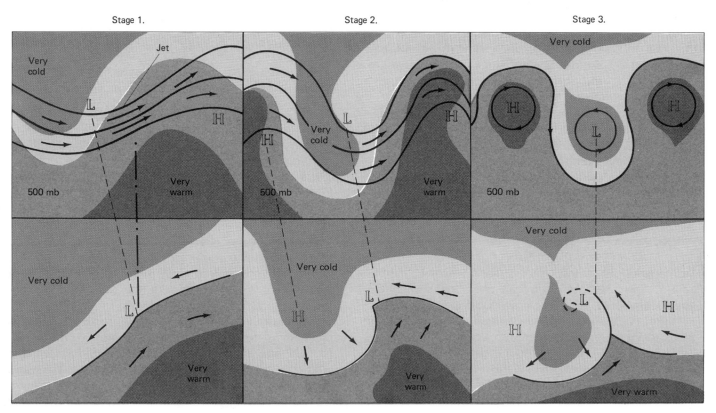

skater effect! This draws cold air southward on the western side and warm air northward on the eastern side of the surface low.

The increasing temperature contrasts immediately intensify the upper-tropospheric troughs and ridges. For instance, the isobaric surfaces of the upper trough (which is located west of the surface low) sink even lower than before, because the air is growing colder and denser. Conversely, warm air brought under the upper ridge raises the isobaric surfaces even further there. This is shown in Figure 17.24 Stage 2.

The chain reaction is completed when we see that the more the upper troughs and ridges intensify, the greater the rising motion that is induced above the surface low and so the more the surface low is intensified by the ice-skater effect, and so on.

Eventually the chain reaction comes to an end. The cold air sweeps completely around to the eastern side of the low; the warm air is displaced aloft and the storm occludes. Once the cold air has moved around the low, the upper trough no longer slopes to the west with height. At this point (see Figure 17.24 Stage 3) the surface low and upper low line up vertically, which means an end to the strong rising motion above the low center. Thus, the chain reaction is terminated and the low slowly begins to die.

In this final stage the extratropical low has a cold core. (This is exactly the opposite of the structure of a fully developed *tropical* cyclone [i.e., hurricane].) Surface lows at this stage of development generally slow down and may even come to a standstill. Why? Since the upper low lies directly above the surface low, there are no upper-tropospheric winds to "steer" the surface low.

Cyclones will develop by this process of baroclinic instability as soon as there is a large enough north–south temperature contrast (no temperature contrast — no upper troughs or ridges and no jet stream). Thus, once again, you see that temperature differences drive the winds, but then the winds act to reduce the temperature differences.

# FRONTOGENESIS

Frontogenesis is another process that involves a chain reaction. This chain reaction can only begin when the large-scale winds bring warm and cold air together.

Figure 17.25 shows such a situation. This wind pattern tends to increase the temperature contrast at all levels. The temperature changes produce changes in the height of the isobaric surfaces, making them sink where the air is getting colder and rise where the air is getting warmer.

Stage 1.

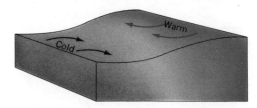

Stage. 2

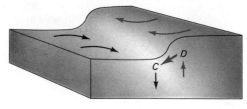

Induced secondary circulation

Stage 3.

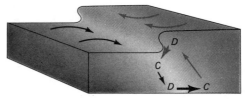

Induced secondary circulation

FIGURE 17.25

The ideal wind and temperature patterns for frontogenesis. Stage 1, large-scale wind field increases temperature contrast; Stage 2, secondary circulations are induced; Stage 3, secondary induced circulations are fully developed. They help to further increase temperature contrasts, make fronts slope, and produce clouds in the warm air. C and D represent regions of convergence and divergence, respectively.

This naturally increases the slope of the isobaric surfaces aloft, causing air to accelerate downslope.

In short, a secondary wind circulation is induced in the frontal zone, as is shown in Figure 17.25 Stages 2 and 3. Vertical motions are induced, with rising air on the warm side of the frontal zone and sinking air on the cold side.

These secondary circulations further increase the temperature gradient, and so a chain reaction begins. The increased temperature gradient then increases the slope of the isobaric surfaces, which further intensifies the induced secondary circulations, and so on.

Two realistic features come as by-products of this explanation of frontogenesis, thereby reinforcing the theory. First, the front slopes upward toward the cold air, with the cold air wedged under the warm air. Second, the warm air above the front rises, while the cold air below sinks.

Furthermore, the chain reaction is needed to explain why fronts exist in the first place. There are so many other processes such as turbulence that tend to mix air, thereby *reducing* any contrasts, that it requires a very active process to build up strong temperature contrasts. In fact, as soon as there are no longer the proper general wind conditions, fronts rapidly dissipate.

## SOME CASE STUDIES

"Untried virtue is no true virtue." The weather features and processes discussed in this chapter take on meaning only when you have studied the actual weather day after day. Here we only have room to present a few textbook cases. The weather situations that are discussed were chosen because of their unusual features, and they tend to represent extremes of weather. The one "typical" storm presented here is actually one of the rare storms that fits the ideal pattern very closely. But this is a good way to begin to learn the art of weather analysis and forecasting.

### July 12, 1966 — One Day in the Life of a Heat Wave

July 1966 was one of the hottest months on record for the eastern two-thirds of the United States. There were two distinct heat waves, the first made memorable because it hit during the July 4 weekend. The second came during the week of July 10–17 and was especially bad in the Midwest. Huron, South Dakota, reported an all-time high of 44°C on July 10, but the effect was far more serious in St. Louis where there were several days when the temperature readings climbed higher than 40°C and markedly increased the mortality rates.

The weather map for July 12, 1800 GMT, (shown in Figure 17.26) tells the story all too well. A large high pressure area had come to rest over the southeastern United States. The clockwise flow of air around the high pumped hot, humid air up from the south. All areas shaded in red had temperatures of 35°C or higher.

Since this map described the weather a few hours before the hottest time of day, most stations (especially in the west) had highs several degrees higher. More than 1 million square kilometers had highs of 38°C or more later that afternoon.

This map shows that the heat wave had ended in the Northern Plains States. A cold front that was creeping southward dropped temperatures by 10 to 15°C in the Dakotas. You will see also that there was a local cool area in Ohio as the result of thunderstorms.

The 500 mb chart for that morning (Figure 17.27) shows a closed, high pressure area over Missouri. The high at 500 mb is located northwest of the surface high, as you should expect. Why? Because highs slope toward the region of warmest air with increasing height, and on both maps the warmest air is located northwest of the high centers.

The 500 mb chart is useful for telling how the weather will change, because you can use it to apply the steering rule. Any time there is a closed low or high at the 500 mb level directly above the surface weather system you know that there will be only very slow changes in the weather, because there are no upper winds to steer the weather. The surface high is found under a region of weak easterly winds. According to the steering rule it should move very slowly to the west, and this is precisely what happened. By the next morning the high was centered over Mississippi.

No wonder they had such an extended heat wave. That high had been there for two weeks because there were no upper winds to steer it away. The longer the high remained there, the more hot, humid air it pumped up from the Gulf of Mexico.

The only big change in the weather for July 13 should be the eastward sweep of the cold front. Notice that the front is located under a region of reasonably fast westerly winds. Using the steering rule which tells us that systems tend to move at slightly more than half the speed of the 500 mb winds above, the cold front should be located a little more than 500 nautical miles farther to the east by the morning of July 13. In reality, the front moved about 700 nautical miles, in part because the 500 mb winds sped up during this time. The 500 mb flow was also helpful in showing that the front would not advance far southward; Nebraska had to wait more than 24 hours before it got some relief from the heat.

### January 17, 1977 — A Record-Breaking Cold Wave

One look at the weather map for the morning of January 17, 1977 (Figure 17.28) tells you that heat waves do not last forever. As you have already read in Chapter 8, this was one of the coldest days in one of the coldest months on record throughout the eastern two-thirds of the United States. A large high pressure area from northern Canada had settled over the middle of the United States. Frigid air plunged southward, dropping temperatures below freezing as far south as the Gulf of Mexico.

This air remained frigid during the day, since the snow-covered ground reflected much of the sunlight. Then, at night, under mostly clear skies and light winds, the tem-

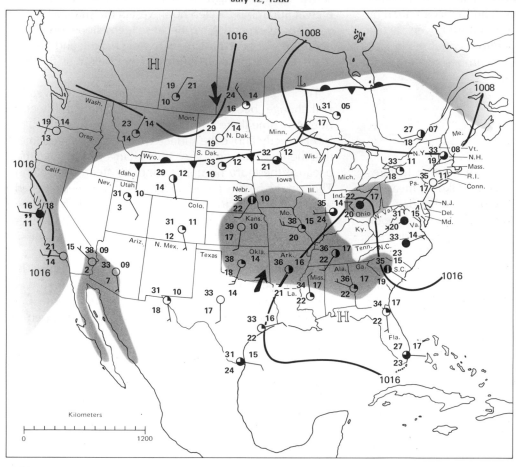

Surface weather map
1800 GMT
July 12, 1966

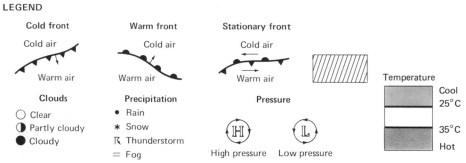

LEGEND

**Cold front**

Cold air

Warm air

**Warm front**

Cold air

Warm air

**Stationary front**

Cold air

Warm air

**Temperature**

Cool
25°C

35°C
Hot

**Clouds**

○ Clear

◐ Partly cloudy

● Cloudy

**Precipitation**

• Rain

* Snow

Ⱪ Thunderstorm

= Fog

**Pressure**

Ⓗ High pressure

Ⓛ Low pressure

FIGURE 17.26
Surface weather map for 1800 GMT, July 12, 1966. Pressure interval for isobars is 8 mb. Heavy red arrow indicates general movement of warm air; heavy black arrow indicates movement of cold air. All times indicated are Greenwich Mean Time (GMT), which is five hours ahead of EST and four ahead of EDT.

peratures dropped below −20°C as far south as northern Georgia.

To see why this weather was so unusual you need only one glance at the 500 mb chart (Figure 17.29). Strong winds over the center of the country were blowing straight from northern Canada. Normally at this time of year the winds come from slightly north of west. In this map the northerly component was, to say the least, not

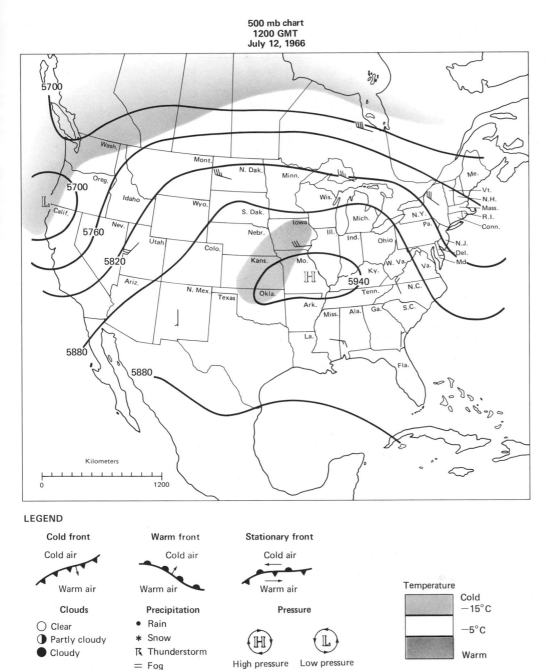

500 mb chart
1200 GMT
July 12, 1966

LEGEND

**Cold front**

Cold air

Warm air

**Warm front**

Cold air

Warm air

**Stationary front**

Cold air

Warm air

**Clouds**

○ Clear

◑ Partly cloudy

● Cloudy

**Precipitation**

• Rain

✳ Snow

R Thunderstorm

= Fog

**Pressure**

Ⓗ High pressure   Ⓛ Low pressure

**Temperature**

Cold
−15°C

−5°C

Warm

FIGURE 17.27

The 500 mb map for 1200 GMT, July 12, 1966. Contours represent height of 500 mb level in meters above sea level. Contour interval is 60 m. Notice that winds tend to blow parallel to contours and are faster when contours are closer.

slight. According to the steering rule, all the weather systems in the United States should have been coming straight from northern Canada, which they were.

In fact, the strong northwesterly winds aloft persisted almost without a break throughout the entire month. (Notice the closed low aloft over Maine.) The result was record-breaking cold. Now if you are smart enough to ask why the northwesterly wind was so persistent in the first

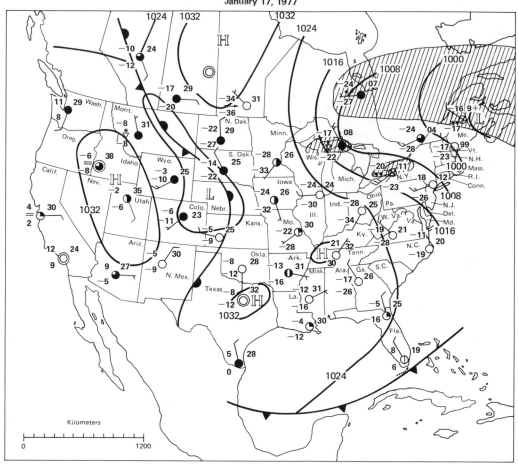

Surface weather map
1200 GMT
January 17, 1977

LEGEND

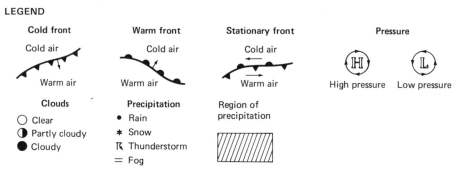

| Cold front | Warm front | Stationary front | Pressure |
|---|---|---|---|
| Cold air | Cold air | Cold air | |
| Warm air | Warm air | Warm air | High pressure   Low pressure |

**Clouds**
○ Clear
◐ Partly cloudy
● Cloudy

**Precipitation**
• Rain
∗ Snow
℞ Thunderstorm
= Fog

Region of precipitation

FIGURE 17.28
The surface weather map for 1200 GMT, January 17, 1977.

place, you will not get a good answer. No one has one yet.

The satellite picture for this time (Figure 17.30) is also quite informative. The cold front appears as a rather poorly defined, broad cloud band stretching eastward from Mexico across to Cuba and over the Atlantic Ocean. Rows of stratocumulus clouds can be seen over the Gulf of Mexico and south of Texas and Louisiana—a sign of cold air pouring over much warmer waters. Another sign of the cold air over warm waters is the open convection cells that can be seen in a region about 1500 km east of the United States. For a refresher on cloud streets and convection cells, look back to Chapter 2.

It is almost cloudless over Kansas so that much of the

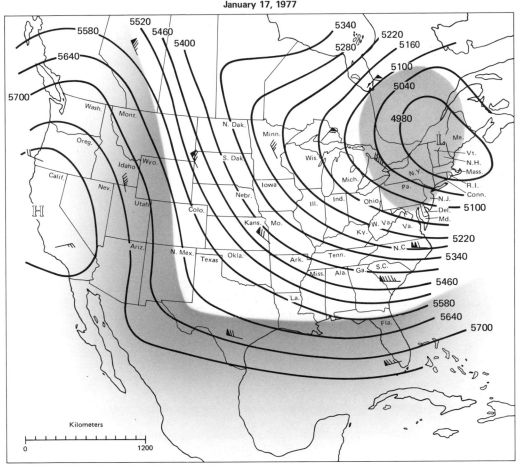

**500 mb chart
1200 GMT
January 17, 1977**

**LEGEND**

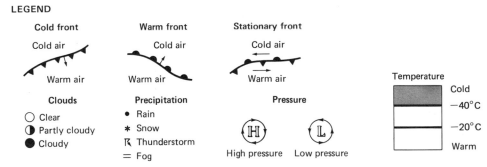

**Cold front**

Cold air

Warm air

**Warm front**

Cold air

Warm air

**Stationary front**

Cold air

Warm air

**Clouds**

○ Clear
◐ Partly cloudy
● Cloudy

**Precipitation**

• Rain
* Snow
R Thunderstorm
= Fog

**Pressure**

Ⓗ High pressure  Ⓛ Low pressure

**Temperature**

Cold
−40°C
−20°C
Warm

FIGURE 17.29
The 500 mb map for 1200 GMT, January 17, 1977.

whiteness that you see is due to snow and not clouds. Geography buffs may notice that, far from the cold, the central valley of California is outlined almost perfectly by fog. Finally, you should notice the difference between the west and east coasts of Central America. Clouds end abruptly at the mountain tops, and the west side of the continent is almost clear. It should be easy for you to guess the wind direction at this time.

**December 5–6, 1977—A Classic Storm**
This storm has already been briefly described in Chapter 1 and shows all the classic features of an ideal extratropical cyclone. From Figures 17.31 and 17.32 you can see that

FIGURE 17.30
The satellite picture for
January 17, 1977.

the cold and warm fronts remain well defined on both maps. As the cold front swept eastward, it generated a number of thunderstorms, some of which produced hail the size of golfballs and larger. Several tornadoes even touched down in Tennessee and South Carolina on the afternoon of December 5.

This storm had the typical large area of precipitation in the cold air north of the warm front. The one "imperfect" feature of this storm was that the clearing did not occur so rapidly after the cold front passed, except down in Texas. The very wide area of cloudiness prevented this classic storm from appearing in all its beauty on the satellite photos until somewhat later (see Figure 1.14).

The movement and development of the storm were also classic. The storm began to develop right around 0000 GMT on December 5 when a region of strong winds at the 500 mb level passed directly over the weak surface low (see Figure 17.33). The low then deepened both at the surface and at the 500 mb level (see Figure 17.34) as cold air was drawn southward on the western side and warm

air northward on the eastern side of the lows at both levels.

The surface low moved in the direction of the 500 mb winds above. According to the steering rule it should have moved somewhat over 700 nautical miles, whereas it actually moved 850. This could also have been predicted. As the low intensified, the upper winds strengthened (because the temperature contrasts also increased), thus making the storm accelerate.

### January 25–27, 1978 — A Killer Storm

This may have been the greatest of all the great storms during the winter of 1977–78. Bad as the previous winter had been, with its unrelieved cold, this winter was worse because of the monstrous storms. On the morning of January 25, 1978 there were two rather innocuous lows, one at each end of the Mississippi River, as you can see in Figure 17.35. Nevertheless, these set the stage for an explosive development of the major blizzard that fol-

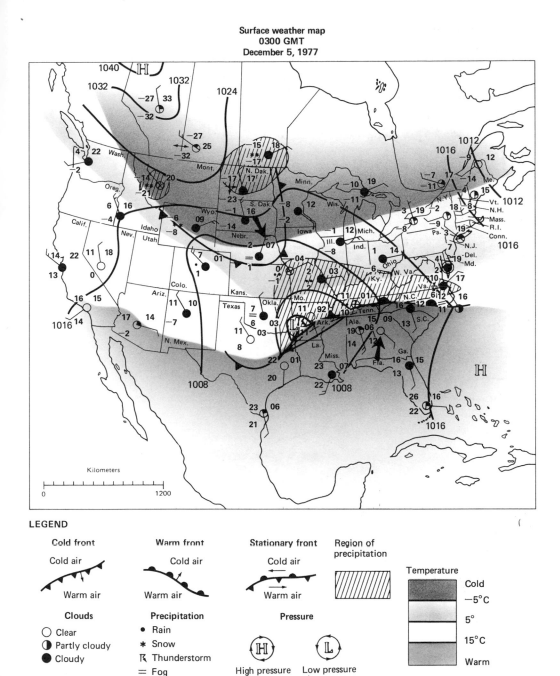

**Surface weather map**
**0300 GMT**
**December 5, 1977**

**LEGEND**

| Cold front | Warm front | Stationary front | Region of precipitation |
|---|---|---|---|
| Cold air | Cold air | Cold air | |
| Warm air | Warm air | Warm air | |

**Temperature**

Cold
−5°C
5°
15°C
Warm

**Clouds**
○ Clear
◐ Partly cloudy
● Cloudy

**Precipitation**
● Rain
✳ Snow
Ⴉ Thunderstorm
= Fog

**Pressure**

Ⓗ High pressure     Ⓛ Low pressure

FIGURE 17.31
The surface weather map for 0300 GMT, December 5, 1977.

lowed within hours. The job of the low to the south was to pump warm, moist air up the east coast of the United States. At the same time a blob of frigid cP air was being blown southeastward into the middle of the country by the low in the north.

Extratropical lows receive their energy from temperature contrasts which in this case were enormous.

The fully developed blizzard of January 26 is shown in Figure 17.36. Note that the two lows had merged into one with enormous and sharp temperature contrasts. The

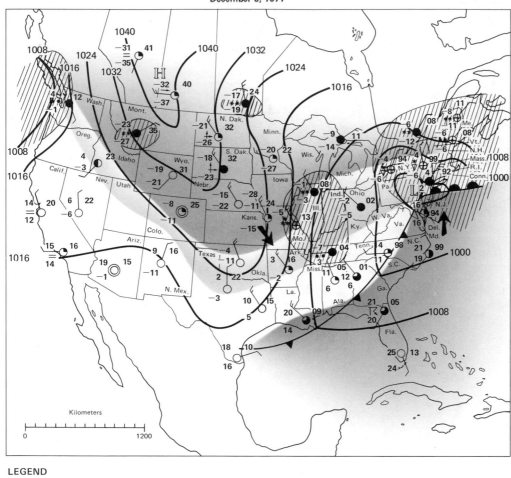

Surface weather map
0300 GMT
December 6, 1977

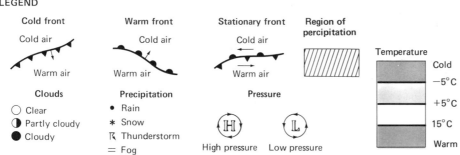

LEGEND

**Cold front**

Cold air

Warm air

**Warm front**

Cold air

Warm air

**Stationary front**

Cold air

Warm air

**Region of percipitation**

**Clouds**

○ Clear

◖ Partly cloudy

● Cloudy

**Precipitation**

• Rain

\* Snow

R Thunderstorm

= Fog

**Pressure**

Ⓗ High pressure

Ⓛ Low pressure

**Temperature**

Cold

−5°C

+5°C

15°C

Warm

FIGURE 17.32
The surface weather map for 0300 GMT, December 6, 1977.

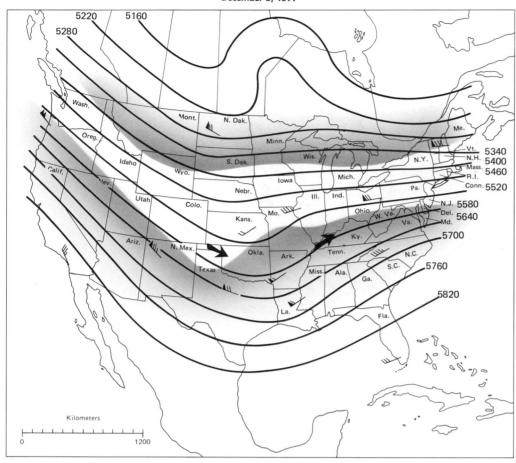

5220   5160
5280

Wash.
Oreg.
Calif.
Nev.
Idaho
Mont.   N. Dak.
Minn.
S. Dak.   Wis.
Wyo.
Iowa
Utah   Colo.   Nebr.
Kans.   Mo.
Ariz.   N. Mex.
Texas
Okla.   Ark.
La.
Mich.
Ill.   Ind.
Ohio   W. Va.
Ky.
Tenn.
Miss.   Ala.
Ga.
Fla.
Pa.
Me.
N.Y.
S.C.
N.C.
Va.

Vt. 5340
N.H. 5400
Mass. 5460
R.I.
Conn. 5520
N.J. 5580
Del. 5640
Md.
5700
5760
5820

Kilometers
0          1200

LEGEND

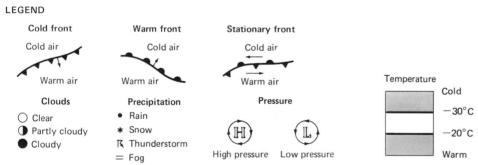

| Cold front | Warm front | Stationary front |
|---|---|---|
| Cold air | Cold air | Cold air |
| Warm air | Warm air | Warm air |

| Clouds | Precipitation | Pressure | |
|---|---|---|---|
| ○ Clear | ● Rain | | |
| ◐ Partly cloudy | * Snow | Ⓗ | Ⓛ |
| ● Cloudy | ℞ Thunderstorm | High pressure | Low pressure |
| | = Fog | | |

Temperature
Cold
−30°C
−20°C
Warm

FIGURE 17.33
The 500 mb chart for 0000 GMT, December 5, 1977.

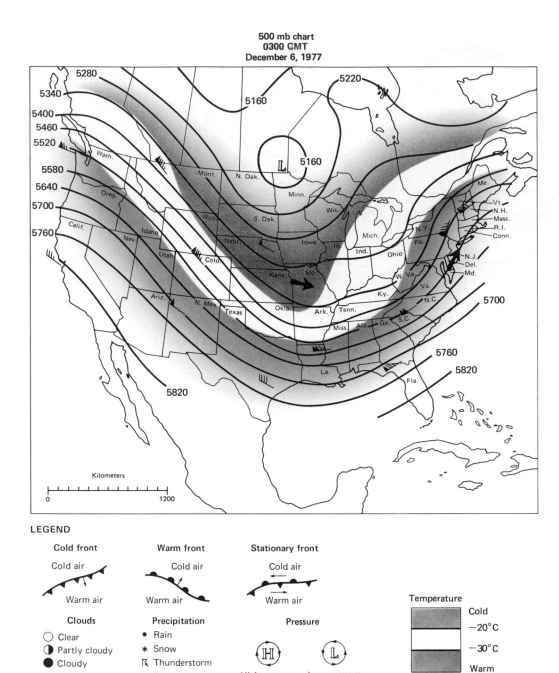

500 mb chart
0300 GMT
December 6, 1977

LEGEND

**Cold front**

Cold air

Warm air

**Warm front**

Cold air

Warm air

**Stationary front**

Cold air

Warm air

**Clouds**

○ Clear
◑ Partly cloudy
● Cloudy

**Precipitation**

● Rain
✳ Snow
⚡ Thunderstorm
= Fog

**Pressure**

Ⓗ High pressure
Ⓛ Low pressure

**Temperature**

Cold
−20°C
−30°C
Warm

FIGURE 17.34
The 500 mb chart for 0000 GMT, December 6, 1977.

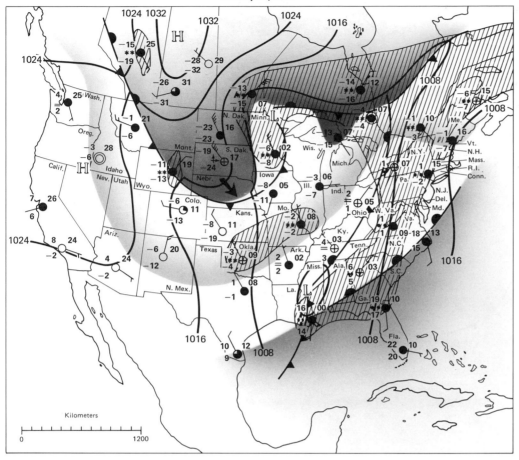

**Surface weather map
1200 GMT
January 25, 1978**

LEGEND

**Cold front**

Cold air

Warm air

**Warm front**

Cold air

Warm air

**Stationary front**

Cold air

Warm air

**Region of precipitation**

**Clouds**
○ Clear
◑ Partly cloudy
● Cloudy

**Precipitation**
● Rain
✱ Snow
℞ Thunderstorm
= Fog

**Pressure**

High pressure    Low pressure

**Temperature**
Cold
−10°C
0°C
10°C
Warm

FIGURE 17.35
The surface weather map for 1200 GMT, January 25, 1978.

## Surface weather map
## 1200 GMT
## January 26, 1978

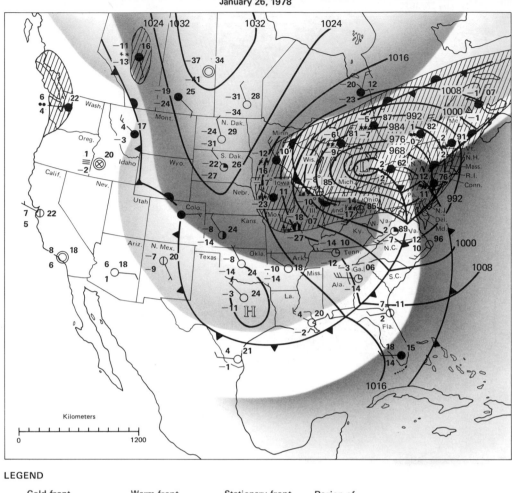

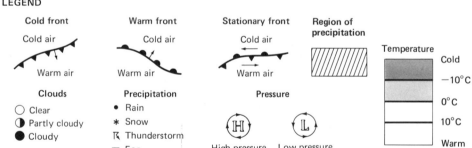

### LEGEND

**Cold front**

Cold air

Warm air

**Warm front**

Cold air

Warm air

**Stationary front**

Cold air

Warm air

**Region of precipitation**

**Clouds**

○ Clear

◐ Partly cloudy

● Cloudy

**Precipitation**

• Rain

＊ Snow

ʀ Thunderstorm

= Fog

**Pressure**

Ⓗ High pressure

Ⓛ Low pressure

**Temperature**

Cold

−10°C

0°C

10°C

Warm

FIGURE 17.36
The surface weather map for 1200 GMT, January 26, 1978.

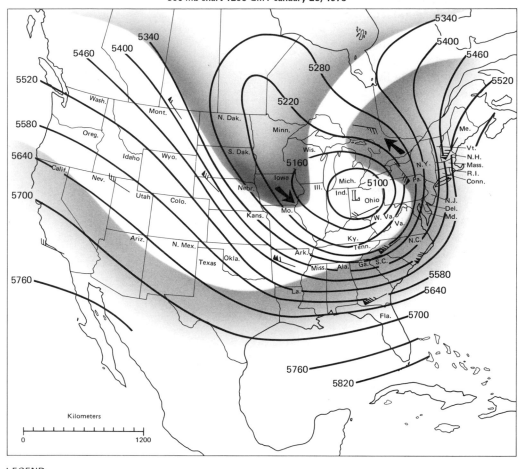

500 mb chart 1200 GMT January 26, 1978

LEGEND

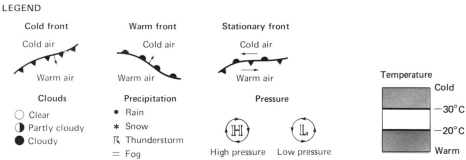

**Cold front**
Cold air
Warm air

**Warm front**
Cold air
Warm air

**Stationary front**
Cold air
Warm air

**Clouds**
○ Clear
◐ Partly cloudy
● Cloudy

**Precipitation**
● Rain
\* Snow
℞ Thunderstorm
= Fog

**Pressure**
Ⓗ High pressure
Ⓛ Low pressure

**Temperature**
Cold
−30°C
−20°C
Warm

FIGURE 17.37
The 500 mb chart for 1200 GMT, January 26, 1978.

storm is by no means simple in structure; count all of its fronts. Some of the coldest air had actually been pulled around to the south of the storm (showing that it had reached its peak intensity) in the Ohio Valley States, while it was much warmer up in Ontario. On the eastern side of the storm temperatures rose above 15°C, and people living along the East Coast had no idea of the blizzard conditions raging less than 1000 km to their west.

Between 20 and 85 cm of snow fell in large parts of all the states from Iowa and Wisconsin to northern Tennessee and West Virginia. In places where there was only 10 cm of snow, the incredible sustained winds of 35 to 60 knots with higher gusts whipped the new snow as well as snow from previous storms into drifts 4 m high. Where the snowfall was greater, so were the drifts which reportedly reached heights of 15 m in parts of Michigan! Think about it. That is as high as a five-story building! Thousands of motorists were stranded and some were even buried.

More than 80 people died during this storm, mostly from exposure to the cold when they abandoned their cars and got lost in drifts. The central pressure of this storm dipped below 960 mb, making it as low as many hurricanes. No one who lived through this storm would take issue with such a classification.

Added to the intensity of this storm is the fact that once it reached southern Canada it stalled and then slowly and aimlessly drifted northward, prolonging blizzard conditions an extra 24 hours. A look at the 500 mb chart (Figure 17.37) is quite revealing in this regard. The storm stalled because a very deep low had formed aloft, almost directly above the surface low. Thus there were almost no winds aloft to steer the storm away. But there were enormous winds all around the storm center, and the 500 mb chart shows the classic temperature pattern for storm development. These extremely rapid winds around the low center did whip the cold front eastward, causing temperatures to drop by 15°C in a few hours along the East Coast and ending the heavy rains there quite abruptly.

By the morning of January 27 the 500 mb chart showed a cold-core low (not shown here) aloft, so the storm slowly weakened as it drifted away. Nevertheless, the prolonged cold it brought kept the snow from melting so that many schools and businesses remained closed for several weeks.

# PROBLEMS

**17-1** Why would the source regions of cP air be highly polluted if they were urbanized?

**17-2** All-time low temperature records are much lower to the west of the Appalachian Mountains than to the east. Explain why. (The answer has nothing to do with the ocean.)

**17-3** Why does mT air from the Gulf of Mexico produce fog in fall and winter, but thunderstorms in spring and summer in the southeast United States?

**17-4** Maritime polar air typically forms in low pressure areas. Explain this.

**17-5** Why do air masses tend to form in highs, but fronts form in lows?

**17-6** Why is thunderstorm activity so restricted in cT air, despite the high-surface temperatures?

**17-7** Where is it warmer in Figure 17.3: in Ottawa, Canada, or New Orleans, Louisiana? Explain this.

**17-8** Judging from a map and from Figure 17.4, estimate the average eastward speed of the cold front. Explain how the temperature drop at each city is consistent with the change in wind direction.

**17-9** Estimate the average slope of the front in Figure 17.5.

**17-10** Explain why fronts have a slope.

**17-11** Draw Figure 17.9 as it would appear in the SH.

**17-12** Why is it often quite warm just before a cold front passes?

**17-13** Why is snow much more likely if the low pressure center passes south of you?

**17-14** A cold front passes by in Portland, Oregon. Explain why there is rarely a dramatic cooling, and why an extended period of showery weather often follows.

**17-15** Explain why freezing rain is reasonably common during winter along the northeast coast of the United States and the east coast of Canada.

**17-16** Memorize Figure 17.11 and know how to use it to forecast the typical weather sequences observed for passing lows.

**17-17** Explain why low pressure areas slope upward toward the cold air regions, and why high pressure areas slope upward toward the warm air regions.

**17-18** Read the Note on page 292. What does the practical outdoor weather-forecasting technique predict on December 5, 1977, for (a) Montreal, Canada; (b) Columbus, Ohio; (c) Oklahoma City, Oklahoma; (d) St. Louis, Missouri; and (e) Birmingham, Alabama? To do this problem, you must compare the wind directions on Figures 17.31 and 17.33 for each city.

**17-19** Explain why coastal lows (see Figure 17.15) develop frequently and often become quite intense during winter.

**17-20** Explain why lows develop frequently to the east of the Rocky Mountains and often grow quite intense as they move eastward during winter.

**17-21** What kind of fog is occurring in Boise, Idaho, in Figure 17.28? Explain.

**17-22** Judging from Figure 17.29, do you think the eastern United States remained cold? Or did it warm up rapidly?

**17-23** Describe the weather at San Francisco, Kansas City, Montreal, and Jacksonville as shown in Figure 17.31. Then describe the way the weather changed over the next 24 hours (use Figure 17.32).

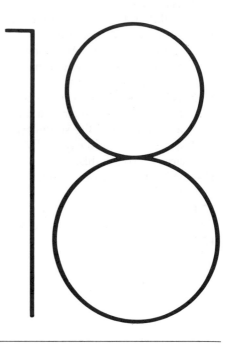

# HURRICANES AND THE TROPICAL BACKGROUND

A group of about fifteen friends decided to hold a hurricane party on August 17, 1969 during hurricane Camille. Ignoring the order to evacuate the Gulf Coast, they felt safe on the third floor of a sturdy hotel building. But they must have realized that something was wrong when they looked outside and saw a struggling woman floating by, level with their window! Miraculously, the woman lived to tell the tale. For all the rest, it was their last party.

Hurricanes are the deadliest storms on earth. They can produce damage in almost every conceivable manner. Their tremendous winds can snap trees like twigs and raise mountainous waves at sea. Born over the tropical waters, they produce utter chaos when they plow into coastal regions, not so much because of their high winds but because of the enormous waves and high tides they produce. Then, even though the wind quickly dies down once hurricanes move inland, they can still create havoc because no other storms produce as much rain. Hurricanes have caused unprecedented flooding even in areas that are accustomed to torrential downpours.

Hurricanes are defined as intense tropical cyclones that have a maximum wind speed of 65 knots or more. The word *hurricane* may originate from the name of the Mayan Storm God, Hunraken, or perhaps from the Carib word, urican, for big wind. The name hurricane is used for storms when they form in the Atlantic Ocean, but the same storms are called typhoons in eastern Asia, baguios in the Philippines, cyclones in India, and willy-willies in Australia. Generally, the typhoons and baguios are far more intense than their cousins in the other oceans because they have a better chance to develop more completely over the Pacific Ocean.

This chapter begins with a discussion of some of the weather records and destruction caused by hurricanes. The life cycle of hurricanes is then considered, from the birth of a hurricane in the background of the more typical tropical weather patterns to their death over land or in the higher latitudes. In the final section of this chapter the most destructive aspect of hurricanes—the storm surge—is considered.

## DESTRUCTION BY HURRICANES

A hurricane warning service was established around 1900 by President McKinley after Willis Moore of the Weather Bureau had demonstrated that hurricanes and storms had sunk more ships than all the wars in history. McKinley was prompted to say, "I am more afraid of a West Indian Hurricane than I am of the entire Spanish Navy." The Spanish Navy would have returned the compliment. Even as late as World War II, hundreds of lives were lost at sea because of typhoons, the worst of which claimed the lives of 790 sailors, 3 destroyers, and 146 airplanes.

This book began with a brief account of the 1938 New

England hurricane. That storm produced all the forms of damage and destruction of which hurricanes are capable, but many other hurricanes have claimed more lives. The cyclone that struck East Pakistan in November 1970 killed 300,000 people. One terrible thing about this storm was that it was an almost exact duplicate of an earlier East Pakistan cyclone that had struck in 1876.

Most lives lost in hurricanes are claimed along coastal lowlands. The cyclones of 1970 and 1876 are just two of a long list of astoundingly destructive storms in that general area. The coastline of that part of the Bay of Bengal is rather flat with an extremely dense population. In addition to this, the orientation of the coastline is almost ideal for creating huge tidal surges. In 1737, a cyclone produced a storm surge estimated to have been as high as 12 m! This wall of water advanced on Calcutta, and the storm also claimed about 300,000 lives.

Another low-lying coastal area that is extremely susceptible to such hurricane damage is the Gulf Coast of the United States from Texas to Florida. The most lethal American hurricane smashed into Galveston, Texas, in September 1900. In the years before the storm, the residents of Galveston had, for "aesthetic" purposes, gradually removed a 5 m high sand dune that might have protected them from what followed. When the hurricane did strike in 1900 its floodwaters were easily able to rush unimpeded into the city, drowning 6000 people. One eyewitness claimed that in the center of the city the water from the storm surge rose "4 feet in as many seconds."

It is partly because of the large death toll of that storm that extensive precautions are taken every time a hurricane takes aim against the Gulf Coast. Thousands of people living in the coastal sections are evacuated inland. When Carla in 1961 and Camille in 1969 smashed into the coast their enormous energies were directed against mostly empty dwellings. In Camille alone the evacuation may have saved 100,000 lives, since the storm surge was estimated to have been 7 m above normal *high* tide from the watermarks left inside a building at Pass Christian, Mississippi, and thousands of buildings were utterly destroyed. One movie taken during Camille shows a building collapse almost as fast as a falling house of cards. (See Figure 18.1).

There are also many miraculous stories of people who survived the incredible fury of hurricanes. The famous "Labor Day Storm" of September 2, 1935 was perhaps the most intense Atlantic hurricane of the twentieth century. It passed directly over Long Key, Florida, where it was described by J. E. Duane, a cooperative observer of the Weather Bureau. Duane wrote

During this lull (the eye of the hurricane passed directly overhead) the sky continued to clear to northward, stars shining brightly and a very light breeze continued; no flat calm. About the middle of the lull, which lasted a total of 55 minutes, the sea began to lift up, it seemed, and rise very fast; this from the ocean side of camp. I put my flashlight out on sea and could see walls of water which seemed many feet high. I had to race fast to regain entrance of cottage but water caught me waist deep, although writer was only about 60 feet from doorway of cottage. Water lifted cottage from its foundation and it floated. 10:10 P.M. Barometer now 27.02 inches, wind beginning to blow from SSW.

10:15 P.M. The first blast from the SSW full force. House now breaking up—wind seemed stronger than at any time during the storm. I glanced at barometer which read 26.98 inches, dropped it in water and was blown outside into sea; got held up in broken fronds of cocoanut tree and hung on for dear life. I was then struck by some object and knocked unconscious.

Sept. 3: 2:25 A.M. I became conscious in tree and found I was lodged about 20 feet above ground.

From Tannehill, *Hurricanes*

Only the sturdiest structures can withstand the full fury of the hurricane. Huge cement breakwaters weighing 1 million kilograms have been knocked aside like children's blocks, tree trunks are snapped, and even the sturdiest houses are pulverized. There are even several eyewitness accounts of church steeples that were silently (except for the ringing church bells) lifted from the rest of the structure and then smashed to the ground. The apparent silence is easy to account for—the incredible roar of the wind completely overwhelms all other noises, except perhaps for the ringing of the bells.

When hurricanes move inland the wind quickly dies down. Within a day of landfall the winds inevitably fall below hurricane intensity, but the rains can keep on falling for several days.

Such rain has proved to be a blessing on countless occasions. Land suffering under prolonged drought has had its moisture replenished and crops saved by hurricane rains. The governments of Honduras and Japan have on several occasions complained that the United States possibly diverted much needed hurricane rains from their countries by seeding the hurricanes while they were at sea.

But the hurricane rains can be excessive. After hurricane Camille moved inland, it dropped as much as 83 cm of rain and caused record floods on the James River of Virginia. Then, in 1972, Agnes deluged the eastern United States. The rains from Agnes relieved a prolonged drought in Georgia but, as you have already read in the section on floods in Chapter 10, Agnes did far more bad than good. One final irony is that in 1974 hurricane Fifi smashed into Honduras killing 10,000 people, mostly

FIGURE 18.1
Damage from hurricane Ca-
mille along the Gulf Coast.
Photos show an apartment
complex before and after the
hurricane hit.

from inland flooding and landslides. Fifi also destroyed most of the corn crop that had earlier been suffering from a drought. In an ironic turnabout, this time the government of Honduras accused the United States of seeding the storm so that it produced too much rain!

The average rainfall intensity in a hurricane is roughly 10 cm/day. This rainfall is not steady since it is produced by thunderstorms. The rainfall intensity also varies considerably from one storm to another and can be excessive. The 200 cm that fell on La Reunion in 24 hours was excessive. The 244 cm that fell at Silver Hill, Jamaica, in four days was excessive. Even the mere 58 cm that fell at Taylor, Texas, in 24 hours in September 1921 was excessive and flooded normally dry San Antonio with up to 3 m of water.

## THE BIRTH OF HURRICANES: THE TROPICAL BACKGROUND

Every year thousands of tourists head for the tropics to spend their vacations splashing and swimming in the warm tropical waters. Here the trade winds blow, keeping the temperature on the islands from rising too high. When it happens to get a bit too warm in the afternoon a thunderstorm will often cool things off. Some of the most delightful weather occurs here, since the thunderstorms do not last long and soon the sun comes out again.

However, there is a potential danger in these warm, idyllic, tropical waters. Warm water is the fuel that hurricanes run on. In fact, hurricanes form only over warm waters that are at least 26°C and preferably over 27°C—the warmer the ocean temperature, the more intense the hurricane is likely to become!

Thus, hurricanes tend to form during the late summer and early autumn months when the tropical waters are at their warmest. Figure 18.2 shows the breeding grounds and seasons of the hurricanes. You should notice that the South Atlantic Ocean is the one ocean in which *no* hurricanes ever seem to form. The mystery is partly dispelled when you observe that the water temperatures in the South Atlantic are usually too cold to kindle a hurricane (see Figures 16.13 and 16.14).

You will also notice from Figure 18.2 that no hurricanes ever form within 3° latitude of the equator. Most hurricanes form between 5 and 25° latitude. Hurricanes need the Coriolis force, but there is no Coriolis force on the equator. The Coriolis force does not provide the hurricane with any energy, but it enables the storm to become the efficient, organized engine that it is.

Warm water and the Coriolis force are not the only prerequisites for hurricane formation. If they were, hurricanes would form near Hawaii. However, hurricanes rarely form or pass anywhere near Hawaii. Hurricanes

FIGURE 18.2
Regions where hurricanes originate. Each dot signifies the point where a storm first reached hurricane intensity. Arrows show typical paths taken by hurricanes.

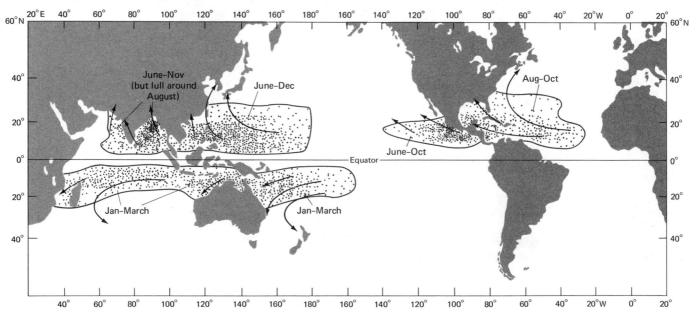

can only form when the wind is uniform at most heights through the troposphere. Therefore, because there are usually easterly winds at the surface, hurricanes do not form where there are upper-tropospheric westerly winds or where the upper tropospheric easterly winds are very strong. In the mid-Pacific near Hawaii, there are northeasterly trade winds at the surface, but westerly winds aloft (see Figure 18.3). Similarly, the strong summertime upper-tropospheric easterlies south of Asia and India suppress hurricane formation during the height of the summer monsoon in July and August.

This last prerequisite for hurricane formation is very significant and shows in more than one way how hurricanes differ from *extra*tropical cyclones. The *extra*tropical cyclone develops only when there is a strong increase of the westerly winds with height. Because of the geostrophic relationship this implies that there must be a strong temperature gradient at ground level. Hurricanes, on the other hand, develop only when there are no significant wind changes with height, which implies that the temperature near ground level must be reasonably constant over a wide area!

But hurricanes never develop from a completely uniform background. They always develop from some preexisting disturbance in the tropical atmosphere, no matter how subtle and harmless it may seem. We still need to learn much more about these preexisting disturbances, but we do know some of their important features.

Tropical weather analysis is a subtle and difficult field. For instance, analyzing pressure patterns is very helpful for understanding extratropical weather systems where a falling barometer usually indicates approaching rain. But this rarely works in the tropics.

When most tropical weather disturbances pass by they hardly affect the barometer at all. Atmospheric pressure rises and falls twice a day with an almost alarming regu-

FIGURE 18.3
Streamline analysis of the winds in the tropics, at 200 mb, for January and July.

Streamlines 200 mb--June to August

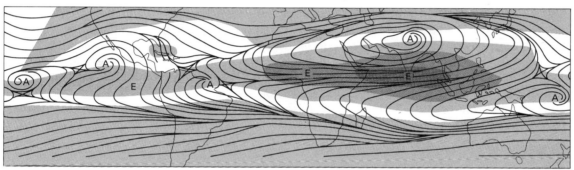

Streamlines 200 mb—December to February

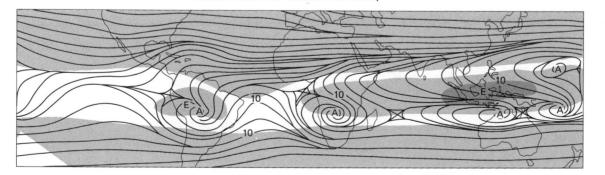

larity in the tropics. Twice each day the pressure rises 1 to 2 mb, to a high around 10 A.M. and 10 P.M. local time (see Figure 18.4). These variations are due to tides in the atmosphere produced by the sun's heating in the upper atmosphere. These tidal pressure variations also occur outside of the tropics, but they are normally masked by the larger pressure changes of the extratropical highs and lows.

Temperature analysis is also almost useless in the tropics. Almost all the air (except in the deserts) is mT air, and temperature variations are very small. Even when the rare cold fronts do penetrate the tropics, they no longer have the large temperature contrasts that they had at higher latitudes. Indeed, they are often distinguishable only by an increase in cloudiness and a shift in the wind direction.

How then can we analyze tropical weather systems? The simplest thing to do is to look at the satellite pictures.

The dominant weather system of the tropics is the ITCZ, as you have seen in Chapter 16. Like the ITCZ, a hurricane is actually an organized system of thunderstorms. Therefore, any disturbance that enhances thunderstorm activity (such as the ITCZ) may eventually produce a hurricane. Generally, sections of the ITCZ do not develop into a hurricane because they are located too close to the equator. Occasionally, however, waves form on the ITCZ, bringing it far enough from the equator to get a hurricane started. The surface winds of these waves in the ITCZ also have the appropriate cyclonic sense of rotation, as is shown in Figure 18.6, because the ITCZ forms in a region of low pressure. Hurricanes tend to form on the poleward side of these waves, and there are times

when four or five waves can be seen on the ITCZ from satellite pictures, with a hurricane forming on the trough of each wave (see Figure 18.5)!

It is possible to generalize our findings about the ITCZ to many other tropical weather systems. The most favored regions for thunderstorm development are marked by convergence of the surface winds and divergence of the winds in the upper troposphere. Meteorologists, therefore, analyze the wind field by drawing **streamlines** (as in Figure 18.3) in order to detect these regions of convergence and divergence. Streamlines are simply lines that are parallel to the wind direction.

A second important feature of the ITCZ is that the trade wind inversion is generally missing. You should recall that the trade wind inversion is produced by the sinking air of the Hadley cell and is a dominant feature in many tropical soundings. Sometimes there is no actual inversion, but only a relatively stable layer with a small lapse rate. In any case, this stable layer or the trade wind inversion is largely responsible for suppressing thunderstorm activity.

The trade wind inversion is strongest on the eastern side of the oceans at about 20° latitude where it is reinforced by cold upwelled ocean waters and the more stable eastern side of the subtropical highs (see Figure 18.7). Therefore, the trade wind inversion gets weaker and is found higher above sea level as you move westward or toward the ITCZ, thus allowing larger cumulus clouds and even thunderstorms to form.

It is not surprising then that most tropical disturbances

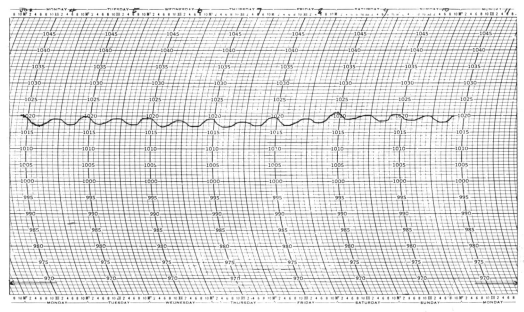

FIGURE 18.4
Typical tropical barograph trace at Barbados shows variations due to tides in the atmosphere.

FIGURE 18.5
Tropical storms developing on the ITCZ. From west to east, Kate, a tropical depression, and Lisa. Lisa later smashed into La Paz, Mexico, killing over 100 people.

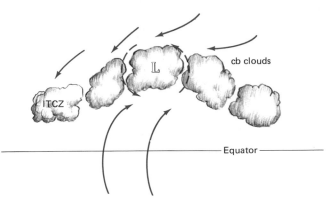

FIGURE 18.6
The winds rotate cyclonically (counterclockwise in the NH) about a poleward bulge or wave in the ITCZ. The solid arrows depict the winds, while the dashed arrows are not winds but merely help to indicate the sense of rotation.

are characterized by a higher and weaker trade wind inversion, which indicates a reduced sinking tendency of the air. The possibility of hurricane development is therefore indicated by a region of reduced stability. This illustrates another very important use of atmospheric soundings.

There is one type of tropical disturbance that has been studied extensively. This is the so-called "easterly wave," so named because it moves from east to west with the tropical easterlies. In the early 1950s some meteorologists thought that all hurricanes developed from easterly waves, but we now know this is not true. Easterly waves are found in the Atlantic and Western Pacific and are not even responsible for all hurricanes there. An easterly wave is shown in Figure 18.8.

The easterly wave usually has a pressure change that is less than 3 mb, but it is easily recognizable from the associated wind patterns and the soundings. To the west of the trough line the surface winds are divergent and the trade wind inversion is intense and close to the ground. All this implies sinking motion, and the area is mostly clear. To the east of the trough line the opposite situation prevails and, as a result, there is enhanced cloudiness. If hurricanes develop, they will, of course, develop to the east of the trough line.

Easterly waves often develop as disturbances over the mountains of Africa. These disturbances occasionally produce tremendous squall-line thunderstorms in Africa but, as they move westward over the cold waters of the Canaries current, they die down. Then they rejuvenate once they reach the warmer mid-ocean waters and, from that point on, hurricanes may develop.

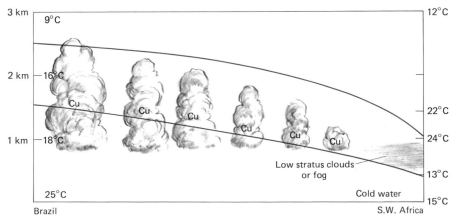

FIGURE 18.7
The trade wind inversion across the South Atlantic Ocean at 22° S latitude during July. The inversion is particularly strong on the eastern side of the ocean. It gets weaker and higher toward the west (where it is not usually an inversion, but only a stable layer), allowing progressively larger cumulus clouds to develop.

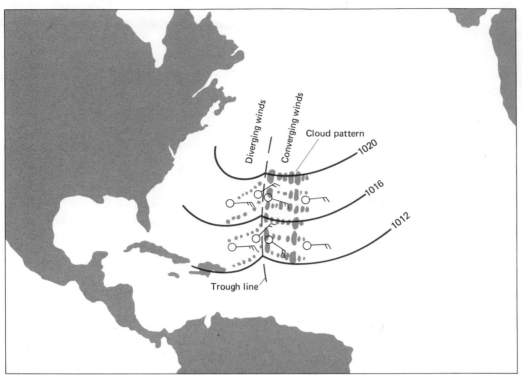

FIGURE 18.8
An easterly wave. This wavelike disturbance moves slowly to the west. Notice how the clouds tend to line up in bands parallel to the wind. Large clouds form only to the east of the trough line when the surface winds converge.

Disturbances only at the surface are generally not sufficient for hurricane development. There must also be a disturbance of the upper troposphere as well. One of the most common upper-tropospheric disturbances that leads to hurricane development is the dying, old remains of an upper-tropospheric cyclone wave from the westerlies (see Figure 18.9). Such waves are often abandoned by the westerlies when they move too deeply into the tropics. They have two features that promote the development of hurricanes.

First, divergence usually prevails on the eastern side of these old, abandoned troughs. This, of course, helps induce rising motion and enhances thunderstorm development. Second, these old troughs usually have cold cores. This means that the atmospheric lapse rate is slightly steeper and more unstable under these troughs, because the surface temperatures are so uniform in the tropics. The increased instability also enhances thunderstorm activity. This is one ironic feature in hurricane formation. Mature hurricanes, as you will soon see, have warm cores—yet they are often born with cold cores.

These old upper-tropospheric troughs do not actively generate hurricanes until they line up with disturbances near the surface. Conversely, easterly waves will often not intensify until they drift under one of these old upper-tropospheric lows. There is a case on record of a preexisting disturbance consisting of a conglomerate of thunderstorms that for several days drifted innocently northward from the Pacific coast of Colombia to the Gulf of Mexico. As soon as it moved under an old upper-tropospheric cyclone, it suddenly blossomed into hurricane Agnes in 1972.

## THE GROWTH OF HURRICANES: A CHAIN REACTION

Meteorologists closely watch all of these tropical disturbances for any signs of a developing hurricane. Most of these preexisting disturbances never amount to much. Perhaps only 10% of them begin to intensify. Once they begin to intensify a chain reaction starts to take over, so that almost 70% of the storms that attain tropical storm designation (maximum winds of 35 knots or more) go on

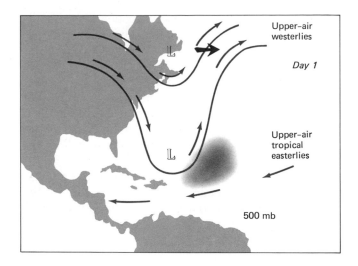

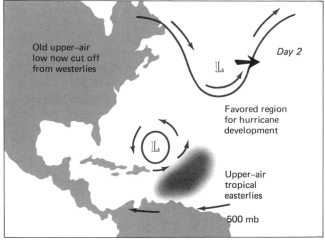

FIGURE 18.9
Hurricanes tend to develop to the southeast of old, abandoned (i.e., cutoff) upper-air lows from the westerlies. The favored region usually coincides with the border of the westerlies and the normal tropical easterlies.

to develop into full-fledged hurricanes. In a sense the tropical atmosphere does not "know" the meaning of moderation. What is this chain reaction?

First, thunderstorms sprout up in the disturbance region. The thunderstorms release large quantities of latent heat into the mid and upper troposphere, warming it and increasing its pressure aloft. This produces divergence aloft that decreases the total weight of air in the column and then causes the sea-level pressure to fall.

As the sea-level pressure falls a well-defined low pressure area is established. The low-level winds converge and begin to spin faster around the low pressure center because of the Coriolis force and the ice-skater effect, but in the lowest kilometer convergence is enhanced by friction slowing the winds, which then turn into the low

center. These winds pick up vapor from the warm ocean waters below and as they approach the low pressure center they rise, releasing the vapor and latent heat. The center of the low grows warmer than before, causing more divergence aloft, and so on. Thus a chain reaction has been established (see Figure 18.10).

Once the sea-level pressure of the hurricane falls significantly, there is an additional source of heat. The surface air would normally cool adiabatically as it moves toward the lower pressure of the hurricane center; instead, its temperature is maintained by contact with the warm ocean waters below. Thus, the core of the hurricane is heated by both latent heat and sensible heat.

You may now be able to see why hurricanes cannot develop when there are any significant vertical differences in the wind. If there were such differences, then the latent heat carried aloft would be swept away and could not lead to a lowering of the central pressure. Therefore, a vital link in the chain reaction would be broken. Again, you should notice how completely this differs from the development of an *extratropical* cyclone, which intimately depends on large vertical wind differences and the tilting with height of the entire system.

The intensity of the growing hurricane soon reaches a limit. This limit is set by two factors. Friction, of course, reduces the kinetic energy and acts as an increasing drag on the winds as the winds get faster. The other important limitation on the hurricane's intensity is set by its fuel supply (i.e., the latent heat of condensation), which is set by the temperature of the ocean water. Rough estimates indicate that each increase of 1°C of the ocean temperature allows a decrease of 12 mb for the central pressure of the hurricane. Of course, the lower the central pressure, the stronger the winds and greater the precipitation tend to be.

Oddly enough, friction—which normally wears things down—actually is indirectly responsible for the hurricane's intensity. The low-level convergence in the hurricane is largely due to friction, and this convergence produces the vertical motion that releases the latent heat. This latent heat drives the storm. Remove the friction and you have removed one of the vital links in the process. Thus, even though friction may limit the ultimate wind speed, it is in a strange way partly responsible for the large speeds in the first place.

## THE MATURE HURRICANE

What does a mature hurricane look like? All hurricanes rotate cyclonically and are intense, low pressure systems (see Figure 18.11). Hurricanes are generally smaller than

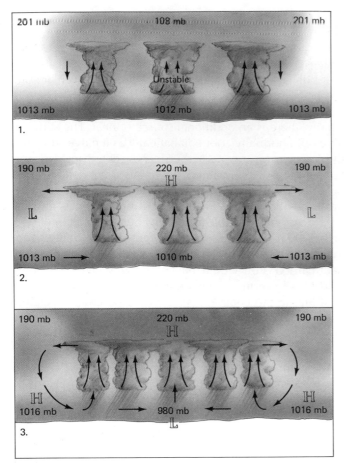

FIGURE 18.10
The chain reaction leading to hurricanes. 1. A disturbance region, perhaps with cold air aloft, favors thunderstorm activity. 2. The thunderstorms warm the upper troposphere, creating a high aloft. Air begins diverging aloft, dropping pressures at the surface below. 3. Convergence into the surface low further enhances the thunderstorm activity.

extratropical cyclones. A typical hurricane has a diameter of about 650 km, but hurricanes can range anywhere from 100 to 1000 km in diameter.

Near the surface the temperatures within hurricanes are almost constant. However, since the central pressure is lower, this means that the air in the center is *potentially* warmer. Above ground level not only does the *potential* temperature increase inward toward the center, but the actual temperature does as well. Central temperatures aloft may be 10°C or even more warmer than the surrounding air. Thus, the hurricane is a warm-core low (this contrasts sharply with the structure of extratropical cyclones).

By drawing isobaric surfaces you have seen (in Chapter 16) that warm-core lows decrease in intensity with height. In fact, above the 400 mb level, hurricanes become high

pressure areas. This, of course, accounts for the divergence aloft and the predominance of anticyclonic circulation beyond about 100 km from the eye near the tropopause.

The hurricane is a warm-core low, the strongest winds are usually found a relatively short distance above the ground (again in contrast with extratropical lows), and the wind speeds generally increase toward the center of the hurricanes. This is another example of the so-called ice-skater effect.

In fact, by the time the air has come within 10 or 15 km of the center of the hurricane, it is spinning so rapidly that it simply cannot be forced any farther toward the center. This air is then forced to rise and as it ascends it also spreads outward. It is this effect that leads to the formation of the eye of the hurricane.

The **eye** of the hurricane is a more or less circular region located in the center of the hurricane that is generally less than 50 km in diameter. As you read earlier in this chapter, the weather in the eye is almost calm by comparison and largely free of clouds. During the day the sun typically shines and there is a strange feeling of warmth, probably due to the relatively light winds (10 to 15 knots). Immediately surrounding the eye is a ring of the most intense weather of the hurricane.

The air at the **outer edge** of the eye is probably dragged upward and outward by the surrounding air. This creates a further pressure deficit in the eye, which induces air from above to sink into the eye. That is why it is mostly clear and warm inside the eye. Eyes can also form inside intense extratropical cyclones for much the same reason (though they are extremely rare); however, they are more common inside dust devils, waterspouts, and tornadoes (see Chapter 20).

This central region of almost calm winds is surrounded by a ring of the most intense winds. The strongest winds of the hurricane are located just outside the eye on the right side of the storm (right with respect to its forward motion). The maximum winds for a fully developed hurricane typically vary between 100 and 150 knots. The ring that has hurricane force winds varies between 20 and 50 km and is rarely wider.

The eye is surrounded by a formidable wall of cumulonimbus clouds, known as the **eye wall.** Not only are the strongest winds found within the eye wall, but the most intense precipitation is also. The eye wall consists of an almost continuous ring of intense thunderstorms.

Extending out from the eye wall are usually two spiral bands that give the hurricane the appearance of a galaxy in space. The spiral bands consist of many individual thunderstorms. Both the precipitation and winds are more

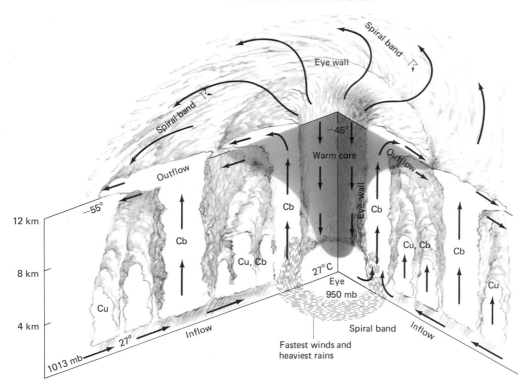

Cirrus and Cirrostratus cloud shield

FIGURE 18.11
The structure of a mature hurricane.

intense in the bands than in the surrounding region of the storm. The bands rotate with the storm at speeds from 10 to 30 knots. The full explanation for the spiral bands has yet to be given, but we do know that they are needed to allow the low-level winds to converge as far as the eye wall.

Meteorologists have also found that you can estimate the strength of a hurricane merely by looking at satellite pictures of these spiral bands. When the spiral bands are tightly wrapped (more like a spool of thread), that indicates an intense hurricane. After all, the spiraling winds help to wrap up the spiral bands. When the spiral bands begin to straighten out, that is one sign of the death (or weakening) of the hurricane.

Central pressures in hurricanes are quite low and often give the places that they pass over the lowest pressures ever recorded there. The typical central pressure of hurricanes is about 950 mb, but an all-time low of 878 mb has been recorded in a typhoon.

There is one other feature of interest about hurricanes. As a hurricane approaches a region, it occasionally spawns tornadoes in its outer fringes. These tornadoes are generally less intense than the squall-line tornadoes of the midlatitudes (see Chapter 19), but they are still quite severe and often strike before people have completed their preparations for the approaching hurricane.

## THE MOVEMENT AND DEATH OF HURRICANES

The extensive cirrus and cirrostratus cloud shield, which is ejected from the upper layers of the hurricane by the diverging winds, is often the first sign of an approaching hurricane. Large ocean waves that are generated by the high winds can provide an even earlier warning sign to people living along the coast, because the waves inevitably move faster than the storm itself. But the meteorologist provides us with the most reliable and most advanced warning of the impending danger.

The simplest forecasting rule is that hurricanes (like extratropical lows) are steered by the winds of the upper troposphere. At first, hurricanes move from east to west, since they are generated in the tropical easterlies. The typical speed of a hurricane in the tropics is about 10 knots because the tropical winds are relatively slow. The motion is often erratic, especially in those places where the background wind is very light. Hurricanes have been known to stop in their tracks and then reverse direction completely. Therefore, there are places that have been struck twice by the same hurricane. (This has happened several times in Japan, Taiwan, and the Philippines.)

Occasionally, two hurricanes approach each other closely and then revolve around one another like planets, with the smaller, less intense hurricane moving more rapidly. This odd behavior is known as the Fujiwara effect (see Figure 18.12).

As hurricanes move around the large subtropical highs they eventually turn poleward. After they move poleward of about 30° latitude they enter the zone of prevailing westerlies and then they curve from west to east (meteorologists call this recurving; see Figure 18.2). After they "hook up" with the stronger westerlies they usually move more rapidly and predictably.

The life span of a typical hurricane is about 10 days—from its birth as an innocuous tropical weather disturbance to its death. Death takes place after the hurricane leaves the region of warm waters and moves over land or over colder waters (see Figure 18.13). If the large-scale winds allow the hurricane to remain over tropical waters, it could persist indefinitely; a few hurricanes have been known to last almost a month. But sooner or later the hurricane does get blown away from the tropical waters.

I have already emphasized that death does not occur the instant that hurricanes pass over land. Hurricanes Agnes and Camille both produced some of their worst flooding several days after they had moved inland. True, the wind speeds rather quickly subsided so they could no

FIGURE 18.12
The Fujiwara effect. When two hurricanes approach each other, they tend to revolve around each other like planets. The smaller, less intense hurricane will move more rapidly.

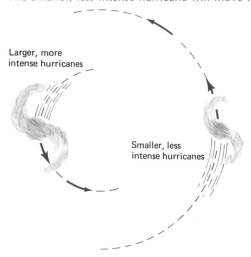

Larger, more intense hurricanes

Smaller, less intense hurricanes

FIGURE 18.13
Hurricanes Ione and Kirsten over the Pacific Ocean near Mexico on August 24, 1974. Notice the closed convection cells in the clouds to their north, indicating cold waters. These hurricanes soon died when they moved over the colder water.

longer be classified as true hurricanes, but then again some hurricanes have become rejuvenated after crossing the land and passing over warm waters. It is more accurate to say that once hurricanes pass over land or over cold waters they undergo a metamorphosis.

Hurricane Agnes (see Figure 18.14) is one example of this metamorphosis. When it first passed over land the winds almost immediately died down, but it still had a warm core. Then it joined with a midlatitude trough in the westerlies and underwent a complete transformation to become a *cold*-core *extratropical* cyclone by the time it reached Pennsylvania (see Figure 18.15). By this time the only hint it gave of its noble origin was the enormous rains it produced. It finally died as an extratropical low. This has also happened to many other hurricanes such as the hurricane of 1938 (which still had an eye when it passed over northern New Hampshire) and hurricane Hazel in 1954. Figure 18.16 shows hurricane Blanche on July 27, 1975 shortly before a cold front caught up with it.

## THE STORM SURGE

The tide is the pulse of the ocean. For countless centuries people living near the shore have watched sea level rise and fall each day. In some places the water rises and falls

FIGURE 18.14
Sequence of satellite pictures of hurricane Agnes.

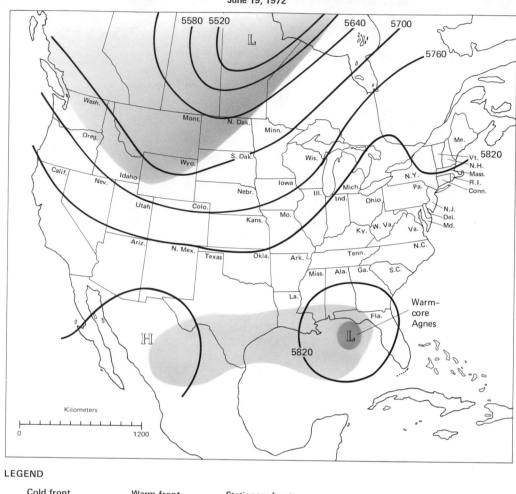

**500 mb chart**
**1200 GMT**
**June 19, 1972**

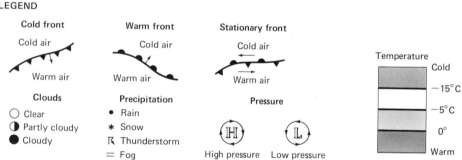
FIGURE 18.15
Transformation of Agnes from a warm-core system into a cold-core system, as seen on the 500 mb charts of June 19 and June 24, 1972.

approximately twice each 25 hours, whereas in other places it rises and sinks approximately once each 24 hours. In some places the seas rise and fall only a small fraction of a meter, while at others (such as the Bay of Fundy) the difference can exceed 15 m!

These astronomical tides are different for each location, but they are so regular that they can be predicted for years in advance. In the days of sailing ships, departures and arrivals were often timed to take advantage of the tides. Indeed, in many cases the ships *had* to sail with the tides.

**500 mb chart
1200 GMT
June 24, 1972**

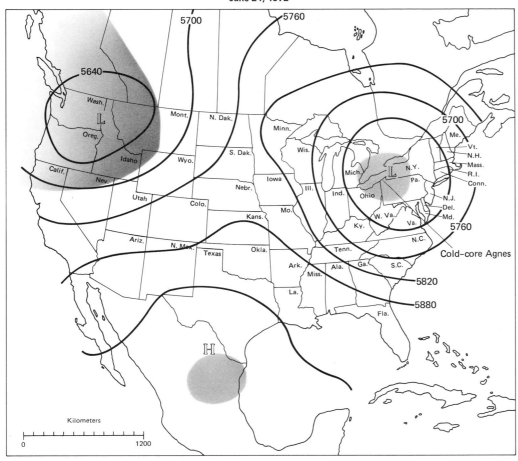

**LEGEND**

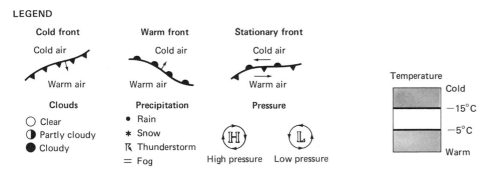

The weather is the factor that destroys the regularity of the tides. Unusual weather can either raise or lower sea level. Of course, most of the time, there is not much concern when the weather lowers sea level. Some ships may be grounded in the unusually shallow water, but otherwise nothing serious results. On the other hand, when the weather causes sea level to rise it can have disastrous results in low-lying coastal areas. The sharp rise in sea level that accompanies some storms is known as the storm surge. The storm surge will produce the greatest

FIGURE 18.16
Hurricane Blanche about to join up with an extratropical cyclone on July 27, 1975. Note the cold front over the Great Lakes approaching from the west.

damage when it coincides with, and therefore rides upon, the normal high tide.

Most intense low pressure areas that pass by a coastline raise the level of the sea by a meter or less. Rises of more than 2 m are rare and generally disastrous. Although we have no accurate records of exactly how high sea level can be raised, there are well-documented rises of 6 or 7 m, as in Camille, and some unverified accounts that go as high as 12 m, for the 1737 cyclone in Calcutta. Many of the most spectacular storm surges have occurred during hurricanes.

The weather can affect sea level in several different ways. One of the well-known effects is the so-called inverted barometer effect. It is easy to illustrate. Fill a pan with mud or wet clay. Then spread your fingers and press down on the mud. The mud will be depressed beneath your hand but will ooze up between your fingers where the pressure is lower.

Similarly, if you built a giant tube and placed it over the ocean and then began pumping the air out of the tube, the water would rise exactly as it does in a barometer (see Figure 18.17). This is precisely what happens in the hurricane, except that the hurricane needs no giant tube to keep the pressure low. As a result of this, we find that *for every millibar you lower the air pressure, the water will rise 1 centimeter.* Since most hurricanes have pressures that are 50 to 75 mb lower than the surroundings, the inverted barometer effect should raise sea level by some 50 to 75 cm. Even the most intense hurricane could only lift

sea level 1.5 m by this effect alone. Obviously there are other effects that must add to the inverted barometer effect to produce the more spectacular rises in sea level.

Every intense storm has strong winds. Not only do these storms create large ocean waves (which may reach 15 m

FIGURE 18.17
The inverted barometer effect. As air is pumped out of a container, the pressure falls and water gets "sucked" up. For each millibar of pressure that is lowered, water rises 1 cm. A low pressure area produces the same effect, but not so abruptly.

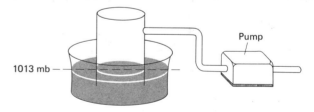

A. Before pumping

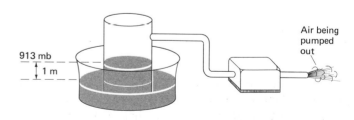

B. After pumping

in hurricanes and 30 m in extratropical lows), but they actually push the water downstream. When the winds blow against a coastline they pile the water against the shore. Actually the water piles highest at some small angle to the right of the wind (in the NH) because of the Coriolis force. Generally, the shallower the water, the higher it will be piled by the wind. Conversely, when the wind blows out to sea, it will drag the waters away from the coast and depress sea level.

One of the most notorious cases of this piling effect occurred during an extratropical low of hurricane intensity that inundated the coastline of England and the Netherlands at the end of January 1953 (see Figure 18.18). Much of Holland is below sea level but is protected from the sea by large dikes. For hundreds of years the Dutch had worked to keep out the sea, but during this storm the waters rose almost 3 m and huge waves pounded away at the dikes. When the dikes finally gave way in a few places the sea poured in reclaiming 25,000 square kilometers. Two thousand people were drowned and 600,000 were left homeless (see Figure 18.19).

There is one last effect of storm weather that can add to the raising of sea level. This is the so-called resonance effect. There are times that a storm can move along a coastline in such a way that the wind blows away from the shore at first but, after the storm center passes, the wind direction reverses (see Figure 18.20). Thus at first the waters will be blown away from the coast. Then, when the center of the storm passes by and the wind drops down, the water will begin to slosh back as if in a giant bathtub. As the water rushes back on its own, the wind suddenly begins blowing toward the shore to give the waters an extra push. In this way the water will rise even higher than if a sustained wind had been blowing toward the shore all along!

In much the same way it is possible to make a swing in the playground go very high with only a series of small but well-timed pushes. In like manner opera singers have been known to crack crystal glass if they sing at a frequency that matches the natural frequency at which glass vibrates.

Such a resonance effect has proven disastrous several times on the Gulf Coast and it has also caused flooding

FIGURE 18.18
Weather map of January 31, 1953, showing areas of coastal flooding (dark shading) and sustained winds above 30 knots (red shading).

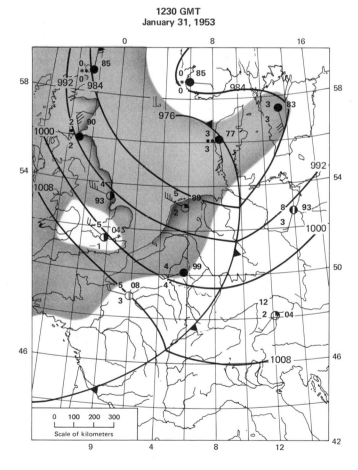

**1230 GMT**
**January 31, 1953**

FIGURE 18.19
View of flooding in the Netherlands from the storm of January 31, 1953.

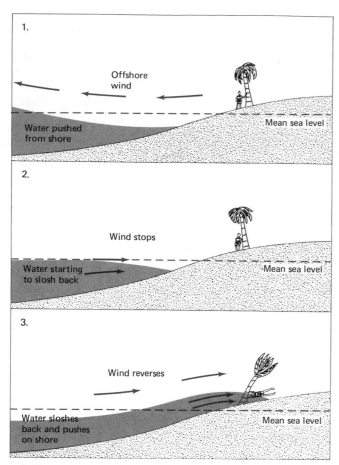

FIGURE 18.20
Disastrous rises in sea level can be produced when the wind reverses from offshore to onshore.

around Lake Okeechobee in Florida and Lake Ponchartrian in Louisiana.

There is a final example of a resonance effect which is somewhat more complicated but quite real. The mound of water raised by the inverted barometer effect would act like a long ocean wave if the low pressure area suddenly disappeared. This wave would travel outward much like the waves that emanate from a rock that is thrown in the water. Now when the low pressure area moves it has the same effect, and the raised mound of water begins to travel outward. Resonance will occur when the low pressure area moves with the same speed as the wave. This only happens in shallow water (because the wave moves too fast in deeper water). The region where this effect has the most serious consequences is in the shallow water at the mouth of the Bay of Bengal along the shore of India and Bangladesh (East Pakistan).

# PROBLEMS

**18-1** Name a beneficial aspect of hurricanes.

**18-2** The next time a hurricane strikes land, compare its rains with the *annual* average rainfall where you live.

**18-3** Explain why hurricanes are mainly late summer and early autumn storms.

**18-4** Explain the lull in hurricane activity around India from mid-July to mid-September. Use Figures 16.7 and 18.3 to justify your answer.

**18-5** Explain, using Figures 16.14 and 18.2, why hurricanes that form off the western coast of Central America frequently dissipate before they get too far.

**18-6** Hurricanes and extratropical lows are produced by two completely different weather systems. Describe the two distinct weather patterns.

**18-7** Why are satellites so very helpful in tropical weather analysis?

**18-8** Explain why hurricanes and other tropical weather systems generally move from east to west until they leave the tropics.

**18-9** Describe the chain reaction leading to the formation of a hurricane.

**18-10** Estimate how much the temperature of the air moving into the center of the hurricane is raised by contact with the ocean. Assume no temperature change, but a pressure decrease from 1010 to 960 mb.

**18-11** Using the fact that we can add the rotation speed of the hurricane to its speed of movement to find the wind speed, explain why the strongest winds of a hurricane are found on its right side, with respect to its direction of motion.

**18-12** Explain why there is always a *high* pressure area at the tropopause directly above hurricanes.

**18-13** See if you can simulate the Fujiwara effect by creating little whirlpools with two knives or spoons in a pot of water.

**18-14** Why do hurricanes quickly die down when they pass over land?

**18-15** Explain why hurricanes and other intense lows produce such high tides at coastal locations.

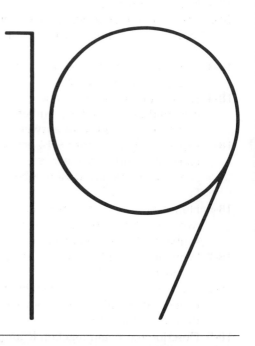

# THUNDERSTORMS, HAILSTORMS, AND TORNADOES

Phaethon, the son of Apollo and a mortal woman, wanted to prove his divine origin to his friends. He was allowed to visit his father, who then promised him any wish. Phaethon requested permission to drive the chariot of the sun across the sky for just one day. Apollo implored him to ask for anything else, "It is no simple task to hold the horses in rein." But Phaethon refused to change his wish.

No sooner did he take off into the sky than he regretted his decision. Phaethon quickly panicked and dropped the reins, and the uncontrolled horses then left the steep path and wandered erratically across the sky. In places the chariot dipped too close to the earth, burning the ocean — according to mythology, it was on this day that the Sahara was created. Because of the imminent danger, the other gods appealed to Zeus for salvation. Zeus then had no choice but to act. He hurled a thunderbolt at the poor boy and struck him dead. Thus, the earth was saved by the power of lightning, and Phaethon fell to earth as a shooting star.

We now classify such stories as mythology and observe that such primitive beliefs have been replaced by today's religions. The religions that now guide people's beliefs are no longer so graphic, simple, or crude. They deal more with subtle, inner issues such as guilt, salvation, brotherhood, and the like. The "outer" issues now belong to the domain of the "natural" world and can be explained by the laws of science. Everything in the universe has become rational.

Even so, a vestige of the irrational remains in all of us. It may take nothing more than a thunderstorm to bring out our most primitive fears. Many of us in today's urban civilization lead lives that appear largely sheltered from raw nature, but thunderstorms can strike with all their fury in the most densely populated places. When lightning strikes nearby and makes a deafening thunderclap, it can start an entire busload of atheists praying.

In this chapter we consider the thunderstorm and the wonders which accompany it. Thunderstorms naturally produce lightning and thunder and, when severe enough, can produce hail and even tornadoes.

A thunderstorm consists of a cumulonimbus cloud and develops when the air is unstable and sufficiently humid. There are several different types of thunderstorms, and we will analyze the two main categories — the so-called air-mass thunderstorm and the more severe frontal or squall-line thunderstorm. Since severe thunderstorms can produce large damaging hail and tornadoes, we will look at some of the warning signs for such weather. In this regard we discuss weather radar which is one of the most helpful tools for understanding and providing short-term warnings of these violent storms. Finally, lightning and thunder are discussed.

# SOME RECORDS

In many inland locations thunderstorms show a definite preference for the late afternoon hours, usually shortly after the hottest time of the day. As most storms go they are not very large, since they are only about 10 km in diameter on the average. Even the supersized thunderstorms are rarely 40 km in width, but they may approach 20 km in height. Because the thunderstorms are moved by the winds at midtropospheric heights, they quickly blow over in most cases.

Shortly before the storm strikes it may be calm and muggy, but the sky grows very dark. Then, suddenly, the wind picks up and can become quite strong and gusty, the temperature drops sharply, lightning flashes, and a heavy downpour begins. If the storm does produce hail, the hail usually falls in the first few minutes, and it may fall with or without rain. Rainfall is often especially intense for the first 5 or 10 minutes, and the whole storm typically lasts from 30 minutes to an hour.

Toward the end of the storm the rain tapers off and may remain quite light for the last 5 or 10 minutes. The wind subsides and, more often than not, the temperature begins to rise again, although not quite so dramatically as it fell. The sky becomes light and, if the sun peers through, you may be lucky enough to see a rainbow. Of course, with a severe thunderstorm by that time you may be dead.

Thunderstorms occur most frequently in the tropical areas of the world, partly because of the intense sunlight year round. Very few places get more than 100 thunderstorms a year, but at Bogor on the island of Java thunder is heard on 322 days a year. In the United States there are two areas most favored for thunderstorms—one in Florida and one in New Mexico (see Figure 19.1a). Africa has more thunderstorms than anywhere else on earth (see Figure 19.1b), and the electrical effects of these African thunderstorms are felt around the world.

Tornadoes are generated by severe thunderstorms and often can be seen extending down from the base of the cloud, much like a giant twisting funnel (see Figure 19.2). In the midlatitudes, tornadoes tend to be located in the southwest corner of the thunderstorms. Wherever they touch ground tornadoes cause incredible damage because of the tremendous winds and extremely low pressures associated with them. A typical tornado is only a few hundred meters in diameter and, since it moves more or less with the cumulonimbus cloud above, it rarely lasts for more than a minute or two in any one place. But this is all the time it needs—with winds that may exceed 100 m/s (200 knots), destruction is almost instantaneous.

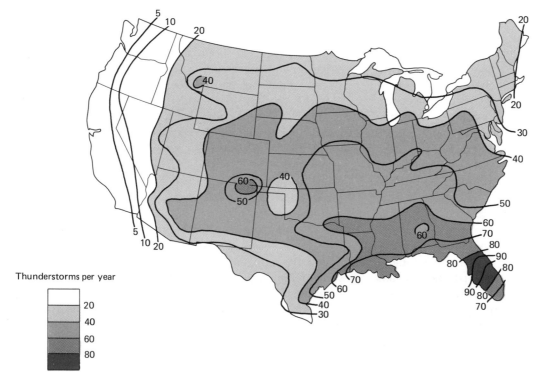

**Annual frequency of thunderstorms in the United States**

Thunderstorms per year

| | |
|---|---|
| | 20 |
| | 40 |
| | 60 |
| | 80 |

FIGURE 19.1a
Annual frequency of thunderstorms in the United States.

THUNDERSTORMS, HAILSTORMS, AND TORNADOES

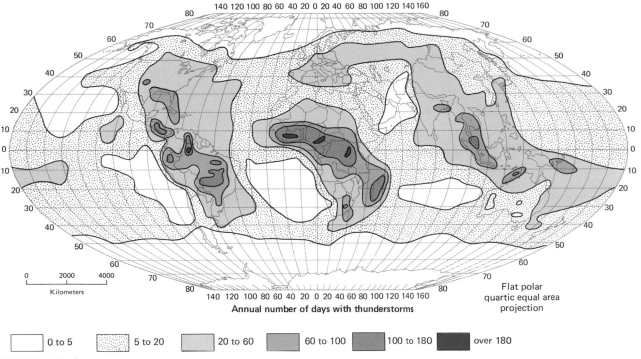

FIGURE 19.1b
World distribution of thunderstorms.

Almost all tornadoes rotate in a counterclockwise direction in the NH, but recently several clockwise tornadoes have been documented and at least one has been filmed. Sometimes two tornadoes, one cyclonic and one anticyclonic, extend down from the same parent cloud.

Sometimes the funnel cloud cannot be seen, either because it occurs at night or because it is well embedded within the rain area of the thunderstorm. Nevertheless, the tornado always gives some warning, since it is accompanied by an extremely loud roaring noise and unusually intense lightning. Another warning sign of a possible tornado is the mamma clouds that form at the base of the more severe thunderstorms. Many of the more severe thunderstorms are also immediately preceded by a low-lying, dark, **arcus** cloud that sometimes looks like a long, rolling cylinder and sometimes like a wedge (see Figure 19.3). Often you can see the air near the arcus cloud rise because of the ragged edges which appear to get sucked up into the parent cloud above.

Large thunderstorms are common in the tropics where they provide almost all the precipitation. On the other hand, in the middle and high latitudes much of the precipitation, especially that which falls during the winter half of the year, is not associated with thunderstorms but with the large extratropical cyclones and their stratiform clouds. Outside of the tropics and subtropics, thunderstorms are quite rare in the winter and are most common in the late spring and summer.

In the United States, large hail and tornadoes tend to be more common in late spring but drop off considerably by July. This is because, as you will see, they are produced by severe thunderstorms which are found in conjunction with low pressure areas. Lows can still be intense during spring, but they usually weaken considerably and drift far poleward during the summer.

Severe thunderstorms mete out death by three principal means. The first is lightning, which kills over 200 people annually in the United States alone. Tornadoes account for more than 100 deaths annually, and floods, which result from the torrential downpours produced by some of these storms, cause almost 200 deaths annually in the United States. Worldwide statistics are no doubt considerably higher. There are also records of people killed by hail; one such storm in India claimed 230 lives. However, for the most part, deaths from being struck by hail are extremely rare, although large hail has almost certainly caused a number of fatal plane crashes (see Figure 19.4).

These storms also cause enormous damage to crops and to buildings. The damage from hail actually exceeds the damage from tornadoes by about 15%, and together they account for over $200 million annually in the United States.

FIGURE 19.2

The Union City (Oklahoma) tornado of May 24, 1973. Note the pendant cloud around the tornado. This is produced by adiabatic cooling due to lower pressure near the tornado.

FIGURE 19.3

The arcus cloud at the leading edge of a severe thunderstorm.

THUNDERSTORMS, HAILSTORMS, AND TORNADOES

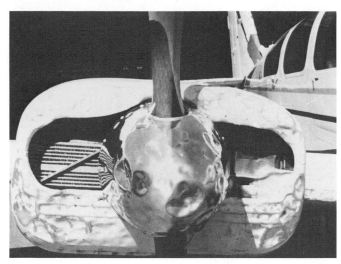

FIGURE 19.4
Airplane battered by hail.

The National Weather Service now has an excellent warning system for severe weather, and this has saved thousands of lives. It is a remarkable tribute to the Service that deaths over the past 30 years have actually decreased, despite the large rise in population in the most susceptible areas. Property damage has, of course, risen tremendously during this time, partly because there are many more houses and buildings, and partly because of inflation. Nevertheless, people on the Great Plains are often safe in their storm cellars while the houses above them are being demolished.

The worst tornado in U.S. history was the famous Tri-State tornado. It occurred on March 18, 1925 and traveled for 350 km through Missouri, Illinois, and Indiana. This is one of the longest known paths for any tornado, and its death toll of 689 is the highest in the record books. This tornado carved a path of total destruction that was as much as 1.5 km wide in places.

Most tornadoes do not pass through towns simply because most of the land is agricultural—but the Tri-State tornado was *not* an average tornado. It passed directly through nine towns and destroyed about 3000 homes. By today's standards the property damage would have been astronomical, but even then it totaled almost $17 million. There is no way that this property damage could have been avoided, but many of the lives could have been saved if only there had been a warning system at the time.

Very few people reported seeing anything like a funnel shaped vortex cloud in the course of this storm. Apparently the storm cloud was so close to the earth there was no room for a pendant cloud. At Princeton, Indiana, the tornado was described as blackness moving across the south part of the city with the air full of tree limbs. One person in another part of the path mentioned a "fog" rolling towards him and described the cloud as a turbulent, boiling mass. . . .

Snowden D. Flora, *Tornadoes of the United States*

Tornadoes are rather unusual events but, when they do occur, they often come in bunches. For instance, on the day of the Tri-State tornado, 7 other tornadoes were also seen raging in the vicinity. Other famous tornado outbreaks include the Palm Sunday outbreak. On April 11, 1965 (Palm Sunday) a total of 31 tornadoes struck in 6 states and killed 256 people. This, however, was a mere baby when compared with the all-time record outbreak of April 3-4, 1974 when 148 tornadoes that killed 337 people spread over 11 states (see Figure 19.5). In this outbreak most of the tornadoes fortunately missed the larger towns and cities; if they had not been bypassed, the death toll might have been much higher than it was.

### The Jumbo Tornado Outbreak of April 3–4, 1974

Let me give a little background material on the April 3–4 disaster. It was preceded by a smaller outbreak of 20 tornadoes which occurred on April 1 (April Fools day). At that time a low pressure area that had come from the Pacific Ocean was beginning to cross the Rocky Mountains. Then events began to proceed in classic fashion.

As the low crossed the Rockies it largely lost its identity (refer to the ice-skater effect of Chapter 17), but by April 2 it had re-formed on the eastern side of the Rockies as a weak low centered over Cheyenne, Wyoming. This low began to move to the east and soon started drawing cold air southward on its western side. The storm then rapidly intensified following the usual pattern, and heavy snow began to fall on the northwestern side of the storm late on the same day, continuing until April 3. Southwest of the low center, in Texas, Arkansas, and Oklahoma, strong winds and dry conditions behind the cold front led to dust storms.

Meanwhile, on the eastern side of the storm, conditions were quite different. Warm moist air was being sucked up from the south. This air was extremely unstable and gave all the warning signs of a major tornado outbreak. A severe storm warning was issued for the entire area that was soon to be struck. Then, rather suddenly, about 3:00 P.M. on April 3rd, all hell began breaking loose as tornadoes formed along not just one, but three separate lines, known as squall lines. The surface weather map of this storm is shown in Figure 19.6.

The worst damage occurred in the town of Xenia, Ohio. A 14-year-old boy scout actually took a movie of this tor-

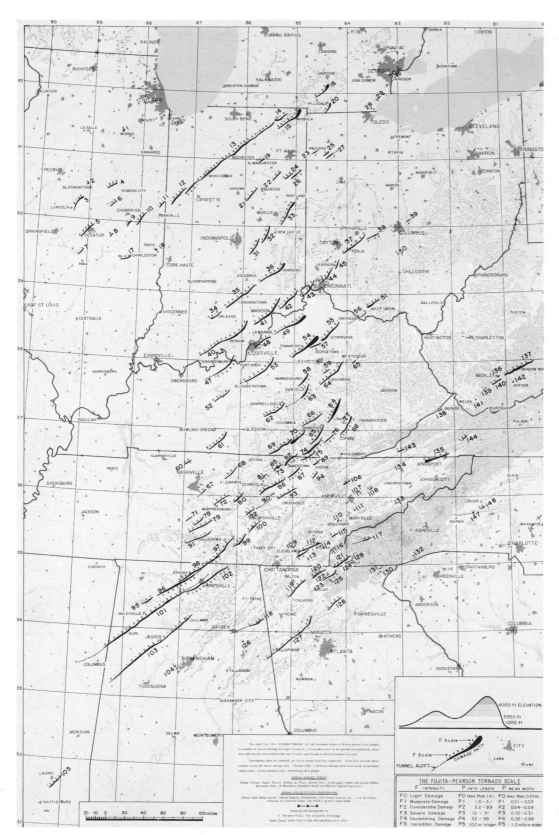

FIGURE 19.5
Jumbo tornado outbreak of
April 3–4, 1974.

THUNDERSTORMS, HAILSTORMS, AND TORNADOES

**Surface weather
map 2100 GMT
April 3, 1974**

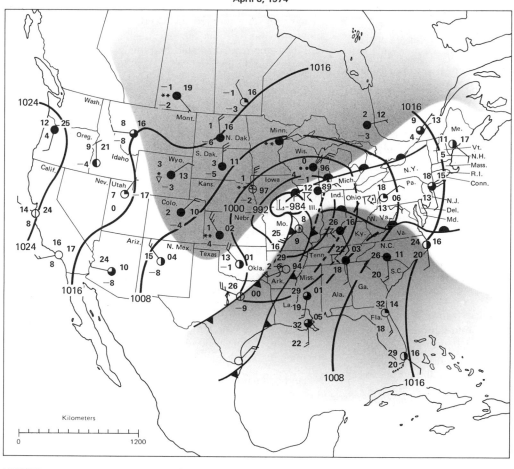

**LEGEND**

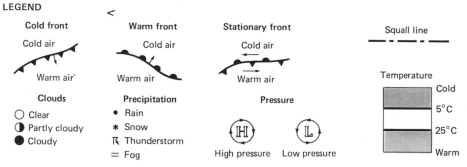

Cold front — Cold air / Warm air

Warm front — Cold air / Warm air

Stationary front — Cold air / Warm air

Squall line

**Clouds**
○ Clear
◑ Partly cloudy
● Cloudy

**Precipitation**
● Rain
∗ Snow
↟ Thunderstorm
= Fog

**Pressure**
Ⓗ High pressure   Ⓛ Low pressure

Temperature
Cold
5°C
25°C
Warm

FIGURE 19.6
Surface weather map for 2100 GMT, April 3, 1974. Squall lines are denoted by heavy dot-dash lines. This complex killer storm has three squall lines, two cold fronts, and two warm fronts!

nado as it was approaching him. Small funnel clouds touched ground now and again, lifting parts of houses and other assorted objects. However, this was merely a preliminary action. All at once, a very wide and almost pitch-black major funnel descended to the ground, causing total devastation. Shortly thereafter the boy scout wisely departed the scene.

**Some Interesting Facts and Figures**

The lifting power of tornadoes is incredible, and there are many records of entire houses being lifted and moved as much as 70 m without being destroyed. Locomotives have been lifted off their tracks, bridges separated from their foundations, and entire sections of small rivers sucked up, thereby exposing the dry river beds. In these

SOME RECORDS 337

cases it has been known to rain fish and frogs (not cats and dogs). In one tornado more than 200 fully loaded coal cars were lifted off their tracks, even though each car weighed more than 40 tons.

Straw has been driven into planks of wood, large furniture has been blown several miles, sometimes without suffering any damage, chickens have been stripped of their feathers and been left standing stiffly at attention and dead. There is even the improbable story that a rooster was blown into a jug with only his head protruding.

Although people blown about by tornadoes are usually killed, there are a number of humorous incidents in which people escape death and even injury. A *reliable* account of two Texans is told in which one named Al went to the door of his house to see what was happening and was blown over the treetops. His friend, Bill, went to the door to see what had happened to his friend and he likewise

found himself, also, sailing through the Texas atmosphere but in a slightly different direction from the course his friend was taking.

Both landed about 200 feet from the house with only minor injuries. Al started back and found Bill uncomfortably wrapped in wire. He unwound his friend and both headed for Al's house, crawling because the wind was too strong to walk against. They reached the site of the house only to find that all the house, except the floor, had disappeared. The almost incredible part of the story is that Al's wife and two children were huddled on a divan, uninjured. The only other piece of furniture left on the floor was a lamp.

In another story

A small tornado near Topeka, Kansas, picked up a farm hand, carried him one hundred feet, and let him down in as good shape as when he took off, except that he was covered with mud. His employer intimated that this was the fastest move he had ever seen the man make.

Snowden D. Flora, *Tornadoes of the United States*

Tornadoes can occur almost anywhere in the world, but they are most prevalent in the midlatitudes. The United States has by far the largest number and most intense tornadoes because of its particular geography, which freely allows contrasting air masses to clash (see Figure 19.7).

Hail was one of the ten biblical plagues and it is apparent that the Egyptians and Hebrews were both well aware of the damage it can cause.

23. And Moses stretched forth his rod toward heaven: and the Lord sent thunder and hail, and the fire ran along upon the ground; and the Lord rained hail upon the land of Egypt.

24. So there was hail, and fire mingled with the hail, very grievous, such as there was none like it in all the land of Egypt since it became a nation.

25. And hail smote throughout all the land of Egypt all that *was* in the field, both man and beast; and the hail smote every herb of the field, and brake every tree of the field.

*King James Bible*, Exodus, Chapter 9

FIGURE 19.7
Tornado frequency in the United States. Units are relative.

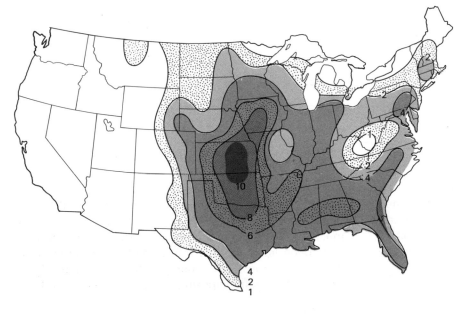

THUNDERSTORMS, HAILSTORMS, AND TORNADOES

Another interesting account of a tremendous hailstorm is given by our old friend, Benvenuto Cellini

One day when we found ourselves a day's distance from Lyons (it was nearly two hours before sunset), we heard the crackling of thunder and noticed how very clear the sky was: I was a bow's shot in front of my companions. After the thunder we heard such a tremendous, fearful noise reverberating in the skies that I was convinced it must be the Day of Judgement. I paused for a while, and there was a fall of hail, without a drop of water. The hail was bigger than pellets shot from a blow-pipe, and when it hit me it was very painful: little by little its size increased, till it was like the bullets from a crossbow . . . . The hailstones grew to the size of large lemons. I sang a Miserere and while I was praying to God in this devout way a hailstone fell that was so large that it smashed a very thick branch from the pine under which I thought I was safe . . . .

In the same way one of them fell on poor old Lionardo Tedaldi who, as he was kneeling down like me, was forced onto his hands. . . . The storm continued some while, and then stopped: we had all been given a pounding. . . . Then a mile in front we found such a spectacle of ruin so much greater than our own misfortune that it defies description.

All the trees were stripped and smashed; all the animals around had been killed, as well as a good number of shepherds. We saw a mass of stones which were so large that it was impossible to get both your hands round them.

*Autobiography of Benvenuto Cellini*

It is rather difficult to believe that all during this time the sky remained clear (perhaps Cellini stopped looking up), but the other details of this story seem quite accurate. Hailstones the size of lemons or larger are certainly not unheard of, especially since the largest *single* hailstone was the size of a melon and weighed nearly 1 kg. However, most accounts of hailstones weighing 2 kg or more are probably descriptions of groups of hailstones that have frozen together on the ground.

Hailstones have been known to cover the ground uniformly to depths up to 30 cm and have been known to blow into drifts as deep as 2 m in places. There are even times when these drifts have lasted a month or more into midsummer before completely melting.

On July 2, 1953, a single hailstorm left a path (**swath**) of desolation 8 to 15 miles wide and 100 miles long, in which an estimated 3 million bushels of wheat were destroyed. This storm cost the farmers and insurance companies $6 million in crop damage alone. Other storms, such as the one Cellini described, have killed many small animals and even some horses and cows. Needless to say, windows are regular targets of large hail.

Despite the fact that hailstorms and tornadoes are both produced by severe thunderstorms, the two do not always occur in the same places. As mentioned earlier in Chapter 12, hail is more likely on the high ground of the western Great Plains. By comparing Figures 19.7 and 19.8, you can see that the region of most frequent hailstorms lies northwest of the region of most tornadoes. Hail is also quite frequent on the highlands of India and the plateau of Southern Africa.

Usually the rainfall from thunderstorms is intense but sufficiently short-lived so that no one place gets too much rain. Nevertheless, when thunderstorms happen to stall over a particular place, enormous rainfall totals can result and may produce disastrous floods. Virtually all the record-breaking rainfalls up to a few hours in duration result from such thunderstorms. The Rapid City and Big Thompson floods have already been mentioned. To this list must be added the Johnstown, Pennsylvania, flood of July 21, 1977 in which 76 people drowned. Once again the sudden damage was caused when a thunderstorm stalled over a river valley and dumped all of its rain in one place.

# AIR-MASS THUNDERSTORMS

The air-mass thunderstorm is usually the less intense type of thunderstorm. It is so named because it occurs in a more or less disorganized manner, far from any fronts or any other well-defined atmospheric disturbances. It is produced almost entirely by atmospheric instability in an atmosphere with warm, moist air, and it generally contains the seeds of its own destruction. Very often the thunderstorm consists of several distinct cells, each somewhat less than 10 km in diameter.

The air-mass thunderstorm was investigated in detail during the Thunderstorm Project of 1946–47. The thunderstorm cell was found to go through a life cycle that lasts somewhat more than an hour before it ultimately destroys itself. This cycle is broken down into three distinct stages—the cumulus stage, the mature stage, and the dissipating stage. (This life cycle has already been described in a somewhat crude way in Chapter 2; see pages 25–26.)

The cumulus stage is the early growing stage of the thunderstorm. The air throughout the entire cloud rises, and updrafts of 5 to 20 m/s are typical. During this stage raindrops and ice pellets develop within the cloud, but no precipitation falls from the cloud because of the updrafts. Eventually, the updrafts cause the precipitation particles

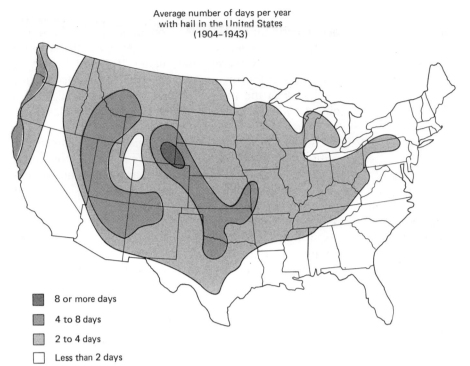

Average number of days per year
with hail in the United States
(1904–1943)

■ 8 or more days

■ 4 to 8 days

▨ 2 to 4 days

☐ Less than 2 days

FIGURE 19.8
Hailstorm frequency in the United States.

to grow rapidly so that their weight soon begins to drag the air downward in parts of the cloud.

With the creation of downdrafts the storm enters the mature stage. In this stage both the updrafts and downdrafts reach their maximum intensity, and lightning and thunder are produced. The downdrafts are often intensified when some of the drier surrounding air at higher levels gets sucked in and mixes with the humid, rain-filled cloud air. This process is known as **entrainment** and it reduces the bouyancy of the warm cloud air. Evaporation results and causes the air to cool and become even denser so that any sinking tendency is enhanced. At ground level heavy rain begins to fall under the downdraft area. Hail can also fall from some of the larger storms at this time.

The weight of liquid water within the cloud and the effect of entrainment thus helps begin the destruction of the storm by opposing the updrafts. The destruction of the storm is accelerated once the thunderstorm has sucked up most of the nearby warm, moist air. Then there are no more updrafts, and downdrafts are present everywhere. This is the final, dissipating stage. The downdrafts themselves are weakened because the cloud has already been largely flushed of its enormous weight of rain by the earlier, stronger downdrafts of the mature stage. Only light rain falls during this stage from the weakened downdrafts. The entire dying cloud begins to collapse slowly.

When the downdraft first reaches the ground the tem-

perature may drop 5 to 10°C or more within a minute or two. Then, after the thunderstorm has ended, the temperature gradually rises and it may get almost as hot and muggy as it was before the storm began (see Figure 19.9). In fact, air-mass thunderstorms rarely provide more than an hour or two of relief from the sticky, warm conditions that produced them in the first place.

**Note:** Some thunderstorms seem to produce downdrafts of extraordinary intensity. The thunderstorm that struck New York City on June 24, 1975 passed directly over Kennedy Airport. A jet carrying more than 130 passengers approached the runway just as the blast from the downdraft hit. The plane was jerked downward and collided with one of the approach towers. All but a few on board were killed.

Such enormous downdrafts are not common but may possibly be identifiable. They seem to come from clouds that are rather elongated in their direction of motion. For this reason they have been named *spearhead echos* by Ted Fujita (who discovered them). Apparently, they are caused when the updraft of the thunderstorm rises into the dry air of the stratosphere. Mixing of the updraft air with this dry stratospheric air causes intense cooling by evaporation, and the downdraft rushes headlong toward the ground.

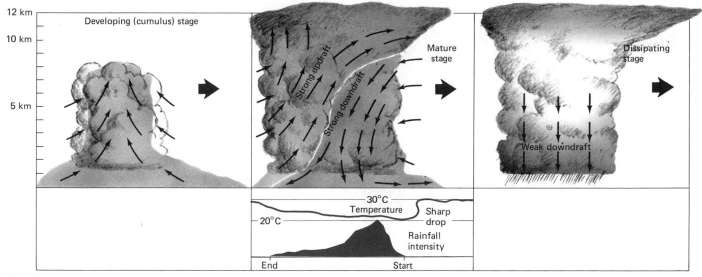

**FIGURE 19.9**
Life cycle of an air-mass thunderstorm. Wind speed is proportional to length of arrows. Double arrows indicate direction of movement of thunderstorm. The typical rainfall rates, temperatures, and pressures are shown under the storm in the mature stage. Weather changes should be read from *right* to *left,* because motion of storm is from left.

# SQUALL-LINE THUNDERSTORMS

All thunderstorms result from atmospheres with unstable lapse rates and are fueled by warm, moist air, but squall-line thunderstorms get an extra boost from the larger scale weather patterns. It is this extra boost that helps make the squall-line thunderstorm the giant it is. At times it almost seems as if there is a conspiracy by the atmosphere to do everything possible to add to the severity of these storms. The term **squall line** merely refers to a line of thunderstorms that forms in a favored region.

The weather patterns that help produce squall-line thunderstorms must satisfy several conditions. First, there is a "tongue" of especially warm, moist air advancing from the south at lower levels of the atmosphere. This shows up very well on the 850 mb chart (at about 1500 m). Second, there are strong winds usually from the west or southwest (except in the tropics where they may be easterly) consisting of cool, very dry air at upper levels of the troposphere. This shows up very well on the 500 mb chart (at about 5500 m). The winds at this level should be divergent so that they help suck up air from below. Third, there is generally a surface low pressure area nearby with an advancing cold front (except for tropical squall lines). These features are shown in Figure 19.10, and an actual squall line is shown in Figure 19.11.

In the hours preceding the onset of the squall-line thunderstorm, the lower layers of the atmosphere are gradually warming, while the air aloft is cooling. This increases

**FIGURE 19.10**
Conditions leading to squall-line development in the United States. For squall lines in Africa or India there would be no cold front at the surface, and 500 mb winds would be easterly.

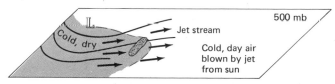

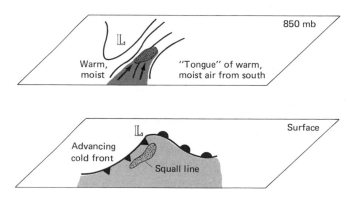

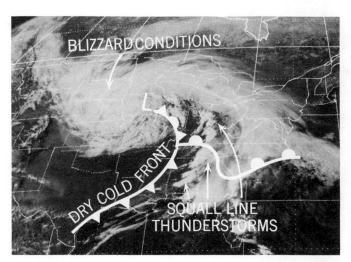

**FIGURE 19.11**
Satellite picture of the squall line of April 18, 1978. The individual thunderstorms are distinctly visible. There are no clouds associated with the cold front for this storm. Like many spring storms, snow was falling northwest of the storm center, at the same time that severe thunderstorms were raging in the warm unstable air south of the storm center.

the natural instability of the air. Thus, even when the air is initially stable, the large-scale winds at different levels soon destabilize it.

Example 19.1
In Figure 19.12 show that the large-scale winds are destabilizing the atmosphere.

Reasoning: Over the point marked $X$, $T = 13°C$ at 1500 m, while $T = -9°C$ at 5500 m for a difference of 22°C in 4 km. Thus, the lapse rate is 5.5°C/km, which is absolutely stable. In one hour, because of the winds, the temperature will rise to 15°C at 1500 m and fall to −15°C at 5500 m for a difference of 30°C in 4 km.

Solution
Thus, the lapse rate will increase to 7.5°C/km; this is unstable for saturated air! Therefore, we have a case in which the (large-scale) winds have set up a favorable situation for (small-scale) thunderstorms.

Vertically rising motion also tends to destabilize the atmosphere, as you will find out shortly. Rising motion is assisted by the divergence of the winds aloft which literally sucks air up from below. The rising tendency is also aided by the approaching cold front, which acts like a wedge, forcing the warm, moist air aloft. It is important to note

that the presence of the cold front is not absolutely necessary—first, because it is not present for the squall-line thunderstorms of India and Africa and, second, because even when one is present the squall line typically propagates hundreds of kilometers ahead of it. However, no one will deny that a cold front can help to initiate and further intensify the situation. Nowhere else in the world do the squall-line thunderstorms match the intensity of the American thunderstorms.

The strong winds aloft serve an extra function. They actually cause the thunderstorm to tilt somewhat. This also adds to the severity of the storm because then the precipitation particles do not fall back into the updraft. Instead, the warm, moist tongue of air at lower levels assures an almost continuous source of fuel, while the cold downdraft wedges under the warm, moist air, forcing it upward. Both of these factors combine to add to the lifetime of the squall-line thunderstorms, which may well live for several hours. (This extra time is of vital importance for the production of large hailstones.) Some of these tilted thunderstorms reach enormous size and are known as **supercell** thunderstorms. (Figure 19.13 shows a supercell storm.)

**FIGURE 19.12**
The winds can destabilize the atmosphere. Here cold air is being brought (advected) toward $X$ at the 500 mb level, while directly beneath at the 850 mb level warm air is being advected toward $X$. The arrows show how far the wind moves in one hour.

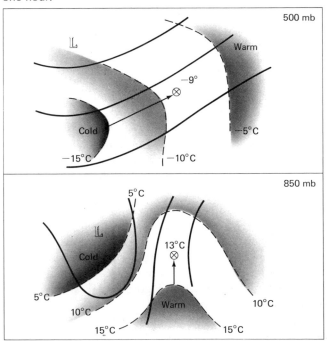

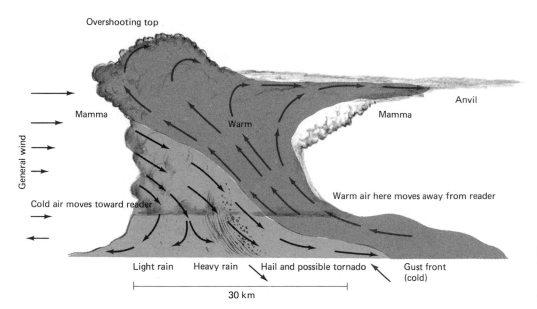

Overshooting top

Mamma

General wind

Warm

Anvil

Mamma

Cold air moves toward reader

Warm air here moves away from reader

Light rain    Heavy rain    Hail and possible tornado    Gust front (cold)

30 km

FIGURE 19.13
The structure of a large (supercell) squall-line thunderstorm.

The structure of an individual squall-line thunderstorm is shown in the Figure 19.13. As the anvil of the storm moves overhead, blocking the sun, the temperature begins to fall. The full fury of the thunderstorm may not hit for another hour or even more, but when it does the temperature drops suddenly, the wind gusts with fury, and you are quickly drenched with heavy rain or pounded with heavy hail. The weather sequence is actually quite similar to that of the air-mass thunderstorm, except that the squall-line thunderstorm tends to be more severe.

Individual squall-line thunderstorms tend to move more or less in the direction of the midtropospheric winds, but at a somewhat slower speed than the winds. In the United States this means that they usually move from the southwest to the northeast. Perhaps they are initiated by a pulse of air that may be sent out by the advancing cold front. Therefore, a number of storms tend to form in a line so that the squall line may consist of nothing more than a long line of thunderstorms.

One strange feature of the entire squall-line and the extremely large thunderstorms—the supercell storms—is that they tend to move at an angle to the right of the general winds. At first this astounded meteorologists, but the explanation was soon found to be rather simple. Since the warm air comes from the south, the southernmost storm in the line begins to deprive all the thunderstorms to its north of their vitally needed warm air supply. Furthermore, the cold downdrafts wedge southward, forcing the warm air from the south to rise sooner. (This largely accounts for the drifting of the supercell.) Thus, the storms to the north begin to dwindle, and the only place that a new

storm can form is even farther to the south. All this takes place while the thunderstorms are moving; the entire picture looks something like the example shown in Figure 19.14. Oddly enough some individual supercell storms move to the left of the general winds.

Pictures have been taken in which the entire cumulonimbus cloud is seen rotating slowly until the storm gathers force and its spinning progressively moves lower and lower. Ultimately, this movement extends from the cloud base as a tornado. These developments are now finally being seen within thunderstorms for the first time by special radar known as doppler radar. Although now only in the developmental stage, it is possible that we will soon be able to pinpoint the touchdown of tornadoes as much as one hour in advance.

## WEATHER RADAR

Radar was invented in the years before World War II. During the early tests of this highly secret weapon, it became necessary to overcome some very annoying problems. Chief among these problems was the fact that clouds and rain blocked enemy planes and ships from view. It didn't take long to realize that in the years to come when the world was no longer at war that the chief use of radar would be to detect and portray the weather.

Now, although radar is very complicated, it will be easy for us to tackle since we have already answered why the sky is blue; radar has a similar explanation. A radar unit is a machine that sends out microwaves (i.e., long

Motion of squall line ➡

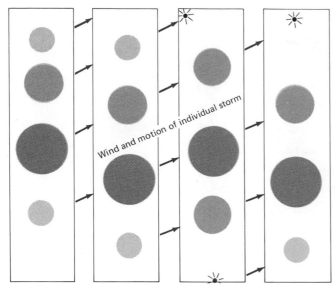
Wind and motion of individual storm

FIGURE 19.14
Individual thunderstorms move to the left of the squall line. The storms dissipate when they have lost their warm air supply, because new storms have sprouted up to their south. The individual storms appear red when they have an available warm air supply; dark, when they do not.

wavelength electromagnetic waves). These waves travel with the speed of light until they hit an object which scatters them. Some of the waves get scattered directly back to the radar unit where they are then recorded on a scope. The minute amount of time it takes for the signal to return is directly proportional to the distance of the scattering object. All this information is then displayed on a scope where an electrical signal makes the scope bright in the appropriate places.

You might ask, what kinds of objects can scatter radar signals. The principle is the same as it was for light scattering. Any object that is large enough can scatter radar signals. Air molecules can scatter light waves but they are far, far too small to scatter the much longer microwaves emitted by the radar unit. The early researchers of radar understood this and used microwaves whose wavelength was about 1 cm.

Unfortunately, some of the larger cloud droplets and raindrops are large enough to scatter 1 cm microwaves. The solution was simple—since you cannot make raindrops smaller, you must use longer wavelengths to bypass the raindrops. Military and aviational radar now uses longer wavelength microwaves, but not so long that the

radar will bypass flocks of birds or swarms of insects. Meteorological radar uses either 3 cm or 10 cm wavelengths. These are large enough to bypass the cloud droplets but not the raindrops or snowflakes.

Meteorological radar therefore sees only those clouds that produce precipitation. Radar is useful in telling exactly where it is raining or snowing at any given instant in time. It also gives a very good estimate of the intensity of the precipitation, because a greater returning signal means more and larger raindrops or snowflakes.

Thus radar is quite useful in predicting when and where rain will begin once there is rain approaching the general area. For instance, meteorologists may know a day or more in advance that a general area will get thunderstorms, but they do not know the precise locations that will be struck until the thunderstorms have formed. At this point the radar takes over, showing how individual thunderstorms are moving and what lies in their path.

Since radar provides a good estimate of rainfall intensity, it is also very useful in flash-flood forecasting. The radar data is quickly combined with hydrological information about the rivers and streams in a particular area, and how much water can soak into the soil before it must pour into the streams. The flash-flood forecasts that have been made with the aid of radar provide people with an extra hour or two of warning. This has doubtless saved thousands of lives.

Meteorological radar is designed so that it can sweep around like a cannon on an armored tank. It can either sweep around in a horizontal circle, or it can sweep up and down. When it sweeps horizontally, the return signal is portrayed on a PPI (plan-position indicator) scope. This looks like a regular map or the normal radar that you see in movies (see Figure 19.15). It can have an effective horizontal range up to about 400 km before the curvature of the earth hides too much of the troposphere from view. The signal from the PPI scope tells the location of precipitation, its intensity, and where it is moving.

The RHI (range-height indicator) scope portrays a cross section of the atmosphere when the radar sweeps vertically. The RHI scope is useful for telling the height of the clouds, which is often a good measure of their intensity. It is also useful for telling the height at which falling snow is melting, as in Figure 19.16. During the winter there are many times when it is difficult to forecast whether it will be raining or snowing. The RHI scope is sometimes useful in this regard because of a strange phenomenon.

Snowflakes reflect radar signals poorly compared to raindrops of the same size. However, as the falling snow begins to melt the outside of the snowflake's ice crystals contain a thin coating of liquid water and then look like

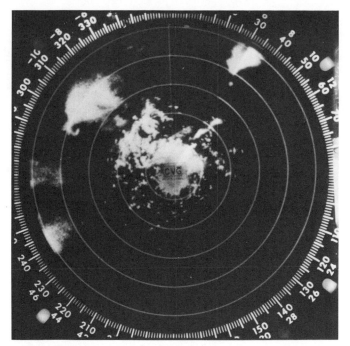 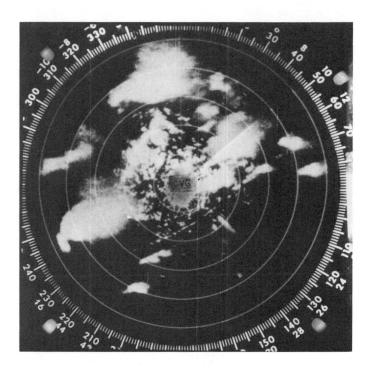

**FIGURE 19.15**
Two radar (PPI) photos about half an hour apart, on April 3, 1974 from Evansville (near Cincinnati). The thunderstorm of the first frame has moved to the northeast, and another thunderstorm has formed by the second frame. All these thunderstorms have a hook shape, which indicates tornadoes. Circular markers are at intervals of 10 nautical miles.

**FIGURE 19.16**
RHI photo showing the bright band produced by melting snowflakes.

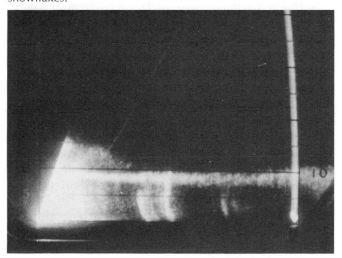

giant raindrops to the radar. Thus, the level at which the snow melts can often be seen as a bright band on the RHI scope. As long as this bright band remains well above ground level, there is no danger of snow; but whenever the bright band begins to descend toward ground level, snow is imminent. At such times you can imagine the excitement among the meteorologists at the National Weather Service—many of them are lovers of snow!

From the PPI scope you can see when you have airmass thunderstorms and when there is a well-defined squall line. There is, however, another less obvious use for radar that at first surprised meteorologists. Radar, it seems, can provide warning of a tornado.

In 1945, a meteorologist observed that a thunderstorm that had produced a tornado had a rather funny shape on the PPI scope (as in Figure 19.15). It had a hooklike echo extending from the southwest corner of the storm. Unfortunately, the significance of this discovery was not appreciated for eight years. By 1953, meteorologists had realized that hook echos are *characteristic* of tornado situations, and often a hook echo will appear as much as half an hour before the tornado strikes earth.

**Warning:**
Occasionally, the hook echo is nothing more than a false alarm, and in other cases tornadoes will not produce hook echoes. Nevertheless, most of the time hook echos and tornadoes occur together, and many lives have been saved because of this warning sign revealed by the radarscope.

The explanation of the hook echo is rather straightforward: the air is rising so rapidly in the region of the storm that the cloud droplets simply do not have enough time to grow large in the lower parts of the cloud, and the small droplets do not reflect the radar signal well.

**The Radar Summary Chart**
Because of the limited range of any individual radar unit, **radar summary** charts are produced. The summary charts are not actual radar pictures, but merely a composite summary of all the individual radar pictures across the country. The chart shows all the general areas in which precipitation is falling, with their directions and speeds. It also distinguishes between squall lines and general areas. It will show some of the larger individual thunderstorms and tell whether they are intensifying or diminishing.

As an excellent, though by no means typical, example of a radar summary chart look at Figure 19.17, which shows the state of the nation at 2135 GMT on April 3, 1974. You can clearly see the three major squall lines that are depicted by the long, narrow rectangles. The symbol (570/) indicates the height of the top of an individual thunderstorm in hundreds of feet. Thus, 570 stands for 57,000 feet, which is unusually high, and yet there are several storms that are even higher on this chart. The result of these monstrous storms is TRWXXA—thunderstorms (TRW) of extreme intensity (XX) with hail (A).

**Doppler Radar**
In the past few years great strides have been taken with the new doppler radar. **Doppler radar** works on the same principle that you experience everyday when a car, truck, train, or plane passes by. As the car (for example) approaches, the sound is always high pitched. As soon as it has passed, the pitch drops a few notes. The faster it moves, the more the pitch drops.

**Note:** Believe it or not, this same phenomenon, known as the **Doppler effect,** is observed in astronomy. Stars moving toward the earth appear bluer (higher pitched), while stars receding from the earth appear redder (lower pitched). Since the light of the most distant galaxies in space is redder than it should be, scientists have concluded that the universe is expanding.

But let's get back to doppler radar. When a signal is sent out by the radar it will come back with a higher pitch (i.e., with shorter wavelength) if the scattering object is approaching the radar unit. The more rapidly the object approaches, the higher the pitch of the return signal. Therefore, it is actually possible to tell the *component* of velocity for objects moving toward or away from the radar unit.

The objects of importance to meteorologists are raindrops and snowflakes. Since these are carried along by the winds, doppler radar is actually able to "see" into clouds and tell the winds within clouds. Unfortunately, the doppler radar cannot tell anything about the component of the wind speed at right angles to the unit. Therefore, in order to obtain a complete picture of the winds within clouds, it is necessary to have two or more nearby doppler radar units.

Doppler radar units are still so expensive that they are only being used on an experimental basis in a few places. Two such units were installed in Oklahoma in the hopes that some severe weather would soon pass through the general vicinity. For several years no severe weather occurred there, much to the relief of the general public. Finally, a very large thunderstorm that produced a tornado crossed the area and, for the first time, meteorologists were able to "see" inside a severe thunderstorm. The picture they obtained is shown in Figure 19.18. It remains to be seen whether this picture is characteristic of severe thunderstorms in general.

In Figure 19.18 we can see a type of wind structure known as a double vortex. It is actually quite similar to the flow inside a falling raindrop! This type of flow pattern acts to minimize the frictional drag that the surrounding wind exerts on the thunderstorm and may explain why thunderstorms are able to move more slowly than the surrounding winds. It also helps to explain why tornadoes tend to form in the southwest corner of the thunderstorms. That corner naturally contains the cyclonic or counterclockwise vortex that is the stronger of the two, since tornadoes occur near low pressure areas (cyclones). On rare occasions the other vortex may be strong enough to produce an anticyclonic tornado, and sometimes two tornadoes extend from the same cloud. The double-vortex flow pattern makes this result seem natural.

Through the use of radar meteorologists have also discovered other, previously unknown features of the atmosphere and its storms. It was from radar pictures that we first realized that hurricanes contain spiral rainbands. The fine structure of extratropical lows is now being revealed by radar. Even the annoying and sometimes misleading signals picked up by radar (often called angels) are now being used to advantage to locate temperature inversions and waves in the atmosphere. However, radar's greatest use is that it can provide the last-minute warning of impending danger from severe thunderstorms and tornadoes.

## THUNDERSTORMS AND ATMOSPHERIC STABILITY

Because thunderstorms occur when the atmospheric lapse rate is unstable, the thermodynamic diagram is an excellent tool for analyzing and forecasting severe weather. It is quite remarkable that such a complicated phenomenon can be made so simple to predict.

Albert Showalter developed a simple index (called the **Showalter Stability Index — SSI**) to forecast the possible intensity of thunderstorms. We know that the air is unstable if the lifted parcel is warmer than the surroundings it is lifted into. Showalter found a simple relation between the degree of instability and the resulting weather.

The Showalter Stability Index, SSI, is obtained in the following way. First, take a parcel of air at 850 mb. Lift it dry adiabatically until it becomes saturated, and thereafter lift it moist adiabatically until it reaches the 500 mb level. Then, subtract the final temperature of the lifted parcel from the temperature of the surroundings at 500 mb. Thus,

$$SSI = T(\text{air at 500 mb}) - T(\text{lifted parcel})$$

The air is unstable when the SSI is negative. Table 19.1 shows how the resulting weather depends on the SSI. For instance, if the SSI = −5, you can expect tornadoes and large hail. Needless to say, the SSI does not often get as low as −5, and usually it is positive.

---

Example 19.2
What type of weather would you expect from the sounding shown in Figure 19.19.

Technique: Using the thermodynamic diagram, lift a sample of air from 850 to 500 mb. Next, take the SSI and use Table 19.1 to tell you what kind of weather to expect.

The first thing to do is probably to reread the section on thermodynamic diagrams, which you may have forgotten by now. From Figure 19.19 you should be able to see that

TABLE 19.1
The Relation of the Showalter Stability Index to the Intensity of Thunderstorms

| SSI | Conditions |
|---|---|
| 3 to 0 | Light showers |
| 0 to −3 | Thunderstorms, possible small hail |
| −3 to −6 | Intense thunderstorms, possible large hail and tornadoes |
| −6 or less | *Severe* tornadoes and very large hail |

this air rises dry adiabatically until it reaches 820 mb. At this point, $T = T_d = 19°C$. Then continue lifting moist adiabatically until the air reaches 500 mb, at which time its temperature is 1°C. Subtracting this from the actual temperature at 500 mb, we find that SSI = −7.

Solution
Begin praying immediately. When the SSI = −7 this indicates severe tornadoes.

---

I remember the first time I saw such a sounding. Even though I was assured that a tornado actually did occur within a few hours, I still couldn't believe it. True, the sounding does contain warm, moist air in the lower levels and cold dry air in the upper troposphere and this is ideal for severe thunderstorms, but what troubled me is the fact that the sounding contains an inversion. Everyone knows that an inversion implies stable conditions. What was even more puzzling to me is the fact that most severe weather events are preceded by soundings like this one. What could the explanation be?

The argument goes something like this — *if* you could get the air from 850 mb over the inversion and up to 500 mb, *then* it certainly would continue to rise on its own. Now, a few pages earlier you read that when severe weather develops there is usually a warm tongue of air at lower levels coming up from the south, so that the air at 850 mb should be getting warmer. At the same time the cool, dry air at 500 mb should be getting cooler. These temperature changes act to eat away at the inversion. By heating the air near the ground, the intense sunlight is another factor that contributes to breaking up the inversion.

The inversion is also weakened by the general rising tendency of the air and any disturbance such as an approaching cold front that can initiate vertical motion. This happens in the following way. The humid air near the ground quickly becomes saturated as it rises and then cools slowly at the moist adiabatic rate. At the same time, the dry air aloft cools more rapidly at the dry adiabatic

**Echo Coverage Symbols on the Radar Summary Chart**

| Symbol | Meaning | Called |
|---|---|---|
|  (line symbol) | A line of echoes | Line |
|  (area symbol) | An area of echoes | Area |
| ⊕ | Over 9/10 coverage | Solid |
| ⑪ | 6/10 to 9/10 coverage | Broken |
| ⊘ | 1/10 to 5/10 coverage | Scattered |
| ⊙ | Less than 1/10 coverage | Widely scattered |
| ○ | Isolated cell | Cell |
| ✹ | Strong cell detected by two or more radars | Cell |

**Symbols Indicating No Echos**

| Symbol | Meaning |
|---|---|
| NE | No echo (equipment operating but no echoes observed). |
| NA | Observation not available. |
| OM | Equipment out for maintenance. |

**Weather Symbols**

| Symbol | Meaning |
|---|---|
| R | Rain |
| RW | Rain showers |
| A | Hail |
| S | Snow |
| IP | Ice Pellets |
| SW | Snow Showers |
| L | Drizzle |
| TW | Thunderstorm |
| ZR, ZL | Freezing precipitation |

**Intensity Trend**

| Symbol | Meaning |
|---|---|
| + | Increasing |
| − | Decreasing |
| NC | No change |
| NEW | New |

## Intensity and Trend of Precipitation

Type of precipitation is further annotated to show *intensity* and *intensity trend*. Intensity follows the precipitation symbol, and a solidus (/) separates intensity from intensity trend.

**Echo Intensity**

| Symbol | Echo Intensity | Estimated Precipitation |
|---|---|---|
| − | Weak | Light |
| (None) | Moderate | Moderate |
| + | Strong | Heavy |
| ++ | Very strong | Very heavy |
| X | Intense | Intense |
| XX | Extreme | Extreme |
| U | Unknown | Unknown |

## Heights of Echo Bases and Tops

Heights in hundreds of feet MSL (mean sea level) are entered above and/or below a line to denote echo tops and bases, respectively. Examples are:

| 450 | Average tops 45,000 feet. |
| $\frac{220}{80}$ | Bases 8,000 feet; tops 22,000 feet. |
| 330 | Top of an individual cell, 33,000 feet. |
| \650/ | Maximum tops, 65,000 feet. |
| A350 | Tops 35,000 feet reported by aircraft. |
| 35 → | Individual echo movement to the northeast at 35 knots. |
| → | Line or area movement to the east at 20 knots. |

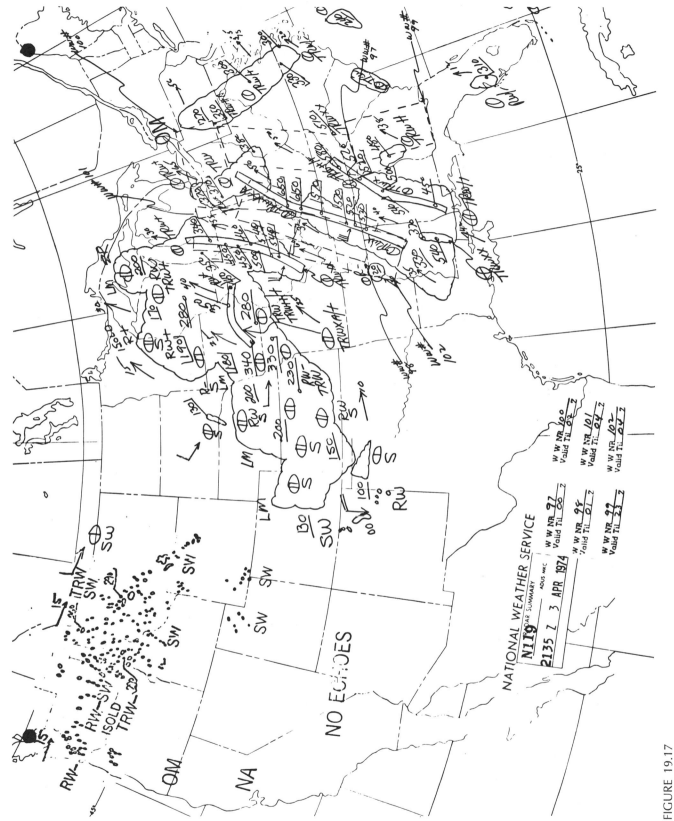

FIGURE 19.17
The radar summary chart for April 3, 1974 at 2135 GMT. The codes for interpreting the various symbols are also included.

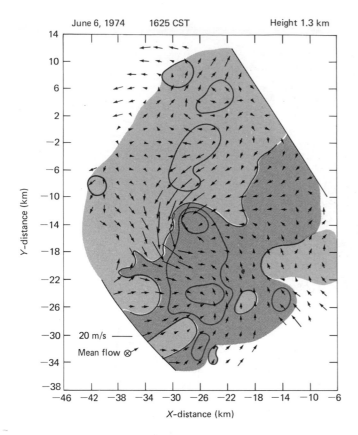

June 6, 1974     1625 CST          Height 1.3 km

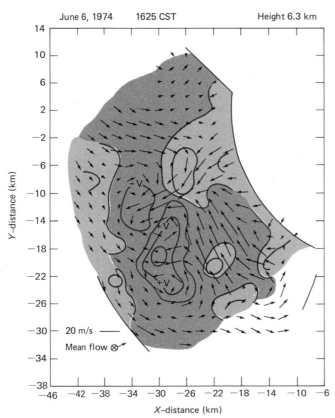

June 6, 1974     1625 CST          Height 6.3 km

THUNDERSTORMS, HAILSTORMS, AND TORNADOES

**FIGURE 19.18**
Winds at two levels within a severe thunderstorm in central Oklahoma at 1625 CST on June 6, 1974. Red shading indicates updrafts, dark shading downdrafts, contours are in meters per second (m/s). The symbol $+V$ indicates a cyclonic vortex; $-V$ indicates an anticyclonic vortex. Notice how the main updraft slopes westward with height, and that downdrafts dominate around the fringe of the storm.

rate. Thus the top of a rising layer soon becomes much colder than the bottom or, in other words, the lapse rate has become much steeper than before the layer rose. This process is depicted in Figure 19.20. A situation like this is called **convectively unstable**. The potential for convective instability exists any time the equivalent potential temperature of an air layer decreases with height. (Thus there is some use for the equivalent potential temperature discussed on page 201 in Chapter 13, after all!)

What finally happens is that the air near the ground breaks through the inversion in isolated places and the atmosphere then acts the way water acts behind a dam which has collapsed. Yes, the inversion actually keeps the warm, moist, and *potentially* unstable air from rising until the last possible moment, when the floodgates are literally thrown open.

The dry air aloft also plays a role in increasing the severity of the thunderstorms. Downdrafts are intensified because the dry air that is entrained into the sides of the cloud cools enormously when raindrops evaporate into it. The updrafts are also intensified by contrast with this cool air. This dry air also helps in making hailstones larger: first, by adding to the thunderstorm's intensity and, second, as we know already from Chapter 12, hailstones will not melt as rapidly in dry air. Thus, another important reason why hail is not so common in Florida thunderstorms is that there is seldom dry air aloft there.

# LIGHTNING AND THUNDER IN HISTORY

According to many ancient religions all over the world, lightning is manufactured by one of the gods and is hurled to earth to cut down some mortal who has perhaps been irreverent. If you were unfortunate enough to be struck by lightning, the ancients interpreted this as a proof that you had incurred the wrath of the gods, and you were therefore considered unclean.

Of course, if you were dead this might not matter too much to you, but sometimes it is possible to revive a per-

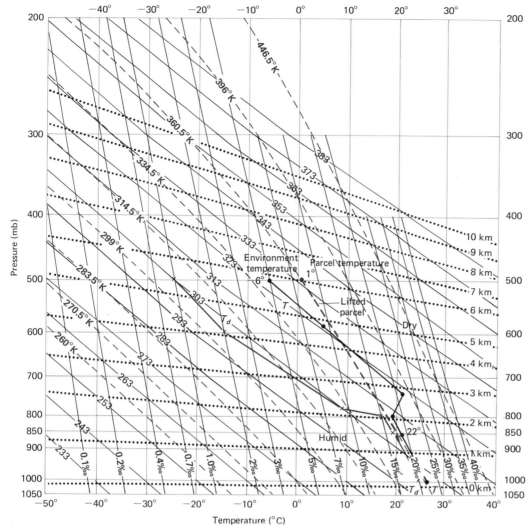

FIGURE 19.19

A sounding that typically precedes severe thunderstorms, hail, and tornadoes. Such severe weather will occur only when large-scale weather conditions lead to the breakdown of the inversion. Note that for this sounding a parcel lifted from 850 mb (lines with arrows) will arrive at 500 mb about 6°C warmer than its surroundings.

son who has only been stunned by the shock. Indeed, many people have died who might have been saved by artificial respiration. Animals killed by lightning were also considered to be unclean and were not to be touched, even if there was famine in the land.

Such primitive ideas still persist. The witch doctors of many tribes have the specific job of driving away thunderstorms, and it is assumed by members of the tribe that they have the power to direct lightning bolts. In fact, between 1926 and 1930, there were three accusations leveled against witch doctors of the Kgatla tribe in the Bechuanaland Protectorate that they had directed lightning at the huts of certain natives. In one of the trials a witch doctor was accused of murder and, believe it or not, he actually pleaded guilty! As a punishment for the crime, he was severely branded in the mouth with a large piece of burning wood.

In the past, there have been other practices even more foolish. During the Middle Ages many people thought that ringing church bells would keep the lightning away. Unfortunately, lightning is attracted to the highest point in an area, so that the steeples of large cathedrals acted as ideal lightning rods. Needless to say, cathedrals were often struck by lightning, and it does not take too much imagination to realize how poorly the unfortunate bell ringers fared—they fared quite poorly! Literally hundreds

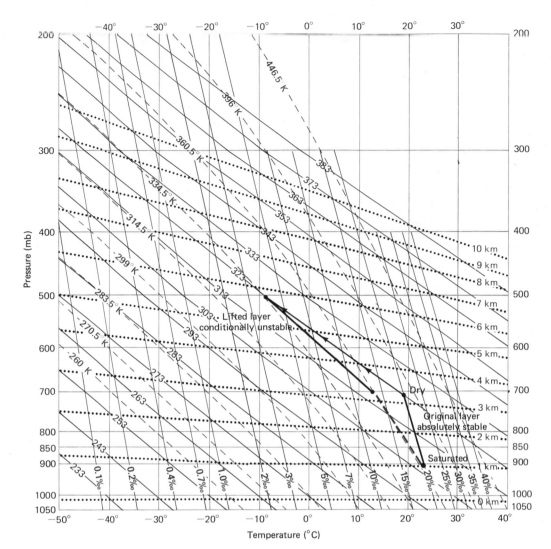

FIGURE 19.20
Lifting a layer of air that is saturated at the bottom and dry at the top quickly destabilizes it.

of bell ringers were struck and killed by lightning as they were in the very act of "protecting" the townfolk.

The tendency for lightning to strike high places such as cathedrals has had another impressive effect. Besides the fact that a number of cathedrals were burnt to the ground, these edifices were also used in former days to store huge quantities of gunpowder. During thunderstorms they therefore literally became time bombs, and there are several cases of enormous explosions. The first recorded case occurred in 1769 when the church of St. Nazaire in Brescia went up with 100 *tons* of gunpowder. This explosion destroyed one-sixth of the city and sent 3000 souls to their eternal salvation, or elsewhere.

Ironically, in another part of the globe, Benjamin Franklin had by this time invented the lightning rod. In a letter to the Royal Society of London written in 1750 he announced his discovery.

Fix on the highest parts of the edifices upright rods of iron, made sharp as a needle, and gilt to prevent rusting and from the foot of these rods a wire down the outside of the building *into the ground*, or down the shrouds of a ship and down her side, till it reaches the water. Would not these pointed rods probably draw the electrical fire silently out of a cloud before it came nigh enough to strike and thereby secure us from that most sudden and terrible mischief.

B. F. J. Schonland, *The Flight of Thunderbolts*

By 1753, Franklin made this account available to the general public and gave instructions on how to construct a lightning rod in *Poor Richard's Almanac* (see Figure 19.21). The evidence quickly mounted in favor of Franklin's lightning rod. But, despite the evidence, some people remained disbelievers, and for the next 50 years many buildings, ships, and lives were needlessly lost.

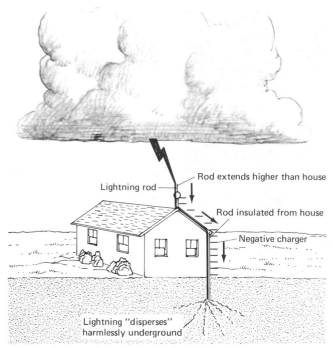

FIGURE 19.21
Principle of the lightning rod. The lightning flash is conducted down the rod (which is insulated from the house) and disperses harmlessly underground. The rod must extend higher than the house to "attract" the lightning.

# LIGHTNING

Lightning (see Figure 19.22) is nothing more than a gigantic electric spark. From the time of the ancient Greeks something had been known about the nature of electricity, but we do not know if anyone realized the electrical nature of lightning. However, the Greeks found that when amber was rubbed together with a piece of fur sparks would be created, and the fur would rise. Today we see the same thing when we rub a balloon across our arms. In fact, during dry winter weather merely walking across a wool rug can produce sparks when you then touch a wall or door. At times this can cause annoying and even mildly painful shocks.

A spark is nature's way of releasing an electric charge that has been built up too much. How are these charges built up in the first place? Since all atoms are made up of protons (positive charges) and electrons (negative charges) you *might* think that every object should be charged. Normally, however, there are exactly the same number of electrons and protons, so that the charges cancel and the object is neutral. But when substances such as amber and fur are rubbed together the amber acquires a small excess of electrons (the Greek word for amber *is*

elektron), and the fur acquires a small excess of protons.

In 1746, in Leyden, Holland, the first device that could store electrical charges was invented by Pieter van Musschenbroch. This was called the Leyden jar by Franklin. It was then the Age of Reason and all the intelligentsia at the time found science intriguing and fun. Soon many people began experimenting with their own electrical condensers for the purpose of making large sparks. Around 1750, Benjamin Franklin built his own electrical condenser in Philadelphia and began experimenting with it.

The sparks made by such machines were used to perform such "noble" experiments as killing small and even medium-sized animals. Franklin himself fared little better because he got careless one day while he was trying to execute a turkey. He received such a shock from his own condenser that he was knocked unconscious and nearly died. But Franklin did gain something from his experiments with electricity, since he soon realized that lightning is nothing more than a giant spark. This ultimately led to his invention of the lightning rod and his discovery of the electrical nature of the thunderstorm. And finally, he did get to kill a turkey with his condenser.

At present, we still don't fully understand how lightning is produced because there are so many complicated processes involved, and because the interior of a thunderstorm does not provide scientists with ideal laboratory conditions. Current research now indicates that the main process by which electrical charges are built up within thunderclouds is much the same as the process by which the Leyden jar becomes charged. This process is known as charging by **induction**.

We can imagine a simple experiment. Rub some amber with cat's fur; the amber now has an excess of electrons. Then, bring the amber into contact with a large metal ball. At first the metal ball may have no net electric charge but, when the amber is brought close, some of the electrons within the ball flee from the amber. This happens because like charges repel, and the electrons in the ball are repelled by the electrons in the amber. Once the amber rod touches the ball, some of the electrons rush off the amber to the nearby protons on the ball. This results in a net negative charge on the ball, and the process can be repeated until a significant charge builds up on the ball (see Figure 19.23).

This situation is quite analogous to one from everyday life. Imagine that men are positive charges and that women are negative charges (metaphorically, of course!). At first there are an even number of men and women at a Ball. But then a busload (amber) of women are brought in. Obviously, the men at the Ball will welcome the extra women; but the women already there will be somewhat repelled by the situation. As more and more busloads of

FIGURE 19.22
Lightning.

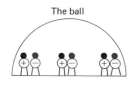

The ball

Amber bus

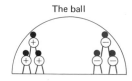

The ball

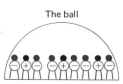

The ball

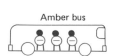

Amber bus

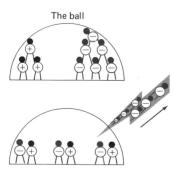

The ball

FIGURE 19.23
Electrical charging by induction. When the net charge grows excessive on the ball, sparks may fly (i.e., discharge occurs).

THUNDERSTORMS, HAILSTORMS, AND TORNADOES

women are brought to the Ball, the imbalance increases until it becomes intolerable. Finally, the atmosphere will become quite "charged" and sparks may fly. Relations will reach the breaking point and most of the excess women will be **conducted** out of the Ball at a shockingly rapid rate. If you do not like the story as it stands, reverse the roles of the players but the effect will remain the same.

Now, let's see how this works in a thunderstorm. Under normal conditions the earth has a negative charge, whereas the upper atmosphere has a positive charge. Since all clouds form in the region between earth and the upper atmosphere, the little raindrops, **graupel** (ice pellets), and hailstones all will act like the metal ball when the charged amber rod is brought near. The positive charges drift toward the bottom of each graupel (etc.), so as to be near the negatively charged ground, while the negative charges of the graupel are attracted to the positive charge of the upper atmosphere and drift toward the top of each graupel. At this point, the graupel still has *no* net charge, but it has been **polarized.**

The graupel acquires a net charge when it collides with other graupel. Since the larger particles fall faster than the small ones, the larger particles fall on top of the small ones. Thus, the large graupel acquires a net negative charge when some of the electrons from the top of the small particles move toward the positive charges of the large graupel. Similarly, the small graupel acquire positive charges.

The main reason that graupel are probably more effective in charging the thunderstorm is because the graupel have a tendency to bounce off one another after collision, while drops tend to coalesce. After all, if two raindrops with no net charge coalesce, the resulting larger raindrop will still have no net charge. There is, of course, other evidence that the graupel is more important in charging the thunderstorm. For instance, it is found that the main centers of positive and negative charges almost always develop in the subfreezing part of the cloud.

Now we come to the final step. The rising air of the thunderstorm then helps to sort out the different sized graupel. The large particles fall faster and thus move toward the bottom of the cloud, while the smaller graupel are carried aloft into the upper part of the cloud by the strong updrafts. Thus, a large accumulation of positive charges develops in the upper parts of thunderstorms, whereas a large accumulation of negative charges accumulates in the lower parts of the thunderstorms at about the level where the temperature is −10°C.[1]

---

[1] There is also a small pocket of positive charges near the bottom of most thunderclouds, which we will not explain here.

The induction mechanism is self-reinforcing and acts like a chain reaction. The growing accumulation of positive charges near the cloud top and of negative charges lower in the cloud increases the polarization of all particles in between. This in turn increases the amount of negative charges that are transferred from the small to the large graupel, which leads to a further accumulation of positive charges near the cloud top and of negative charges near the cloud bottom, and so on. It takes only a few minutes for the charges to build up the gradient of voltage or **electric field** to the breaking point, and then the air, which is normally an insulator, literally breaks down in a channel and becomes a wire. This occurs when there is a voltage gradient of 100,000 volts/m over a distance of roughly 50 m in raindrop-filled air. The lightning flashes and discharges the cloud (see Figure 19.24).

Most lightning bolts occur entirely within the cloud. Only about one flash in four actually strikes the ground. When it does the region of large negative charges near the bottom of the cloud creates such a strong electric field that during the lightning stroke a considerable negative

FIGURE 19.24
One theory of thunderstorm electrification. 1. Graupel neutral, but charges are oriented with positive charge on bottom. 2. Fast-falling, large graupel gets negative charge from top of small graupel upon collision. 3. Particles sort by size; large graupel accumulate on bottom so cloud is charged. 4. Discharge occurs when charge concentration (technically, the electric field strength) becomes excessive.

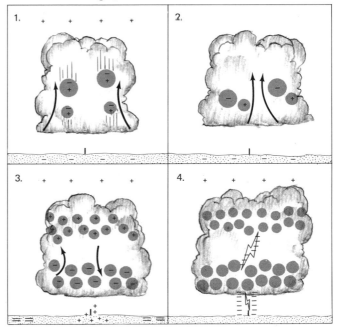

charge is transferred into the ground. Thus lightning is mostly responsible for maintaining the negative charge of the earth and the positive charge of the upper atmosphere. The positive charge of the atmosphere reaches a maximum around 3:00 P.M. London time (actually GMT), which corresponds to midafternoon in Africa. Recall that the majority of the world's thunderstorms occur each day at about this time in Africa.

In spite of the enormous power of each lightning bolt, it takes about 1 million lightning bolts to transfer 1 gram of electrons into the ground! But the lightning flash is so brief that the mere 20 coulombs of electric charge that it transfers is a poor indicator of its power. Lightning flashes can build up a voltage of several hundred million volts! The electric current of an average flash is 20,000 amperes. This information should impress you if you have ever gotten a shock at 110 volts and perhaps 20 amperes in your home.

The air gets superheated by the lightning flash and temperatures reach as high as 30,000°C—which is far hotter than the surface of the sun. Thunder (see Chapter 20) results when this superheated air expands explosively.

### Saint Elmo's Fire

Sometimes the electric field does not build up to the point where lightning is produced, but it is still strong enough in the immediate vicinity of pointed objects to produce what is known as **Saint Elmo's fire**. Saint Elmo is the patron saint of all sailors, and often during thundery weather the top of the masts of ships will glow as if they are on fire—but this fire won't burn. This is Saint Elmo's fire which has long been regarded with reverence. It can also sometimes be seen on land on the top of tall steeples and, on occasion, airplane pilots have found themselves enveloped in its bluish or reddish glow. You can see something resembling Saint Elmo's fire if you take a woolen blanket or sweater on a cold dry night and rub it vigorously in a completely dark room.

Saint Elmo's fire is technically known as a series of corona discharges. This means that the object emits an almost continuous barrage of sparks which produce a hissing noise that is often completely masked by the other noises of the thunderstorm. For this reason, Saint Elmo's fire is often *thought* to be silent.

There is no doubt that witnessing such a spectacle can cause feelings of religious awe. In fact, we have good reason to believe that the burning bush that Moses saw on Mount Horeb was a bush enveloped in Saint Elmo's fire.

2. And the Angel of the Lord appeared unto him in a flame of fire out of the midst of a bush: and he looked, and behold, the bush burned with fire, and the bush was not consumed.

3. And Moses said, I will now turn aside and see this great sight, why the bush is not burnt.

4. And when the Lord saw that he turned aside to see, God called unto him out of the midst of the bush . . . .

*King James Bible,* Exodus, Chapter 3

One technique of artificial lightning prevention employes a principle similar to that of Saint Elmo's fire.

FIGURE 19.25
Lightning can spread along the ground. This is why it is not wise to stand under a tall object such as a tree during a thunderstorm.

Small, needle-shaped pieces of aluminum chaff are fired into the cloud with the hope that they will produce a series of sparks or corona discharges that will then keep the overall electrical field in the cloud from growing excessive.

**What to Do When Lightning Strikes**

If you have the misfortune of being outside during a thunderstorm, seek shelter inside a car if possible. There you can safely weather out the storm, even if the car happens to get struck. If there is no car or house around, avoid being the tallest object yourself, and don't stand too near the tallest object, such as a tree. Lightning may well strike the tree and then run down the trunk and along the ground where it can electrocute you (see Figure 19.25). The golfer, Lee Trevino, was struck by lightning in this way and is lucky to be alive. If you are swimming in the water, get out. If you are in the mountains, get off the high, exposed ridges and seek shelter in some depression in the rocks. When you are indoors turn off all electrical devices (especially the TV) when possible. Finally, if you happen to feel the hair rising on your head, dive to the earth and pray that Saint Elmo will spare you.

Some final advice is important enough to repeat. If someone near you happens to be struck by lightning, you may be able to revive them. Try artificial respiration (assuming you know how to do it) and perhaps you may save a life.

# PROBLEMS

**19-1** How long does rain from an average thunderstorm last?

**19-2** How long does a typical tornado take to pass?

**19-3** Why do tornadoes tend to occur in bunches?

**19-4** Why are *severe* thunderstorms, hail, and tornadoes most common in spring?

**19-5** Look at Figure 19.6. Compare the weather at Duluth, Minnesota; Fort Worth, Texas; and Louisville, Kentucky. Aren't springtime lows interesting?

**19-6** Why do the highest incidences of thunderstorms, hailstorms, and tornadoes occur in different places in the United States?

**19-7** Explain how air-mass thunderstorms are self-destructive, and why this is not necessarily so for squall-line thunderstorms.

**19-8** Using Figures 17.26 and 17.27, compute the SSI for Oklahoma City at 1800 GMT on July 12, 1966. Use the surface temperature at 970 mb instead of the 850 mb temperature. What is your forecast?

**19-9** Explain why a halo is sometimes seen before a thunderstorm passes.

**19-10** In each of the following cases start with a layer of air 100 mb thick, from 800 to 700 mb, at a uniform temperature of 20°C. Keep the layer 100 mb thick as it moves up or down. Calculate the final lapse rate for these cases: (a) if the entire layer rises dry adiabatically, and the pressure is lowered by 100 mb; (b) if the entire layer sinks, and the pressure is increased by 100 mb; and (c) if pressure is decreased by 100 mb, but the top of the layer rises dry adiabatically while the bottom rises moist adiabatically.

**19-11** What is so important about choosing the proper wavelength for weather radar?

**19-12** What is the height of the tallest cloud top in Figure 19.17?

**19-13** Why are two doppler radars needed to obtain a complete picture of the winds within a thunderstorm?

**19-14** Explain the hook echo that is sometimes seen on the PPI scope.

**19-15** Explain the bright band that is sometimes seen on the RHI scope.

**19-16** How can dry air aloft add to the severity of a thunderstorm?

**19-17** Explain how a thundercloud can get charged by induction.

**19-18** Explain the principle of the lightning rod.

**19-19** Is it a good idea to stand on a mountain top or under a solitary tree on a flat plain during a thunderstorm? Explain.

# LOCAL WINDS

April 19, 1976 was the second day of a record-breaking heat wave in New York City. Thousands of New Yorkers who had suffered from the heat on the previous day rushed to the beaches of Long Island. There they managed to escape the 36°C of mid-Manhattan—but they got more relief from the heat than they had anticipated. A chilling afternoon sea breeze sprung up and dropped temperatures down near 15°C along the beaches, causing a mass, frenzied stampede back to the torrid city.

The sea breeze is one of the most widely known examples of a local or small-scale wind. Many of these local winds appear to be quite simple in principle but are strongly influenced by the larger scale winds and, therefore, can become quite complicated. It is for this reason that the subject of local winds was postponed so long.

Almost every place on earth has its own unique weather. Differences in topography or even in the nature of the underlying surface produce local winds that can sometimes (as in the case above) have tremendous effects on the local weather conditions. But even when the small-scale winds *seem* not to have any observable effect on the weather, meteorologists now realize that they help distort the weather of the entire globe and render weather unpredictable more than about two weeks ahead (see Chapter 21).

In this chapter we will look at and analyze several types of small-scale winds. The chapter begins with the sea- and land-breeze circulations and then moves to the vari-ous local mountain winds such as the chinook, katabatic winds, and mountain and valley breezes. Various desert winds such as the Khamsin, dust storms, and dust devils are then considered. Finally, a brief discussion of other small-scale phenomena such as sound and gravity waves is presented.

## SEA BREEZE BY DAY: LAND BREEZE BY NIGHT

Most of the world's coastlines are subject to the sea and land breezes. The basic principle of the sea breeze was already presented in Chapter 15. Repeating it briefly, during the day the land (and thus the air over the land) gets heated considerably, while the air over the sea hardly changes. Then, the warm, light air over the land rises and is replaced by the cooler air from the sea.

A complete circulation cell forms with the sea breeze confined to the lowest kilometer. Above this, the wind direction is reversed, and the air returns seaward. Then as the air passes over the ocean it sinks, and any small clouds that were carried by this wind soon dissipate in the environment of sinking air (see Figure 20.1).

Thus during the daytime clouds form over the land, while it is usually quite clear just offshore along a strip 10

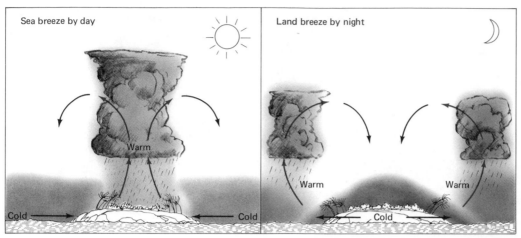

FIGURE 20.1
The sea breeze and the land breeze. Day clouds over land, night clouds over water.

to 100 km wide. At times this is beautifully illustrated by satellite photos, as you can see in the pictures of India and Florida (Plates 11a and b). The land over Florida is mostly covered by cumulus clouds (except over Lake Okeechobee and the St. John River), but along the coast it is completely clear.

At night the wind direction reverses, because the air over the land becomes cooler than the sea air. Now it is the sea air that rises and the cooler land air that drifts out from the land as a land breeze. At night, clouds tend to form over the sea; if you happen to walk along the beach during a summer night, lightning can often be seen in the distance over the ocean. Because the temperature contrasts are generally much weaker at night, the land breeze tends to be weaker than the sea breeze.

The sea breeze usually begins a few hours after dawn. I will never forget my first week as a lifeguard on Atlantic Beach, Long Island. The weather was remarkably consistent the entire week. When I arrived on the beach at 8:00 A.M. it was clear and calm and the temperature was 23°C. Within the next hour a cool sea breeze of about 10 knots brought the temperature down to 21°C. The sea breeze would continue all day until about 5:00 P.M. after which it calmed down again.

The sea breeze averages between about 10 and 20 knots and it tends to be strongest on warm spring days when the ocean is still quite cold. The case mentioned at the beginning of the chapter is an extreme example of the cooling powers of the sea breeze, but it is quite common for afternoon temperatures along the seashore to be 5 to 10°C cooler than inland on sunny spring and summer days. When the ocean water is relatively warm, the cooling along the coast is not so large, and this means that

tropical and subtropical sea breezes sometimes provide only a moderate degree of cooling.

The sea breeze begins along the coast in the morning and progresses both landward and seaward as the day progresses. At times the temperature, wind, and humidity change sharply when the sea breeze roars in. In these cases the leading edge of the sea breeze behaves like a front and, indeed, it is then called a sea-breeze front (see Figure 20.2). On the inland side of the sea-breeze front it is hot, but a sharp transition to much cooler air sometimes occurs within a kilometer. The sea-breeze front typically penetrates 10 to 20 km inland, but there are times when it has penetrated as much as 200 km inland!

The sea breeze is affected by the Coriolis force. Early in the morning it tends to blow straight onto shore but, as the day goes on, the Coriolis force gradually twists the winds. By late afternoon the sea breeze usually strikes shore obliquely, with land to its left (in the NH). There are even places where the sea breeze and land breeze undergo a continuous clockwise rotation (in the NH) over 24 hours, as is shown in Figure 20.3.

The sea breeze can either be strengthened or weakened by the larger scale winds. When the geostrophic wind blows toward the sea, it suppresses the sea breeze and does not allow it to penetrate as far inland as normal. If the offshore geostrophic wind is large enough, it may suppress the sea breeze completely. Generally, the larger the land–sea temperature contrast, the larger the offshore geostrophic wind must be in order to suppress the sea breeze completely.

I remember one day on the beach when the normal sea breeze was completely suppressed and the wind was blowing from the land. Not only was it far hotter than

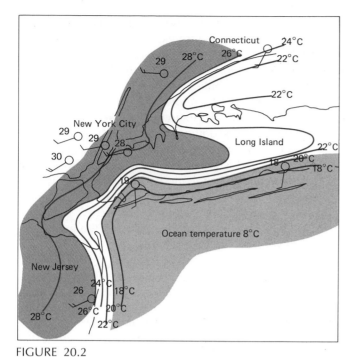

FIGURE 20.2
Spring sea breeze with sea-breeze front on Long Island at 1600 EST, April 19, 1972. The temperature along the beaches of Long Island was probably 15°C or less. Notice how strong westerly winds prohibit any sea breeze in New Jersey. Solid lines are isotherms at 2°C intervals.

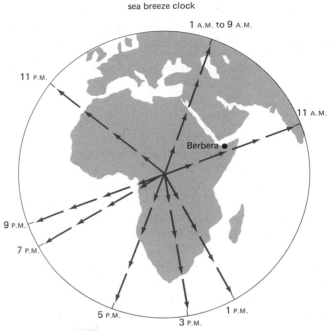

FIGURE 20.3
The "sea-breeze clock" in Berbera (Somali Republic) during September. Wind direction (shown by clock hands) veers almost continuously, except from midnight to early morning, making a complete circle every 24 hours. The winds at many coastal locations have a similar tendency.

usual along the beach, but the insects that are normally kept inland by the sea breeze were blown onto the beach. That day, as I proudly sat on the lifeguard tower I got stung by the largest bumblebee I have ever seen.

There are other wind systems that are analogous to the sea breeze. Naturally, large lakes have similar wind circulations, and the lake breeze at Chicago is one well-known example. But even stranger analogs to the sea breeze exist. You have already read about the forest breeze in Chapter 9. This is the cool breeze that emerges from the forest's fringe onto the heated plain. There is even a phenomenon known as a snow breeze in which a cool breeze blows from snow-covered surfaces toward warmer bare ground.

## MOUNTAIN WINDS

There is a saying, "the mountains make their own weather," and any mountaineer knows that there are many different winds in the mountains. After the winter snows have melted, the mountain absorbs the sun's rays

and heats the air on the slopes directly facing the sunlight. This heated air glides upslope in a layer about 100 to 200 m thick, so that within an hour after dawn a gentle wind starts blowing up the mountain slope. This wind increases in strength until the early afternoon when it reaches a typical peak velocity of 4 to 8 knots. As you have seen in Chapter 2, this rising air current often produces a cumulus cloud above the mountain top. The upslope breeze weakens in the late afternoon and usually ends an hour or so before sunset.

At night the mountain cools and the wind direction reverses. Now, a cold downslope wind blows all night. The downslope wind has approximately the same average speed—but it is much more irregular and gusty. Mountaineers have sometimes been disturbed by these nightly wind gusts that can occur at roughly 5-minute intervals all night. This behavior has been verified by scientists and leads to the conclusion that on many mountains the cold air at night oozes downhill rather like porridge than water. Friction keeps the cold air from flowing until there is a sufficiently thick layer, whereupon it bursts downhill. Then a new layer of cold air must build up before the air can flow again.

After sunset at 21:01 Central European Time, a weak air current developed which died down again within a minute. At 21:06 there was another air movement and another at 21:11, followed by gusts at regular intervals of 5 minutes which persisted through the night disturbing our sleep in the most unpleasant way. An occasional gust would be missing but the next one would appear punctually on time.

J. Kuttner, from Geiger, *Climate Near the Ground*

The winds are also similar in an enclosed valley. During the day warm air moves up the valley and up the valley walls as well, while at night the circulation is reversed (see Figure 20.4). An interesting picture taken shortly after dawn shows this wind at work (see Figure 20.5), breaking up the fog which had formed in the valley overnight. Here the fog is breaking up in the middle of the valley first, since the sinking air is there.

One exception to the daily cycle of upslope and downslope winds is the glacier wind. The cold air from mountain glaciers blows downslope at all times, suppressing plant growth for perhaps 100 or more meters beneath the bottom of the glacier. This glacier wind is strongest in the afternoon, when the temperature contrast between the icy glacier and the warm surroundings is the greatest.

The winds in the mountains are not always so gentle. Sometimes they can blow at hurricane speeds. In fact, as you have read earlier the windiest places on the earth's surface are found on or near mountains.

One of the potentially destructive mountain winds is the so-called **katabatic** (or fall) wind. Its principle is similar to that of the nightly downslope breeze, but the ka-

tabatic wind is far more intense because it is produced by a large pool of cold air from extensive highland regions. The katabatic wind certainly does not flow like porridge when fully developed. Instead, it flows like water cascading down a narrow chute.

Katabatic winds are therefore very common around the edges of Greenland and Antarctica, where frigid air from the ice sheets pours down to the neighboring warmer ocean waters. Wherever the slope of the land is gentle, the katabatic wind is also gentle but, where the slope of the land is steep, the katabatic wind can be furious. It is the katabatic wind that is responsible for the windiest average on earth at Cape Dennison, Antarctica. Gales exceeding 100 knots are *common* there and the wind chill factor is severe, to say the least (see Figure 20.6).

Picture drift so dense that daylight comes through dully, though, maybe, the sun shines in a cloudless sky; the drift is hurled screaming through space, at a hundred miles an hour, and the temperature is below 0°F. Shroud the infuriated elements in the darkness of a polar night, and the blizzard is presented in a severer aspect. A plunge into the writhing storm-whirl stamps upon the senses an indelible and awful impression seldom equalled in the whole gamut of natural experience. The world a void, grisly, fierce and appaling. We stumble and struggle through the Stygian gloom; the merciless blast—an incubus of vengeance—stabs, buffets and freezes, the stinging drift blinds and chokes. In a ruthless grip we realize that we are . . . poor windlestraws on the great sullen roaring pool of time.

*Weather*, 1972

The katabatic wind reaches its greatest intensity at the

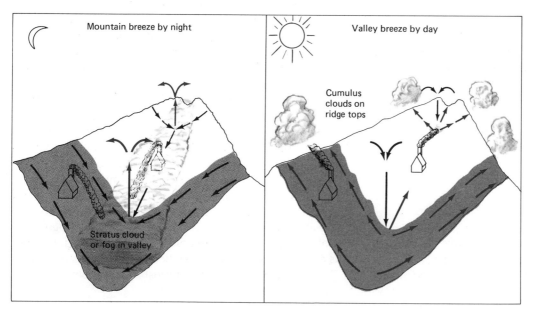

FIGURE 20.4
Mountain breeze by night, valley breeze by day. Notice the sinking air aloft directly above the valley floor during the day.

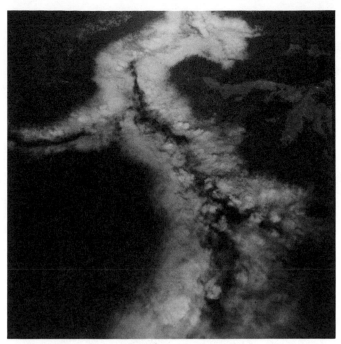

FIGURE 20.5
Stratus clouds in the Redwood Creek Valley breaking up in the morning (1000 PST, November 4, 1971) beccause of sinking air in valley center.

FIGURE 20.6
Standing against the wind at Cape Dennison, Antarctica (Mawson Institute of Antarctic Research). Wind speed at this time was a steady 45 m/s.

shoreline. Once it passes over the sea it loses its impetus, but still can make coastal seas choppy. Usually it can penetrate only a few kilometers seaward where it abruptly terminates but, occasionally, it penetrates over 100 km offshore before gradually losing its identity.

You might wonder how the katabatic wind can descend so far and still remain cold. As the air descends the slopes it will "try" to warm dry adiabatically, but it is often cooled significantly by the underlying surface that consists largely of snow and ice (see Figure 20.7). The high winds often whip up snow high into the air (the "drift" you just read about in the above quote). This suspended snow cools the air significantly and adds to its weight.

But even if the air did warm at the dry adiabatic rate, it would still often arrive at sea level much colder than the warm air over the sea, simply because it is so cold to begin with.

The large-scale winds and pressure patterns have a great influence on the intensity of the katabatic winds. When no general wind is present, a cold air dome forms above the plateau, and cold air often "dribbles" down the edges of the plateau. The large-scale winds can either hold this dome away from the edge or help to push it against the edge and cause the cold air to cascade or "stampede" over the edge.

The katabatic wind will also be enhanced when a low pressure area lies to your left or a high to your right as you look down to the sea (in the NH). This large-scale pressure pattern acts to blow the winds away from the mountain base so that air from above tends to get "sucked" down from the highlands (see Figure 20.8). As you will soon read, the chinook also forms under the same general pressure conditions.

There are several well-known examples of katabatic winds in Europe. One of these is the **Bora** (or north wind, even though it blows from the northeast) that rushes from the highlands of Yugoslavia down to the shores of the Adriatic Sea. It is most notable at Trieste, where wind gusts commonly range between 50 and 100 knots. During the Bora it is considered wise for small ships to harbor elsewhere. Agriculture also suffers greatly from the Bora's desiccating effect.

The Bora is most common during the winter half of the year whenever a large high pressure area containing polar air sits over the snow-covered mountains. Then, at night, the frigid air grows even colder and heavier and begins to flow downhill. At this time a small but significant low pressure area typically forms over the much warmer waters of the Adriatic Sea. Thus, the Bora like many katabatic winds, is actually a combination of a downslope wind and a land breeze (see Figure 20.9). This correctly

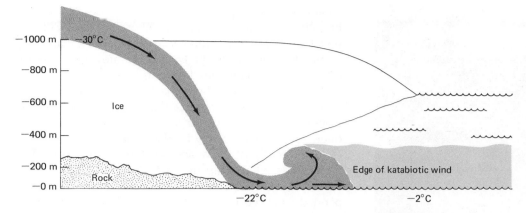

FIGURE 20.7
The katabatic wind at Cape Dennison. As cold air descends the slope, it accelerates. Note the sharp leading edge of the wind. On descending, the air warms somewhat more slowly than the dry adiabatic rate, since it is cooled by the snow it picks up.

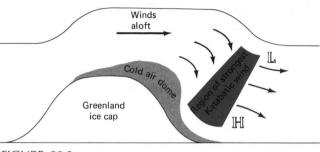

FIGURE 20.8
The large-scale wind conditions favoring a strong katabatic wind. Notice how the dome of cold air is pushed over the edge of the ice cap by the winds aloft, while the pressure patterns below also help to suck the cold air down.

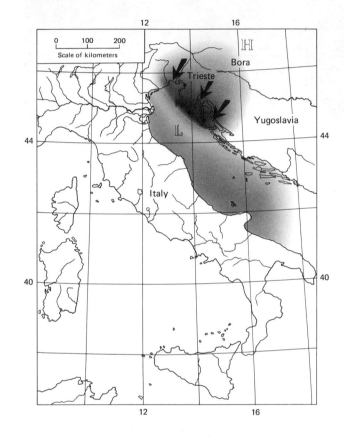

FIGURE 20.9
Conditions favoring development of the Bora. It is strongest early on winter mornings when the land is especially cold.

implies that the Bora is strongest in the hours near dawn and weakest during the afternoon.

The **Mistral** is another notable example of a katabatic wind. This is the cold, dry wind of winter and early spring that blows down the Rhone River Valley past Arles and Marseilles and out into the Mediterranean Sea. The Mistral forms under similar but not identical conditions to the Bora. A large high pressure area situated to the northeast drives polar air through the narrow Rhone River Valley. It is intensified because of a funneling effect but, even so, it is generally not quite as strong as the Bora. The Mistral is

strongest when there is also a low pressure area just to the south over the Mediterranean (as with the Bora). Lows occur in that general area with greatest frequency—you guessed it—during winter and early spring (see Figure 20.10).

These winds are known to affect the tempers of the people living there. Houses built where the Mistral is strongest often have no doors or windows on the north side where the wind strikes with full force. But even without windows, there is plenty of light and, indeed, it is the Mistral that is commonly associated with clear, deep,

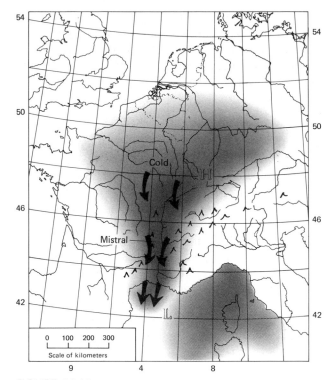

FIGURE 20.10
Conditions favoring the development of the Mistral.

blue skies — those under which the artistic genius of Van Gogh burst forth.

Now that I have mentioned a possible link between a wind and human behavior, it is appropriate to discuss the chinook. You should recall that the chinook (also called the foehn, Santa Ana, or Zonda) is the warm, dry downslope wind that was discussed in Chapter 13. It is well known that many people become tense and some even grow quite disturbed when these winds are blowing (see Chapter 25).

You already know why the chinook is warm and dry; in passing over the mountains the air loses vapor, but acquires latent heat released by condensation (see page 197). You also know that the chinook cannot occur freely, because warm winds are light and will not sink naturally. The chinook therefore *must* be forced by the large-scale wind and pressure patterns. In this respect it differs from the katabatic winds that certainly can be enhanced by the large-scale winds, but may occur naturally.

Several conditions must be satisfied in order to produce a chinook. First, there must be a strong wind blowing across a mountain ridge. Second, as you look downstream from the mountain ridge there must either be a low pressure area to your left or a high pressure area to your right to help "suck" down the air. Finally, the chinook is most noticeable when the air aloft is potentially *much*

warmer than the surface air it replaces (see Figure 20.11). This means that the lapse rate in the lower troposphere is small.

As the air passes over the ridge it often produces a crest cloud that covers the mountain top. Since the air has been forced up, it then rebounds downward after passing over the ridge. As with a wave in the bathtub or a swing, this air then "overshoots" and may drop a kilometer or more beneath its original level, perhaps reaching the ground where it will be quite warm and dry. (This overshooting effect is sometimes sufficient by itself in stable situations, to produce the warming of the chinook.) Some of the air can also be forced to the ground by friction when the giant rotor, discussed in Chapter 2, forms.

This should bring to mind the subject of mountain wave clouds. You should expect that the warm air will tend to rise again because of its lightness. The resulting vertical oscillations of air often produce several mountain waves. Mountain wave clouds are often a sign that the chinook is blowing.

The chinook is usually gentle and produces a slow warming, with somewhat dry air. Sometimes, however, it can be truly fearful, producing strong gusty winds (partly from the rotor) that may exceed 100 knots. When this is combined with the extremely low humidity and occasional sharp temperature rises, the drying effect can be startling. It is reported that 75 cm of snow disappeared in one half day when the temperature rose 19°C in 7 minutes (for a total rise of 45°C in a few hours) at Kipp, Montana, on December 1, 1896 after a chinook set in. It may be appropriate to mention that the world's highest temperature occurred during a chinook in Azizia.

The area just east of the Rocky Mountains, especially near Boulder, Colorado, is famous for such windstorms. I remember arriving in Boulder late in June 1969 at the tail

FIGURE 20.11
Conditions needed for a noticeable chinook. Notice how the winds aloft oscillate vertically, bringing potentially warm air closer to ground level on the lee side.

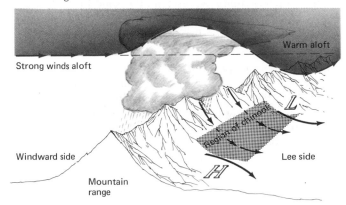

end of an unusually late season windstorm. The sky was an incredible clear, deep blue, as is characteristic of such windstorms. However, the plants did not look so beautiful. All the leaves on the trees had withered, and many never did recover. Many branches and even whole trees were blown down and lay in the street with crisp, desiccated leaves. But the saddest were the roses—mummified just as they had begun to bloom.

The chinook is more beneficial during the winter, when the warming it produces often clears the snow and provides welcome relief from winter's harshness. This account of an Alberta winter chinook shows off its good side.

That night the wind dropped. We awoke to find the thermometer standing at −52°F. The steam rose in little clouds from the stables. . . . Little pools of mist marked the spots where cattle stood in huddled bunches . . . exactly as a mist will gather over a pool on a chilly summer's night . . . . Our main bunch of horses were on a pasture 4 miles from home when the blizzard struck. . . . When the snow finally cleared [10 days later] dead horses were to be found with their manes and tails eaten off by their starving companions, lying in every sheltered corner . . . . And then, almost as suddenly as it had commenced [the chinook began and] the siege was raised; the snow vanished like magic, grass coming up green and fresh as fast as it had disappeared. Horses that had seemed about to die fattened overnight.

From W.G. Kendrew, *Climates of the Continents*

The chinook can produce extreme and astonishingly rapid temperature variations when there is a cold air mass resting against the mountain slopes. The gusty, variable chinook winds can cause the cold-air mass to slosh back and forth, producing alternate rises and falls in the temperature.

This once happened at the edge of the Black Hills of South Dakota. The temperature rose 27°C (from −20 to 7°C) in two minutes, from 7:30 to 7:32 A.M. on January 22, 1943 at Spearfish, South Dakota (which rests against the Black Hills). When the warm air replaced the cold air, streets and windshields were instantly coated with a layer of frost (the warm air cooled below the frost point when it touched the still cold, solid objects).

When I left, the temperature was between −21 and −24°C which was expected. About halfway south (to Spearfish) the windshield on the car frosted so suddenly and heavily that I was well on the way toward the ditch before I could get stopped. When I got out to clean the windshield of the car it felt like a warm spring day.

Cedric A. Barnes, *Monthly Weather Review*, 1943

Then from 9:00 to 9:27 A.M. the temperature fell by

32°C (from 12 to −20°C), when the cold air sloshed back into Spearfish. As you can see from Figure 20.12, this was just one of several oscillations. The largest 24-hour fall in temperature took place in Browning, Montana, when the temperature dropped from 7 to −49°C.

The chinook occurs several days a month during the winter-half of the year and, therefore, has a warming effect on the climate. You can see this by comparing temperature statistics from Havre, Montana, which is near the Rockies, to those from International Falls, Minnesota, which is too far east to feel the chinook. Both of these cities are located at the same latitude, but the average temperature of December to February is almost 6°C higher at Havre because of the chinooks.

The influence of the chinook shows up far more clearly in the all-time high temperature records for these months. For instance, in February, Havre has recorded 22°C, while the all-time high in International Falls is only 12°C for the month. But do not think that it never gets cold at Havre. When the chinook was *not* blowing it has gotten down to −49°C at Havre, which is colder than it has ever gotten at International Falls (−45°C)!

### The Low-Level Jet and Nighttime Thunderstorms
Mountains are involved in another strange wind called the **low-level jet**.

A few hundred meters above the surface of the Great Plains this strong jet of wind often blows from the south on summer nights. This jet occurs when the general wind blows from the south. The Rocky Mountains act to funnel

FIGURE 20.12
Thermograph record showing sharp temperature changes at Rapid City, South Dakota, on January 22, 1943. The actual changes were more abrupt since the thermograph responds only slowly. The abrupt temperature changes cracked a number of plate glass windows.

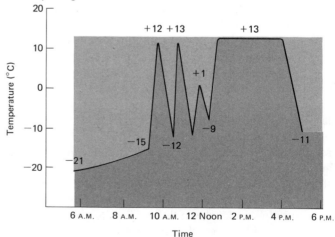

it, but that doesn't explain why it is fastest at night.

At night the mountains to the west cool more than the Plains. The isobaric surfaces then drop over the mountains, which helps to strengthen a southerly geostrophic wind.

Furthermore, the air near the ground is more stable at night (when the ground cools). Therefore, it doesn't mix well with the air higher up which is faster. This is why the *winds at ground level tend to be slower at night*. Conversely, the winds several hundred meters above ground level at night are no longer slowed down by being forced to mix with the slower ground-level winds. Having lost this restraint they accelerate, often reaching speeds of 30 m/s.

The main reason that the low-level jet has been described here is that it is intimately related to the occurrence of thunderstorms at night on the Great Plains. In several places throughout this book you have read that for most places over land thunderstorms are most common during the late afternoon. This is not true on the Great Plains.

On the Great Plains thunderstorms are most common around midnight. This is partly due to the fact that thunderstorms frequently form over the eastern slopes of the Rocky Mountains late in the afternoon and then move eastward, reaching the Great Plains several hours later.

But meteorologists have also shown that thunderstorms are enhanced and are even created in regions where the winds of the low-level jet converge (and rise), as they do at night over the Great Plains.

# WINDS OF THE DESERT

Pretty soon we see something coming that stood up like an amazing wide wall, and reached from the desert up into the sky and hid the sun, and it was coming like a nation, too. Then a faint breeze struck us, and then it come harder, and grains of sand begun to sift against our faces and sting like fire, and Tom sung out:

"It's a sand-storm — turn your backs to it!"

We done it; and in another minute it was blowing a gale, and the sand beat against us by the shovelful, and the air was so thick with it we couldn't see a thing. In five minutes the boat was level full, and we was setting on the lockers buried up to the chin in sand and only our heads out and could hardly breathe.

Mark Twain, *Tom Sawyer Abroad*

Desert winds are seldom pleasant. Not restrained by a plant cover, they whip across the naked land, scorching

by day and chilling by night. When they blow fast enough (usually over 25 knots), they pick up the loose dusty earth and blow it for hundreds or even thousands of kilometers. Satellite pictures have shown the blowing dust travel from the Sahara clear across the Atlantic.

The Khamsin, or Scirocco, is the first desert wind I will describe. The Khamsin (from the Arabic word for fifty) blows all along the desert lands bordering the Mediterranean Sea, from Morocco to the Middle East. Under normal conditions northerly winds prevail here, blowing from the Mediterranean Sea and providing some moderation in the heat.

But the Khamsin is not an ordinary wind — it comes *from* the desert and it knows no moderation. It has produced the highest temperatures ever recorded in the lands bordering the Mediterranean, and this is odd because the Khamsin occurs most frequently during the fall and spring but almost never during the summer. During the Khamsin the RH (relative humidity) often drops below 20% and may dip below 10% and temperatures can exceed 40 to 50°C. Even at night the temperature remains excessive during the Khamsin, so there is little relief from the heat. And added to this is the fine dust or sand that all too often fills the air and the reversed electrical field.

The Khamsin is nothing more than the wind and weather of the warm sector of low pressure areas that form over the Mediterranean Sea. (It is not common during summer because lows rarely form then over the Mediterranean.) Since the warm sector of these lows consists of Sahara air, it does have a "local" quality all its own. Figure 20.13 shows the weather during the Khamsin of May 21, 1970. In this figure the low pressure area over the eastern Mediterranean is drawing air up from the south through Saudi Arabia, Egypt, and Israel, and the temperatures are as high as 50°C! See if you can find the few places cooled by sea breezes in this figure.

Occasionally the Khamsin crosses the Mediterranean Sea and arrives in southern Europe. In Spain it is called the **leveche** and is still very hot and dry, because it has only a short passage across the sea. In Italy and Greece, where it is called the **gharbi** because of its longer run over the sea, it is not quite so hot, but it is far more humid and brings muggy, rainy weather. Occasionally this rain is reddened by dust that had been transported all the way from the Sahara.

At times these Khamsin winds even cross over the mountains of southern Europe and then produce foehn winds on the north slopes. When this happens in northern Sicily,

the air is misty, the sky yellowish to leaden, filled with heavy vapors, through which the sun can be seen only as a pale disk if at all .... Everyone stays at home as much as possible and does

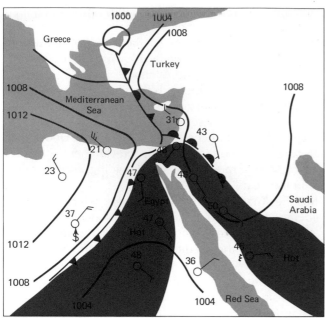

**FIGURE 20.13**

Khamsin (or Sharav) of May 21, 1970 at 1200 GMT (2 P.M. local time). Note all the blowing sand and sandstorms. The cold front in Egypt is bringing the oppressive heat to an end. All areas shaded in red are 45°C or more! Where do you see evidence of sea breezes?

nothing. When the scirocco is specially hot, its scorching breath does great injury to the vegetation; the leaves of the trees curl up and fall off in a few days, and if it sets in when the olive trees and vines are in blossom a whole year's harvest may be lost.

From W. G. Kendrew, *Climates of the Continents*

Most Khamsins end with the arrival of cold fronts. Indeed, this is precisely what is happening in the Khamsin of Figure 20.13, and you can see that the western part of Egypt is already cooling off. But the relief brought by the cold front can be a mixed blessing, since the northwest winds behind the cold front are often strong enough to produce violent sandstorms.

In fact, the following weather sequence taken from the bible is realistic. After the plague of hail had moistened the soil of Egypt and the surrounding deserts, locusts bred.

12. And the Lord said unto Moses, Stretch out thine hand over the land of Egypt for the locusts that they may come up upon the land of Egypt, and eat every herb of the land, *even* all that the hail hath left.

13. And Moses stretched forth his rod over the land of Egypt, and the Lord brought an east wind upon the land all that day, and all *that* night; *and* when it was morning, the east wind brought the locusts . . . .

16. Then Pharaoh called for Moses and Aaron in haste; and he said I have sinned against the Lord your God, and against you.

17. Now therefore forgive, I pray thee, my sin only this once, and entreat the Lord your God, that he may take away from me this death only.

18. And he went out from Pharaoh and entreated the Lord.

19. And the Lord turned a mighty strong west wind which took away the locusts and cast them into the Red Sea . . . .

But the relief brought by the west wind was only temporary, for another plague on Egypt immediately followed this one—the plague of darkness.

22. And Moses stretched forth his hand toward heaven; and there was a thick darkness in all the land of Egypt three days.

23. They saw not one another, neither rose any from his place for three days . . . .

*King James Bible* Exodus, Chapter 10

This was the darkness of a sandstorm produced by the passage of a cold front with its mighty west wind.

The approach of a sandstorm is ominous, as Huck Finn just informed us. In many cases it looks as if a solid, black wall several kilometers high (as in Figure 10.14) were bearing down, ready to engulf and submerge everything in its path. This appearance is often produced at the leading edge of dry, cold fronts where the dust is forced upward over the advancing wedge of cold air. Once it passes over an area it can make the air choking but, of course, it is *not* a solid wall. Even so, the dust tends to permeate everything, and in the dust storms of the 1930s on the Great Plains it was said that the dust entered the houses and even the refrigerators (or ice boxes), despite the fact that all detectable cracks and chinks around doors and windows had been stopped up with oil-soaked rags.

**Dust Devils, Waterspouts, and Tornadoes**

Another common feature of the desert is the dust devil (as you can see on the map of the Khamsin of May 21, 1970; Figure 20.13). **Dust devils** are vortices or whirlwinds in the atmosphere that form when the heat of the ground becomes excessive. Dust devils are similar in appearance to tornadoes and waterspouts, but are generally smaller (see Figure 20.14a and b). Aside from the size, there are a few other important differences.

Dust devils originate under mostly clear skies and grow from the ground up, while tornadoes and waterspouts usually grow downward from the bases of large thunder-

FIGURE 20.14a
A dust devil. Note the hollow central tube.

FIGURE 20.14b
A waterspout with a dry center near the Florida Keys.

storms (occasionally waterspouts grow out of medium-sized cumulus clouds). Most tornadoes and waterspouts rotate cyclonically, whereas only slightly more than half the dust devils rotate cyclonically.

Recently, meteorologists have courageously decided to investigate these phenomena first hand—that is, they now cross right through dust devils and waterspouts with instrumented planes and trucks. They even pursue tornadoes, but not too closely at present because of their great intensity. From these investigations we are beginning to learn about the detailed structure of these whirlwinds.

The dust devil begins when the lapse rate near the ground is greater than the dry adiabatic lapse rate (i.e., when the air is unstable). They typically form shortly after noon on a nearly clear day with light winds. The dust devils often form over favored spots, perhaps just as whirling eddies form near some corner of buildings whenever the wind is blowing. The dust devils are then carried along by the general winds and tend to tilt, since the winds above ground level are somewhat stronger.

Once the air begins to rise, other hot air near the ground is drawn in from the sides. The ice-skater effect then comes into play, and the converging air, which may not have had any observable spinning at first, soon is spinning anywhere from 5 to 35 m/s. The air rises rapidly (roughly 10 m/s) as it spirals inward, dragging sand, dust, and, on occasion, various loose objects hundreds of meters into the air. The sand or dust renders the dust devil visible, and this means that dust devils (sans dust) may well go undetected where there is an adequate cover of plants.

Dust devils are typically 10 m in diameter, 100 m high, and last from 1 to 5 minutes. Sometimes, they can be much larger and longer lasting. They can exceed 50 m in diameter and *visibly* reach up to 1 km, but glider pilots have felt their *invisible* updrafts more than 5 km above the ground. Freak dust devils have lasted for several hours, and there is one story of a dust devil that hovered over a large pile of sand and then proceeded to remain there for 4 hours, removing a total of 4 cubic meters of sand. Finally, someone drove a truck into the dust devil and somehow this dispersed the devil.

Most mature dust devils (as well as tornadoes and waterspouts) have a central eye, much as a hurricane does! In this eye the air descends, rotates far more slowly, and is almost free of dust. Thus, the behavior of the dust devil eye is similar to the hurricane eye, and the explanation is also similar.

The rapid change of rotational speed in the air as you cross into the eye produces a ring in which small eddies may be superimposed on the general whirling. These smaller eddies sometimes add to the rotational speed, much as the Mad Hatter-type rides at the amusement parks do. In tornadoes these extra little vortices are called *suction spots* and can add to the damage (see Figure 20.15).

Fortunately, dust devils are not very often dangerous. I remember one hot, almost breathless day on the beach. Suddenly the calm was broken by the whirl of a dust devil, and beach chairs and umbrellas were lifted 10 to 20 m in the air. Frenzied mothers ran after their children like Olympic sprinters, but within a minute the whole thing had blown over (much to my disappointment). Actually, the mothers had little cause for alarm—dust devils have

been known to pick up kangaroo rats, but there is no known case in which a dust devil swept up a child.

## SOUND WAVES

When the giant meteorite struck Siberia on July 30, 1908, it knocked down the trees for 50 km around, and its tremendous impact was heard more than 1000 km away. It also produced waves that were too low in frequency to be audible, but these waves were recorded because of pressure disturbance they created on barographs as far as England. When Krakatoa exploded in 1883, the sound was heard 5000 km away in Australia, and the low frequency waves it produced traveled around the world several times.

Sounds coming from such huge explosions often have the same peculiar property that radio waves exhibit. The sound is heard in a zone directly around the explosion and again at much greater distances, but there is an intermediate zone of silence. The distant sound is refracted (commonly said to be reflected) from great heights in the atmosphere.

Sound waves (like light waves) are always refracted away from the regions in which they travel fastest. Since sound travels roughly with the average speed of the air molecules, its speed is therefore proportional to the square root of the air temperature in windless air. Thus, the speed of sound (without wind), in meters per second, can be expressed

$$\text{Speed of sound} = v_s = 20\sqrt{T}$$

Therefore, sound travels faster in warmer air.

---

Example 20.1
What is the speed of sound when the temperature is 27°C?

Procedure: As with most scientific laws the absolute (Kelvin) temperature must be used. Therefore,
$$T = 27 + 273 = 300°$$

Equation:
$$v_s = 20\sqrt{T}$$
$$= 20\sqrt{300}$$

Solution
347 m/s

---

FIGURE 20.15
Horizontal slice through a typical whirlwind showing that small vortices or suction spots are produced within dust devils, waterspouts, and tornadoes because of the large wind shear. These can add to the destruction. The length of the arrows are proportional to the wind speed.

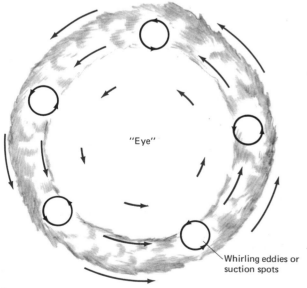

"Eye"

Whirling eddies or suction spots

From this fact we can usually tell the distance of a lightning bolt. The flash itself travels with the speed of light and reaches your eyes almost instantaneously, but the thunder travels with the speed of sound and (from the example above) takes about 3 seconds to travel a kilometer. Thus, for example, if you see lightning and then have to wait 15 seconds to hear the thunder, you know that the lightning flash was 5 km away.

There are also many times you have probably seen lightning without hearing any thunder. This is due to refraction (recall the discussion in Chapter 2). The light from the flash may get refracted slightly by the density and temperature variations of the atmosphere, but the sound from thunder is *strongly* refracted.

Since sound waves travel faster in warm air, they curve away from the warm air. During thunderstorms warm air is almost always near the ground, so that the sound from thunder is refracted or bent away from the ground. For this reason, thunder cannot be heard from clouds more than about 20 km away.

Conversely, sound can be heard over astounding distances in the center of polar highs. Then the wind is almost calm, so there is no turbulence, and the air near the ground is very cold. Thus, any sound moving upward is refracted back to the ground by the warmer air above. This means that the sound waves do not spread out as rapidly in all directions of space, but are channeled along the ground (see Figure 20.16). Under these conditions human voices can sometimes be heard distinctly over 5 km away.

*Sound waves are refracted back to the ground whenever the speed of sound is greater at some height than it is at ground level.* Thus, sound will be refracted back to the ground whenever the temperature of the upper atmosphere is high enough. But the speed of sound is also affected by the wind. Therefore, if there is a wind blowing toward the observer, this adds to the speed of sound in that direction and makes sound refraction back to the ground more likely!

During the summer whenever there is a storm off the coast of Newfoundland, low fequency sound waves known as **microbaroms,** are recorded continuously at New York. These waves are refracted from the top of the stratosphere. Normally the stratosphere is not warm enough to allow the waves to be refracted, but during the summer there are strong easterly waves that blow from the storms near Newfoundland toward New York where they are recorded. The wind, therefore, adds to the sound speed and allows refraction to occur.

During winter storms, microbaroms are generally recorded only twice a day around 11:00 A.M. and 11:00 P.M. (see Figure 20.17). At this time of year the winds at the stratopause blow from the west so that no sound is refracted from that level. The sound that is refracted therefore comes from the thermosphere, but it only comes twice a day because the winds up there are affected by the atmospheric tides. Twice a day (around 11:00 A.M. and 11:00 P.M.) the thermospheric winds blow from the east, allowing both refraction and microbaroms, but around 5:00 A.M. and 5:00 P.M. they blow from the west, *preventing* both refraction and microbaroms; this is depicted in Figure 20.18.

# GRAVITY WAVES

Some properties of gravity waves have already been described in Chapter 2. Cloud billows and mountain wave clouds are two examples of gravity waves that have

FIGURE 20.16
Refraction of sound in (a) a thunderstorm and (b) a cP air mass.

(a) Sound refracted up from ground.      (b) Sound refracted to ground.

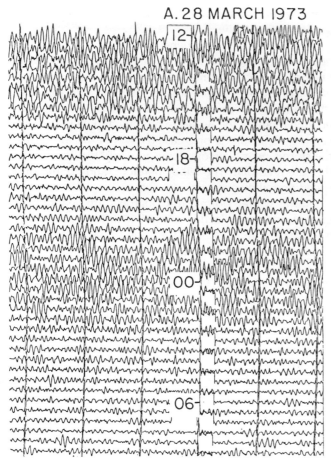

A. 28 MARCH 1973

FIGURE 20.17
Microbaroms at Lamont, showing two peaks and two lulls in 24 hours. This is a reasonably typical pattern during the winter.

FIGURE 20.19
An arcus cloud from a dissipated thunderstorm actually produced another thunderstorm when it passed under a large cumulus cloud.

been rendered visible by condensation. However, most gravity waves are much longer than the waves you see in the cloud billows. Typical wavelengths range from a single kilometer to several hundred kilometers.

I recently saw a time-lapse movie made from a sequence of satellite pictures. The movie showed gravity waves emanating from an intensifying low pressure area over the South Pacific Ocean near New Zealand. The wavelength of these waves was several hundred kilometers. The waves were rendered visible because their vertical motion affected the clouds they passed through. The waves then traveled slowly across the entire Pacific Ocean, taking several days to reach the ITCZ near Colombia, South America. As each wave crest passed through the ITCZ, giant thunderstorms sprouted up from clouds that had previously been rather inactive. Without the satellite pictures there would have been absolutely no way to predict when these thunderstorms would develop, but

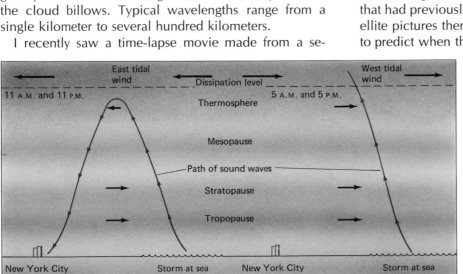

FIGURE 20.18
The tidal winds in the thermosphere cause the semidiurnal (twice daily) variation in the microbaroms during the winter. At 5 A.M. and 5 P.M., microbaroms encounter the dissipation level (above about 115 km) before they can be refracted. At 11 A.M. and 11 P.M., easterly tidal winds help waves refract before they reach the dissipation level.

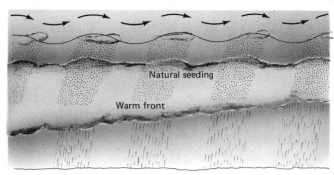

Rainbands

FIGURE 20.20
Gravity waves producing small cirrus clouds, which then seed the clouds below naturally, and lead to the formation of rainbands. (The dotted areas represent the ice crystals.)

with them you could have pinpointed the culprit several days in advance!

Similar gravity waves occasionally travel out ahead of cold fronts and may trigger squall-line thunderstorms when the large-scale atmospheric conditions are right.

Other waves or pulses form when dissipating thunderstorms collapse. The spreading downdraft is often marked by an arcus cloud, and new thunderstorms will often form when two arcus clouds collide or when one passes under an ordinary cumulus cloud, as in Figure 20.19. This process also indisputably shows up on satellite film loops.

Recently, radar studies of extratropical cyclones have revealed that the most intense precipitation often forms in wavelike bands that are embedded in the general area of rains. These bands exhibit many features of gravity waves and, indeed, sensitive barographs show that gravity waves with lengths of 10 to 100 km commonly precede and accompany precipitation from low pressure areas. The waves may increase precipitation by producing cirrus clouds whose ice crystals then seed the clouds below as in Figure 20.20.

# PROBLEMS

**20-1** At coastal cities during spring and summer it is often warmest before noon. Explain why this is true.

**20-2** Explain why cumulus clouds often dissipate when moving inland from the sea at night, but grow when moving inland during the day.

**20-3** Explain why high clouds, such as cirrus, are relatively unaffected by the sea breeze.

**20-4** Sea-breeze fog is blown in from the sea but rarely penetrates more than a few miles. Why?

**20-5** Explain why, in the NH, the sea breeze tends to turn progressively clockwise during the day.

**20-6** In Figure 20.2 there is no sea breeze in New Jersey, but a strong sea breeze in Long Island and Connecticut. Why?

**20-7** I once camped out for several nights in Fundy National Park, New Brunswick, Canada. The campground lies along the Bay of Fundy at the base of a large, elevated plateau. Days were clear and calm, but every night it got cold and very windy around 9:00 P.M. The cold wind continued all night, making it quite miserable. Explain this wind.

**20-8** Why is the weather usually crystal clear during a Bora or Mistral?

**20-9** The chinook may be warm even if there is no condensational heating on the upwind side of mountains. Consider that after air passes over the mountains its oscillation carries it 2 km lower than its original level at the low point. How much of a warning will this produce if the environmental lapse rate is (a) 8°C/km and (b) 2°C/km. From this, generalize and explain how the warming effect of the chinook depends on atmospheric static stability.

**20-10** Explain why the chinook must be a forced phenomenon, and describe the weather conditions most favorable for producing a chinook.

**20-11** Explain why winds a few hundred meters above the surface are weaker during the day and stronger at night.

**20-12** From mid-June to mid-August a huge hot low extends from Saudi Arabia to India. What effect does this have on the winds of the eastern Mediterranean Sea and on the Khamsin?

**20-13** With the answer to the last problem in mind, explain why the highest temperatures ever recorded in Tel Aviv, Israel, are 46.5°C in May, 44.4°C in June, 41.1°C in August, 42.0°C in September, but only 35.6°C in July!

**20-14** Explain why all storms or winds that are rotating extremely rapidly have a clear core.

**20-15** The ice-skater effect has a simple mathematical formulation for dust devils, waterspouts, and tornadoes. The velocity multiplied by the radius ($vr$) is equal to a constant. Compute how fast the wind will be blowing 10 m from the center of the whirlwind if it blows 5 m/s when it is 100 m from the center.

**20-16** What is the speed of sound in air with temperature 0°C?

**20-17** Why is lightning easier to hear when a thunderstorm is approaching than when it has just passed?

**20-18** Thunder is heard 15 seconds after lightning is seen. How far away was the lightning flash?

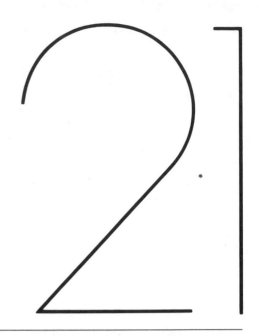

# MODERN WEATHER ANALYSIS AND FORECASTING

While serving his government in the Friends' Ambulance Unit during World War I, a conscientious objector by the name of Lewis F. Richardson (1881–1953) wrote a monumental work called *Weather Prediction by Numerical Process*. In this book, Richardson laid the foundations for numerical weather forecasting long before the computer was invented.

At times the book has a visionary ring to it.

Imagine a large hall like a theatre, except that the circles and galleries go right round through the space usually occupied by the stage. The walls of this chamber are painted to form a map of the globe . . . . A myriad [human] computers are at work upon the weather of the part of the map where each sits but each computer attends to only one equation or part of an equation. The work of each region is coordinated by an official of higher rank. . . .

Four senior clerks in the central pulpit are collecting the future weather as fast as it is being computed . . . .

Messengers carry piles of used computing forms down to a storehouse in the cellar.

In a neighboring building there is a research department, where they invent improvements.

Lewis F. Richardson, *Weather Prediction by Numerical Process*

Richardson estimated that it would take 64,000 human computers to "race the weather" mathematically—in other words, to make a forecast before the weather actually occurred. Those lines were written in 1917, and in a sense they have been fulfilled. Now giant electronic computers are being used instead of humans, and these computers are able to make the staggering, but necessary, number of computations quickly enough to forecast the weather. We now also have a worldwide network of weather stations which gather and relay information, large central computing centers, weather-record storage centers, and meteorological research centers.

However, Richardson's mathematical masterpiece lay idle for over 25 years, partly because its usefulness depended on the existence of the computer. There was a second reason for its neglect by the scientific community. It took Richardson six weeks to make all the necessary computations to forecast the weather for only one point on earth, six hours in advance. His forecast called for a pressure change of 145 mb, which he realized was a "glaring error," but one he was not able to explain. (It was not a careless error.) It took meteorologists several years to analyze Richardson's error. A truly enormous intellectual effort has since gone into producing modern weather forecasts. The material in the first half of this chapter is intended to give you an idea of what goes into modern weather forecasting.

The weather forecasts that result are often good, but are far from perfect. We therefore will try to evaluate weather forecasting and can also consider the basic question of exactly how predictable the weather really is.

# NUMERICAL WEATHER FORECASTING

Numerical weather forecasts are based on the same principle as motion pictures. The movie camera does not photograph continuously. Instead, it takes pictures at small but finite intervals. The shorter the interval between pictures, the more accurate the movie. If the time interval becomes too large or the motion too rapid then the illusion of motion is destroyed, and all you see is a jerky collage of photographs.

So it is with the numerical depiction of the weather. Ideally to make a perfect weather forecast we must know the pressure, temperature, humidity, and winds at every point on the earth. But this is clearly impossible; so we merely record the weather at widely scattered weather stations that are often several hundred kilometers apart. Then for the purpose of the computer, this information is estimated at grid points which are spaced at regular intervals.

The grid-point representation used in the numerical models can be illustrated by showing how it depicts a landscape. The actual landscape shown in Figure 21.1 contains a large mountain, a series of small rolling hills and valleys, and a deep, narrow canyon.

The grid-point view of the same scene obviously introduces some distortions. The mountain is rather well represented, but the rolling hills and valleys have been mysteriously transformed into one low, wide mound; the canyon has completely disappeared.

Such misrepresentations also occur in the movies and under fluorescent lights or "strobe" lights. When the motion is slow enough, it is well represented. However, stagecoach wheels and airplane propellers often seem to spin backwards. This misrepresentation is known as **aliasing** and is simply explained. Imagine the stagecoach wheel has four spokes and in the time between successive pictures the wheel turns forward slightly less than one-fourth of a rotation. Then on film it will *seem* as if the wheel is moving slowly backwards. Actually, the optical illusion is due to the fact that your eye focuses on the nearest spoke in each consecutive frame—you only think you are seeing the same spoke (see Figure 21.2).

The canyon does not show up at all, simply because it is far too narrow and fits between two adjacent grid-points. In the same way, it you take a movie of a lizard catching an insect with its tongue, the movement of the tongue is so rapid that the action may take place entirely between two successive frames. Then, your movie will show the insect mysteriously transferred from the leaf into the mouth of the lizard.

What is the meteorological interpretation of this? Large-scale weather systems such as lows, highs, and Rossby waves are accurately represented by the grid-point view, but small-scale phenomena such as thunderstorms are often completely missed. Intermediate-sized weather systems such as fronts and squall lines, which are comparable in size to the spacing of the grid-points, are subject to the misrepresentation of aliasing, and great efforts have been taken to minimize such aliasing problems (mathematically).

The LFM (*Limited area Fine Mesh*) model has the closest grid-point spacing of any numerical model that is run on a daily basis. But even with the LFM model, thunderstorms cannot be represented at all because the grid points are 130 km apart. One implication of this wide spacing is that computer models have no *direct* way of predicting the small-scale phenomena such as thunderstorms.

Fortunately, *indirect* means of forecasting the small-scale phenomena can often be found! For example, thunderstorms will occur when the atmosphere is sufficiently humid and when the lapse rate is unstable. What the computer does is to predict the areas that will have unstable lapse rates and, thus, *areas* in which thunderstorms are likely. Meteorologists have even found indirect mathematical ways of representing some of the effects that thunderstorms have had on the larger scale weather systems, such as their ability to transport large quantities of heat and water vapor up from the ground and high into the troposphere.

Since most hurricanes are smaller than typical extratropical lows and highs, the grid-point spacing of the LFM model is not quite fine enough to predict hurricane behavior accurately. Thus, when a hurricane threatens to approach land, an extremely fine mesh computer model

FIGURE 21.1
An actual scene and its representation using grid points. The grid-point scene represents the mountain quite well, but it transforms the small rolling hills and valleys into a low, wide mound, completely missing the deep, narrow canyon.

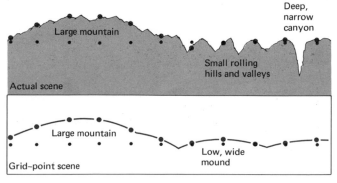

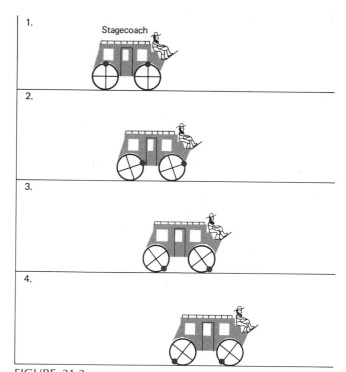

FIGURE 21.2
Aliasing. Notice how the spokes on the wheels seem to be turning slowly backward as the stagecoach moves forward. Actually, the spokes move forward by almost a one-quarter turn between each frame.

with grid points a mere 60 km apart is used in the threatened area. This model has predicted hurricane movement and behavior very well the few times that it has been used to date, and it even can simulate the spiral-band structure of the hurricane.

You might ask why this fine mesh model is not used all the time. After all, it would eliminate many of the misrepresentations of the courser models. The answer is simple — if the very fine mesh model were used on a worldwide basis it would require so many computations that even the fastest computer could not compute the weather as fast as it actually happened. As soon as faster computers become available we *will* use the finer but more demanding models.

What are the mathematical equations that are used in the numerical models? Newton's second law of motion (*F* = *ma*) is the basic law of motion. It is really three equations, since the air motions are three dimensional. Then, there is an equation which states that energy is conserved (the thermodynamic equation), an equation which states that mass is conserved (the continuity equation), and an equation for water vapor. (The ideal gas equation, which has already been used to eliminate density is not directly needed.) These equations must be solved for each grid

point on the entire globe! First, however, all the equations must be specially adopted to express the behavior of the fluid, air. But these equations lie far beyond the scope of this book and therefore are not presented here.

Even without seeing the equations, you can appreciate what an enormous job the computer must perform. *If* the LFM model were used over the entire globe (it is just used over North America), it would have roughly 125,000 grid points at each level. Since it has six levels, it would have a total of 750,000 grid points. Now, six equations must be solved at each grid point, so that there are a total of 4.5 million equations with 4.5 million unknowns that the computer must solve. (Remember the trouble you had in elementary algebra solving two equations and two unknowns.)

But the problem doesn't end here. When these equations are solved they have only predicted the weather 4 minutes ahead. If a larger time step were used, several other types of inaccuracies and distortions, besides those already indicated in Figure 21.1, would be produced. Therefore, in order to predict the weather 24 hours in advance the entire process must be repeated and updated 360 times!

Now you can see why the grid points are not so close together. If we cut the grid-point spacing in half, there would be 4 times as many grid points at each level. Then, a mathematical theorem demands that we also use a time interval that is half as large as before, to avoid misrepresentation due to something known as computational instability.

Thus, cutting the grid-point spacing in half means that there must be 8 times as many computations. This is why the LFM model is not used over the entire globe. It is also why the very fine mesh hurricane model is used only in emergencies over a very limited area. Still, we have come an incredibly long way since Richardson dreamed it all up.

## MODERN SATELLITE METEOROLOGY

A second great revolution in meteorology has been initiated by the satellite. Now, for the first time, we are able to see clouds from both sides. As you have seen on many occasions throughout this book, the satellite gives an overall view of entire weather systems that we never before had seen. Several new weather systems such as tropical cloud clusters (regions several hundred kilometers wide of increased cumulus and cumulonimbus clouds) have been discovered because of satellite pictures.

The satellite provides us with more than just a static, pictorial view of the weather. Photographs are relayed every half hour from the latest series of GOES (*Geostationary Orbiting Environmental Satellites*), and when these pictures are animated in sequence the weather systems "come to life" and move.

The motion of clouds at low and high levels of the troposphere are now used to estimate the winds there. This information is especially useful in data-poor areas such as the oceans and the tropics where there is almost no radiosonde information. In addition, for short-period forecasts (i.e., several hours in advance) the animated sequence is just about the best indicator of the movement of cloud areas.

This pictorial sequence is given day and night. You might ask, "How can we see clouds or, for that matter, anything at night?" The answer is simple. During the day we certainly can use visible light sensors that pick up the light reflected from the earth and clouds. At night, there is infrared "light." You should recall that the earth emits infrared radiation to space. Planck's law of radiation tells us how much radiation is emitted at each wavelength for any particular temperature. The higher the temperature, the more radiation is emitted.

Some wavelengths of radiation are hardly absorbed by the atmosphere. Thus, when they are emitted by the ground they escape to space unless they are blocked by clouds. When it is cloudy, the radiation reaching the satellite at these wavelengths comes from the top of the clouds. Cloud tops usually occur high enough in the troposphere so that they are much colder than the ground (fog and low stratus clouds are exceptions).

This means that we can distinguish cloud areas from ground areas merely from the fact that areas of large radiation will be ground and areas of little radiation will be cloud. Furthermore, we can even estimate the height of the cloud top because the higher the cloud top is, the colder it will be and therefore the less radiation the satellite will receive. This is very helpful because, if intense cumulonimbus clouds are embedded in a ground cloud cover, they can barely be seen in visible light but they stand out dramatically in the infrared. This technique is now so refined that it can even tell the difference between land and water and also outline the warm waters of the Gulf Stream, as you can see in Figures 21.3 and 21.4.

One of the major stumbling blocks in meteorology has always been the lack of information over certain large areas such as the oceans, the polar lands, and the tropics. Some of the more recent satellites are equipped to receive infrared and microwave radiation at a number of different wavelengths. This information is then relayed to earth where it is actually possible to compute the temperature and humidity at all heights in the atmosphere and in the most remote places on earth!

The procedure for determining the temperatures from the satellite radiation data involves another tremendously complicated mathematical process. The basic idea, however, is rather simple and, once again, involves understanding of the laws of radiation. Those wavelengths that are hardly absorbed by the atmosphere are also not emitted by it either. Infrared radiation of such wavelengths either comes from the ground or, if it is cloudy, the cloud tops. Some microwaves even penetrate clouds so that we can still determine the ground temperature from microwave radiation when it is cloudy.

Other wavelengths are readily absorbed and then reemitted by the atmosphere, so that the radiation received by the satellite at these wavelengths can only come from higher regions of the atmosphere. Therefore, radiation at these wavelengths provides temperature information of the higher atmosphere. In similar fashion, wavelengths that are moderately absorbed and reemitted by the atmosphere yield temperature information about intermediate heights in the atmosphere. (See Figure 21.5.)

FIGURE 21.3
Infrared satellite picture for April 13, 1977, showing contrast between the cold coastal waters and the warm land and Gulf Stream waters. Warm surfaces appear dark. These charts are now used as navigational aids in yachting races!

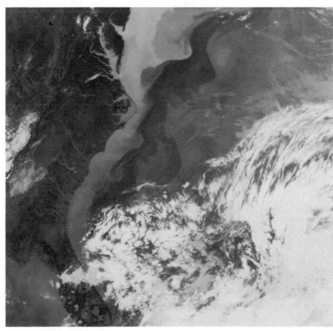

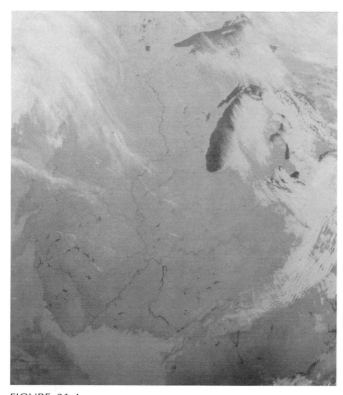

**FIGURE 21.4**
Infrared satellite picture for November 5, 1973, showing contrast between the cold land at night and the warmer water of lakes and rivers. Notice also the clouds that formed over Lake Superior and Lake Michigan. These brought snow squalls south and east of the lakes.

**FIGURE 21.5**
How temperatures are determined from satellites. Poorly absorbed infrared wavelengths come from the ground or near ground level, while highly absorbed wavelengths come from higher levels in the atmosphere. When it is cloudy, infrared radiation comes from the cloud tops, but some microwaves can penetrate from below the clouds.

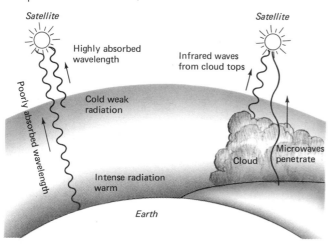

When all the mathematical work is finished, the satellite has provided us with a temperature sounding that is accurate to within 2°C at most heights in the troposphere and lower stratosphere. Because the radiation for each wavelength comes from a layer of finite thickness, the soundings lose some accuracy in those regions where there are sharp temperature changes, such as at frontal inversions and at the tropopause. All in all, the satellite is not as accurate as the radiosonde, but it is available all over the world.

In places where there are no clouds to track by satellite and no radiosonde wind measurements, the temperatures measured by the satellite can be used to infer the winds, since the thermal wind is a good approximation (see Chapter 15) everywhere except at the equator.

## PRODUCTION OF WEATHER FORECASTS

An incredible effort, only briefly outlined here, is a necessary part of every weather forecast. The first step of the job is to obtain the necessary meteorological data. Every 12 hours, at midnight and noon Greenwich time, radiosondes are launched at over 1000 weather stations in the NH, and this information, along with much more surface weather information and satellite measurements, are relayed to the central meteorological offices in cities such as Washington, D.C., and Moscow. The collection of this information for the entire hemisphere is completed within 3 hours. (A similar job is performed in the SH.)

The information is then analyzed to produce the various weather maps and to prepare for use in the large computer models of the atmosphere. Analysis is a simple matter where there is an adequate network of weather stations, but it involves a complicated mathematical procedure, known as *objective analysis,* in areas that have little or no data. Objective analysis must also be used to obtain the values of temperature, pressure, and so on, at the grid points. The analysis of data and the subsequent transmission of weather maps to the various weather stations is accomplished within an hour.

Once the data is fed into the numerical prediction models, the computer begins its race with time. A mathematical atmosphere moves over a mathematical earth. A mathematical sun shines down, heating mathematical continents and oceans. Mathematical heat and vapor are then exchanged between the mathematical surface and atmosphere, and the mathematical weather evolves. The computer produces a picture of the weather to come—12,

24, 36, and 48 hours ahead of the initial picture. All these computations are performed within an hour!

Since the numerical predictions are given only at grid points, objective analysis must be used again to obtain forecasts for particular locations. Local weather effects are then taken into account to some degree by using a complicated statistical procedure known as MOS (*Model Output Statistics*).

When all of this work has been done, the forecasts are then transmitted in two forms from the central meteorological offices. Forecast weather maps are sent out and specific printed forecasts are issued for each weather station. This forecast package is called *Guidance,* and the individual weather stations receive their Guidance within 6 hours of the time the data was first being collected. The entire process involves thousands of people around the world.

In the final step of the process there *is* room for human judgment. Despite all the effort that has gone into the production of the numerical forecasts, they are far from perfect. An experienced forecaster will be aware of the conditions under which the numerical predictions do not perform well; there may be local peculiarities in the weather, sudden and rapid developments in the weather that took place since the original data was collected, or weather data that fit unnoticed between the grid points. For instance, for thunderstorm forecasts the computer can only issue general warnings, but the local meteorologist can supplement this by using the latest updated weather observations, satellite pictures, or weather radar to give specific short-range forecasts. But this subjective improvement in the weather forecasts is found only with the 12- and 24-hour forecasts; the 36- and 48-hour computer forecasts are hard to beat on the average.

# EVALUATION OF THE FORECASTS

Everyone knows that weather forecasts are not perfect. But how accurate are they? This is not a question that can be easily answered. Let me give you an example.

If you forecast no rain for Yuma, Arizona, for every day of the year, you would be over 95% accurate. However, your forecasts would be worthless because everyone living there knows it hardly ever rains. What the people living in Yuma *do* want to know is when it *will* rain. It should then be obvious that it is much easier to predict the weather in Yuma than in most other places. In most places there is no doubt that the weather forecasts will be less "accurate," but the forecasters may be quite skillful.

One good way of assessing the accuracy of weather forecasts is to keep a record of the errors. Thus, if the temperature was forecast to be 27°C but it turned out to be only 24°C, then your error was −3°C. The errors are squared to avoid having negative and positive errors cancel. Then take an average to produce what is known as a **mean square error** (see Table 21.1).

Forecasts have been evaluated on the basis of the mean square error of temperature, pressure, *probability* of precipitation, and so forth. The forecast errors are now much smaller than they were in the early 1950s (i.e., before the numerical predictions), and there has been a significant reduction of error with every improvement in the numerical models.

Meteorologists can now predict temperature 48 hours in advance as accurately as the meteorologists in 1955 could only 24 hours in advance. Unfortunately, there has not been nearly as much improvement in the precipitation forecasts, and the fact they are often given in terms of probabilities (e.g., 10% chance of rain) causes confusion and scorn.

However, it is essential to give probabilities when predicting precipitation. There are many situations in which meteorologists are not 100% confident in their forecasts. For instance, places which lie at the edge of a region that is forecast to have rain may get nothing if the storm's path changes ever so slightly. Furthermore, it is often possible to predict a *general area* of showers and thundershowers more than 24 hours in advance, but showers are spotty and some places may escape without a drop of rain while others get drenched.

The value of probability forecasts becomes apparent when economic decisions have to be made. For instance, if the forecast calls for a 70% chance of thunderstorms with high winds, a painter will probably choose not to paint anyone's house, but the ice cream vendor will most likely still show up.

TABLE 21.1
Mean Square Error

| Forecast Number | Actual Temperature | Forecast Temperature | Error | Error Squared |
|---|---|---|---|---|
| 1 | 24 | 27 | −3 | 9 |
| 2 | 28 | 26 | +2 | 4 |
| 3 | 18 | 15 | +3 | 9 |
| 4 | 19 | 14 | +5 | 25 |
| 5 | 20 | 19 | +1 | 1 |

Total = 48

Mean square error = 9.6
Root mean square error = 3.1

 MODERN WEATHER ANALYSIS AND FORECASTING

The numerical forecasts have been analyzed in great detail, and it is found that errors can be broken down into three basic categories.

1. Errors due to faulty measurement and missing information.
2. Errors due to incomplete knowledge of physical processes such as evaporation and turbulence.
3. Errors due to imperfections in computer (finite difference) mathematics and the large size of the grid-point spacing.

One example of how grid-point spacing affects the accuracy of the numerical forecasts is shown in Figure 21.6. The computer was run several times, each time with a different spacing of the grid points. The different predictions were then compared to the actual outcome of the precipitation 36 to 48 hours ahead. You can see that the model with the smallest grid-point spacing is the most accurate, but even that one it not too good. (This happened to be a rapidly developing, difficult forecast.)

# THE ULTIMATE PREDICTABILITY OF THE ATMOSPHERE

Publications such as the *Farmer's Almanac* list the weather for every day of the coming year in full detail. Their forecast for Monday, February 6, 1978 for the greater Boston area called for sunny and unseasonably warm weather. Some of you may perchance recall what the *actual* weather was that day. Boston suffered through the worst snowstorm in its history, or at least the worst one since the snows of February 27 to March 7, 1717. The total snowfall in Boston measured 70 cm, and some of the surrounding areas received more. Of course, drifts were far, far higher. The storm also produced hurricane-force wind gusts all along the New England coast, with record high tides (4.5 m above mean sea level at Portland, Maine). Nothing moved for almost a week afterward.

This storm represented something of a triumph for the National Weather Service. The storm was a **coastal storm** (see Figure 17.15) and it developed out over the Atlantic. Normally, such storms are extremely difficult to forecast because of a lack of information over the ocean, but the National Weather Service did exceptionally well. They anticipated the storm's rapid development and also predicted correctly that the storm would slow down and stall just south of Boston. All in all, they anticipated the major snowfalls to a good degree and gave remarkably accurate warnings up to 48 hours in advance.

This success is to be sharply contrasted with the *long-range "forecasts."* When these (i.e., the *Farmer's Almanac*) are evaluated, they truly turn out to be *absolutely worthless.* They are issued simply because people (including myself) still buy them. An interesting historical note is that Benjamin Franklin himself published such long-range forecasts in his *Poor Richard's Almanac* from the year 1732, despite the fact that only in the year 1743 did he learn where storms come from (recall the argument in Chapter 1).

In general, the large computer models provide very accurate forecasts at 24 hours and still show considerable skill at 48 hours. However, more in advance than this, the accuracy of the computer forecasts rapidly disintegrates, so that any forecast made more than about two weeks in advance is not likely to be worth much more than a knowledge of the average weather (i.e., climate) conditions for that time of year.

A truly staggering effort has gone into improving our weather forecasts, but somehow the atmosphere yields its

FIGURE 21.6

Predictions of accumulation of precipitation (in millimeters) during the last 12-hour period of a 48-hour forecast. Notice how forecasts become more accurate as grid-point spacing is decreased.

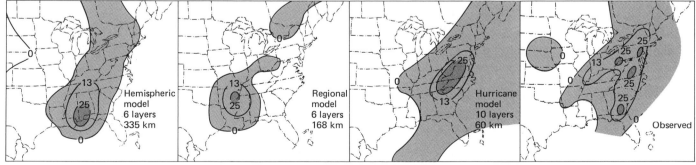

secrets only with great reluctance. The improvement in weather forecasts over the last decade has been disproportionately small compared to the the vast improvements in observations, communications, computer power, mathematical techniques, and understanding of the physics of the atmosphere. Why?

Phenomena such as the tides can be predicted years in advance except for those variations caused by the weather. Why then, can't we predict the weather more than a week or two in advance? The answer lies in the basic difference between the nature of the tides and the nature of the weather.

The tides are largely periodic; if the weather happens to disturb the tide, that disturbance will quickly die down as soon as the weather pattern which caused it has moved away. By contrast, the weather is largely not periodic. To be sure, there is an annual cycle and a daily cycle in the temperature that can be predicted fairly well centuries in advance—*on the average*. But the effects of weather disturbances such as thunderstorms and highs and lows do not die out and are not periodic. The job of the meteorologist is to predict precisely what these disturbances will do and how they will evolve; no one would hire a meteorologist to forecast summer.

The weather disturbances are often produced by instability processes. Thus, thunderstorms form when the lapse rate becomes unstable, and extratropical lows develop as a result of baroclinic instability. Monstrous storms often grow from almost insignificant initial disturbances that have become unstable. A slight difference in the shape or location of that tiny, initial disturbance can therefore potentially lead to a completely different weather pattern within a few days.

What this means is that if we make a mistake in measuring the wind or the temperature, and the like, or if we leave out some *seemingly* unimportant weather feature because it entirely fits between two adjacent grid points, or even if we round off the computations, then the mistake will eventually grow so large that our prediction will be worthless. Experience and theory indicate that small errors tend to double within two or three days. These small errors are inevitable and that means that within about two weeks, large forecast errors are also inevitable. It appears that the details of the day-to-day weather simply *cannot* (even in theory) be predicted more than about two weeks in advance.

One of the important problems facing meteorologists is whether we will be able to predict any changes in the *climate*. In view of the fact that we have virtually no skill in predicting the weather two weeks in advance, this might seem like a ridiculous question. Yet, it is possible that we may be able to forecast changes in the average weather or climate even though we cannot predict the specific daily changes themselves.

In 1956, Norman Phillips published the first numerical study of the general circulation of the atmosphere. His mathematical problem started with an atmosphere at rest. In a few days, temperature differences began to grow appreciable because the model had more sunlight at the equator than at the poles. These temperature differences produced winds. In a few weeks, the model had developed highs and lows with fronts, and there was actually a jet stream.

This early numerical model was quite simple compared to the enormously complicated **General Circulation Models** (GCMs) that are now used to perform tests on and simulate features of the world's climates. For instance, the GCMs have showed with some success how the weather patterns at the height of the last ice age differed from those of today.

The GCMs have recently begun to include such features as albedo-feedback mechanisms, the interaction of the oceans and the ground below with the atmosphere, and even changing the solar intensity. We hope that within a few years these GCMs will provide us with many unanswered questions about the nature of climate changes on earth. As you will see in Chapter 24, many of the theories of climate change are little more than conjectures, and it will prove interesting to see how they stand up "under fire."

# PROBLEMS

**21-1** You can simulate the effect of aliasing by spinning a gyroscope under a fluorescent light. You can also draw consecutive frames, as in Figure 21.2, on consecutive pieces of paper in a pad or in the margins of a book and then flip pages to see a similar effect. Explain aliasing.

**21-2** Why are computer models of weather useless in describing the details of thunderstorms?

**21-3** For the following conditions state which type of satellite picture is more useful: visible or infrared. (a) At night; (b) for fog and low clouds; (c) for thunderstorms embedded in a region of cloudiness; (d) for snow-covered ground; (e) for locating ocean currents; (f) for determining the height of cloud tops.

**21-4** Explain how infrared and microwave radiation as measured by satellites can be used to determine the vertical profile of temperature in the atmosphere.

**21-5** Forecaster A misses the temperature by an average of 4°C, but if he had merely used the climate tables he would have missed the temperature by an average of 6°C. In a different city forecaster B misses the temperature by an average of 1°C, but if he had merely used the climate tables he still would have missed by an average of only 1°C. Who do you think is a better forecaster? (*Note:* All numbers represent root mean square errors.)

**21-6** What is the advantage of issuing probability forecasts?

**21-7** What is the value of all long-range, *detailed* forecasts?

**21-8** Why is it fundamentally impossible to forecast the details of the weather more than a few weeks in advance?

# CLIMATES AROUND THE WORLD

When you travel around the world you find such incredibly different climates that at times it is hard to believe they all occur on the same planet. Each climate determines in large measure the plant life, soil, and even landscape of the area. Furthermore, whether humans have free will or not, the climate determines much about the way we live and perhaps even the way we feel and think.

People adapt rather slowly to changes of climate. A colleague of mine spent almost one year doing research on the ice in the Arctic Ocean and when he returned to New York City at the end of November he had become so *acclimatized* to the polar conditions that he was able to walk around comfortably in short sleeves while everyone else felt cold in overcoats.

When the Spaniards discovered Tierra del Fuego, they were amazed to find that the Indians walked around virtually naked at 0°C.

They were naked except for an evil-smelling sealskin between their legs, and their bodies glistened with a sheen of fish oil which protected them so effectively from the cold that when one of the women suckled her child she did not bother to brush the snow from her naked breast.

Antonio Pigafetta, "Magellan's Voyage, a Narrative Account of the First Circumnavigation" quoted in Ian Cameron, "Magellan."

On the few nights each year when the temperature drops below 18°C in equatorial Africa the native people build fires to keep warm, so limited is their experience with cool weather.

Some people can never get used to changes of climate. Consider the case of Sam McGee.

*There are strange things done in the midnight sun*
  *By the men who moil for gold;*
*The Arctic trails have their secret tales*
  *That would make your blood run cold;*
*The Northern Lights have seen queer sights,*
  *But the queerest they ever did see*
*Was that night on the marge of Lake Lebarge*
  *I cremated Sam McGee.*

*Now Sam McGee was from Tennessee,*
  *Where the cotton blooms and blows.*
*Why he left his home in the South to roam*
  *'Round the Pole, God only knows.*
*He was always cold, but the land of gold*
  *Seemed to hold him like a spell;*
*Though he'd often say in his homely way*
  *That he'd sooner live in hell.*

When Sam realized that he was dying from the cold, he made the narrator swear to cremate him when he was gone.

*Well he seemed so low that I couldn't say no*
  *Then he says with a sort of moan:*

*"It's the cursed cold and it's got right hold
'Till I'm chilled clean through to the bone
Yet taint being dead, it's my awful dread
of the icy grave that pains;
So I want you to swear that foul or fair,
You'll cremate my last remains."*

When Sam died the narrator lived up to his promise

*Some Planks I tore from the cabin floor,
And I lit the boiler fire;
Some coal I found that was lying around,
And I heaped the fuel higher;
The flames just soared, and the furnace roared
Such a blaze you seldom see;
And I burrowed a hole in the glowing coal,
And I stuffed in Sam McGee.*

After a few hours the narrator looked in to see how Sam was doing.

*And there sat Sam, looking cool and calm
In the Heart of the furnace roar.
And he wore a smile you could see a mile,
And he said, "Please close that door.
It's fine in here but I greatly fear
You'll let in the cold and storm
Since Ah left Plum Tree, down in Tennessee
It's the first time Ah've been warm!"*

*Robert W. Service, "The Cremation of Sam McGee"*

The chapter begins by defining climate and presenting a climate classification scheme. Since the scheme contains more than a dozen climate types, you *might* expect that it would be impossible to memorize. However, by keeping in mind a knowledge of the global winds, you will see that there is a somewhat simple, ideal pattern to the world's climates that is reasonably easy to memorize and very easy to apply to the real world. The bulk of this chapter is then spent describing the world's climates. World vegetation patterns are also discussed since they are so closely related to the climate types.

# CLIMATE CLASSIFICATION AND CLIMATES OF AN IDEAL CONTINENT

The climate of a region is its average weather. The two most important variables of climate are temperature and precipitation, while others such as cloudiness, fog, humidity, and wind are also important.

Seasonal variations must also be included in any account of climate. If we merely used annual averages to define the climate we might think that San Francisco, California, and Peking, China, have very similar climates. San Francisco has an annual temperature and precipitation of 13°C and 51 cm, respectively, while Peking has 12°C and 61 cm, respectively.

However, there are a few "minor" differences between these climates. In July, when the average temperature in San Francisco is 15°C, Peking swelters at 27°C. Furthermore, July is the middle of the long dry season in San Francisco, but it is the heart of the rainy season in Peking. In January, it is bone chilling(−4°C) and bone dry in Peking, whereas San Francisco is pleasantly cool (10°C) and in the middle of its rainy season. Finally, San Francisco is known for its frequent fog, while the dry air in Peking's winter months is filled with dust (loess) from the Gobi Desert.

In many places the weather varies considerably from year to year so that it is necessary to choose some standard averaging period. This is usually chosen to be 30 years, because it is a long enough period to permit the year-to-year variations to average out and short enough so that distinct climatic trends can be spotted by comparing different 30-year periods.

Wladimir Köppen, the great climatologist (and father-in-law of Alfred Wegener), devised the first modern climate classification scheme in 1900. In it he documented the amazingly close connection between climate and plant life. In fact, his scheme is based largely on the world patterns of plant life, and many of the climate regions are named after the dominant vegetation of the region.

Basically, there are five types of botanical regions—forests, grassland, desert, tundra, and ice cap.

Ice-cap climates are found wherever the average temperature of the warmest month is below 0°C. Tundra vegetation dominates when the temperature of the warmest month is between 0 and 10°C. Once the temperature of the warmest months reaches 10°C or more, forests will form provided there is adequate moisture. Usually this means adequate precipitation, but trees also grow in deserts along the riverbanks and in oases.

The amount of precipitation necessary to support forests depends on the temperature of the region, since evaporation rates increase with increasing temperature. Therefore, in the tropics the precipitation must be at least 125 cm annually to support forests, while the great boreal forests of Canada and Siberia can survive on 25 cm or less annually.

When the annual precipitation falls below a certain critical value, grasslands replace the forests. Isolated trees may grow, but these are often stunted. As the precipitation drops still lower, even the grasses cannot survive, and desert vegetation takes over (see Figure 22.1).

There are several other climate–vegetation relationships that are important. Tropical vegetation grows if the monthly average temperature remains at least 18°C. Subtropical vegetation usually grows in regions where at least 8 months of the year have a temperature of at least 10°C. Thus, San Francisco is classified as subtropical under this criterion, even though you need to wear sweaters most days of the year.

In this book I use the climate classification scheme developed by Glenn Trewartha which is a variant of Köppen's scheme. The climate map of the world is shown in Figure 22.2, with a brief legend to define each region. The detailed criteria for Trewartha's climate scheme are given in Appendix 2.

Briefly, A climates are tropical climates with all months at least 18°C, C climates are subtropical with 8 or more months at least 10°C, D climates are temperate with 4 to 7 months at least 10°C, E climates are subpolar or boreal forest with 1 to 3 months at least 10°C, and F climates are polar (tundra or ice cap) with no months reaching 10°C. The B climates are the arid or semiarid climates in which the evaporation exceeds the precipitation, whereas H climates are the highland climates that vary sharply with altitude.

In the discussion that follows I freely use the letter symbols when describing the climates. Therefore, you will frequently have to refer back to this section or to Appendix 2.

Most of the difficulty in memorizing the world's climates is due to the irregularities of the world's geography. Therefore, it is instructive to draw an "ideal continent" (i.e., one without irregularities) and then imagine what its climate patterns would be. Then, every time you looked at the climate of the real world and found that a particular region did not match the ideal picture you could immediately say, "This region has an anomalous climate because of a geographical irregularity." That would, of course, constitute only the beginning of an explanation, but it sounds impressive.

The ideal continent I chose (shown in Figure 22.3) runs from pole to pole and is bounded on the east and west by lines of longitude (and, of course, oceans). Since the continent is symmetrical around the equator, we need treat only the climate of one hemisphere, the other being a mirror image.

Rainy tropics (designated Ar) extend in a band from the equator to 5 or 10° latitude, due to the more or less constant presence of the ITCZ (loosely defined over land). The Ar climates will extend farther poleward on the eastern side of the continent because of the tropical easterlies (which blow upslope on the east coast), while on the western shore the Ar climate will be somewhat restricted.

In this band we find the tropical rain forest or jungle. Here temperatures remain almost constant year round and average about 26 or 27°C. The only significant temperature variations occur between day and night in this land of perpetual summer.

Proceeding poleward, we next encounter the wet and dry tropics (Aw). This climate is so named because it has one dry season and one rainy season each year. The rainy season occurs when the ITCZ wanders north of the equator around July. The rest of the time the ITCZ is too far south to affect the region, and sinking air from the Hadley cell dominates. The farther poleward you move, the shorter the rainy season and the drier the climate becomes. Here there are significant monthly variations in temperature with the coldest month invariably January (in the NH) and the warmest month generally the one just prior to the onset of the rains.

The Aw climate grades to semiarid (BS) and then finally desert (BW) from about 20 to 30° latitude in the center and at the western shore of the continent. The vegetation changes accordingly from forest to savannah-type grassland to desert.

The large subtropical highs centered at about 30° latitude over the oceans play a very important role in the climate of the regions from about 15 to 45° latitude, but they bring completely different weather to the eastern and

FIGURE 22.1
The dominant type of vegetation and how it depends on mean annual temperature and precipitation. This diagram is oversimplified because it should include the effects of seasonal variations on weather.

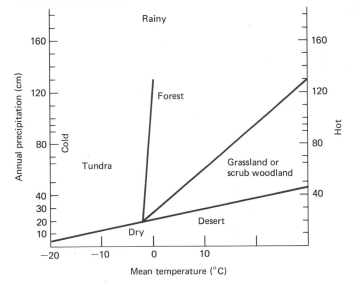

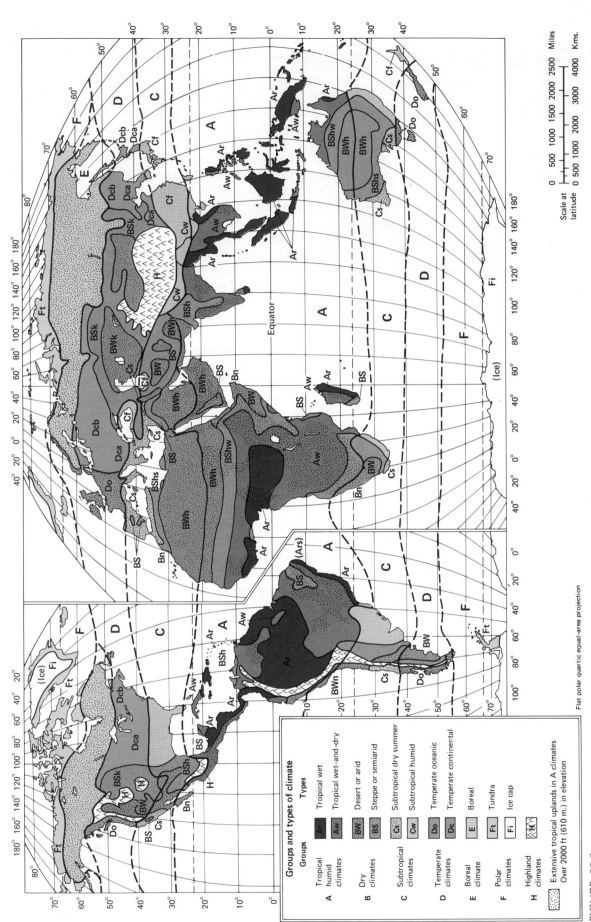

FIGURE 22.2
Climate map of the world.

Groups and types of climate

| Groups | Types |
|---|---|
| **A** Tropical humid climates | Ar Tropical wet |
| | Aw Tropical wet-and-dry |
| **B** Dry climates | BW Desert or arid |
| | BS Steppe or semiarid |
| **C** Subtropical climates | Cs Subtropical dry summer |
| | Cw Subtropical humid |
| **D** Temperate climates | Do Temperate oceanic |
| | Dc Temperate continental |
| **E** Boreal climate | E Boreal |
| **F** Polar climates | Ft Tundra |
| | Fi Ice cap |
| **H** Highland climates | H |

Extensive tropical uplands in A climates
Over 2000 ft (610 m.) in elevation

Scale at latitude

Miles
0   500   1000   1500   2000   2500

Kms.
0   500   1000   2000   3000   4000

Flat polar quartic equal-area projection

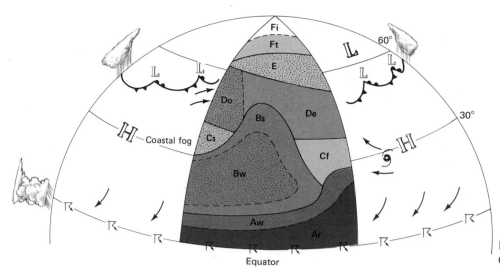

FIGURE 22.3
Climate map for an "ideal" continent.

western coasts of the continents. On the western side of the continent the high brings very dry, stable weather that is hot inland, but cool and foggy along the coast because of upwelling. On the eastern side of the continent the high brings warm but humid weather, with plentiful rain from showers and occasional hurricanes.

During the winter season the subtropical highs weaken and drift toward the equator. The influence of the westerlies is felt increasingly from about 25° latitude poleward. The westerlies bring passing lows and rain to both coasts. On the west coast the temperatures generally remain mild for the given latitude because of the oceanic influence, but on the eastern side of the continent the westerlies help draw in cold cP air from the interior of the continent.

Above about 40° latitude the influence of the subtropical high weakens even during the summer, permitting the west coast to have some summer precipitation and allowing the eastern side of the continent some relief from the heat. During the winter the westerlies dominate almost completely, bringing oceanic air constantly against the west coast and one cold wave after another to the eastern side of the continent.

As a result of these influences the west coast has a dry, summer, subtropical climate (Cs) between about 30 and 40° latitude. The long dry season makes it impossible for forests to survive, except perhaps along the foggy coast or in the mountains. Then from about 40 to 60° latitude the west coast has a temperate oceanic climate (Do) with cool pleasant summers, but relatively mild and *very, very cloudy*, wet winters.

The eastern side of the continent has a humid subtropical climate (Cf) between about 25 and 35° latitude. This has adequate moisture at all seasons, with an un-

pleasantly hot, humid summer. This climate grades into the temperate continental (Dc) climate above about 35° latitude. The Dc climate persists up to 45 or 50° latitude and, while its summers are more tolerable than those of the Cf climate, its winters are dramatically colder, with snow more likely than rain during the winter above about 40° latitude.

If the continent is sufficiently wide, then the center will tend to be semiarid or arid because of the remoteness of a moisture source. This is true especially on the lee or downwind side of any mountain ranges.

From 50 to about 65° latitude, the boreal forests gird the continent. The boreal climate (E), named after this vegetation type, has pleasant summers and frigid winters, and one outstanding feature of this climate is the large annual variation of temperature. Naturally, the boreal forest is located farther poleward on the warm, western coast.

Tundra (Ft) should be found above about 65° latitude and this region will eventually give way to an ice-cap climate (Fi). The ice-cap climate might take over about 75° latitude, but perhaps somewhat farther south along the coasts that have a large moisture supply.

Now that I have presented the climate pattern of an ideal continent only one minor problem remains—the climate pattern of the real world. Actually, the real world should no longer be so mysterious. Compare the climate pattern of the ideal continent with that of the real world and you should see more similarities than differences.

We will now look at the actual climates of the world in greater detail. Most of what you will read in the following pages would probably mean more if you had seen the places first hand and, since most of you will probably never see them all, you may be pleased to know that all you have to do is visit the natural history museums, bo-

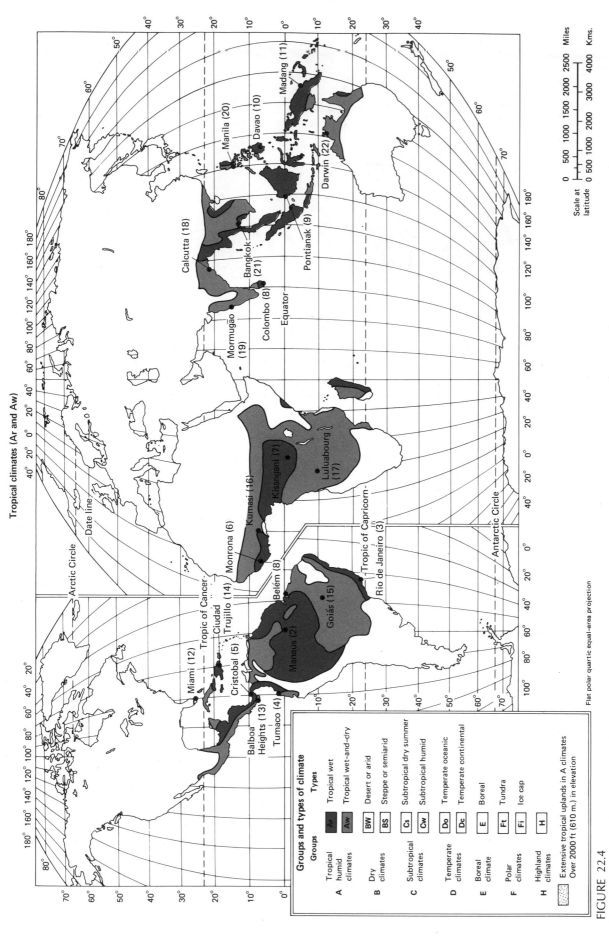

**Tropical climates (Ar and Aw)**

Scale at latitude

| 0 | 500 | 1000 | 1500 | 2000 | 2500 | Miles |

| 0 | 500 | 1000 | 2000 | 3000 | 4000 | Kms. |

Flat polar quartic equal-area projection

**Groups and types of climate**

| Groups | Types |
|---|---|
| **A** Tropical humid climates | Ar Tropical wet |
| | Aw Tropical wet-and-dry |
| **B** Dry climates | BW Desert or arid |
| | BS Steppe or semiarid |
| **C** Subtropical climates | Cs Subtropical dry summer |
| | Cw Subtropical humid |
| **D** Temperate climates | Do Temperate oceanic |
| | Dc Temperate continental |
| **E** Boreal climate | E Boreal |
| **F** Polar climates | Ft Tundra |
| | Fi Ice cap |
| **H** Highland climates | H |

Extensive tropical uplands in A climates
Over 2000 ft (610 m.) in elevation

FIGURE 22.4

Locations of tropical (Ar and Aw) climates. Climate statistics for cities shown are given in Tables 22.1 and 22.2.

tanical gardens, or better zoos to get a good feeling for many of the world's settings. I recommend them to you.

## TROPICAL CLIMATES

In most of the lowlands surrounding the equator, the world's greatest tropical rain forests (commonly called jungles) are found (see Figure 22.4). Here in the rainy tropics (Ar) the rising air of the Hadley cells, known as the ITCZ, is never too far away so that most of the year is rainy and humid. There are perhaps only one or two cases where cold fronts have reached the equator, but by that time the air had been so thoroughly modified that the temperature dropped only a few degrees (see Figure 22.5). The equatorial regions maintain an almost constant level of warmth all year round.

The tropical rain forests are among the least known regions of the world. It is therefore no wonder that weather and climate records for these regions are rather poor and perhaps not entirely representative. Nevertheless, a coherent picture of the climate definitely emerges.

The weather in the rain forest can best be described by one word—monotonous. Day after day it is almost the same. Mornings are very humid but clear, with temperatures around 20 to 23°C. By late morning the sun has heated the earth enough so that cumulus clouds begin to form in the deep layer of humid air. These clouds build up to peaks each afternoon and, by shading the surface, keep the temperature from rising too high. These clouds frequently attain cumulonimbus size and then provide the tropics with its rain. Afternoon temperatures commonly reach a high of 30 to 33°C. The afternoon showers are usually brief, but may be incredibly intense with violent thunder and strong wind gusts. Then, as the sun dips down in the western horizon, the sky starts to clear again, peace and tranquility return, and the temperature starts down to its morning low.

In the rainy tropics the monthly mean temperatures generally vary less than 3°C over the year. The mean temperature hovers around 27°C at sea level, and night is the only cool time. In the midlatitudes people are so used to a cold season each year that they find it hard to believe that there can be places of unrelieved warmth.

One of the amazing things about equatorial weather is that extreme heat waves are unheard of! Almost no place within the rainy tropics has ever recorded a temperature as high as 38°C, but almost every afternoon of the year has a high of at least 30°C. The diurnal temperature *range* for most Ar stations is nearly 10°C, whereas the range be-

FIGURE 22.5
Satellite picture of 1700 GMT, July 17, 1975, showing cold front in South America approaching the equator. The cold air behind the front decimated that year's coffee crop in Brazil.

tween the all-time high and all-time low is less than 20°C for many stations.

Almost all tropical rain comes from showers or thundershowers. In most places, these occur in the afternoon but, in some unusual localities, the bulk of the showers may even occur at night. The rainy seasons are generally those times of the year when the thunderstorms are more intense and more likely. Often, the showers will occur at almost the exact time of day for a number of days. The best forecasts are those that use the persistence method— that is, simply predict for tomorrow what happened today. It is very difficult to forecast the day that the pattern will be broken, but the satellites are finally helping the tropical meteorologist anticipate these changes.

On the ideal continent the equator would get two rainy seasons each year as the ITCZ passed back and forth. Although there are a few places such as equatorial Africa that do have two rainy seasons, many places do not. (Recall that the ITCZ is broad and diffuse over land.)

The Ar climates receive at least 150 cm of rain annually and usually receive over 200 cm. Along some coastlines and mountain slopes, amounts can easily exceed 500 cm. In fact, the rain forest is not able to survive wherever relatively dry conditions (less than 6 cm per month) persist for more than two months in a row. Climate statistics for cities with Ar climates are given in Table 22.1.

TABLE 22.1
The Climate of Selected Ac Cities

| | | Jan | Feb | Mar | Apr | May | Jun | Jul | Aug | Sep | Oct | Nov | Dec | Year | Record High and Low Temperatures |
|---|---|---|---|---|---|---|---|---|---|---|---|---|---|---|---|
| 1.ª Belém, Brazil (1°S) | Tᵇ | 25 | 25 | 25 | 26 | 26 | 26 | 26 | 26 | 26 | 26 | 26 | 26 | 26 | 37 |
| | ΔT | 8 | 8 | 7 | 8 | 8 | 9 | 9 | 9 | 10 | 10 | 11 | 9 | 1.3 | 16 |
| | P | 31.8 | 35.8 | 35.8 | 32.0 | 25.9 | 17.0 | 15.0 | 11.2 | 8.9 | 8.4 | 6.6 | 15.5 | 244 | |
| 2. Manaus, Brazil (3° S) | T | 26 | 26 | 26 | 26 | 26 | 27 | 27 | 28 | 28 | 28 | 28 | 27 | 27 | 38 |
| | ΔT | 7 | 7 | 7 | 7 | 8 | 8 | 8 | 9 | 9 | 9 | 8 | 7 | 1.7 | 17 |
| | P | 24.9 | 23.1 | 26.2 | 22.1 | 17.0 | 8.4 | 5.8 | 3.8 | 4.6 | 10.7 | 14.2 | 20.2 | 181 | |
| 3. Rio de Janeiro, Brazil (23° S) | T | 26 | 26 | 26 | 24 | 23 | 22 | 21 | 21 | 21 | 23 | 24 | 25 | 24 | 39 |
| | ΔT | 6 | 6 | 6 | 6 | 7 | 7 | 7 | 7 | 7 | 6 | 6 | 6 | 5 | 8 |
| | P | 12.4 | 12.2 | 12.9 | 10.6 | 7.9 | 5.3 | 4.1 | 4.3 | 6.6 | 7.9 | 10.4 | 13.7 | 108 | |
| 4. Tumaco, Colombia (2° N) | T | 26 | 26 | 26 | 27 | 27 | 26 | 26 | 26 | 26 | 26 | 26 | 26 | 26 | 32 |
| | ΔT | 4 | 4 | 4 | 4 | 4 | 4 | 4 | 4 | 4 | 4 | 4 | 4 | 1.0 | 18 |
| | P | 42.8 | 29.7 | 24.3 | 37.0 | 44.1 | 30.4 | 19.5 | 18.5 | 18.5 | 15.0 | 12.4 | 17.7 | 310 | |
| 5. Cristobal, Canal Zone (9° N) | T | 27 | 27 | 27 | 28 | 27 | 27 | 27 | 27 | 27 | 27 | 26 | 27 | 27 | 36 |
| | ΔT | 4 | 4 | 4 | 5 | 6 | 6 | 5 | 6 | 6 | 6 | 5 | 5 | 1.1 | 19 |
| | P | 8.6 | 3.8 | 3.8 | 10.4 | 31.7 | 35.2 | 39.5 | 38.7 | 32.2 | 40.0 | 56.5 | 29.7 | 330 | |
| 6. Monrovia, Liberia (6° N) | T | 27 | 27 | 28 | 27 | 26 | 24 | 24 | 25 | 25 | 26 | 26 | 26 | 26 | 36 |
| | ΔT | 10 | 10 | 11 | 10 | 9 | 4 | 4 | 6 | 6 | 8 | 9 | 10 | 4 | 17 |
| | P | 0.5 | 0.3 | 11.2 | 29.7 | 34.1 | 91.8 | 61.6 | 47.3 | 76.0 | 64.1 | 20.9 | 7.4 | 444 | |
| 7. Kisangani, D. R. Congo (0° N) | T | 25 | 25 | 25 | 25 | 25 | 24 | 24 | 24 | 24 | 24 | 24 | 24 | 24 | 36 |
| | ΔT | 11 | 10 | 9 | 10 | 9 | 10 | 9 | 9 | 10 | 10 | 10 | 11 | 1.4 | 16 |
| | P | 8.4 | 10.4 | 17.6 | 14.3 | 15.5 | 8.9 | 11.2 | 22.7 | 19.1 | 24.9 | 17.1 | 7.1 | 177 | |
| 8. Colombo, Sri Lanka (7° N) | T | 26 | 26 | 26 | 27 | 27 | 26 | 26 | 26 | 26 | 26 | 26 | 26 | 26 | 37 |
| | ΔT | 8 | 8 | 8 | 7 | 5 | 4 | 4 | 4 | 5 | 6 | 6 | 8 | 0.8 | 15 |
| | P | 8.9 | 6.9 | 14.7 | 23.1 | 37.1 | 22.3 | 13.5 | 10.9 | 16.0 | 34.8 | 31.3 | 14.7 | 234 | |
| 9. Pontianak, Indonesia (0° N) | T | 27 | 27 | 28 | 28 | 28 | 27 | 28 | 28 | 28 | 28 | 28 | 27 | 28 | 36 |
| | ΔT | 7 | 7 | 8 | 8 | 7 | 8 | 8 | 8 | 8 | 8 | 7 | 7 | 1.0 | 20 |
| | P | 27.4 | 20.8 | 24.2 | 27.7 | 28.2 | 22.1 | 16.5 | 20.3 | 22.8 | 36.6 | 38.9 | 32.3 | 318 | |
| 10. Davao, Philippines (7° N) | T | 26 | 27 | 28 | 28 | 27 | 27 | 27 | 28 | 28 | 27 | 27 | 26 | 27 | 36 |
| | ΔT | 8 | 9 | 10 | 10 | 8 | 8 | 8 | 9 | 9 | 9 | 9 | 8 | 1.5 | 18 |
| | P | 12.2 | 11.4 | 13.2 | 14.7 | 23.4 | 23.0 | 16.5 | 16.5 | 17.0 | 20.1 | 13.5 | 15.5 | 197 | |
| 11. Madang, New Guinea (5° S) | T | 27 | 27 | 28 | 27 | 27 | 27 | 27 | 27 | 27 | 28 | 28 | 27 | 27 | 37 |
| | ΔT | 7 | 7 | 7 | 8 | 7 | 8 | 8 | 9 | 8 | 7 | 7 | 7 | 0.7 | 17 |
| | P | 30.8 | 30.3 | 37.9 | 42.9 | 38.4 | 27.4 | 19.3 | 12.2 | 13.5 | 25.4 | 33.8 | 36.8 | 349 | |

ª In Tables 22.1 to 22.9 numbers next to each city also appear on the world maps that outline the particular climate region.

ᵇ In Tables 22.1 to 22.9: $T$ = average temperature (°C); $\Delta T$ = temperature range (°C); $P$ = average precipitation (cm); * = less than 0.1 cm.

The world's largest tropical rain forest is located in the Amazon and Orinoco River basins in South America. The largest branch of the African rain forest is located in the Congo River basin, but this rain forest is limited by the African highlands to the east and the Sahara to the north, both blocking a plentiful supply of moisture. The third great tropical rain forest is found in southeast Asia and the nearby islands. Generally the tropical rain forests fall within 10° of the equator, but occasionally they can be found as high as 25° latitude.

The rain forest (see Figure 22.6) consists primarily of very tall, evergreen trees whose crown area is typically 30 to 60 m above the ground. Although there are several distinct layers of life below the crown area, most of the action of the jungle takes place well above the ground. In fact, the ground is bare in many places and it is possible to

FIGURE 22.6

The tropical rain forest showing all the lianas hanging from the canopy. Except for openings like this where the sun penetrates the canopy, the forest floor is relatively clear of undergrowth.

move about quite freely. Very little sunlight penetrates to the floor so that the entire forest is pervaded by a dim, dark green hue. Because the ground is often flooded, the trees send out giant buttresses to keep themselves well anchored.

The relatively bare appearance of the forest floor has led the inexperienced to starve in the rain forest, while countless tropical flowers and nourishing fruits grow in the canopy above. It is possible to climb up into the canopy because hundreds of vines, known as **lianas** (often thicker than a man's arm), hang down from the treetops. These lianas and other creeping plants stretch from treetop to treetop, tying the entire rain forest together. The well-known explorer of rain forests, Ivan Sanderson, has walked and crawled distances up to ½ km through the canopy, 40 m above the ground.

This description of the tropical rain forest contrasts sharply with the phony image of the jungle sometimes portrayed in the movies. Often in Hollywood's "jungles," entire safaris have to hack their way through dense underbrush, which in reality only exists at the edge of the rain forest or in savannah woodlands and never in the heart of the jungle. At the edges of the jungle, such as the riverbanks, there *is* an almost impenetrable wall of greenery that extends from the ground to the treetops since this area is exposed to the sunlight. This "jungle wall" has kept both human and beast out.

There is also a zone known as the cloud forest that only exists on *extremely* rainy mountain slopes above the elevation of the true rain forest. Here, trees are far shorter and there *is* a virtually impenetrable undergrowth.

Rain forest animals tend to be small and live in the canopy. Conversely, tropical insects tend to be huge, and beetles over 15 cm long have been captured. Poisonous insects and animals abound, and colorful tropical birds live and play in the canopy. Indeed, one of the main characteristics of the tropical rain forest is the tremendous diversity of life. Thousands of species of plants live close together, and many of the plants take root in the canopy and either live off the trees (**saprophytes**) or merely live on them (**epiphytes**). Many of the plants in the rain forest have leaves so large that a person can hide under them.

In such a climate the pace of life and death is startlingly rapid. As soon as something dies, it rots or is eaten. The constant high humidity and warmth makes paper soggy and many things such as shoes moldy. No wonder the forest floor is mostly bare—the lack of sunlight keeps most plants from growing and the rapid decay process keeps dead leaves and even entire tree trunks from accumulating and littering the forest floor.

The wet and dry tropics (Aw) are found on the poleward side of the rainy tropics. Aw climates have a very warm hot season and often a distinct cool season. There are also distinct wet and dry seasons. The climate of the wet and dry tropics can generally be broken down into three well-defined seasons. First, at the time of the low sun there is a distinct cool season that is almost rainless. Then, as the sun gets higher in the sky the hot and dry season sets in. This is followed by the rainy season that is slightly cooler than the hot season. Then, after the rains end it may grow hot again for a few weeks, but soon the cool season arrives once more.

The monsoon climate of India is perhaps the best known example of Aw climate. To give you an idea of a typical Aw year, we will look more closely at India's climate, which is similar to most other Aw climates but perhaps somewhat more extreme than most.

Begin the year in January. Throughout India the weather is mostly clear with deep blue skies, because the ITCZ is far to the south and India is situated under the sinking air of the Hadley cell. This is the coolest time of the year throughout India, although in the extreme south the average temperature is still 25°C. India does not have any extremely cold weather largely because the Himalaya mountains block out the cold Siberian air to the north. Occasionally, a low pressure area will pass around the south side of the Himalayas and bring some rain to the northern part of India. Many Indians depend on these meager winter rains for a winter wheat harvest that often means the difference between survival and starvation.

By March it has become noticeably warmer, and every Indian knows that the hot season is almost upon them. The hottest time is May in the south of India and July in the northwest. This heat occurs just before the monsoon rains set in and it is notorious. At this time coastal locations may get some relief from the heat because of sea breezes, but even along the seacoast the warm month averages 30°C. Inland, the warmest month averages as much as 36°C.

During the daytime in the hot season the earth feels like a giant furnace and temperatures regularly rise over 40°C. Humans suffer quietly and do as little as possible in the heat of the day. Plants also fare poorly, rapidly withering and turning grayish brown. The earth cracks and hardens.

The one consoling feature is that the air is dry, so that the native Indians can keep their houses somewhat cooler by using grass or bamboo screens, known as tatties, which are sprinkled with water. When the wind blows through the tatties the water evaporates and cools the air. But this method fails during the most unbearable few days immediately preceding the rains when the wind drops down.

Beginning in February the continent of Asia warms as the sun moves north. The ITCZ leaves its winter position over the South Indian Ocean and heads northward toward India. It finally reaches the southern tip of India about the beginning of May and the rains often set in quite suddenly. Once the ITCZ has passed north of a point the wind direction switches to the south, and extremely warm, humid air pours into India (glance back at Figure 16.7).

The ITCZ continues its northward progress, finally reaching the Himalayas in July. The rainy season persists until the ITCZ begins moving back south in September. By the end of October, the ITCZ has finally left southern India.

Almost all the rain comes from showers and thundershowers and is therefore quite intense. When I was a child I had the impression that rain fell continuously during the summer monsoon. This is not true. Since the rain is showery, it tends to be erratic.

There are also some weak low pressure areas known as subtropical lows that are regions of generally rising air and enhanced thunderstorm activity. Much of the monsoon rainfall occurs when these subtropical lows slowly pass by. They bring several days of extremely rainy weather during which continuous rain is sometimes falling. But after they pass by there may be days with mostly clear skies and little if any rain.

The monsoon is also not so reliable as you might think. During some years the monsoon starts early and finishes late, bringing too much rain and flooding. Other years it comes late and then leaves too soon. If the weak subtropical lows are too weak or entirely missing, the monsoon can come and go without leaving enough rain, and then the crops fail and mass starvation results. The farther northwest you go the more unreliable the rains become (somewhat like the Sahel).

Near the end of the rainy season the cyclones (hurricanes) form with increasing frequency in the Bay of Bengal. You have already seen what can happen when these monsters smash into shore. During the rains the temperature falls several degrees, which is fortunate because otherwise the wet bulb would reach dangerous levels. In fact, in the Black Hole of Calcutta there have been times when the wet bulb rose above body temperature and some of the prisoners died. In spite of the warmth (for the temperature still averages 27°C), there are times during the rainy season when people still light fires indoors simply to dry out the air and their possessions!

During the rains everything springs back to life and the earth turns green with amazing rapidity. The ground soaks up the moisture until it becomes saturated and finally muddy and impassable. The insects also multiply rapidly swarming and breeding over the rain-soaked ground, spreading disease as rapidly as they can. After the rains stop these insects add greatly to human misery in the few hot weeks before the weather turns cool and dry again.

The wet and dry tropics are generally located between 5 and 20° latitude. Precipitation varies from place to place, but usually ranges between 100 and 200 cm annually, with at least 3 dry months (less than 6 cm per month). The annual temperature *range* typically lies between 3 and 10°C. Climate statistics for some Aw stations are shown in Table 22.2.

The dominant type of vegetation in the Aw region is the so-called **savannah** woodland. This is an open type of woodland–grassland combination that should be familiar

TABLE 22.2
The Climate of Selected Aw Cities

| | | | Jan | Feb | Mar | Apr | May | Jun | Jul | Aug | Sep | Oct | Nov | Dec | Year | Record High and Low Temperatures |
|---|---|---|---|---|---|---|---|---|---|---|---|---|---|---|---|---|
| 12. | Miami, Florida (26° N) | T | 20 | 20 | 21 | 24 | 26 | 27 | 28 | 28 | 28 | 25 | 22 | 21 | 24 | 38 −2 |
| | | ΔT | 9 | 10 | 9 | 8 | 8 | 8 | 7 | 8 | 7 | 8 | 8 | 9 | 8 | |
| | | P | 5.5 | 5.0 | 5.3 | 9.1 | 15.5 | 22.9 | 17.6 | 17.1 | 22.2 | 20.8 | 6.9 | 4.2 | 152 | |
| 13. | Balboa Heights, Canal Zone (9° N) | T | 27 | 27 | 28 | 28 | 27 | 27 | 27 | 27 | 27 | 26 | 26 | 27 | 27 | 36 17 |
| | | ΔT | 9 | 10 | 9 | 9 | 8 | 7 | 7 | 8 | 7 | 7 | 7 | 8 | 2 | |
| | | P | 2.5 | 1.0 | 1.8 | 7.4 | 20.3 | 21.3 | 18.0 | 20.0 | 20.8 | 25.7 | 25.9 | 12.2 | 177 | |
| 14. | Ciudad Trujillo, Dominican Republic (18° N) | T | 24 | 24 | 25 | 25 | 26 | 27 | 27 | 27 | 27 | 26 | 25 | 24 | 26 | 37 15 |
| | | ΔT | 10 | 10 | 9 | 9 | 8 | 9 | 9 | 9 | 8 | 8 | 8 | 10 | 3 | |
| | | P | 6.1 | 3.6 | 4.8 | 9.9 | 17.3 | 15.8 | 16.3 | 16.0 | 18.6 | 15.2 | 12.2 | 6.1 | 142 | |
| 15. | Goiás, Brazil (16° S) | T | 24 | 25 | 25 | 25 | 24 | 23 | 23 | 24 | 25 | 26 | 25 | 25 | 24 | 40 5 |
| | | ΔT | 13 | 14 | 14 | 16 | 17 | 18 | 18 | 18 | 18 | 17 | 14 | 14 | 3 | |
| | | P | 31.7 | 25.1 | 25.9 | 11.7 | 1.0 | 0.8 | 0.0 | 0.8 | 5.8 | 13.5 | 23.9 | 24.1 | 165 | |
| 16. | Kumasi, Ghana (7° N) | T | 25 | 26 | 26 | 27 | 26 | 25 | 24 | 25 | 26 | 26 | 25 | 25 | 26 | 38 11 |
| | | ΔT | 12 | 12 | 10 | 10 | 8 | 8 | 7 | 8 | 9 | 9 | 10 | 12 | 3 | |
| | | P | 2.0 | 5.8 | 14.5 | 12.9 | 19.0 | 20.1 | 10.9 | 7.9 | 17.3 | 18.0 | 9.4 | 2.0 | 140 | |
| 17. | Luluabourg, D. R. Congo (6° N) | T | 25 | 26 | 26 | 25 | 24 | 23 | 23 | 24 | 25 | 25 | 25 | 25 | 25 | 34 14 |
| | | ΔT | 9 | 9 | 9 | 10 | 11 | 12 | 12 | 11 | 10 | 9 | 8 | 8 | 3 | |
| | | P | 13.7 | 14.2 | 19.6 | 19.3 | 8.4 | 2.0 | 1.3 | 5.8 | 11.5 | 16.5 | 23.1 | 22.6 | 158 | |
| 18. | Calcutta, India (23° N) | T | 18 | 21 | 26 | 29 | 30 | 29 | 28 | 28 | 28 | 27 | 22 | 18 | 25 | 44 7 |
| | | ΔT | 14 | 14 | 13 | 12 | 10 | 6 | 6 | 6 | 8 | 8 | 12 | 14 | 12 | |
| | | P | 1.0 | 2.8 | 3.6 | 5.1 | 12.7 | 28.4 | 30.7 | 29.2 | 22.8 | 10.9 | 1.3 | 0.5 | 149 | |
| 19. | Mormugão, India (15° N) | T | 26 | 26 | 28 | 29 | 28 | 26 | 26 | 26 | 27 | 27 | 27 | 26 | 27 | 37 15 |
| | | ΔT | 9 | 9 | 8 | 5 | 5 | 4 | 4 | 4 | 6 | 6 | 8 | 9 | 3 | |
| | | P | * | * | * | 1.8 | 6.6 | 75.3 | 79.3 | 40.4 | 24.2 | 9.7 | 3.3 | 0.5 | 241 | |
| 20 | Manila, Philippines (15° N) | T | 25 | 26 | 27 | 28 | 29 | 28 | 28 | 28 | 27 | 27 | 26 | 25 | 27 | 38 14 |
| | | ΔT | 9 | 10 | 11 | 11 | 11 | 8 | 7 | 7 | 7 | 8 | 8 | 9 | 4 | |
| | | P | 2.3 | 1.3 | 1.8 | 3.3 | 13.0 | 25.4 | 43.9 | 42.9 | 36.2 | 19.6 | 14.7 | 6.7 | 212 | |
| 21. | Bangkok, Thailand (14° N) | T | 26 | 28 | 29 | 30 | 30 | 29 | 29 | 29 | 28 | 28 | 27 | 26 | 28 | 40 10 |
| | | ΔT | 14 | 13 | 12 | 12 | 11 | 9 | 9 | 9 | 9 | 9 | 10 | 12 | 4.5 | |
| | | P | 0.5 | 2.8 | 2.8 | 5.8 | 13.2 | 15.2 | 17.5 | 23.4 | 35.6 | 25.2 | 4.6 | 0.3 | 147 | |
| 22. | Port Darwin, Australia (12° S) | T | 29 | 29 | 29 | 29 | 28 | 26 | 25 | 26 | 28 | 30 | 30 | 29 | 28 | 41 10 |
| | | ΔT | 7 | 7 | 8 | 9 | 10 | 10 | 10 | 10 | 10 | 9 | 9 | 8 | 5 | |
| | | P | 38.6 | 31.2 | 25.4 | 9.7 | 1.5 | 0.3 | * | 0.3 | 1.3 | 5.1 | 11.9 | 23.9 | 149 | |

to you if you have ever seen nature films taken on the Serengeti Plains of Africa (see Figure 22.7). This is where lions, hyenas, and countless other grazing animals roam freely in one of nature's last large-game preserves. Trees grow in clumps that appear like islands in a sea of grass.

The extended dry season prevents the forest from being continuous. The longer the dry season, the more isolated the clumps of trees and the more dominant the grasses. Trees and grasses alike do not easily survive a long dry season with its all too frequent fires. The grasses dominate because they rapidly grow and reach maturity soon after the rains begin, while the trees must live several years, surviving dry seasons before they produce seeds.

The awesome solemnity of a dry season can be seen by watching these nature films of the Serengeti Plains. The films are almost always taken during the dry season, because traveling is quite difficult in any other season when the insects are also intolerable. After a month or two of dry weather the grasses and leaves of trees begin to wither, and the animals start to migrate in search of food and water. Often the waterholes and smaller streams completely dry out so that only the fittest animals survive. After several rainless months every living thing pines for the rains to return.

After leaving the rain forest and entering the wet and dry tropics the vegetation more or less continuously thins out, except along the east coast of most continents. The open grassy areas widen and the trees become more and more isolated as the dry season becomes longer. Then, on the poleward boundary of the Aw regions you often approach the deserts and even the grasses die out and give way to the sparse, thorny vegetation or the sands of the desert.

# DRY CLIMATES

The dry (B) climates are simple to define. They are areas in which the evaporation exceeds the precipitation (see Figure 22.8). However, the distinction between semiarid and desert climate is somewhat arbitrary, although deserts usually have more than twice as much evaporation as precipitation. In all arid and semiarid areas the limited amount of water rapidly evaporates into the air. When a river runs through a dry area, it loses a great deal of its water through evaporation and seepage into the rocks. Water vapor can even be sucked from seemingly dry ground where there are no rivers or lakes.

There are two main types of deserts. The first are the subtropical deserts that are generally located between 20 and 30° latitude. These are the result of the sinking air of the Hadley cells. The ITCZ rarely gets farther poleward than 20° latitude, while the midlatitude lows rarely penetrate below 30° latitude. These deserts are found from the center to the western coast of the continents, but not near the east coasts (except in the Horn of Africa) because of the easterly winds.

The second type of desert occurs mainly in the midlatitudes in the heart of the continents surrounded by mountain ranges and therefore remote from any source of water. They are not found poleward of 50° latitude largely because the evaporation rate is so low in the cool air that there is almost always adequate moisture.

There are only a few parts of some deserts that are almost rainless. Most desert areas receive between 10 and 30 cm of precipitation annually, and on the tropical borderline they can receive as much as 40 cm. Most of the scanty precipitation comes from showers, but on the poleward side of the subtropical deserts and in the midlatitude deserts some winter rain or snow may fall from passing lows.

Desert rains are notoriously unreliable, and from studying the desert a general climatological rule can be formulated.

**Rule:** The smaller the annual average precipitation, the more irregular and unreliable it is.

The rule is simple to illustrate. Port Sudan in the Sahara averages 10 cm per year; but in 1910 they only received 2 cm, while in 1925 they received 42 cm. Many places in the heart of the desert can go several years without any rain and then one large shower may produce 10 cm in an hour. Then, no more rain may fall for several more years. From this you should be able to figure out that during most years in the desert the precipitation is below average!

The lack of reliability is not too serious in the desert itself since no one expects rain there, but it has been devastating to people living on the fringe of the desert in the semiarid (BS) climates (recall the Sahel drought described in Chapter 10).

We can look at the dry areas of the southwest United States for another example of the unreliability of rainfall. The once mighty Colorado River, now dammed, brings the Southwest its most important source of water and an important source of hydroelectric power as well. Its water is apportioned between the various states and Mexico so thoroughly that only a trickle now runs into the sea. Unfortunately, the apportionments were based on rainfall from 1900–1920 that research has now shown to be

FIGURE 22.7
The Serengeti Plains. Savannah or open woodland is characteristic of the Aw climates.

among the rainiest in the past 1000 years. Furthermore, damming the river has significantly increased water losses from increased evaporation and groundwater flow.

Between 1975 and 1977 the Southwest experienced a serious drought in which precipitation fell by 25 to 50%. Hundreds of thousands of acres of irrigated farmland had to go without water so that people might drink. The drought finally ended with a vengeance late in 1977; nonetheless, the population of the Southwest is too large for the available water supply, and there may come the time when people will have to leave or create additional water themselves.

Summer days in the desert can be truly frightful. In the subtropical deserts afternoon temperatures almost always exceed 40°C, and large areas have daytime highs that often exceed 45°C. Remember that the air temperature is taken in the shade, but most objects in the desert are exposed to the sun. Even the sand which has a high albedo can reach a temperature of 80°C on sunny afternoons. Let no one tell you that the desert's heat is not so bad because it is "dry heat." That is pure nonsense. A temperature of 45°C is brutal and the sunlight makes it even worse.

I once passed through Baker, California, when it was 49°C and I have never felt more uncomfortable in my life. When I put my hand out of the car window to feel the air go by, it felt like an oven.

After a brief, cooler period near dawn the desert quickly heats up in the morning. The mean temperature of the hottest month in the subtropical deserts is often well over 30°C, and many of those places average slightly over 35°C. This is impressive when you consider that there are seldom any weather stations in the most inhospitable parts of the desert. Greenland Ranch in Death Valley, California, has a mean temperature of 38°C during July, which is just about the warmest month anywhere on earth.

The midlatitude deserts seldom get so hot. Summer temperatures still average near 30°C, while daytime highs frequently reach 40°C. Altitude effects are also quite important, and valley stations such as Lukchun in the heart of Asia and the Dead Sea (both of which are below sea level) have temperatures that astound the highlanders.

The misery of the desert does not end with the scorching temperatures. The extreme heat of the air near the

Dry climates (B)

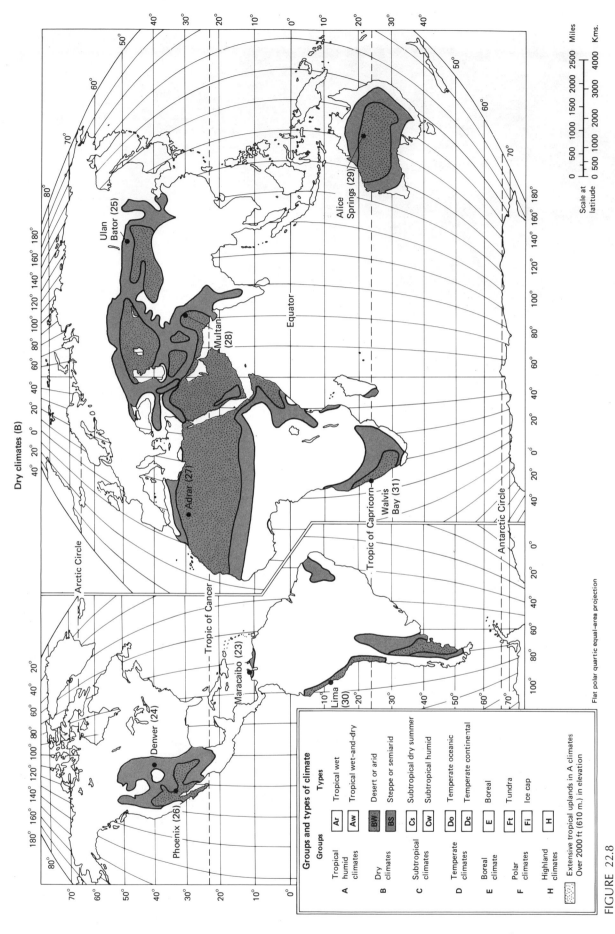

Groups and types of climate

| Groups | | Types |
|---|---|---|
| A | Tropical humid climates | Ar | Tropical wet |
| | | Aw | Tropical wet-and-dry |
| B | Dry climates | BW | Desert or arid |
| | | BS | Steppe or semiarid |
| C | Subtropical climates | Cs | Subtropical dry summer |
| | | Cw | Subtropical humid |
| D | Temperate climates | Do | Temperate oceanic |
| | | Dc | Temperate continental |
| E | Boreal climate | E | Boreal |
| F | Polar climates | Ft | Tundra |
| | | Fi | Ice cap |
| H | Highland climates | H | |

Extensive tropical uplands in A climates
Over 2000 ft (610 m.) in elevation

Scale at
latitude

Flat polar quartic equal-area projection

FIGURE 22.8

Location of the dry (Bw and Bs) climates. Climate statistics for cities shown are given in Table 22.3.

ground produces instability. Ordinarily is it far too dry to produce rain, but the unstable air does produce windy conditions, with occasional whirling dust devils. Since there is little vegetation, the wind is relatively unimpeded and also picks up sand and dust easily. This hot, blowing air scorches you at the same time that the blowing sand stings and sticks.

The relative humidity of the desert air is very low—certainly less than 20% and often less than 10%. When this is combined with the high temperature of the desert, it produces the highest evaporation rates in the world *from any moist surface* (such as human skin). No human being can survive without water for two summer days in the desert. The moisture leaves with every breath and through every pore in the skin, quickly dehydrating and even mummifying the unprepared. In such a climate it is best to travel at night and rest during the day. Even so, a human being requires 8 liters of water a day.

The traditional dress of the Arabs of the desert is no mere accident. It is designed to protect the body from sun, wind, sand, heat, and even cold of night. It also helps retain some moisture. It may seem strange to go about so completely covered up, but in the desert that is wise policy.

Cumulus clouds do form in the desert and occasionally rain can be seen falling from their bases. But, more often than not, in the summer this rain evaporates completely before reaching the ground (this is called **virga**). The few times that the rain reaches the ground in sufficient quantities flash floods may result.

One less known feature of the desert is the nighttime cold. The clear, dry air allows tremendous cooling of the air near the ground during the night, and diurnal temperature ranges in the desert generally exceed 15°C. There are even some parts in the eastern Sahara that have diurnal temperature variations as high as 30°C during some months.

During the hottest few months the nights are the only comfortable time, but at other times of the year the nights are positively cold. Then, the desert traveler who is prepared for the heat must endure the cold of night. Many desert travelers claim that it is this feature that they have found to be truly unbearable. Winter in places like the Gobi Desert of Mongolia is notorious for its totally inhospitable, bone-chilling cold, and even the subtropical deserts have occasional frost.

Coastal deserts such as the Atacama of Chile and the Namib of Southwest Africa are cooled by the upwelled coastal waters found on the western coasts of the continents in subtropical latitudes. The cold (10 to 15°C) water often cools the air to the dew point, creating fog which then is carried inland by a sea breeze. It also stabilizes the lower atmosphere, thereby making precipitation even more unlikely. When the west coasts also have mountain ranges—as is the case in Southwest Africa and North and South America—the subtropical high tends to get locked in place. The net result of all these factors is an almost rainless climate. You have already read that the Atacama Desert is the driest place in the world.

On rare occasions the winds and the high pressure area over the Pacific Ocean near Chile weaken and the upwelling stops. Warm water creeps southward from Peru and Ecuador, killing the plankton and disrupting the entire food chain because it is poor in basic nutrients. This phenomenon is known as El Niño, because it tends to occur around Christmas. (Niño stands for the Christ child in Spanish.) The El Niño of 1972 so decimated the anchovy population that it still hadn't recovered as of 1978, and this has had a devastating effect on the Peruvian economy.

Statistics for the semiarid and desert stations are presented in Table 22.3. They include the subtropical deserts —the Sahara, Arabian, Kalihari, Australian, Mohave, and so on; the coastal deserts—the Namib, Atacama, and the like; and the midlatitude deserts—for example, the Gobi and other Asian deserts, the Great Salt Lake Desert, and the Patagonian (the only midlatitude coastal desert).

The entire desert landscape is distinctive. The angular shapes of the terrain, the gray, yellow, and red colors, and the relative absence of plants are trademarks of the desert. Vegetation has little or no leaves in order to preserve water, and many desert plants have extensive root systems and are coated with a waxy layer which further curtails water loss. Finally, some plants are covered with thorns or little prickles to ward off thirsty animals.

Most of the desert consists of rocky plains. Only 25% consists of shifting sand dunes, which may be as high as 300 m. Very little of the desert is actually lifeless, and in all but the most barren places the sparse desert plants wait for the few rains. When one of these rainfalls happens to come at the right time, your eyes may be greeted by the miraculous sight of the desert in bloom (see Figure 22.9). But even in bloom the desert is inhospitable.

# SUBTROPICAL AND TEMPERATE CLIMATES

Poleward of 25° latitude there is a definite change in the climate. Here there is a distinctly cool or cold season and the annual temperature variations exceed the diurnal vari-

TABLE 22.3
The Climate of Selected BS and BW Cities

| | | Jan | Feb | Mar | Apr | May | Jun | Jul | Aug | Sep | Oct | Nov | Dec | Year | Record High and Low Temperatures |
|---|---|---|---|---|---|---|---|---|---|---|---|---|---|---|---|
| 23. Maracaibo, Venezuela (11° N) BSh | T | 28 | 28 | 29 | 29 | 30 | 30 | 30 | 30 | 29 | 29 | 28 | 28 | 29 | 39 19 |
| | ΔT | 9 | 9 | 9 | 9 | 10 | 10 | 11 | 11 | 10 | 9 | 9 | 9 | 2 | |
| | P | 0.3 | * | 0.8 | 2.0 | 6.9 | 5.6 | 4.6 | 5.6 | 7.2 | 15.0 | 8.4 | 1.5 | 58 | |
| 24. Denver, Colorado (40° N) BSk | T | −1 | 0 | 3 | 9 | 13 | 19 | 23 | 22 | 17 | 11 | 4 | 0 | 10 | 40 −34 |
| | ΔT | 15 | 15 | 14 | 15 | 15 | 16 | 16 | 16 | 17 | 17 | 16 | 15 | | |
| | P | 1.5 | 1.7 | 3.1 | 4.9 | 6.7 | 4.9 | 4.5 | 3.3 | 2.9 | 2.9 | 1.9 | 1.1 | 39.5 | |
| 25. Ulan Bator, Mongolia (48° N) BSk | T | −26 | −23 | −12 | 0 | 10 | 14 | 16 | 14 | 10 | −1 | −15 | −24 | −5 | 36 −44 |
| | ΔT | 14 | 15 | 15 | 15 | 15 | 14 | 12 | 14 | 15 | 15 | 14 | 13 | 42 | |
| | P | * | * | 0.3 | 0.5 | 0.8 | 2.5 | 7.4 | 4.9 | 1.0 | 0.5 | 0.5 | 0.3 | 19.6 | |
| 26. Phoenix, Arizona (33° N) BWh | T | 11 | 12 | 16 | 21 | 25 | 30 | 33 | 32 | 29 | 22 | 15 | 12 | 21 | 48 −9 |
| | ΔT | 15 | 16 | 17 | 18 | 18 | 19 | 15 | 14 | 16 | 17 | 17 | 16 | 22 | |
| | P | 1.8 | 2.0 | 1.8 | 0.8 | 0.3 | 0.3 | 2.0 | 2.8 | 1.8 | 1.3 | 1.3 | 2.0 | 18.3 | |
| 27. Adrar, Algeria (28° N) BWh | T | 12 | 14 | 18 | 24 | 31 | 35 | 37 | 36 | 31 | 25 | 18 | 13 | 25 | 51 −4 |
| | ΔT | 17 | 18 | 18 | 18 | 19 | 18 | 18 | 17 | 17 | 16 | 16 | 17 | 25 | |
| | P | * | * | 0.3 | * | * | * | * | * | * | 0.5 | 0.5 | * | 1.5 | |
| 28. Multan, Pakistan (30° N) BWh | T | 13 | 16 | 21 | 27 | 33 | 37 | 34 | 33 | 31 | 26 | 19 | 14 | 25 | 50 −2 |
| | ΔT | 14 | 15 | 15 | 15 | 16 | 14 | 9 | 10 | 14 | 17 | 16 | 15 | 24 | |
| | P | 1.0 | 1.0 | 1.0 | 0.8 | 0.8 | 1.5 | 5.0 | 4.6 | 1.3 | 0.3 | 0.3 | 0.5 | 18.1 | |
| 29. Alice Springs, Australia (24° S) BWh | T | 29 | 28 | 25 | 20 | 16 | 13 | 12 | 14 | 18 | 23 | 27 | 29 | 22 | 47 −7 |
| | ΔT | 15 | 15 | 15 | 15 | 14 | 15 | 16 | 17 | 17 | 17 | 16 | 15 | 17 | |
| | P | 4.3 | 3.3 | 2.8 | 1.0 | 1.5 | 1.3 | 0.8 | 0.8 | 0.8 | 1.8 | 3.0 | 3.8 | 25.1 | |
| 30. Lima, Peru (12° S) BWn | T | 22 | 23 | 23 | 21 | 19 | 17 | 16 | 16 | 16 | 17 | 19 | 21 | 19 | 34 9 |
| | ΔT | 9 | 9 | 9 | 9 | 7 | 6 | 6 | 6 | 6 | 7 | 8 | 9 | 7 | |
| | P | 0.2 | * | * | * | 0.5 | 0.5 | 0.8 | 0.8 | 0.8 | 0.3 | 0.2 | * | 4.1 | |
| 31. Walvis Bay, S. Africa (23° S) BWn | T | 19 | 20 | 19 | 18 | 17 | 16 | 15 | 15 | 15 | 15 | 16 | 18 | 17 | 40 −4 |
| | ΔT | 8 | 10 | 11 | 11 | 12 | 12 | 13 | 12 | 10 | 9 | 8 | 8 | 5 | |
| | P | * | 0.5 | 0.8 | 0.3 | 0.2 | * | 0.1 | 0.2 | * | * | * | * | 2.3 | |

ations. During the winter season the weather tends to be irregular because of passing highs and lows. These systems are weak and uncommon at 25° latitude, but become reasonably common above 30° latitude. The climates of the west and east coasts are so different that they will be discussed separately. We begin with the west coast.

Moving poleward along the west coast from the subtropical deserts, the climate begins to change at about 25 or 30° latitude. Above that latitude belt there are some winter rains, but the summer reverts back to rainlessness (rare summer hurricanes have crept up the coast of California as in 1977). The farther poleward you travel the longer the rainy season becomes so that by about 40° latitude it starts to become the dominant season.

The dry summer subtropics (Cs) are located along the west coasts between about 30 and 40° latitude and, although winter rains do occur, the summer is dominant. The Cs climate is more commonly known as a Mediterranean climate because it is characteristic of almost all the land that borders the Mediterranean Sea. Many people consider this type of climate to be the best in the world (see Figure 22.10). There is plenty of sunshine even in winter, and temperatures are rarely excessive. It is generally a climate of moderation and thus perhaps it is no wonder that the ancient Greeks so religiously espoused a philosophy of moderation.

During the summer the weather is dominated by the sinking air of the Hadley cell and the southward flowing air of the subtropical highs. This air is stable and sinks so that the skies remain clear and blue (except in places like Los Angeles). Upwelling is the rule along the coast, caus-

FIGURE 22.9
The Arizona desert in bloom.
Only about 25% of desert
areas approach a lifeless state.

ing fog and cool temperatures (from 15 to 20°C), with a small diurnal range. Upwelling is rare in the Mediterranean Sea so that the coastal temperatures are much warmer there. A few kilometers inland the temperatures are much higher (from 25 to 30°C), with a large diurnal range since the air is much like desert air during the summer.

Rain is extremely rare in Cs areas during the summer. Many places such as Jerusalem have recorded absolutely no rain during July (NH) for over 100 years.

During the winter the westerlies take over, bringing upslope winds and passing lows and an increased oceanic influence. The modifying influence of the ocean keeps the winter temperatures mild, and this is one of the most attractive features of the Cs climate. Winters tend to average anywhere from 7 to 15°C, and frosts present only an occasional problem to agriculture inland.

Because of the oceans, the annual temperature range is very small along the coast and not too large inland. One of the strange things about Cs climates is experienced when you walk the streets of San Francisco; most likely in July you'll be needing the same sweater you used in January.

Winter storms bring the Cs climate virtually all of their precipitation. Snow is freakish along the coast because the ocean keeps the air above freezing most of the time. The rains last a day or two at a time, but there are many embedded showers that can be especially intense on hilly areas. However, even in the winter there is a lot of sunshine in the Cs climate because the subtropical high is never too far away.

The high mountains have the same basic weather pattern, but snow is much more likely. Mountains can, of course, get showers in the summer that simply do not occur in the lowlands.

The first snow that whitens the Sierra usually falls about the end of October or early in November to a depth of a few inches, after months of the most charming Indian summer weather imaginable . . . . The first general winter storm that yields snow that is to form a lasting portion of the season's supply, seldom breaks on the mountains before the end of November. Then, warned by the sky, cautious mountaineers, together with the wild sheep, deer, and most of the birds and bears, make haste to the lowlands or foothills . . . . The first heavy fall is usually from about two to four feet in depth. Then, with intervals of splendid sunshine, storm succeeds storm, heaping snow on snow, until thirty to fifty feet has fallen . . . .

Even during the coldest weather evaporation never wholly ceases, and the sunshine that abounds between the storms is sufficient to melt the surface more or less through all the winter months.

John Muir, *The Mountains of California*

The total precipitation for Cs climates usually ranges

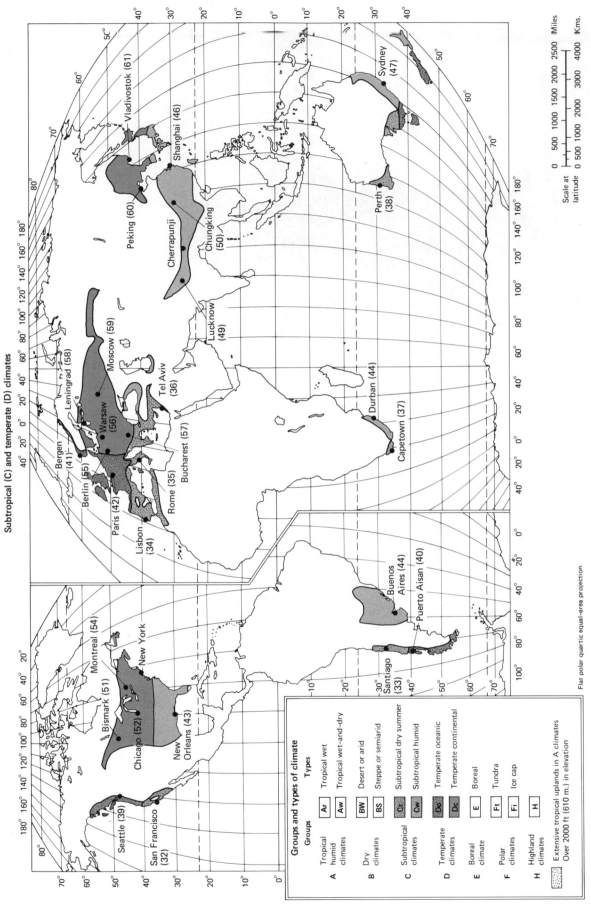

FIGURE 22.10

Locations of the subtropical (Cf, Cs, and Cw) and temperate
(Do and Dc) climates. Climate statistics for cities shown are
given in Tables 22.4, 22.5, 22.6, and 22.7.

between 40 and 80 cm. This relatively small rainfall together with the long dry summer determines the vegetation. The land supports only a scrubby type of tree and bush cover known as **chapparal** or **maquis** (as in Figure 22.11) and the land is not dominated by grasses because the potentially best growing season, is too dry.

Toward the end of the winter rains the land turns green and flowers bloom everywhere. In the warmest parts of the Cs climate winter is, amazingly enough, the green season, while in the cooler parts the greenest time of year is spring. After that, as the dry season progresses, the earth gradually turns browner and grayer.

Toward the end of the dry season the possibility of fires greatly increases. Fire is one of the scourges of the Cs climate and is one of the reasons why trees are not more common. It takes several years for the natural vegetation to recover from the fires and, if a particularly dry winter happens to occur after the plants have reestablished

FIGURE 22.11
Mediterranean scrub woodland.

themselves, there can be enormous destruction. (This is precisely what happened in the early summer of 1977 all through the southwest United States before hurricane Doreen brought desperately needed rain.) Then, after the fires have destroyed the scrubby plant cover, there is an increased danger that the next winter's rains will cause mudslides and landslides in the hilly country.

The Cs climates are located on all western coasts of the continents, roughly between 30 and 40° latitude, as well as all around the Mediterranean Sea. It is surprising that these areas which are so widely separated and completely isolated from each other all have such remarkably similar climates and vegetation (see Table 22.4).

Above 40° latitude the west coast is usually fairly well forested, and the rainy winter season has become noticeably cooler and longer. The west coasts between about 40 and 60° latitude are the locations of the temperate oceanic, or Do, climate. The subtropical high still exerts a strong influence on the summer weather, but it no longer completely dominates the picture, so that rain can fall during the summer in Do locations.

Consider a typical year in the Do climate. We take off for London, England, during January. As the plane drops through the clouds we see the wet, drizzly surface below. People tell us that it has been drizzling like this for three days, but at first we think little of it. After a few more days, during which the sun never penetrates the clouds, we begin to realize that the drizzle is remarkably persistent. In the Do climate during the winter, it rains about two of every three days, the temperature hovers about 5°C, and it is more than 75% cloudy. The air almost always feels damp and fog is reasonably common. This is the season when the westerlies dominate and bring one low pressure area after another. Most of these lows are deeply occluded, which accounts for the drizzle. The annual precipitation averages between 50 and 100 cm in the lowlands, which is very low considering the amount of time it falls. It is understandable, however, since most of it falls as drizzle.

Cold spells are rare in Do regions because the cold air must come from the land and hence from the east. The winter temperatures average above freezing because of the nearby ocean and, therefore, are remarkably warm considering the high latitude. This makes rain more common than snow in the lowlands, but slushy snow is common enough.

The weather is more severe on the exposed mountain slopes. Here precipitation is more intense because the westerly winds are forced to ascend rapidly over the mountains. Exposed mountain slopes in Do climate regions commonly get above 250 cm annually and often

TABLE 22.4
The Climate of Selected Cs Cities

| | | Jan | Feb | Mar | Apr | May | Jun | Jul | Aug | Sep | Oct | Nov | Dec | Year | Record High and Low Temperatures |
|---|---|---|---|---|---|---|---|---|---|---|---|---|---|---|---|
| 32. San Francisco, California (38° N) | T | 11 | 12 | 12 | 13 | 14 | 15 | 15 | 15 | 17 | 16 | 14 | 11 | 14 | 41 −7 |
| | ΔT | 6 | 6 | 6 | 7 | 7 | 7 | 6 | 6 | 8 | 7 | 5 | 5 | 6 | |
| | P | 10.1 | 8.8 | 9.6 | 3.3 | 1.3 | 0.3 | * | * | 0.5 | 1.8 | 3.3 | 10.4 | 47 | |
| 33. Santiago, Chile (33° S) | T | 21 | 21 | 19 | 15 | 11 | 10 | 9 | 10 | 12 | 15 | 18 | 20 | 15 | 37 −4 |
| | ΔT | 12 | 13 | 15 | 16 | 12 | 10 | 12 | 13 | 15 | 15 | 13 | 12 | 12 | |
| | P | 0.3 | 0.3 | 0.5 | 1.3 | 6.3 | 8.3 | 7.6 | 5.6 | 3.0 | 1.5 | 0.8 | 0.5 | 36 | |
| 34. Lisbon, Portugal (39° N) | T | 11 | 11 | 12 | 15 | 19 | 21 | 22 | 22 | 20 | 17 | 14 | 12 | 17 | 39 −2 |
| | ΔT | 6 | 6 | 6 | 7 | 9 | 9 | 9 | 9 | 8 | 7 | 5 | 6 | 11 | |
| | P | 8.3 | 8.0 | 7.8 | 6.1 | 4.3 | 1.8 | 0.5 | 0.5 | 3.7 | 7.8 | 10.7 | 9.1 | 69 | |
| 35. Rome, Italy (42° N) | T | 8 | 9 | 11 | 14 | 19 | 23 | 24 | 24 | 21 | 17 | 13 | 10 | 16 | 40 −7 |
| | ΔT | 8 | 9 | 11 | 12 | 12 | 14 | 13 | 13 | 11 | 11 | 8 | 8 | 16 | |
| | P | 8.4 | 7.4 | 5.1 | 5.1 | 4.8 | 1.8 | 1.0 | 1.8 | 7.1 | 10.9 | 11.2 | 10.4 | 75 | |
| 36. Tel Aviv, Israel (32° N) | T | 13 | 14 | 16 | 18 | 21 | 24 | 26 | 26 | 25 | 23 | 19 | 15 | 21 | 46 −2 |
| | ΔT | 10 | 10 | 10 | 10 | 10 | 10 | 9 | 9 | 10 | 11 | 11 | 9 | 13 | |
| | P | 12.4 | 6.9 | 5.1 | 1.8 | 0.3 | 0.0 | 0.0 | 0.0 | 0.3 | 1.0 | 10.4 | 15.5 | 54 | |
| 37. Capetown, S. Africa (34° S) | T | 21 | 21 | 20 | 18 | 15 | 13 | 13 | 13 | 14 | 16 | 18 | 20 | 17 | 41 −2 |
| | ΔT | 10 | 10 | 10 | 10 | 9 | 8 | 9 | 8 | 9 | 9 | 9 | 10 | 8 | |
| | P | 1.5 | 0.8 | 1.4 | 4.8 | 7.9 | 8.4 | 8.9 | 6.6 | 4.3 | 3.0 | 1.8 | 1.0 | 51 | |
| 38. Perth, Australia (32° S) | T | 23 | 23 | 22 | 19 | 16 | 14 | 13 | 13 | 14 | 16 | 19 | 22 | 18 | 44 −1 |
| | ΔT | 12 | 12 | 11 | 11 | 9 | 8 | 8 | 9 | 9 | 9 | 10 | 11 | 10 | |
| | P | 0.8 | 1.0 | 2.0 | 4.3 | 13.0 | 18.1 | 17.0 | 14.5 | 8.8 | 5.6 | 2.1 | 1.3 | 88 | |

over 500 cm. A great percentage of this falls as snow since the cool sea air only has to be lifted a small distance before the temperature drops below freezing. Above an altitude of about 1 km, snow is far more common than rain during the winter, and these windward slopes often receive extremely high snowfall totals. Paradise Ranger Station, on Mount Rainier, the world's snowiest place, is just such a spot.

During the spring the weather warms slowly because of the conservative nature of the ocean. Cloudiness and precipitation finally diminish, but fog is especially common along the coastline. When it finally does clear up people rejoice and feel rejuvenated.

For the most part, summer is truly delightful. Temperatures average between 15 and 20°C. This means that you will need to wear a sweater on most evenings in Do climates. Daytime temperatures are pleasant and, as the moist ocean air is quickly heated over the land, cumulus clouds form and occasionally produce showers.

Heat waves are rare but they do occur. Afternoon temperatures then rise to 30 or 35°C. This may not seem very hot to most people, but the British have become so acclimatized to the normally cool weather that these puny heat waves actually cause an increase in the mortality rates. It is no wonder that the vivid descriptions of the "unbearable" heat of tropical lands were so often written by the British.

In autumn the cloudiness returns and fog is common inland. Autumn is sometimes the rainiest season because of the warmth of the oceans. Once the general cloudiness sets in (usually in October) it is time to leave, since the sun won't shine much again until the following May.

The cool summer temperatures and the mild, wet winter produce a rich green landscape that is well forested (wherever the forests have not been cleared). Ireland is one example of a Do region that well deserves its reputation for its beautiful greenery.

On the mountainous Do coasts where the precipitation is more abundant, some of the world's densest and grandest rain forests are found. Here the famous coast redwoods (see Figure 22.12) which attain heights over 100 m thrive, as do other giant trees such as the Douglas fir.

In the Do climate all plant life is extremely sensitive to changes in altitude. At sea level the length of the growing season for plants is about half the year. The growing season corresponds roughly to the time during which the

FIGURE 22.12
California coastal redwoods.

average temperature is at least 10°C. Even though the average temperature lies above this figure for about half the year, it never rises far above it. Therefore, any small change in altitude is sufficient to greatly diminish the number of growing days. By 1 km the growing season is cut to less than 60 days (see Figure 22.13). Any slight global cooling trend can therefore have a serious impact on the length of the growing season and potentially could lead to famine.

In short, if you have a summer vacation, Do regions are some of the nicest places to spend it. Simply be careful about picking a mountainous coastline, since it may be too wet even in summer. Climate statistics for representative Do stations are given in Table 22.5.

Now we move to the *eastern* side of the continents. Here, too, the summer weather is dominated by the subtropical high, especially on the equatorward margins; and the winter weather is dominated by the variable westerlies, especially on the poleward margins. There are, however, enormous differences between the east and west coast climates in the midlatitudes.

It is true that the same basic pressure systems govern the climate of both coasts. The differences in climate result from some profound differences in the behavior of these systems on each coast. On the east coast the air

FIGURE 22.13
The length of the growing season in Do climates and how it diminishes with altitude. The growing season corresponds roughly to the time the average temperature is above 10°C.

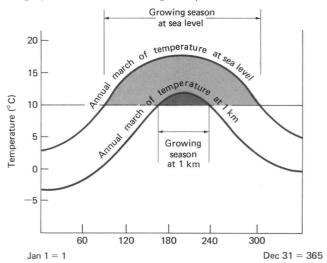

TABLE 22.5
The Climate of Selected Do Cities

| | | | Jan | Feb | Mar | Apr | May | Jun | Jul | Aug | Sep | Oct | Nov | Dec | Year | Record High and Low Temperatures |
|---|---|---|---|---|---|---|---|---|---|---|---|---|---|---|---|---|
| 39. | Seattle, Washington (47° N) | T | 4 | 5 | 7 | 10 | 13 | 16 | 18 | 18 | 15 | 11 | 8 | 6 | 11 | 38 −18 |
| | | ΔT | 4 | 6 | 7 | 9 | 9 | 9 | 10 | 10 | 8 | 7 | 6 | 4 | 14 | |
| | | P | 12.5 | 9.7 | 7.9 | 6.1 | 4.6 | 3.3 | 1.5 | 1.8 | 4.4 | 7.1 | 12.2 | 14.0 | 85 | |
| 40. | Puerto Aisen, Chile (42° S) | T | 14 | 14 | 12 | 9 | 7 | 6 | 5 | 5 | 7 | 9 | 11 | 13 | 10 | 34 −8 |
| | | ΔT | 7 | 7 | 7 | 7 | 5 | 4 | 4 | 5 | 7 | 7 | 7 | 7 | 9 | |
| | | P | 19.8 | 19.8 | 21.1 | 19.0 | 37.4 | 26.4 | 28.2 | 28.2 | 16.5 | 19.8 | 17.8 | 20.1 | 273 | |
| 41. | Bergen, Norway (60° N) | T | 2 | 2 | 3 | 6 | 9 | 12 | 14 | 14 | 11 | 8 | 4 | 3 | 8 | 32 −16 |
| | | ΔT | 4 | 4 | 6 | 6 | 7 | 7 | 7 | 6 | 6 | 5 | 4 | 4 | 12 | |
| | | P | 20.0 | 15.2 | 13.7 | 11.2 | 10.9 | 11.7 | 13.2 | 18.6 | 23.4 | 23.4 | 20.3 | 20.6 | 200 | |
| 42. | Paris, France (49° N) | T | 3 | 4 | 6 | 11 | 13 | 17 | 19 | 18 | 15 | 11 | 6 | 3 | 11 | 41 −17 |
| | | ΔT | 6 | 8 | 10 | 11 | 11 | 11 | 11 | 11 | 10 | 8 | 7 | 6 | 16 | |
| | | P | 3.9 | 3.1 | 4.1 | 4.4 | 5.3 | 5.8 | 5.6 | 5.6 | 5.1 | 5.8 | 4.6 | 4.4 | 58 | |

from the subtropical high comes from *the tropics,* warm and laden with moisture. When this air passes overland it is heated and easily produces showers. The high is also not so persistent as it is on the west coast, and "minor" disturbances such as hurricanes are all too common. Furthermore, on the east coast the westerlies bring continental air rather than the maritime air they bring to the west coast. Therefore, at a given latitude the east coast has much larger annual temperature variations.

Both the humid subtropical (Cf) and temperate continental (Dc) climates have the same basic annual weather sequence. Nevertheless the two climates do not feel the same—the trademark of the Cf is its enervating, hot, humid summer, while the trademark of the Dc climate is its cold winter. We will begin by spending a year in a Cf location and then seeing how it compares to a year in a Dc location.

The Cf climates are located on the eastern side of all the continents, roughly between 25 and 35° latitude. We arrive in Montgomery, Alabama (in the heart of Cf country), on January 1. It is a sunny, pleasant afternoon with a temperature of 18°C. The next day a thunderstorm pours down on Montgomery and the temperature begins to fall. Within 24 hours the temperature hits a low of −5°C. Such cold waves pass through Montgomery a number of times every winter. They occur with increasing frequency and severity the farther north you go. Montgomery is far enough south so that the cold waves alternate with many pleasantly warm winter days, and people there are rarely confined indoors for long.

After the cold wave sets in, we observe the typical weather sequence that indicates an approaching low. This means that a day or so of rain will soon follow the cirrostratus and altostratus clouds, snow being the exception as far south as Montgomery. After the rain ends it often gets warm again for a number of days, but sometimes another cold wave will follow immediately.

The Cf region of the United States has the worst cold *waves,* but the Cf of China is colder on the average because in China there is almost no relief from the constant cold, dry north winds coming from the Siberian high during winter. The severity of the Chinese Cf winter is easily illustrated by the temperature of the coldest month. Shanghai has a January average of only 4°C, while other Cf stations at the same latitude—namely, Port Macquarie, Australia (12°C), Rosario, Argentina (11°C), and Montgomery, Alabama (9°C)—have far warmer average winter temperatures.

Every part of the Cf regions have seen some snow. This was proven again recently when Miami, just south of the Cf–Aw border, received its first snowfall ever in January 1977. Some snow falls every winter on the poleward margins of the Cf; but it rarely amounts to much, and large snowfalls do occur once every several decades. Even the smallest snowfalls in Cf regions are sufficient to paralyze the unprepared residents.

As spring approaches in Montgomery the weather rapidly begins to warm up and plants come to life again. Temperatures at this time are "just right" and the weather can be exhilarating. True, there are a few hot days that give a hint of the hot summer to come, but these are not common until late April. Cold waves come with decreasing frequency, but they are sometimes announced by severe thunderstorms, hail, and even tornadoes associated with squall lines.

By the end of May the weather in Montgomery is unpleasantly hot and the long, hot, humid summer lies ahead. Low pressure areas with their cold fronts have retreated poleward and only occasionally ooze as far south as the Cf regions during late spring and early summer. When they do reach these regions they provide relief for a day or two at most and then are almost immediately replaced by more heat and humidity. In China the rains that occur with these June lows are called the "plum" rains, because they come at the time of the ripening of the plums.

After the plum rains it gets plumb hot. Day after day the weather forecast reads hazy, hot, and humid with a chance of showers or thundershowers. The high temperature varies between 30 and 35°C, while the nighttime low ranges between 20 and 25°C. The mean temperature of the warmest months ranges between 25 and 30°C, making it as hot as (or slightly hotter than) the equator, and just about as humid.

I have had the great misfortune of spending several of my summers in the southeast United States and can personally testify to the sustained, sultry oppressiveness of the summer weather there. When the local residents defend their weather by saying that it really isn't so unpleasant, my only consolation is to tell them that, as a meteorologist, I *know* that it's even hotter (and more miserable I might add) than almost anywhere on the equator.

Even on the seacoast there is little relief from the intense heat and humidity. On the east coast, upwelling is an almost unknown word, and the ocean temperatures may vary between 25 and 30°C. Such high water temperatures mean that it has to get very hot inland to start a sea breeze, but hurricanes have no trouble intensifying in such waters.

As if to add insult to injury, a familiar feature of the Cf summer is the heat wave. Heat waves occur several times each summer and generally last from about two days to two weeks. They occur when the jet stream and the westerlies retreat poleward and the subtropical high expands over the land. This suppresses thunderstorms somewhat and sends afternoon temperatures above 35°C in many places. Rarely do the temperatures exceed 40°C, but then again they don't have to. These heat waves are associated with some of the world's most unhealthy weather which is reflected in the mortality tables. During such heat waves it is wise not to overeat or exercise too strenuously. It is even wiser to leave.

Most of the summer precipitation comes from showers and thundershowers that are intense but short-lived, so that the skies are seldom overcast for long. Although hurricanes represent widely spaced events in time, the enormous rainfall that they bring the few times they do pass by significantly increases the average precipitation of the later part of the summer. The occasional hurricanes add one more interesting twist to the Cf and Dc regions.

In most Cf regions the precipitation is spread rather evenly over the entire year, with the summer usually being somewhat rainier than the other seasons. Precipitation averages between 80 and 160 cm; with no season too dry. Only in China is there a marked dry season during the winter because of the Siberian high (then the climate is called Cw).

The summer rains of China easily compensate for the dry winter; monthly totals exceeding 25 cm are common. This seasonality of the rains is reflected in the river levels. Remember the disastrous floods of the Hwang Ho and Yangtse Kiang rivers. Fed also by the water from the melting snows, the water level commonly rises more than 50 m (the height of Niagara Falls) in the famous Wind Box Gorge of the Yangtse River from the winter to the height of the rainy season!

As fall approaches the weather finally begins to cool. Warm air is still pumped up from the tropics, but it cools as it passes over the land. Late fall and early winter are therefore times of frequent morning fog. Even so, the weather at this time is pleasant, and the autumn is a particularly lovely time of year in Cf regions. All in all, Cf is a lovely climate—if you can get away during the summer. The climate statistics for some Cf stations are given in Table 22.6.

The year-round rains keep the land forested wherever it has not been cleared. In China the forest has been so completely cleared that it is difficult for scientists to reconstruct its original composition. In the United States it is a mixed forest, meaning that it consists of evergreen and deciduous trees (for example, oak and maple are **deciduous**—they lose their leaves). In Australia, eucalyptus trees dominate. Many varieties of eucalyptus also grow in the Ar and Aw regions of Australia. Some of these rival the coast redwoods in height and breadth (see Figure 22.14).

Moving poleward from the Cf climate, we next encounter the temperate continental climate—the Dc. The Dc regions are found only in the NH between about 35 and 50° latitude on the eastern side of the continents. Moving into the heart of the continent, Dc can extend as far as 60° latitude.

It is in the Dc climate that we begin to understand the

TABLE 22.6
The Climate of Selected Cf and Cw Cities

| | | | Jan | Feb | Mar | Apr | May | Jun | Jul | Aug | Sep | Oct | Nov | Dec | Year | Record High and Low Temperatures |
|---|---|---|---|---|---|---|---|---|---|---|---|---|---|---|---|---|
| 43. | New Orleans, | T | 12 | 13 | 16 | 20 | 24 | 27 | 28 | 28 | 26 | 21 | 16 | 13 | 20 | 39 |
| | Louisiana (30° N) | ΔT | 8 | 9 | 9 | 9 | 8 | 9 | 8 | 8 | 9 | 7 | 8 | 9 | 16 | −14 |
| | Cf | P | 9.7 | 10.2 | 13.5 | 11.7 | 11.2 | 11.2 | 17.1 | 13.5 | 12.8 | 7.1 | 8.4 | 10.4 | 13.7 | |
| 44. | Buenos Aires, | T | 23 | 23 | 21 | 17 | 13 | 9 | 10 | 11 | 13 | 15 | 19 | 22 | 16 | 40 |
| | Argentina | ΔT | 12 | 11 | 11 | 11 | 9 | 9 | 8 | 9 | 10 | 11 | 11 | 12 | 14 | −6 |
| | (35° S) Cf | P | 7.9 | 7.1 | 10.9 | 8.9 | 7.6 | 6.1 | 5.6 | 6.1 | 7.9 | 8.6 | 8.4 | 9.9 | 147 | |
| 45. | Durban, S. Africa | T | 25 | 25 | 24 | 22 | 20 | 18 | 18 | 19 | 20 | 21 | 22 | 24 | 22 | 42 |
| | (30° S) | ΔT | 9 | 9 | 10 | 10 | 11 | 12 | 12 | 11 | 9 | 9 | 9 | 9 | 7 | 4 |
| | Cf | P | 10.9 | 12.2 | 12.7 | 7.6 | 5.1 | 3.3 | 2.8 | 3.8 | 7.1 | 10.9 | 12.2 | 11.9 | 101 | |
| 46. | Shanghai, China | T | 4 | 6 | 10 | 14 | 21 | 26 | 28 | 28 | 25 | 19 | 12 | 6 | 16 | 40 |
| | (31° N) | ΔT | 8 | 9 | 10 | 10 | 10 | 8 | 9 | 9 | 10 | 11 | 9 | 8 | 24 | −12 |
| | Cf | P | 4.8 | 6.1 | 8.4 | 9.2 | 9.7 | 17.8 | 14.7 | 14.0 | 13.2 | 7.4 | 5.3 | 3.8 | 115 | |
| 47. | Sydney, Australia | T | 22 | 22 | 21 | 18 | 15 | 13 | 11 | 13 | 15 | 18 | 20 | 21 | 17 | 46 |
| | (34° S) | ΔT | 7 | 7 | 7 | 7 | 7 | 7 | 7 | 9 | 9 | 8 | 8 | 8 | 11 | 2 |
| | Cf | P | 8.9 | 10.2 | 12.7 | 13.5 | 12.7 | 11.7 | 11.7 | 7.6 | 7.3 | 7.1 | 7.3 | 7.3 | 118 | |
| 48. | Cherrapunji, | T | 12 | 13 | 15 | 18 | 21 | 20 | 20 | 20 | 20 | 19 | 16 | 13 | 16 | 31 |
| | India (25° N) | ΔT | 8 | 10 | 9 | 7 | 5 | 4 | 4 | 4 | 5 | 6 | 8 | 8 | 8 | 1 |
| | Cw | P | 1.8 | 5.4 | 18.5 | 66.5 | 128.0 | 269.5 | 244.3 | 178.0 | 109.9 | 49.3 | 6.9 | 1.3 | 1080 | |
| 49. | Lucknow, India | T | 16 | 18 | 24 | 30 | 33 | 32 | 30 | 30 | 29 | 26 | 22 | 18 | 25 | 48 |
| | (27° N) | ΔT | 15 | 16 | 17 | 17 | 17 | 14 | 7 | 7 | 9 | 13 | 13 | 14 | 17 | 1 |
| | Cw | P | 2.0 | 1.8 | 0.8 | 0.8 | 2.0 | 11.4 | 30.5 | 29.2 | 18.8 | 3.3 | 0.5 | 0.8 | 102 | |
| 50. | Chungking, China | T | 9 | 10 | 14 | 20 | 23 | 27 | 28 | 30 | 25 | 20 | 15 | 10 | 19 | 44 |
| | (29½° N) | ΔT | 5 | 6 | 7 | 8 | 8 | 8 | 9 | 8 | 7 | 6 | 5 | 5 | 21 | −2 |
| | Cf | P | 1.8 | 2.0 | 3.8 | 9.7 | 14.5 | 18.1 | 14.2 | 12.0 | 14.7 | 10.9 | 4.8 | 2.0 | 109 | |

meaning of the word, *winter*. Although snow is only occasional at the southern border of the Dc climate, it generally covers the northern borders of Dc regions all winter.

I was always baffled by the fact that the duration of the snow cover increases so sharply over a short distance. For instance, on Cape Cod, Massachusetts, snow covers the ground an average of only 20 days per year, but less than 100 km to the north the snow lasts on the ground more than 100 days per year (see Figure 22.15).

It is easy to explain this sharp transition. In Cape Cod the average temperature during the winter is 0°C, which means that most afternoons the temperature rises above freezing and melting occurs. A mere 100 km farther north the mean temperature is −3°C, so that on many days the temperature never reaches freezing and almost no melting occurs. Then, add to this the fact that the snowfall is somewhat larger in the north, and you can see how it can last a long time.

Winter weather is generally governed by cold north and northwest winds from the large continental high pressure areas. Passing cyclones tend to bring rain to the southern margins and snow to the northern margins of the Dc climate. Occasionally, warm air will penetrate far poleward (especially in America), but such relatively warm days become more and more unusual the farther north you go.

The weather becomes progressively more severe toward the north and, despite the name, *temperate*, it was the Russian Dc winter that decimated the armies of Napoleon and Hitler. On the southern margins of the Dc climate the coldest month averages between 0 and 5°C, while on the poleward margins the coldest month averages only −10 to −15°C and as low as −20°C in eastern Asia.

The severe winter sometimes brings with it an astounding pristine beauty. The white snow glistens on the ground and on the trees, and after the occasional freezing rain storms (ice storms) the trees look as though they are covered with the most dazzling jewelry.

Winter cold waves push the temperature as much as 10

FIGURE 22.14
Eucalyptus forest in Australia. These forests extend from Ar to Cf climate zones and rival the redwood forests in height.

or 20°C below normal, making the climate feel polar. But in a few months spring approaches, and life slowly begins to sprout up after the long winter. Perhaps the best portrayal of this transition from the bleakness of winter to the pastoral beauty and fertility of spring can be seen in the movie, *Dr. Zhivago.*

The scene is not always so pastoral since spring rains can melt the snows of winter, turning the earth swampy and occasionally producing disastrous floods. One other unhappy irony of the spring is that it is the season when the suicide rate is highest. When everything around them is springing back to life, depressed people feel left behind.

Then the spring, which always seems to take too long in coming, rapidly phases into summer. The Dc summers are hot (27°C) at their southern margins, but cool (17°C) at their northern margins. Heat waves still occur, but they are less frequent and shorter as you move north. Cold fronts occasionally pass by and bring a few days of pleasant or even chilly weather.

Most summer precipitation falls from showers and thundershowers, although occasionally rain falls from a few passing lows. As in Cf regions summer is usually the wettest season, except along coastal margins where summer showers are suppressed by the cool surface waters (e.g., the northern island of Japan and the Maritime Provinces of Canada). Precipitation varies from 40 cm in the interior to about 120 cm near the coast, although it can be higher in the mountains.

When fall approaches, the temperature begins to drop rapidly. Relatively warm, humid air still occasionally

Number of days with snow cover

FIGURE 22.15
Duration of winter snow cover in the United States. Notice the rapid transition from southern to northern New England.

TABLE 22.7
The Climate of Selected Dc Cities

| | | Jan | Feb | Mar | Apr | May | Jun | Jul | Aug | Sep | Oct | Nov | Dec | Year | Record High and Low Temperatures |
|---|---|---|---|---|---|---|---|---|---|---|---|---|---|---|---|
| 51. | Bismark, N. Dakota (47° N) | $T$ | −13 | −10 | −4 | 6 | 12 | 18 | 22 | 21 | 14 | 8 | −2 | −9 | 5 | 46 −43 |
| | | $\Delta T$ | 12 | 12 | 12 | 13 | 14 | 13 | 15 | 15 | 15 | 15 | 12 | 12 | 35 | |
| | | $P$ | 1.0 | 1.0 | 2.0 | 3.1 | 5.1 | 8.6 | 5.6 | 4.3 | 3.1 | 2.3 | 1.5 | 1.0 | 39 | |
| 52. | Chicago, Illinois (42° N) | $T$ | −4 | −3 | 3 | 10 | 16 | 22 | 24 | 24 | 19 | 13 | 5 | −2 | 10 | 41 −31 |
| | | $\Delta T$ | 8 | 8 | 8 | 11 | 11 | 11 | 11 | 11 | 11 | 11 | 9 | 8 | 28 | |
| | | $P$ | 4.8 | 4.1 | 6.9 | 7.6 | 9.4 | 10.4 | 8.6 | 8.1 | 6.9 | 7.2 | 5.6 | 4.8 | 82 | |
| 53. | New York, New York (41° N) | $T$ | 0 | 0 | 4 | 10 | 16 | 21 | 25 | 23 | 19 | 13 | 7 | 1 | 12 | 41 −27 |
| | | $\Delta T$ | 7 | 8 | 9 | 9 | 9 | 9 | 9 | 9 | 9 | 9 | 7 | 7 | 25 | |
| | | $P$ | 8.4 | 8.4 | 8.7 | 8.4 | 8.7 | 8.7 | 10.4 | 10.7 | 8.7 | 8.7 | 8.7 | 8.4 | 107 | |
| 54. | Montreal, Quebec (45½° N) | $T$ | −11 | −9 | −4 | 5 | 13 | 18 | 21 | 19 | 15 | 8 | 1 | −7 | 5 | 36 −37 |
| | | $\Delta T$ | 8 | 7 | 8 | 7 | 9 | 9 | 9 | 8 | 8 | 7 | 6 | 7 | 32 | |
| | | $P$ | 9.4 | 8.1 | 9.4 | 6.1 | 7.9 | 8.9 | 9.6 | 8.6 | 8.9 | 8.4 | 8.6 | 9.4 | 103 | |
| 55. | Berlin, Germany (52° N) | $T$ | −1 | 0 | 4 | 8 | 12 | 16 | 18 | 17 | 14 | 9 | 5 | −1 | 9 | 36 −26 |
| | | $\Delta T$ | 5 | 6 | 8 | 9 | 10 | 11 | 11 | 10 | 9 | 8 | 6 | 6 | 19 | |
| | | $P$ | 4.8 | 3.3 | 3.8 | 4.3 | 4.8 | 5.8 | 7.9 | 5.6 | 4.8 | 4.3 | 4.3 | 4.8 | 59 | |
| 56. | Warsaw, Poland (52° N) | $T$ | −3 | −2 | 3 | 8 | 13 | 17 | 19 | 18 | 15 | 9 | 4 | −1 | 8 | 37 −30 |
| | | $\Delta T$ | 5 | 6 | 7 | 9 | 10 | 11 | 11 | 10 | 9 | 7 | 6 | 5 | 22 | |
| | | $P$ | 3.0 | 2.8 | 3.3 | 3.8 | 4.8 | 6.6 | 7.6 | 7.6 | 4.8 | 4.3 | 3.5 | 3.5 | 56 | |
| 57. | Bucharest, Romania (44° N) | $T$ | −3 | −2 | 4 | 11 | 16 | 20 | 23 | 22 | 18 | 12 | 5 | −1 | 10 | 41 −28 |
| | | $\Delta T$ | 7 | 8 | 11 | 12 | 13 | 14 | 14 | 13 | 12 | 11 | 8 | 7 | 26 | |
| | | $P$ | 3.0 | 2.8 | 4.8 | 5.1 | 6.4 | 8.4 | 7.1 | 4.8 | 3.8 | 3.8 | 4.8 | 4.3 | 58 | |
| 58. | Leningrad, U.S.S.R. (60° N) | $T$ | −8 | −8 | −4 | 4 | 10 | 15 | 18 | 15 | 11 | 5 | −1 | −6 | 5 | 33 −38 |
| | | $\Delta T$ | 6 | 7 | 8 | 8 | 9 | 8 | 8 | 7 | 7 | 4 | 4 | 4 | 26 | |
| | | $P$ | 2.5 | 2.3 | 2.3 | 2.5 | 4.1 | 5.1 | 6.3 | 7.1 | 5.4 | 4.6 | 4.1 | 3.6 | 49 | |
| 59. | Moscow, U.S.S.R. (56° N) | $T$ | −11 | −9 | −5 | 3 | 12 | 17 | 19 | 17 | 11 | 4 | −2 | −8 | 4 | 36 −42 |
| | | $\Delta T$ | 7 | 8 | 9 | 9 | 10 | 11 | 12 | 11 | 9 | 7 | 6 | 6 | 30 | |
| | | $P$ | 3.8 | 3.5 | 2.8 | 4.8 | 5.6 | 7.4 | 7.6 | 7.4 | 4.8 | 6.9 | 4.3 | 4.1 | 63 | |
| 60. | Peking, China (40° N) | $T$ | −4 | −2 | 5 | 14 | 20 | 24 | 26 | 25 | 20 | 13 | 4 | −3 | 12 | 43 −23 |
| | | $\Delta T$ | 9 | 10 | 12 | 13 | 12 | 12 | 10 | 10 | 11 | 10 | 9 | 9 | 30 | |
| | | $P$ | 0.3 | 0.5 | 0.5 | 1.5 | 3.5 | 7.6 | 23.8 | 16.0 | 6.6 | 1.5 | 0.8 | 0.3 | 63 | |
| 61. | Vladivostok, U.S.S.R. (43° N) | $T$ | −14 | −11 | −4 | 4 | 11 | 17 | 19 | 18 | 14 | 9 | −2 | −11 | 3 | 33 −30 |
| | | $\Delta T$ | 7 | 7 | 7 | 7 | 8 | 6 | 6 | 6 | 6 | 8 | 7 | 7 | 33 | |
| | | $P$ | 0.8 | 1.0 | 1.8 | 3.0 | 5.3 | 7.4 | 8.4 | 11.9 | 10.9 | 4.8 | 3.0 | 1.5 | 60 | |

makes its way up from the subtropical high, cooling near the surface and resulting in morning fog and all too frequent pollution. Cold waves increase in frequency and severity, as every living thing prepares for winter.

Sensing the approaching winter, the leaves of deciduous trees commit the ultimate sacrifice so that the trees may live to beget many more generations of leaves. The leaves relinquish their vital nutritive fluids which actually ooze back into the branches giving the trees the energy to sprout new leaves in the spring to come.

The death of the leaves is an astoundingly beautiful spectacle to watch (see Plate 12). As their fluids drain away, the leaves turn from green to various shades of red, orange, yellow, and purple, so that when sunlight strikes the forest it makes it appear lustrous and phosphorescent. What a life it must have been that the death should be so magnificent.

During the fall, temperatures drop anywhere from 20 to 40°C. The days become much shorter and, after the color display, the leaves turn brown and are blown off the trees by the ever-increasing winds. The land takes on a dreary and barren appearance which lasts half the year. Unless you enjoy the winter sports, perhaps it is best to follow the birds south for the winter. Read the climate statistics for Dc stations in Table 22.7, and then decide for yourself.

The forest is dominant in the Dc climate. It is predominantly deciduous and thus is bare about half the year. Farther inland as the precipitation decreases, openings form in the forest and these are filled with tall prairie grasses (see Figure 22.16). Even farther inland the grasses dominate, and finally they too give way to semiarid vegetation.

# BOREAL AND POLAR CLIMATES

North of the so-called temperate climates you encounter the great boreal forests (or Taiga) of Russia and Canada (see Figure 22.17). These forests form an almost continuous band which is found in a zone that slopes between 50 and 60° latitude on the eastern shore and to between 60

FIGURE 22.16
Tall prairie grasses on the Great Plains. Once widespread, these have been largely destroyed.

and 70° latitude on the western shore of the continents. The boreal (E) climate gets its name from these great boreal forests of evergreen trees (pines, spruce, hemlocks, and firs; see Figure 22.18).

The main features of the E climate are similar to those of the Dc climate. The main difference is one of degree or, actually, of many degrees! There are still four seasons, but the summer is considerably shorter and doesn't always feel like summer. Only 1 to 3 months have a mean temperature that exceeds 10°C.

Still, the summer is surprisingly warm considering the latitude. The warmest month averages anywhere from 10 to 18°C, and heat waves have pushed the thermometer above 35°C at many stations. Generally, the summer temperatures are quite cool and the days are very long.

When summer ends the temperature begins to drop almost as rapidly as if someone had turned on a giant freezer. The E climates have the widest range of temperature on earth. The annual temperature range is at least 30°C and typically 40°C or more. Of course, you remember Verkhoyansk, Siberia, where the annual temperature range is 65°C! In fact, in the 31 days from October 1 to November 1, the mean temperature at Verkhoyansk falls 22°C, which is more than the *annual* temperature range at most places. In the boreal forest the winter comes on with a suddenness that feels like a smack in the face.

And what a winter it is! A few blizzards (purgas in Russia) alternate with many crystal clear, absolutely frigid days in which cP air is manufactured. Days are very short with the sun barely rising above the southern horizon for a few hours. The warmest parts of the boreal forest may be as high as −10°C during January, but generally it lies between −20 and −30°C in North America and between −30 and −40°C in Siberia. The towns located in the "protected" river valleys such as Verkhoyansk and Oimekon (which may be 1°C colder than Verkhoyansk itself) are especially cold and may be plagued by ice fog for days on end.

Winter often turns out to be the dry season because the cold air simply cannot hold much vapor and, even when it snows, the snow is so light that it doesn't amount to much. Snowfall amounts are often smaller than the totals in the Dc regions farther south. You should recall that the extreme coldness of the ground causes the sublimation of ice crystals on the ground and thus the ground remains white all winter.

After long months of bitter cold, winter gives way to a brief, marshy spring. During this time biting insects such as mosquitos and the notorious black flies (sometimes known as May flies) appear in such tremendous numbers that there is no escaping them. They even bite the very

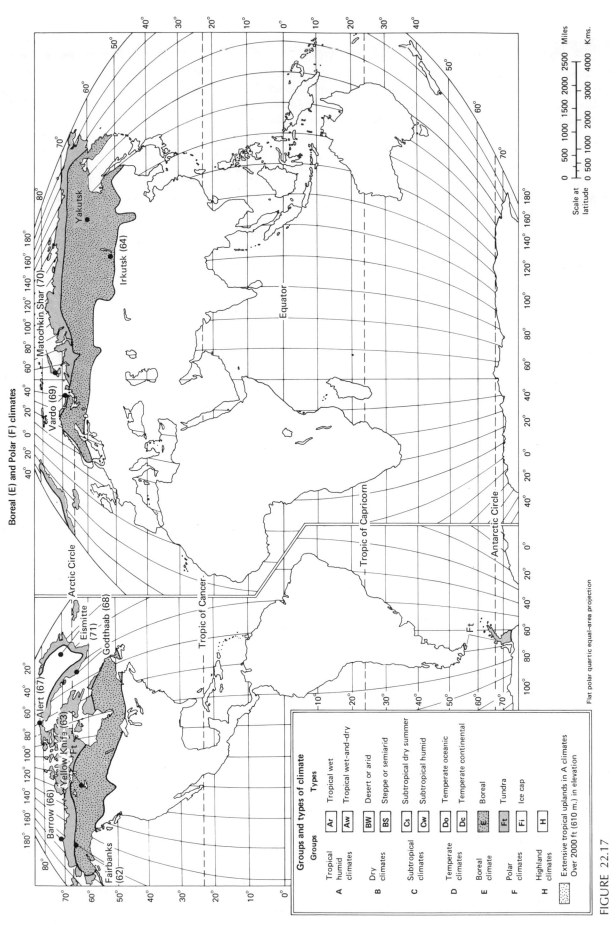

Boreal (E) and Polar (F) climates

**Groups and types of climate**

| Groups | | Types |
|---|---|---|
| **A** Tropical humid climates | Ar | Tropical wet |
| | Aw | Tropical wet-and-dry |
| **B** Dry climates | BW | Desert or arid |
| | BS | Steppe or semiarid |
| **C** Subtropical climates | Cs | Subtropical dry summer |
| | Cw | Subtropical humid |
| **D** Temperate climates | Do | Temperate oceanic |
| | Dc | Temperate continental |
| **E** Boreal climate | E | Boreal |
| **F** Polar climates | Ft | Tundra |
| | Fi | Ice cap |
| **H** Highland climates | H | |

Extensive tropical uplands in A climates
Over 2000 ft (610 m.) in elevation

Scale at latitude

Flat polar quartic equal-area projection

FIGURE 22.17

Locations of boreal (E) and polar (Ft and Fi) climates. Climate statistics for cities shown are given in Table 22.8.

FIGURE 22.18
The boreal forest in Canada.

birds that feed on them. There are times that black flies have been known to swarm so thickly around an animal or person that they completely mask the poor, suffering creature below. The flies have also been known to whip caribou up into such a frenzy that they run blindly until they drop dead from exhaustion.

In the past the black flies bred only once during a summer and lasted a few long weeks before dying. It now

TABLE 22.8
The Climate of Selected E, Ft, Fi Cities

| | | Jan | Feb | Mar | Apr | May | Jun | Jul | Aug | Sep | Oct | Nov | Dec | Year | Record High and Low Temperatures |
|---|---|---|---|---|---|---|---|---|---|---|---|---|---|---|---|
| 62. | Fairbanks, Alaska (65° N) E | T | −24 | −19 | −12 | −2 | 8 | 15 | 16 | 13 | 7 | −4 | −16 | −23 | −4 | 37 −54 |
| | | ΔT | 11 | 13 | 15 | 13 | 13 | 13 | 12 | 12 | 11 | 9 | 10 | 10 | 40 | |
| | | P | 2.3 | 1.3 | 1.0 | 0.8 | 1.8 | 3.5 | 4.5 | 5.5 | 2.7 | 2.3 | 1.5 | 1.3 | 28.7 | |
| 63. | Yellow Knife, N.W.T., Canada (62° N) E | T | −26 | −24 | −17 | −7 | 4 | 14 | 16 | 14 | 8 | −1 | −13 | −23 | −5 | 32 −51 |
| | | ΔT | 8 | 9 | 10 | 11 | 11 | 10 | 9 | 8 | 7 | 6 | 6 | 7 | 42 | |
| | | P | 2.0 | 1.5 | 1.8 | 1.0 | 1.8 | 1.5 | 3.8 | 3.5 | 2.5 | 3.3 | 2.5 | 2.0 | 27.4 | |
| 64. | Irkutsk, U.S.S.R. (52° N) E | T | −21 | −19 | −10 | −1 | 7 | 14 | 16 | 14 | 8 | −1 | −11 | −19 | −2 | 37 −50 |
| | | ΔT | 9 | 10 | 11 | 12 | 12 | 11 | 11 | 11 | 11 | 11 | 9 | 9 | 37 | |
| | | P | 1.3 | 1.0 | 0.8 | 1.5 | 3.3 | 5.6 | 7.9 | 7.1 | 4.3 | 1.8 | 1.5 | 1.5 | 37.8 | |
| 65. | Yakutsk, U.S.S.R. (62° N) E | T | −43 | −37 | −23 | −9 | 5 | 15 | 19 | 16 | 6 | −9 | −29 | −41 | −12 | 36 −64 |
| | | ΔT | 4 | 6 | 10 | 12 | 12 | 11 | 11 | 10 | 9 | 7 | 5 | 4 | 62 | |
| | | P | 2.3 | 0.5 | 1.0 | 1.5 | 2.7 | 5.3 | 4.3 | 6.6 | 3.0 | 3.5 | 1.5 | 2.3 | 34.8 | |
| 66. | Barrow, Alaska (71° N) Ft | T | −27 | −24 | −22 | −18 | −7 | 2 | 5 | 4 | −1 | −9 | −19 | −25 | −11 | 26 −49 |
| | | ΔT | 8 | 8 | 8 | 8 | 7 | 7 | 7 | 6 | 5 | 5 | 7 | 8 | 32 | |
| | | P | 0.5 | 0.5 | 0.3 | 0.2 | 0.3 | 1.0 | 2.0 | 2.3 | 1.5 | 1.3 | 0.5 | 0.5 | 10.9 | |
| 67. | Alert, N.W.T., Canada (83° N) Ft | T | −31 | −30 | −28 | −25 | −14 | −3 | 4 | 1 | −8 | −19 | −27 | −30 | −13 | 19 −47 |
| | | ΔT | 6 | 6 | 6 | 6 | 6 | 5 | 4 | 4 | 5 | 5 | 5 | 6 | 35 | |
| | | P | 0.5 | 0.8 | 0.8 | 0.8 | 1.3 | 1.5 | 1.3 | 2.8 | 2.5 | 2.2 | 0.5 | 1.0 | 16.0 | |
| 68. | Godthaab, Greenland (64°N) Ft | T | −10 | −9 | −7 | −4 | 1 | 5 | 7 | 6 | 3 | −1 | −6 | −9 | −1 | 24 −29 |
| | | ΔT | 5 | 5 | 6 | 6 | 7 | 8 | 8 | 8 | 6 | 5 | 5 | 5 | 17 | |
| | | P | 3.5 | 4.3 | 4.1 | 3.0 | 4.3 | 3.5 | 5.6 | 7.9 | 8.4 | 6.4 | 4.8 | 3.8 | 60 | |
| 69. | Vardo, Norway (70° N) Ft | T | −5 | −5 | −3 | −1 | 4 | 7 | 9 | 8 | 6 | 2 | −1 | −3 | 2 | 27 −24 |
| | | ΔT | 4 | 4 | 4 | 4 | 5 | 5 | 5 | 5 | 4 | 3 | 3 | 4 | 14 | |
| | | P | 6.4 | 6.4 | 5.9 | 3.8 | 3.3 | 3.3 | 3.8 | 4.3 | 4.8 | 6.4 | 5.3 | 6.1 | 60 | |
| 70. | Matochkin Shar, Novaya Zemlya (73° N) Ft | T | −17 | −17 | −16 | −14 | −6 | 1 | 5 | 4 | 2 | −3 | −11 | −15 | −6 | 20 −41 |
| | | ΔT | 8 | 8 | 8 | 8 | 7 | 6 | 6 | 6 | 6 | 5 | 5 | 7 | 22 | |
| | | P | 1.5 | 1.5 | 1.5 | 1.0 | 0.8 | 1.0 | 3.5 | 3.8 | 3.8 | 1.5 | 1.5 | 1.0 | 22.6 | |
| 71. | Eismitte, Greenland (71° N) Fi | T | −41 | −47 | −40 | −31 | −20 | −16 | −11 | −18 | −22 | −36 | −43 | −38 | 29 | −3 −65 |
| | | ΔT | 11 | 11 | 12 | 13 | 11 | 10 | 10 | 10 | 11 | 11 | 11 | 11 | 36 | |
| | | P | 1.5 | 0.5 | 0.8 | 0.5 | 0.3 | 0.2 | 0.3 | 1.0 | 0.8 | 1.3 | 1.3 | 2.5 | 10.9 | |
| 72. | South Pole Station, Antarctica (90° S) Fi | T | −29 | −40 | −54 | −59 | −57 | −57 | −59 | −59 | −59 | −51 | −39 | −28 | 45 | −15 −81 |
| | | ΔT | 3 | 4 | 5 | 6 | 6 | 7 | 7 | 6 | 6 | 4 | 3 | 2 | 30 | |
| | | P | * | 0.2 | 0.0 | 0.0 | 0.0 | 0.0 | 0.0 | 0.0 | 0.0 | * | 0.0 | * | 0.3 | |
| 73. | McMurdo Sound, Antarctica (78° S) Fi | T | −3 | −8 | −19 | −21 | −24 | −23 | −26 | −28 | −24 | −20 | −9 | −4 | −15 | 6 −51 |
| | | ΔT | 5 | 5 | 6 | 7 | 7 | 7 | 10 | 10 | 9 | 7 | 7 | 4 | 23 | |
| | | P | 1.3 | 1.8 | 1.0 | 1.0 | 1.0 | 0.8 | 0.5 | 0.8 | 1.0 | 0.5 | 0.5 | 0.8 | 10.9 | |

seems that a new strain has evolved that breeds repeatedly from the first thaw in spring to the first frost in fall. Let us only pray that this breed doesn't spread too far.

These insects are the plague of the tundra as well; in Siberia one torture of days gone by was to strip a man stark naked outdoors and let the insects kill him. The ordeal took a day or two at most. Needless to say, things like that don't happen anymore.

Even in midsummer, passing lows give rain and there is also some rain from showers. Summer is the wet season in the boreal forest, even though most stations receive less than 10 cm in the wettest month. The annual precipitation averages anywhere from 15 to about 100 cm. Climate statistics for stations with an E climate are given in Table 22.8.

Moving farther north we reach the tundra (Ft) which

FIGURE 22.19
Tundra in bloom in the Colorado Rockies.

borders the Arctic Ocean. Generally lying above the Arctic Circle, the tundra may have a few dwarfed trees at most, but generally the dominant vegetation consists of low shrubs, mosses, and lichens. For the most part the tundra looks rather bleak, but if you look closely you will see many tiny, delicate flowers growing during the short summer (see Figure 22.19).

The warmest month on the tundra is below 10°C, but is at least 0°C. Even with the cool temperatures there is an astounding profusion of life during the few summer months. Countless millions of birds arrive here from warmer climes in the spring for their summer feast of insects and fish, and vast herds of caribou once roamed here (and in the boreal forests to the south) before their decimation by "civilized" humans. Rodents scurry along

the ground and, believe it or not, have always constituted the main dish in the diet of wolves! This is all accompanied by the symphony of the insects. And in the nutrient-rich ocean the cold waters support a soup of plankton that the whales and other creatures thrive on.

But on the tundra the winter is never far away, since frost or even snow may occur on any given day of the year. Most of the year's meager precipitation falls during the summer and tends to fall as drizzle from weak cyclones. Precipitation averages anywhere from 10 to 50 cm annually but may be larger on exposed mountainous coasts. Despite the low precipitation totals which seem more representative of desert stations, the relative humidity on the tundra is usually very high. Fog is also common along the cold seacoasts during the summer.

One distinctive feature of the tundra is **permafrost**. The ground on the tundra is frozen solid to depths as great as 600 m year round, but each summer the top few meters thaws out. The surface waters cannot penetrate the permanently frozen permafrost so the top meter or two of soil turns to mud in the summer. Low-lying land turns into one giant swamp or marsh, while sloping lands literally flow downhill (this is called **solifluction** or **cryoturbation**).

After the short summer the tundra briefly turns a fiery red and then everything quickly refreezes. Most of the animals either move south or hibernate. Winter temperatures on the tundra average between −20 and −35°C, and the sun never rises for weeks or even months, although twilight lasts for weeks. Extreme cold is prevented by the closeness of the sea (recall the heat conducted through the ice and breaks in the ice) so that the tundra winter may be several degrees warmer than winter in the boreal forest farther south!

The Eskimos are the people of the tundra. Forced by other, more aggressive peoples to live where no one else thought it possible to live, the Eskimo has learned to make use of every single scanty resource. And now, like the peoples of other forbidding places such as the desert and the rain forest, the Eskimo is rapidly abandoning the traditional life of hardship and scarcity for the "modern" life with its many conveniences. Thus, the last of the true ecologists on this earth are rapidly disappearing—of their own free will.

Polar lands such as Greenland and Antarctica are covered with ice that averages 2 to 3 km in thickness (see Figures 22.20 and 22.21). Some of this land ice flows down to the sea in glaciers and breaks off into huge chunks known as icebergs. To give you an idea of the immense quantities of ice on these lands, if we were able to pile all of Greenland's ice on the island of Manhattan, it would result in a pile more than 50,000 km high. Yet Antarctica has more than 10 times as much ice as Greenland!

The polar ice-cap climate (Fi) is largely governed by this ice (see Figure 22.22). Temperatures average below 0°C all year, partly because the ice reflects most of the sunlight, and partly because of the high altitude of the ice surface. Almost all of Antarctica averages well below −30°C during July (their coldest month), and in the heart of the continent at Vostok—where the world's lowest temperature was recorded—the July *average* is −68°C! Even in summer, Vostok averages below −30°C.

During the warmest months some melting of the ice is possible, and this accumulates as shallow lakes on the ice surface or runs through crevasses in the ice down toward the sea.

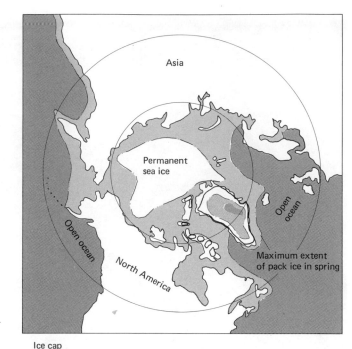

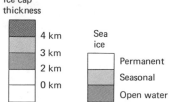

FIGURE 22.20
The extent of sea ice in the NH and the thickness of the Greenland ice cap.

There is a tendency for high pressure to dominate and the air is so pure that the sky often appears violet rather then blue! Fog is common along the seacoast, but inland only an occasional blizzard breaks up the mostly clear skies. The polar lands lie above the major storm belts so that high winds are rare except along steep coastal regions where katabatic winds sometimes roar.

The air is so cold that ice crystals fall from clear skies, and these ice crystals produce halos that are spectacular as well as common. The extreme cold is influential in keeping the mean precipitation below 10 cm over the bulk of the region. Even so, the extreme cold so reduces evaporation that this meager precipitation is adequate to maintain the great ice sheets, despite the fact that every year enormous quantities of ice flow into the sea.

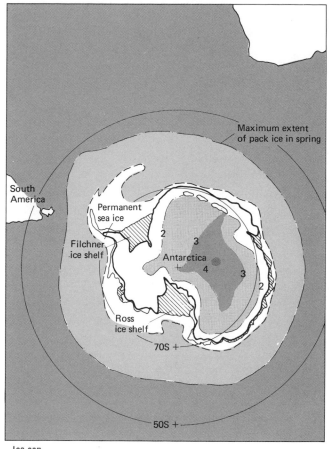

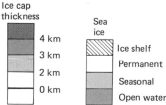

FIGURE 22.21
The extent of sea ice in the SH and the thickness of the Antarctic ice cap.

# MOUNTAIN CLIMATES

Climate changes quite rapidly as you climb a mountain because of the decrease of temperature with height. This change is so dramatic that there are tropical mountains on which the climate at the bottom is Ar, while on the top it is Fi! Climbing such a mountain is like traveling from the equator to the pole (see Figure 22.23).

Consider the climate on Mount Washington in New Hampshire, which is located in a Dc region. Starting from the base at Pinkham Notch at an altitude of 670 m the weather is rather cool. As you begin climbing you are surrounded by tall deciduous trees, but within an hour these give way to pine trees and the air is somewhat cooler. Passing up through the famous Tuckerman's Ravine the pine trees start to look stunted; then there is a zone in which trees the height of a man are densely matted. When you emerge from the ravine onto the Alpine Gardens there are only some dwarf pines, stunted and gnarled with age in the hollows between the rocks; here tundra vegetation dominates. This becomes thinner as you near the summit that seems to be covered by a rubble of loose rocks. The rocks are discolored by a thin layer of lichens, the makeup of the mountains.

The summit (1949 m) is not quite high enough for permanent snow, but it experiences freezing weather and snow in every month of the year. The summit is often shrouded in clouds, duly reported as fog on an average of 310 days a year. Precipitation at the summit is significantly higher than at the base and more of it falls as snow.

Very dramatic changes in climate also occur in the Sierra Nevada Mountains of California. There the climate changes from semiarid to Mediterranean in the foothills on the western slopes of the Sierra. Above 1 km the forests begin to dominate and, as you continue higher, you enter the world of *Sequoia giganteum,* the world's largest living tree. The sequoias thrive at altitudes around 2 km, along with many other giant trees. Above this zone, the forest thins out and gives way to dwarf pines and tundra vegetation (see Figure 22.23). Finally, on the higher peaks in places shaded from the sunlight there are snowfields and even glaciers.

But when you cross the crests and look down the eastern slopes toward Nevada, the change is shocking. Vegetation thins out immediately and bleak desert scenery greets your eyes.

Mountains near the equator show even greater climate ranges. Rain forests at the base give way to midlatitude weather and vegetation, and finally above about 5 km the weather is polar. If hiking paths have been cleared, it takes about two days to hike from equatorial to polar climates. It would take somewhat longer to walk from the equator to the pole. The dramatic changes of climate with altitude have been beautifully documented in the film, *Sky Above, Mud Below,* about an expedition that crossed the island of New Guinea. Climate statistics for stations with H climate (Figure 22.24) are given in Table 22.9.

Temperatures fall at an average rate of 6°C/km in the mountains. At first, precipitation generally increases with altitude, but then it begins to decrease because of the

FIGURE 22.22
View from the ice cap of
Antarctica.

cold. Cumulus clouds often form during the day, bringing showers to the mountains while the lowlands a few miles away tend to be clear.

Quito, Ecuador, is located on the equator at an altitude of 2800 m and has an annual average temperature of 12.5°C; here the difference between the warmest and coldest month is only 0.4°C. During the drier season the diurnal temperature variation is quite large because the air is dry and the sunlight is quite intense at high altitudes. The small annual temperature range at Quito is a direct consequence of its being at the equator. Quito has an annual average temperature that is characteristic of the midlatitudes, but it feels like eternal spring year round.

There is one word of warning. Mountain climates are highly variable and virtually unpredictable. If you decide to hike in the mountains, always be prepared for the worst weather imaginable. Bring warm clothes and something waterproof, if possible (umbrellas are useless, of course). As the saying goes, "The mountains make their own weather" (see Figure 22.25).

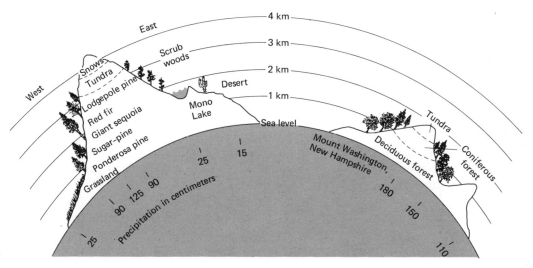

FIGURE 22.23
Profile of vegetation on Mount Washington and in the Sierra Nevada.

**Highland climates (H)**

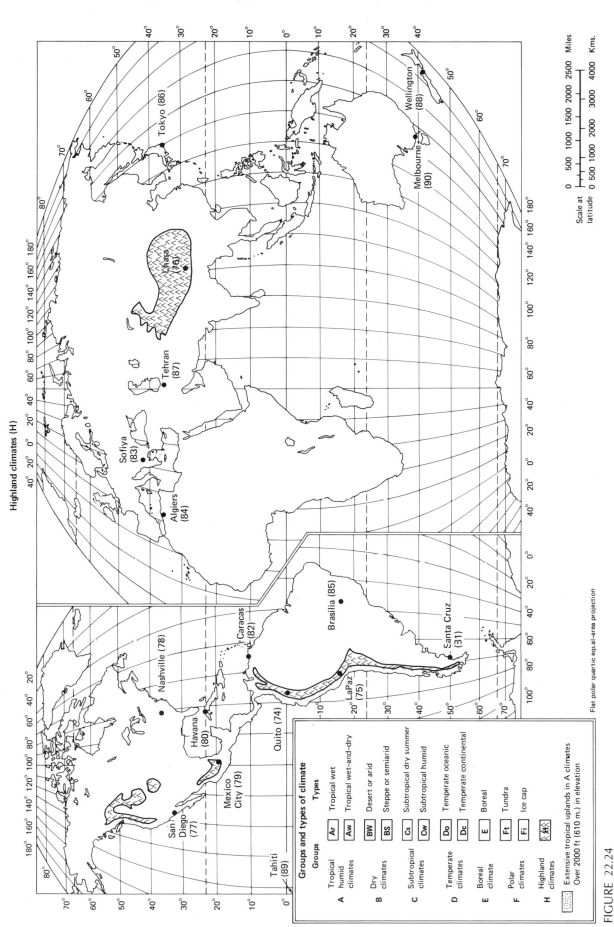

**Groups and types of climate**

| Groups | | Types | |
|---|---|---|---|
| **A** | Tropical humid climates | Ar | Tropical wet |
| | | Aw | Tropical wet-and-dry |
| **B** | Dry climates | BW | Desert or arid |
| | | BS | Steppe or semiarid |
| **C** | Subtropical climates | Cs | Subtropical dry summer |
| | | Cw | Subtropical humid |
| **D** | Temperate climates | Do | Temperate oceanic |
| | | Dc | Temperate continental |
| **E** | Boreal climate | E | Boreal |
| **F** | Polar climates | Ft | Tundra |
| | | Fi | Ice cap |
| **H** | Highland climates | H | |

Extensive tropical uplands in A climates
Over 2000 ft (610 m.) in elevation

Flat polar quartic equal-area projection

Scale at latitude

FIGURE 22.24

Locations of the world's highland (H) climates. Climate statistics for cities shown are given in Table 22.9. Those cities not in the highland regions are part of the climate quiz.

TABLE 22.9
The Climate of Selected H Cities and Climate Quiz

| | | Jan | Feb | Mar | Apr | May | Jun | Jul | Aug | Sep | Oct | Nov | Dec | Year | Record High and Low Temperatures |
|---|---|---|---|---|---|---|---|---|---|---|---|---|---|---|---|
| 74. Quito, Ecuador | T | 13 | 13 | 13 | 13 | 13 | 13 | 13 | 13 | 13 | 13 | 13 | 13 | 13 | 30 |
| (0° S) | ΔT | 12 | 12 | 12 | 12 | 12 | 15 | 15 | 15 | 14 | 14 | 14 | 13 | 0.4 | −4 |
| H | P | 9.9 | 11.2 | 14.2 | 17.6 | 13.7 | 4.3 | 2.0 | 3.0 | 6.9 | 11.2 | 9.7 | 7.9 | 111 | |
| 75. La Paz, Bolivia | T | 12 | 12 | 12 | 11 | 10 | 9 | 9 | 10 | 11 | 12 | 13 | 12 | 11 | 27 |
| (16½° S) | ΔT | 11 | 11 | 13 | 14 | 15 | 15 | 16 | 16 | 15 | 14 | 14 | 11 | 3 | −3 |
| H | P | 11.4 | 10.7 | 6.6 | 3.3 | 1.3 | 0.8 | 1.0 | 1.3 | 2.8 | 4.0 | 4.8 | 9.4 | 57 | |
| 76. Lhasa, Tibet | T | −2 | 1 | 4 | 8 | 12 | 17 | 16 | 16 | 14 | 9 | 4 | 0 | 8 | Not available |
| (30° N) | ΔT | | | | | Data not available | | | | | | | | 19 | |
| H | P | 0.1 | 1.3 | 0.8 | 0.5 | 2.0 | 6.3 | 12.1 | 8.8 | 6.5 | 1.3 | 0.3 | 0.1 | 40 | |

### For the Remaining Cities, Try to Determine the Climate Type

| | | Jan | Feb | Mar | Apr | May | Jun | Jul | Aug | Sep | Oct | Nov | Dec | Year | Record High and Low Temperatures |
|---|---|---|---|---|---|---|---|---|---|---|---|---|---|---|---|
| 77. San Diego, California | T | 13 | 14 | 14 | 16 | 17 | 18 | 21 | 22 | 21 | 19 | 16 | 14 | 17 | 44 |
| (32° N) | ΔT | 11 | 10 | 9 | 8 | 7 | 6 | 6 | 7 | 7 | 8 | 10 | 11 | 9 | −2 |
| Type _____ | P | 4.8 | 3.8 | 3.9 | 2.0 | 0.4 | 0.1 | * | 0.2 | 0.3 | 0.9 | 3.1 | 4.4 | 24.0 | |
| 78. Nashville, Tennessee | T | 3 | 5 | 9 | 16 | 20 | 25 | 26 | 26 | 22 | 16 | 9 | 5 | 15 | 42 |
| (36° N) | ΔT | 10 | 11 | 12 | 12 | 13 | 12 | 12 | 12 | 13 | 14 | 12 | 10 | 23 | −26 |
| Type _____ | P | 12.1 | 11.3 | 12.7 | 10.4 | 10.4 | 8.6 | 9.7 | 8.2 | 7.9 | 5.5 | 8.8 | 11.3 | 117 | |
| 79. Mexico City, Mexico | T | 12 | 13 | 16 | 18 | 20 | 19 | 18 | 19 | 18 | 16 | 14 | 13 | 16 | 33 |
| (19° N) | ΔT | 13 | 14 | 14 | 14 | 13 | 11 | 11 | 10 | 10 | 11 | 12 | 12 | 8 | −4 |
| Type _____ | P | 0.5 | 0.8 | 1.3 | 1.8 | 4.8 | 10.4 | 11.4 | 10.9 | 10.4 | 4.0 | 1.3 | 0.8 | 59 | |
| 80. Havana, Cuba | T | 22 | 22 | 23 | 25 | 26 | 27 | 28 | 28 | 27 | 26 | 24 | 23 | 25 | 40 |
| (23° N) | ΔT | 8 | 8 | 8 | 8 | 9 | 8 | 8 | 8 | 8 | 7 | 7 | 7 | 6 | 6 |
| Type _____ | P | 7.1 | 4.6 | 4.6 | 5.9 | 11.9 | 16.5 | 12.2 | 13.2 | 14.7 | 17.3 | 7.9 | 5.8 | 122 | |
| 81. Santa Cruz, Argentina | T | 15 | 14 | 13 | 9 | 5 | 2 | 2 | 3 | 7 | 9 | 12 | 13 | 9 | 34 |
| (50° S) | ΔT | 12 | 12 | 11 | 10 | 9 | 8 | 7 | 7 | 9 | 11 | 12 | 12 | 13 | −17 |
| Type _____ | P | 1.5 | 1.0 | 0.8 | 1.5 | 1.5 | 1.3 | 1.8 | 1.0 | 0.5 | 1.0 | 1.3 | 2.3 | 15.5 | |
| 82. Caracas, Venezuela | T | 19 | 19 | 20 | 21 | 22 | 21 | 21 | 21 | 21 | 21 | 20 | 19 | 21 | 33 |
| (10½° N) | ΔT | 11 | 12 | 12 | 11 | 10 | 9 | 9 | 9 | 10 | 10 | 9 | 9 | 3 | 7 |
| Type _____ | P | 2.3 | 1.0 | 1.5 | 3.3 | 7.9 | 10.2 | 11.0 | 11.0 | 10.7 | 11.0 | 9.5 | 4.6 | 84 | |
| 83. Sofiya, Bulgaria | T | −2 | 0 | 5 | 11 | 17 | 19 | 21 | 19 | 16 | 11 | 4 | −1 | 10 | 37 |
| (43° N) | ΔT | 7 | 9 | 11 | 12 | 13 | 14 | 14 | 13 | 13 | 12 | 8 | 7 | 23 | −27 |
| Type _____ | P | 3.3 | 2.8 | 4.3 | 5.9 | 8.5 | 8.2 | 6.2 | 5.1 | 5.9 | 5.4 | 4.8 | 3.5 | 64 | |
| 84. Algiers, Algeria | T | 12 | 12 | 14 | 15 | 18 | 21 | 23 | 24 | 23 | 19 | 16 | 13 | 18 | 42 |
| (37° N) | ΔT | 7 | 7 | 7 | 7 | 7 | 7 | 8 | 7 | 7 | 8 | 6 | 6 | 12 | 0 |
| Type _____ | P | 11.2 | 8.4 | 7.4 | 4.1 | 4.6 | 1.5 | 0.1 | 0.5 | 4.0 | 7.9 | 13.0 | 13.8 | 76 | |
| 85. Brazilia, Brazil | T | 22 | 22 | 23 | 22 | 20 | 19 | 18 | 19 | 22 | 23 | 23 | 22 | 21 | 36 |
| (16° S) | ΔT | 8 | 8 | 10 | 11 | 13 | 14 | 15 | 15 | 14 | 10 | 8 | 8 | 5 | 5 |
| Type _____ | P | 22.9 | 19.8 | 12.2 | 8.6 | 3.5 | 0.1 | 0.0 | * | 3.3 | 12.5 | 24.7 | 29.7 | 137 | |
| 86. Tokyo, Japan | T | 4 | 4 | 7 | 13 | 17 | 21 | 24 | 26 | 23 | 17 | 11 | 6 | 15 | 38 |
| (36° N) | ΔT | 9 | 9 | 9 | 9 | 9 | 8 | 8 | 7 | 7 | 8 | 9 | 10 | 12 | −8 |
| Type _____ | P | 4.8 | 6.3 | 10.6 | 13.5 | 14.7 | 16.5 | 14.2 | 15.3 | 23.4 | 20.8 | 9.7 | 5.6 | 157 | |
| 87. Tehran, Iran | T | 2 | 4 | 10 | 16 | 23 | 29 | 30 | 28 | 24 | 18 | 9 | 3 | 16 | 43 |
| (36° N) | ΔT | 10 | 11 | 12 | 12 | 14 | 15 | 15 | 15 | 14 | 13 | 10 | 10 | 28 | −21 |
| Type _____ | P | 4.6 | 3.7 | 4.5 | 3.5 | 1.3 | 0.3 | 0.2 | 0.2 | 0.3 | 0.8 | 2.0 | 3.0 | 24.6 | |
| 88. Wellington, New Zealand | T | 17 | 17 | 16 | 14 | 12 | 10 | 9 | 9 | 11 | 12 | 14 | 16 | 13 | 31 |
| (41° S) | ΔT | 8 | 7 | 7 | 7 | 6 | 6 | 6 | 6 | 6 | 7 | 7 | 7 | 8 | −2 |
| Type _____ | P | 8.1 | 8.1 | 8.1 | 9.6 | 11.7 | 11.7 | 13.7 | 11.7 | 9.6 | 10.2 | 8.9 | 8.9 | 121 | |
| 89. Tahiti, Society Islands | T | 27 | 27 | 27 | 27 | 26 | 25 | 25 | 25 | 26 | 26 | 27 | 27 | 26 | 34 |
| (18° S) | ΔT | 8 | 8 | 8 | 8 | 9 | 10 | 10 | 10 | 10 | 9 | 9 | 8 | 2 | 16 |
| Type _____ | P | 33.6 | 29.2 | 16.5 | 17.3 | 12.4 | 8.1 | 6.6 | 4.8 | 5.8 | 8.6 | 16.5 | 30.2 | 190 | |
| 90. Melbourne, Australia | T | 20 | 20 | 18 | 15 | 12 | 10 | 9 | 11 | 13 | 14 | 17 | 18 | 15 | 46 |
| (38° S) | ΔT | 12 | 12 | 11 | 9 | 8 | 7 | 7 | 9 | 9 | 11 | 11 | 12 | 11 | −3 |
| Type _____ | P | 4.8 | 4.6 | 5.6 | 5.8 | 5.3 | 5.3 | 4.8 | 4.8 | 5.8 | 6.6 | 5.8 | 5.8 | 65 | |

(a)

(b)

FIGURE 22.25
"The mountains make their
own weather." Longs peak in
the morning (a) and in the afternoon (b) of the same day,
August 8, 1976. The morning
was crystal clear and cloudless but by afternoon there
was lightning, thunder and
snow pellets.

# PROBLEMS

**22-1** What do you expect the dominant vegetation type to be in the places listed (a to d) in the following table?

|  | Average Annual Temperature | Annual Range of Monthly Mean Temperatures | Average Annual Precipitation |
|---|---|---|---|
| (a) | 25 | 5 | 100 |
| (b) | 15 | 15 | 90 |
| (c) | −5 | 40 | 60 |
| (d) | 5 | 0 | 110 |

**22-2** By comparing Figures 22.2 and 22.3, which of the following regions would you say had an anomalous climate: (a) the Amazon River Basin, (b) Patagonia (southern South America, east of the Andes), (c) Florida, (d) far eastern Brazil in the San Francisco River Valley, (e) central Australia, or (f) the Horn of Africa (the eastern tip)?

**22-3** Why are the east coasts of continents in the tropics wetter than west coasts? Why is the reverse true above about 40° latitude?

**22-4** Using Figures 16.10 and 16.11, explain how you decided which climates in Problem 22-2 are anomalous.

**22-5** According to Table 22.1, what is the average high and average low daily temperature during the month of April in Rio de Janeiro?

**22-6** (a) What is the coolest month of the year in Monrovia, Liberia (b) In what month do you get the lowest nighttime temperatures? (c) Why is the diurnal range of temperature so small in June and July?

**22-7** Considering its low precipitation, why is Rio de Janeiro classified as an Ar climate?

**22-8** Why is the Amazon rain forest so much rainier and more extensive than the Congo rain forest?

**22-9** Explain why Cristobal, Canal Zone, is so much rainier than Balboa Heights, Canal Zone. A glance back at Figure 17.30 may be helpful.

**22-10** Turn to Figure 16.8 and consider the cloudiness over India. Why should you be able to tell the time of year for each picture without looking at the date?

**22-11** The Serengeti Plains are situated about 3°S latitude. If you want to tour there, when would it be easier and more pleasant to travel—January or July?

**22-12** Explain the characteristic annual cycle of temperature and precipitation in an Aw location.

**22-13** Explain why the precipitation in dry regions is below average for most years.

**22-14** Explain the large diurnal temperature range of the deserts.

**22-15** Why is the world's least rainy place right near the ocean?

**22-16** In what month do the rare rains come: (a) to the northern Sahara and (b) to the southern Sahara?

**22-17** Explain why there are large arid regions in central Asia.

**22-18** Explain why the rainy season begins sooner and ends later in San Francisco than in Los Angeles.

**22-19** Explain why Cape Town, South Africa, is rainy when Durban, South Africa, is not rainy, and vice versa.

**22-20** Why do all subtropical west coasts have cold water, frequent fog, and virtually no summer rain?

**22-21** Contrast the rainy seasons of the Cs climates with those of the Do climates.

**22-22** Explain why September is the warmest month in San Francisco.

**22-23** Compare extreme temperatures at various Cs (and Do) stations with the extremes recorded at various Cf (and Dc) stations. Explain these differences and then give the general wind conditions that can produce record cold temperatures along the west coasts of continents.

**22-24** Virtually all of eastern Asia has more rain in summer than in winter. One notable exception is the coast of Vietnam, midway between Hanoi and Saigon, that has a precipitation maximum in late autumn. Explain this.

**22-25** Explain the typical annual progression of temperature and precipitation in Cf China. Include in your explanation why at the same latitude eastern China is colder than eastern United States on average during the winter, but lower extreme temperatures have been recorded in the United States.

**22-26** Explain why the snowfall in Port Churchill, Canada, is greatest during autumn rather than winter.

**22-27** Explain why most precipitation in Dc regions is showery during the summer but cyclonic during the winter.

**22-28** Why is the winter frequently slightly warmer on the tundra than in the boreal forests farther south?

**22-29** Why are there no Dc or E climates in the SH?

**22-30** At what altitude do you expect tundra to begin in the mountains: (a) at the equator and (b) near Seattle, Washington?

**22-31** Complete the climate quiz of Table 22.9

**22-32** Name your three favorite climates. Which are your least favorite? Why? Compare your answers with the opinions of various friends.

**22-33** For which of the following vacations will you need to bring more than just a light sweater: (a) London in July; (b) Peking, China, in June: (c) Tel Aviv, Israel, in October; (d) San Francisco in July; (e) Perth, Australia, in July; (f) Manilla, Philippines, in January.

**22-34** You want to travel light. For which of the following vacations would you need to bring an umbrella: (a) Cherrapunji, India, in June; (b) Cherrapunji, India, in January; (c) Seattle, Washington, in July; (d) Rome, Italy, in July; (e) Santiago, Chile, in January; (f) Moscow, U.S.S.R., in July?

# THE MARK OF CLIMATE ON THE EARTH

In 1837, a young zoologist named Jean Louis Rodolphe Agassiz agreed to accompany his friend, Jean de Charpentier, to the Diablarets Glaciers in Switzerland. Charpentier believed that the mountain glaciers of the Swiss Alps had once covered large parts of northern Europe. Agassiz was going along merely to show his friend how ridiculous and absurd this idea was. Imagine a glacier covering much of England, Scandinavia, the Alps, and Russia!

Instead, like Saul of Tarsus, it was Agassiz who was converted and became the chief spokesman for the new idea, greatly expanding its concepts. There, before his eyes, was the indisputable evidence written in the rocks that great glaciers were once far more widespread than now, and that great climatic changes have taken place on earth.

The history of the earth lies before our eyes like an open book. High in the mountains of Italy, fossils of seashells are found in many rocks. Leonardo da Vinci realized that these mountain tops were once found at the bottom of the ocean in the distant past. He correctly reasoned that a biblical flood could not account for these fossils because the earth was not flooded long enough for snails to crawl half the width of Italy. Besides, in a flood caused by rains, everything would have been washed down into the sea and certainly would not have been buried deep in the rocks high in the mountains.

But da Vinci stood alone and for a time the biblical in-terpretation won out. The science of geology, more than any other branch of science, has suffered at the hands of various religious dogmas. Even so, by the year 1800, the new ideas were beginning to win support. Geology was finally becoming a legitimate science.

Anyone who has hiked around mountain glaciers knows that they grind up the ground beneath them. As these glaciers slowly creep downhill, they act like plows, pushing masses of rocks and rubble that can accumulate into large driftlike piles known as **moraines**. Moraines are seen all over the Swiss Alps and the Sierra Nevada Mountains of California. The strange thing is that they are also found far from any mountains, in many parts of northern Europe and the northern United States and Canada. Could it be that gigantic glaciers once covered North America and Europe much as they now cover Greenland and Antarctica?

Yes! These moraines are the signatures of ice ages long past. The important point to us is that this and many other features of the earth's surface are related to the climate. Moraines constitute but one example of many, as you will see.

In this chapter we will begin by looking at how climate affects the features of the landscape, and then see how it also affects the chemistry of the soils. The impact of climate on agriculture and animal life is then discussed.

The subject matter of this chapter can be extremely complicated; therefore, it is only possible for us to "scratch the surface." There are many complex interactions that exist among the climate, various geological processes, the chemistry of the soil and underlying rocks, the forms of the landscape, and the animal and plant life. At the risk of oversimplifying, we will stress how the climate affects all these factors.

# CLIMATE AND LANDFORMS

Heat is constantly produced deep within the earth by the decay of radioactive elements. As this heat escapes from the bowels of the earth, various motions of the molten or plasticized rock are produced. We now know that these internal motions have moved the continents across the face of the earth. They also produce earthquakes and volcanoes and have built the great mountain ranges.

But as soon as one part of the earth's surface stands higher than its surroundings, it is attacked and eventually worn down. Prehistoric mountain ranges, once as high as the Himalayas, have been leveled over the course of time, and from their rubble new mountain ranges have arisen.

The reduction of the landscape begins when the rocks are exposed to the air and surface waters. These rocks are broken down and disintegrated by processes collectively known as **weathering**. The loosened or disintegrated material is then carried down to the sea or lowlands by processes known collectively as **erosion**. Weathering and erosion both depend strongly on the climate.

Geologists tell us there are two distinct types of weathering. Rocks can be broken up or weathered *physically*, as they are when water gets into cracks in the rocks and then freezes and expands. Many rocks are split in this way. Rocks can also be broken up or weathered *chemically*, as they are when some of the more soluble minerals in the rocks dissolve in water. Such rocks are left with tiny gaps so that the remaining minerals easily flake off the soft, crumbling rock.

Chemical reactions proceed much more quickly in warm, humid climates where everything including the rocks quickly rots. Chemical reactions take place more slowly in cold climates where rocks would remain long preserved (like frozen foods) if physical processes weren't always at work. Thus, in general, *chemical weathering dominates in warm climates and physical weathering dominates in cold climates.*

Erosion is also strongly affected by the climate. Water is usually the main transporting agent. Rivers and streams carry most of the disintegrated and broken rocks downhill everywhere, except where glaciers plow the earth. Rivers are far more active in wet climates than they are in dry climates. Therefore, erosion proceeds far more rapidly in humid lands than it does in dry lands.

Now let's look at the different landforms and their relation to the climate.

## Landforms of Cold Climates

Yosemite Valley lies hidden from view as you hike up the Merced River into the mountains of California (see Figure 23.1). Leaving the dry lowlands, the land becomes forested as you approach Yosemite, but you still have no hint of what lies ahead. Then, suddenly, after climbing a steep ridge one of the most magnificent scenes on earth greets your eyes. Yosemite Valley is about 1 km wide and more than 10 km long. The almost completely flat valley floor is walled in on two sides by sheer granite cliffs that rise vertically more than 1 km in places. Several of the world's highest waterfalls pour down these cliffs into the valley. There is a serenity here (aside from the constant flow of tourists) and a grandeur that bespeaks a titanic origin.

This monumental valley was carved out of solid rock by the quietly gurgling Merced River and by glaciers that have long since melted. The Merced River had flowed for eons before the ice came, carrying away the loosened and disintegrated rocks and soil to the sea. At that time Yosemite was not so impressive.

Then the climate changed. Snow that fell on the mountain tops did not melt the following summers; instead, it gradually accumulated to great depths. The snow packed down into ice under its own weight, and glaciers were born in the high country. These glaciers then began creeping downhill into the valleys.

The glacier that crept into and buried Yosemite Valley gouged out the rocks in its path. Streams that had once flowed gradually from the sides of the valley now plunge precipitously onto the valley floor. One gigantic dome of rock 1500 m high was torn in half. Half was then reduced to rubble and carried downstream. The other half remains standing today as the famous Half Dome, giving evidence that ice was master here for a short time in the earth's history (see Figure 23.2). Who knows how many more valleys like Yosemite now lie hidden beneath the ice of Greenland and Antarctica?

Many other valleys were carved out by glaciers when ice covered much more of the globe. Geologists are able to distinguish which valleys have been carved out by rivers and which have been carved out by glaciers simply by the shape of the valley. River valleys tend to be V-shaped (as in Figure 23.3) — narrow and steep on the bottom, whereas glacial valleys tend to be U-shaped (as in Figure 23.4) — wide and flat on the bottom.

There are many other indicators that show us where

FIGURE 23.1
Yosemite Valley, California.

glaciers have been at work. The rubble hills or moraines which the advancing glaciers pushed before themselves still stand today in many parts of the earth and may be over a hundred meters high and hundreds of kilometers long. One of these moraines runs right through New York City and down the length of Long Island.

Glaciers leave other signs. They often pick up huge boulders and transport them hundreds of kilometers. Geologists quickly recognize these so-called **glacial erratics** because they are often composed of a type of rock that is not found anywhere in the vicinity. Rocks that glaciers have passed over may be polished or scratched or even chopped off in a characteristic manner.

There are many other factors, besides glaciers, that influence the landscape in cold regions. In colder climates the temperature frequently oscillates around the freezing point of water. When the temperature rises, the ice melts and water seeps into cracks in the rocks. Then, when the

FIGURE 23.2
How glaciers gouged out Yosemite Valley.

FIGURE 23.3
The Black Canyon of the Gunnison in Colorado is a V-shaped valley carved by the Gunnison River.

temperature falls below freezing again, the water in the cracks freezes and expands with such a pressure that the hardest rocks split. When this process is repeated many times (daytime thaw—nighttime freeze), it can mash and plow up the earth.

One result of this process is well known to motorists. Bumps known as frost heaves form on the roads during winter. During the daytime when the ground thaws, water flows into the pores in the soil beneath the road; when this water freezes at night, it expands the pores, forcing the ground and road up a bit. Repetition of the process can ruin the road. On flat, wet ground in the tundra this same process occasionally produces giant mounds of earth and ice known as **pingoes** (see Figure 23.5). Pingoes may be more than 30 m high!

This process also leads to the formation of **ice** and **sand wedges.** The alternate freezings and thawings can widen cracks in the earth that fill with ice and subsequently sand, and these can reach depths of 10 m, as in Figure 23.6.

These ice or sand wedges can form a grid across the land. When viewed from above, they are known as **ice** (or sand) **polygons** and resemble the honeycombs of bee-hives (see Figure 23.7). Some sand polygons were recently discovered when a plane flew over a newly plowed field in New Jersey.

When the soil above the permafrost thaws in summer, it will tend to flow if the land has any slope. Flow patterns, known as **cryoturbation** (or **solifluction**), can be recognized by geologists and identified as having an origin in a cold, tundra climate.

FIGURE 23.4
A U-shaped valley carved by a glacier.

FIGURE 23.5
A large pingo about 30 km south of Prudhoe Bay, Alaska. This is nothing more than an enormous frost heave.

FIGURE 23.6
An ice wedge near Livengood, Alaska.

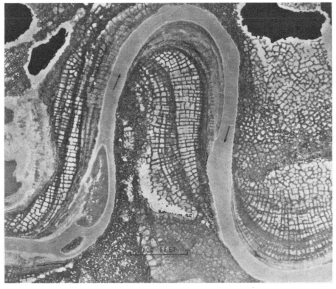

FIGURE 23.7
Ice polygons on a floodplain near Barrow, Alaska.

## Landforms of Arid Climates

Desert landscapes are distinctive and angular. Only about 25% of the desert land is covered by sand and sand dunes; the rest is generally rocky. Large sections of the desert are covered by relatively flat ground across which a few dried out riverbeds meander. Hills tend to be steeply sloped and often resemble cliffs and towers, as in Figure 23.8. The angular features of the landscape are emphasized by the fact that the ground is not covered by vegetation.

The Grand Canyon (see Figure 23.9) is a textbook case of the angularity of desert landscapes. This is one of the most spectacular natural wonders on earth. When you approach from the south, you travel along a level, forested plateau with no hint of what you are to see. Suddenly, the plateau ends at the sheer cliffs of the Grand Canyon. It appears as if the earth has opened up in front of you. The canyon is 1500 m deep and up to 25 km wide.

All the canyon was once level ground. The Colorado River slowly wore it away grain by grain and then washed it down to the sea. Before all the dams were built, Colorado River water was composed of almost 10% red-

dish sand and mud. It has been said that it was too thick to drink, but too thin to walk on.

There are several reasons that deserts have such angular features. In the desert, many of the streams never get very far. After one of the rare flash floods, water pours from the highlands down the slopes, carrying rocks and soil to the desert floor. As soon as the land begins to level out, the current slows down and most of the larger rocks stop moving. These rock deposits eventually build up into a well-known landform known as an **alluvial fan.** The finer mud and sand particles are carried farther downstream onto the desert floor, but the water never gets too far in the desert. It soon sinks into the earth, allowing the mud and sand to accumulate and level the plain.

There are many deserted cliff dwellings (see Figure 23.10) in the southwest United States. These were abandoned around the year A.D. 1300 after an extended and severe drought. We also know that the climate played a role in hollowing out the recessions of the cliffs in which they were built.

These cliffs look as if they were dug out on the bottom. Such cliffs are a common feature of dry climates wherever

FIGURE 23.8
Desert landscape in Arizona, showing angular features and sparse desert vegetation.

a layer of sandstone lies above a layer of rock which is relatively impermeable to water (such as shale). The water table in the arid climates is almost always rather low. The groundwater flows along the bottom of the sandstone and can emerge from springs where the sandstone is exposed.

There are two distinct processes that then create the cliffs. Sandstone consists of sand grains held together by a natural cement. This cement dissolves slowly in water and then the rock crumbles to sand. When the rocks above lose their support they collapse, and a cliff with hollows is created.

The second process has the same effect. Dissolved salts and other material can be left behind when the water above the water table in the rocks evaporates during a dry season. When the water evaporates the salts crystalize in the pores in the rock and (like freezing water) can exert enough pressure to split the rock. Once again this causes the rock to disintegrate and form hollows in the cliffs. These processes are depicted in Figure 23.11.

Steep gullies tend to form after the rare rainstorms in arid regions. This often leads to formations known as badlands, and the most beautiful example of all occurs in Bryce Canyon (see Figure 23.12—better yet, see Bryce itself.) When the rainwater flows down the plateau, it eats into the siltstone, creating deep gullies. More resistant shale and limestone alternate with the siltstone, allowing almost vertical spires to remain relatively unaffected. But the erosion of the siltstone in the spires also proceeds very slowly, because it remains dry in this arid land. If the climate were wetter (as it is on the edge of the next plateau), the siltstone in the spires would be wetter and would crumble long before the gullies were eroded too deeply.

One of the basic themes in the arid regions is that erosion tends to occur more rapidly near the bottom of a rock layer, usually creating cliffs, spires, and steep slopes. This can occur in any climate region whenever a softer, less resistant rock that underlies a harder, more resistant layer gets exposed to the atmosphere. This can lead to cliffs and sometimes waterfalls (such as Niagara Falls), but such features get worn away much more rapidly in the wetter climates. This is nicely illustrated in Figure 23.13.

FIGURE 23.9
The Grand Canyon.

FIGURE 23.10
Cliff dwellings, Mesa Verde.
Ironically, the caves were produced and the dwellings
eventually abandoned as a result of the climate.

## Landforms of Humid Climates

In hot, humid climates the landscape is much less angular. Plants extend their roots down through the soil and
extract minerals from the soil and rocks below. The
everpresent water dissolves the more soluble minerals
such as the salts and carries them downstream. These two
factors combine and help break apart the rocks grain by
grain, rather rapidly from a geological point of view. Even
solid granite rots quickly in warm, humid climates. Thus,
much of the rainy tropics is composed of relatively level
lowlands. Steep slopes are generally found only where
mountain building is somewhat recent or is actively taking place.

Limestone, a soluble rock, is especially soluble in *cold*
water. Therefore, in cold, humid climates limestone gets
worn away faster than any other type of rock. At first
caves are hollowed out and then sinkholes are created
when the cave roofs collapse. Eventually, the limestone
areas get so worn away that they tend to form valleys.

In warm climates even though limestone is still soluble
it is one of the more resistant rocks, and eventually the
limestone areas stand out. After limestone caves collapse
in warm, humid climates they have a tendency to form
sugarloaf-like domes called **karst towers**.

Thus, the strange and exotic-appearing Chinese paintings (Figure 23.14) of these karst towers (Figure 23.15) are
quite realistic and correspond to a landform that is characteristic of humid tropical or subtropical climates.

FIGURE 23.11
Formation of the sandstone cliffs in the arid southwestern
United States. The bottom of the sandstorm collapses (1) when
water running through it dissolves natural cement in the stone,
or (2) when growing salt crystals from the evaporating water
forces grains apart.

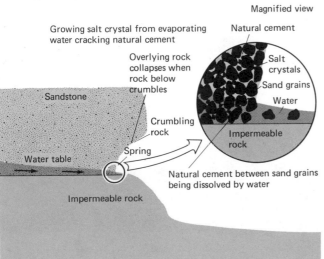

432 THE MARK OF CLIMATE ON THE EARTH

FIGURE 23.12
Bryce Canyon in Utah. Spires formed when water ran downhill and dug gullies. However, if the climate were wetter, the spires would collapse.

## CLIMATE AND THE CHEMISTRY OF THE SOIL

When I was a child I remember digging up the backyard. After I got down past the dark brown layer of topsoil (most of which my parents had bought), there was a transition to a yellow-orange layer. I was told that the topsoil was enriched by the plants and stirred up by the worms, while the soil below had not been affected yet.

But, the soil below is affected. Furthermore, the climate is very important in making the world's soils. The soil in my backyard is characteristic of Dc regions, but very different soils are found in other climate zones. For instance, desert soils often have a somewhat high percentage of sand and, even when not sandy, desert soils are clearly different from other soils of the world.

Soils are the ground-up residue of rocks and minerals. Both the temperature and the humidity affect the chemistry of the soils. The temperature affects the *rate* of chemical change, whereas the moisture supply determines which chemicals are depleted and which are enriched.

### Soils of Humid Climates
The humid regions on earth have an excess supply of water that then flows through the rocks and down the rivers into the sea. The minerals in the ground slowly dis-

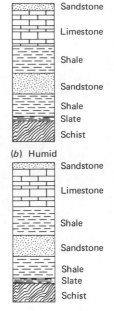

(a) Arid
Sandstone
Limestone
Shale
Sandstone
Shale
Slate
Schist

(b) Humid
Sandstone
Limestone
Shale
Sandstone
Shale
Slate
Schist

FIGURE 23.13
Weathering and erosion in (a) an arid climate and (b) a humid climate. The steeper rock layers are always more resistant. Because of slower erosion and weathering in an arid climate, the landscape features are more angular.

FIGURE 23.14
Chinese landscape paintings appear unrealistic to the unknowing, but they closely resemble limestone sugarloafs (Karst towers). Compare with Figure 23.15.

solve in this water at rates that depend on how soluble the mineral is and how rapidly it dissolves (see Figure 23.16). The highly soluble minerals such as sodium, potassium, and calcium dissolve most rapidly and have made the ocean salty. Even silicon, which makes up a large percentage of the weight of most rocks and minerals, dissolves rapidly enough so that it is slowly washed away in humid areas. Elements such as aluminum and iron are almost insoluble and therefore tend to remain where they are.

The depletion of silicon and the relative enrichment of iron and aluminum are typical characteristics of the soil in hot, humid climates. Consider what happens to solid granite. Granite is composed of minerals that are rich in sodium, potassium, calcium, and silicon. The first stage in the degeneration of granite is the removal of almost all of the first three elements and a small percentage of the silicon. The result is clay.

Clay minerals are somewhat depleted in silicon. In time, even the clay minerals rot, and more of the silicon is lost. The final products are often the iron and aluminum oxides, which are commercially valuable.

Soils in the humid tropics have a large percentage of these oxides as well as a sizable percentage of clays (especially several meters below the surface). Generally, the higher the rainfall (above 200 cm per year), the smaller the percentage of clay that remains (see Figure 23.17). Some of the world's richest bauxite (aluminum-oxide) deposits are found in the tropics.

The humid-tropical iron-oxide and aluminum-oxide soils develop to a relatively great depth. They tend to be very red in Aw and Cf regions because of the iron. In the dry season of the Aw climate they can get baked as hard as rock and are therefore often used in construction, much like bricks. Such soils are called **laterites**.

In the humid regions *outside* the tropics, there is at least one very important difference in soil formation. In cooler regions the dead vegetation does not decay so rapidly; thus it can accumulate on the ground and in the soil. In the tropics all dead matter rots so quickly that it can never accumulate (except in swamps). Dead plant matter is rich in carbon and carbon dioxide. When the carbon dioxide dissolves in water it forms carbonic acid. Other organic acids such as fulvic acid are also produced by rotting vegetation.

These acids are quite concentrated in the topsoil and they act to dissolve the normally insoluble iron and aluminum. Then, after a rainstorm, the iron-and-aluminum-

FIGURE 23.16
Schematic comparison of the rates at which various elements in rock dissolve in water. Note that sodium and calcium have almost entirely dissolved, while iron and aluminum are almost completely intact.

| Sodium | Calcium | Magnesium | Silicon | Iron | Aluminum |
|---|---|---|---|---|---|
|  |  |  |  |  | |

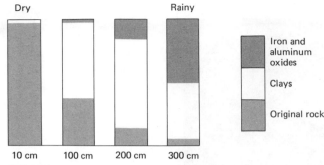

FIGURE 23.17
The percentage of decomposition of original rock into clays and oxides as a function of annual rainfall. The amount of decomposition also depends on the amount of time the rock has been exposed to the elements.

rich solutions percolate down through the soil toward the water table. There the acids are diluted and the iron and aluminum quickly become insoluble once again. Thus, they rapidly accumulate in the soil at some greater depth. This is what I saw when I dug in the earth as a child — the yellow soil layer was enriched in iron and aluminum.

Therefore, the soils in humid regions outside the tropics have distinct layers. The top layer consists of humus (decaying matter) and is deficient in iron and aluminum. Roughly a meter or so below ground level, iron and aluminum accumulate in the form of oxides or clays. Such soils are known as **podzols** (ash soil), after the Russian version in which the lower layer is often ash gray in color. Podzols are found throughout the wetter parts of the Dc climates and under all the boreal forests.

## Soils of Arid Climates

When you travel from the humid to the drier parts of the midlatitude climates the dominant vegetation changes from forest to prairie grassland to semiarid grassland to desert scrub vegetation. There is a parallel transition in the soil. In some places this transition is quite noticeable because the color of the soil changes. I noticed it the first time when I was traveling across South Dakota. In the East where it is more humid the soil is dark, but in the West where it is semiarid the soil is light. Distinct soil layering can be seen in these soils as well (see Figure 23.18).

This color change is an indication of a major difference in the chemistry of the soil. The underground waters that flow into the desert and semiarid regions evaporate and leave behind the soluble minerals which were removed from the soils and rocks of the humid climates. Calcium compounds account for the light color of desert and semiarid soils. Silicon is often deposited as $SiO_2$ (i.e., sand or

FIGURE 23.18
Soil layering can be clearly seen in this prairie soil known as mollisol.

quartz). When sulfur deposits lie nearby, gypsum is produced. The famous White Sands of New Mexico are composed of gypsum. All these deposits fall under the general heading of **evaporites** and are found throughout the world's arid regions.

Because of this tendency for evaporites to accumulate in the world's deserts, you must be very careful before drinking at any water hole. Most of the time desert water does not taste too good, and at times it can be so loaded with poisonous salts that it is deadly. Marco Polo constantly ran into this trouble in his travels across Asia.

. . . but the bread made from wheat grown in the country cannot be eaten by those who have not accustomed their palates to it,

FIGURE 23.19
White Sands, New Mexico. This sand is composed of gypsum.

having a bitter taste derived from the waters, which are all bitter and salty . . . .

Upon leaving Kerman and travelling three days, you reach the borders of a desert extending to a distance of seven days' journey, at the end of which you arrive at Kobiam [Kuhbanan]. During the first three days but little water is to be met with, and that little is impregnated with salt, and is as green as grass and so nauseating that none can drink it. Should even a drop of it be swallowed, frequent calls of nature will result . . . .

Marco Polo, *The Travels of Marco Polo*

The problem of accumulating salts must be constantly watched by farmers who irrigate their fields. If they do not periodically flush their fields, the salts will slowly accumulate as the water evaporates, eventually rendering the soil worthless. Of course, flushing the fields requires enormous quantities of water which are not readily available in the desert.

This is not merely idle chatter; some great societies

have crumbled in good part because of bad irrigation policies. Large regions near the Indus River in India and Pakistan (the Thar Desert), as well as the Fertile Crescent between the Tigris and Euphrates rivers in the Middle East have lost much of their fertility and have even become desertlike when the irrigated soil became too salty.

To summarize, in humid climates the soluble elements are removed from the soil, but in the arid regions the soluble elements accumulate. For intermediate climates (i.e., on the prairie) neither process is dominant; this is the case for some of the world's richest soils, which are called **chernozems** (see Figure 23.20).

### Soil of the Tundra

The tundra also has its own type of soil. At the cool temperatures of the tundra, plant life actually grows faster than dead plant matter decays. Therefore, peat deposits (which are sometimes used for fuel) tend to accumulate on top of the permafrost everywhere except on hillsides. Naturally, this adds to the spongy, swampy conditions of the tundra, but in time to come this may be where our descendents will find their coal and oil.

### World Soils

I will conclude this section by emphasizing that soils develop very, very slowly from the bedrock beneath. During the time it takes to develop a soil (at least hundreds of years in the humid tropics and thousands of years elsewhere), the climate itself may well undergo a drastic change. This is one of the many reasons why the study of soils (pedology) is so complicated.

But even with all these complications, you can see that there is a remarkably close relationship between the world soil map, Figure 23.21, and the world climate map, Figure 22.2.

## CLIMATE AND AGRICULTURE

In Chapter 9 you read about some of the ways that farmers can modify the climate near the ground to improve or save their crops. Thus, oranges are saved from occasional frost damage by using fans on cold, clear, calm nights. Even so, oranges must be raised in warm climates and would not survive or bear fruit in, say, the boreal forest. Getting maximum crop yields depends on choosing the proper crops for each climate.

It is a truly massive job to feed and clothe the people of

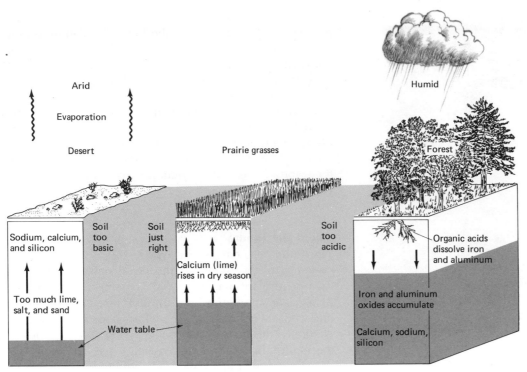

Arid

Evaporation

Desert

Prairie grasses

Humid

Forest

| Sodium, calcium, and silicon | Soil too basic | Soil just right | | Soil too acidic | Organic acids dissolve iron and aluminum |

Too much lime, salt, and sand

Calcium (lime) rises in dry season

Water table

Iron and aluminum oxides accumulate

Calcium, sodium, silicon

FIGURE 23.20
The midlatitude soil profile and how it depends on rainfall.

this world. Millions of people live on the verge of starvation. In many parts of the world a few good harvests in a row inevitably leads to an increase in population. Then, one bad harvest can wipe out thousands of people.

Both good harvests and bad harvests are the results of the weather. Commercial crops and domesticated livestock are quite delicate and are more susceptible to unfavorable changes in the weather than most wild things. Therefore, they need constant caring and wisdom to survive. Of course, we need them, too, and so an entire science has been developed to learn how to exploit the world's climates in order to maximize agricultural production and to avoid disastrous harvests when the climate happens to change slightly.

In the rainy tropics there is an enormous profusion and variety of natural vegetation. This *might* lead you to think that it should be an easy matter to transform the tropical rain forests into breadbaskets. Unfortunately, sad experience has shown that nothing could be further from the truth.

The primitive natives of the rain forest have known this all along. Their method of agriculture is known as slash and burn. They simply burn a clearing in the jungle and then grow their crops there for a year or two. Eventually they leave and burn another area in the jungle.

To you this might seem wasteful and destructive, but it is virtually the only thing to do. Tropical soils are very poor in nutrients because of the rapid rotting and leaching of the minerals. When a swath of jungle is burned, the soil usually contains enough nutrients for one or two harvests at most. Even with fertilizers you fight a losing battle because they too quickly wash away. Furthermore, when you create a large clearing rapid erosion of the soil takes place because of the constant rains.

The situation is not much better in the Aw climate. There the rock hard laterite is often better for the construction industry (i.e., for bricks) than it is for agriculture.

The best way to take advantage of the tropical climate is to grow crops that have always grown there naturally. Because the soil is so poor, it has proven wise to imitate nature and often plant a confusing mixture of many crops in which the wastes of one plant act as the nutrients of another.

Rubber, palm oil, cacao (chocolate), coffee, tea, and a variety of other plants—which are more druglike than foodlike—are the chief agricultural products in the humid tropics. Several of these plants such as coffee do require a short dry season for harvesting and so are found in Aw rather than Ar climates.

The foods that are grown in the tropics are generally

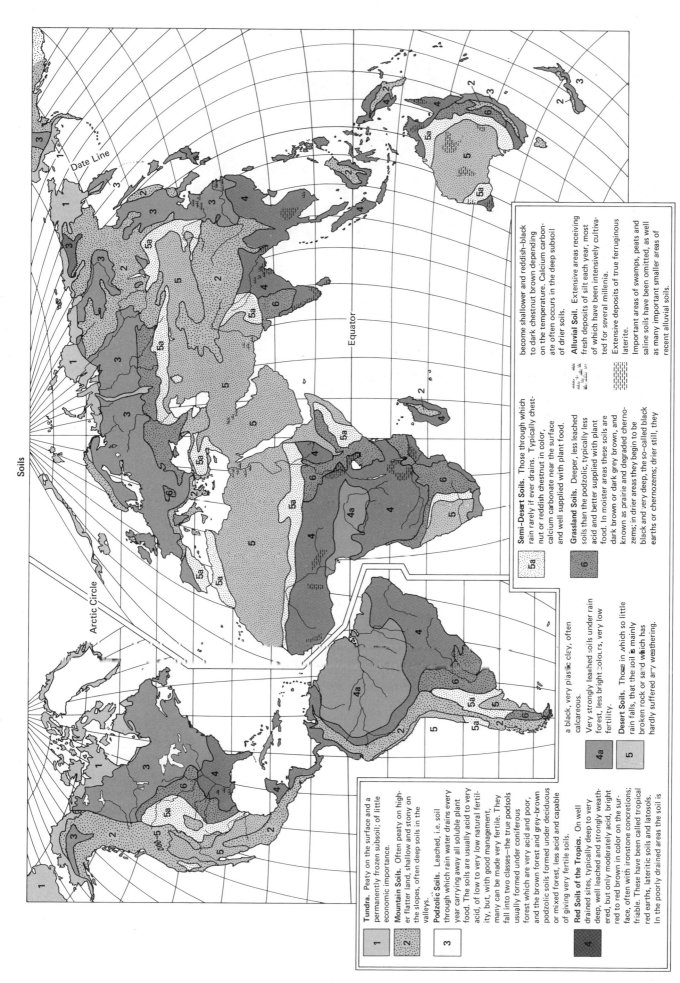

Soils

**Tundra.** Peaty on the surface and a permanently frozen subsoil; of little economic importance.

**Mountain Soils.** Often peaty on higher flatter land, shallow and stony on the slopes, often deep soils in the valleys.

**Podzolic Soils.** Leached, i.e. soil through which rain water drains every year carrying away all soluble plant food. The soils are usually acid to very acid, of low to very low natural fertility, but, with good management, many can be made very fertile. They fall into two classes—the true podsols usually formed under coniferous forest which are very acid and poor, and the brown forest and grey-brown podzolic soils formed under deciduous or mixed forest, less acid and capable of giving very fertile soils.

**Red Soils of the Tropics.** On well drained sites, typically deep to very deep, well leached and strongly weathered, but only moderately acid, bright red to red brown in color on the surface, often with ironstone concretions; friable. These have been called tropical red earths, lateritic soils and latosols. In the poorly drained areas the soil is

a black, very plastic clay, often calcareous.

Very strongly leached soils under rain forest, less bright colours, very low fertility.

**Desert Soils.** Those in which so little rain falls, that the soil is mainly broken rock or sand which has hardly suffered any weathering.

**Semi-Desert Soils.** Those through which rain rarely if ever drains. Typically chestnut or reddish chestnut in color, calcium carbonate near the surface and well supplied with plant food.

**Grassland Soils.** Deeper, less leached soils than the podzolic, typically less acid and better supplied with plant food. In moister areas these soils are dark brown or dark grey brown, and known as prairie and degraded chernozems; in drier areas they begin to be black and very deep, the so-called black earths or chernozems; drier still, they

become shallower and reddish-black to dark chestnut brown depending on the temperature. Calcium carbonate often occurs in the deep subsoil of drier soils.

**Alluvial Soil.** Extensive areas receiving fresh deposits of silt each year, most of which have been intensively cultivated for several millenia.

Extensive deposits of true ferruginous laterite.

Important areas of swamps, peats and saline soils have been omitted, as well as many important smaller areas of recent alluvial soils.

FIGURE 23.21
Soils of the world.

quite high in carbohydrates and sugar, but low in protein. Protein deficiency is therefore a widespread problem in the tropics. Although tropical natives have literally hundreds of ways of preparing and serving bananas, this can never be a sole source of food.

Raising livestock is no easy chore either. Most breeds of domestic cattle do not thrive under constantly warm, humid conditions. The cattle that do best are those with short, shiny, light-colored coats, which enable them to stay cooler. This does not end the problem. In Africa the tsetse fly is rampant throughout the Ar and Aw regions. It injects sleeping sickness into the cattle that proves fatal within three days. Nothing seems to come easily in the tropics.

There is one tropical food that does provide a good source of protein and that feeds more than half the people on earth. That food is rice. Rice is a grain that is grown mainly in Southeast Asia in Aw, Cw and Cf regions. There are several varieties of rice, most of which must be immersed in water in order to grow.

In lowlands the work of planting rice is much easier than it is on the hillsides, where it is essential to build terraces (which look identical to contour lines on a map) to prevent the soils from rapidly eroding and to keep the rice immersed. You can imagine the truly enormous amount

of human labor that has gone into sculpturing these mountainsides without tractors and sometimes without animal power! But all the land must be used to keep the population from starving. (See Figure 23.22)

It is only possible to keep all these terraces wet in a climate that has a long, wet rainy season. Planting begins with the onset of the summer rains. Under ideal conditions the rice is harvested in the drying fields shortly after the rainy season has ended. If the rains end too soon, irrigation is necessary or the rice will shrivel and die.

During the winter, rice cannot be grown in much of Southeast Asia for two reasons. First, rice does not grow once the average temperature drops below 20°C and, second, the tremendous amounts of water in which rice needs to grow are not available during the long, dry Aw or Cw winter. Therefore, during the winter, the people keep busy by planting wheat in their rice fields (except in the colder parts of the Cw region).

Wheat can be grown in some of the same climate *regions* as rice, but the most favorable climate *conditions* for wheat and rice are completely different. Wheat tolerates the cool weather that would kill rice, whereas most varieties of rice need the extremely humid, rainy conditions that are fatal to wheat.

There are many varieties of wheat and they all seem to

grow best in regions where the rainfall averages from 37.5 to 87.5 cm. It is no mere coincidence that wheat grows best in the midlatitude grasslands, because wheat is a grass whose seeds are nutritious. In the past few years biologists have developed new strains of wheat that can tolerate mild droughts. The new varieties have roots which extend deeper into the soil than the older varieties, and so are able to extract water from rather dry soil.

The new varieties of wheat have contributed to the "Green Revolution." They have increased food productivity considerably in many of the world's starvation centers, such as India. And these new varieties are not only more resistant to adverse weather and some pests, but they are also faster growing and more nutritious. Even so, when the weather takes a turn for the worse, the harvests suffer.

When it is too rainy, wheat easily is attacked by a variety of molds (such as wheat rust). In addition, insects easily breed in the wet ground and hungrily attack the edible wheat. Thus, not only does drought destroy crops, but too much rain may prove just as disastrous.

One absolutely horrible example of this was the Great Famine in Ireland in 1848. The Irish subsisted mainly on potatoes which suffer if it is too rainy. The rainy summers of 1845–46 brought poor harvests and widespread hunger, but that was only a prelude of things to come. In 1848 it was also very rainy, but in that year the potato blight (which is fostered by wet conditions) also struck. The potato harvest that year was devastated and an estimated 1.5 million Irish starved to death as a result.

The harvests were also poor in other parts of Europe, promoting widespread economic depression and revolutions. Revolutions seldom result from poor harvests when the government is popular, but no government stays popular too long when people are starving.

In the semiarid climate regions the crop yield is generally so low and unreliable (without irrigation) that grazing is often the best use for the land. However, this must be done carefully because, if the land is overgrazed, it can all too easily degenerate to desert.

Even with irrigation, in the desert it is best to grow crops that do not have a great need for water and that can tolerate sunshine and salty soils. Thus crops such as spinach and pomegranates can be grown profitably. There is also the tamarisk, a tree that serves as a useful windbreak to protect other plants from the desiccating desert winds and that is able to thrive in extremely salty soils.

A world map of the principal uses of the land is shown in Figure 23.23. Compare this and the world map of *natural* vegetation, Figure 23.24, with the world climate map, Figure 22.2.

# CLIMATE AND ANIMALS

Does climate play any role in determining our bodily features or behavior? We know that the skin color of indigenous peoples *tends* to get darker as you approach the equator. But some anthropologists point out that people have migrated across the surface of the earth and intermixed so often that it is now difficult to say what constitutes an "indigenous people."

Skin color differences are so striking that they cannot be denied. Other differences are generally somewhat more subtle, so that people first have to settle the argument of whether they even exist before they can begin to debate if climate plays any role.

One of the undeniable products of climate is the pygmy. Many natives of the world's tropical rain forests are pygmies. They have long been isolated from their more aggressive and larger neighbors living outside the rain forest, and this has allowed evolution a sufficient time to produce the size differences. Their smaller size makes it easier for them to scale the trees and lianas and reach the canopy of the rain forest.

Some climate–animal relationships are universally agreed upon. Insects definitely tend to be larger in tropical regions. There are several varieties of tropical cockroaches, waterbugs, and beetles that are actually larger than small birds.

Conversely, mammals of a particular species tend to be smaller in the tropics. Here the conclusions are somewhat dubious, because many species of large mammals formerly living outside the tropics are now extinct due to our forefathers' efforts. Mammals also tend to get rounder as you approach the poles (see, for example, Figure 23.25).

Large bulk and a round frame minimize exposed surface area in relation to the mass of the animal (recall the discussion on surface tension in Chapter 12). Heat loss is a tremendous problem for small, angular animals.

Eskimos with their shorter legs and stockier bodies are better adapted to cold conditions, whereas many blacks have longer, thinner bodies so as to dissipate more efficiently any excess heat.

There are many examples of animals that have adapted to particular climates. One of the best examples is the camel—the ship of the desert. This animal can go through searing heat for 20 days without drinking, while a man would surely die within 2 days. Under such conditions, the camel hardly sweats, does not urinate at all, and passes such dry feces that they can be used immediately for fuel. But the camel is like the goose who laid the golden eggs—if you cut open a camel there will be no water sack anywhere in its body.

Agriculture

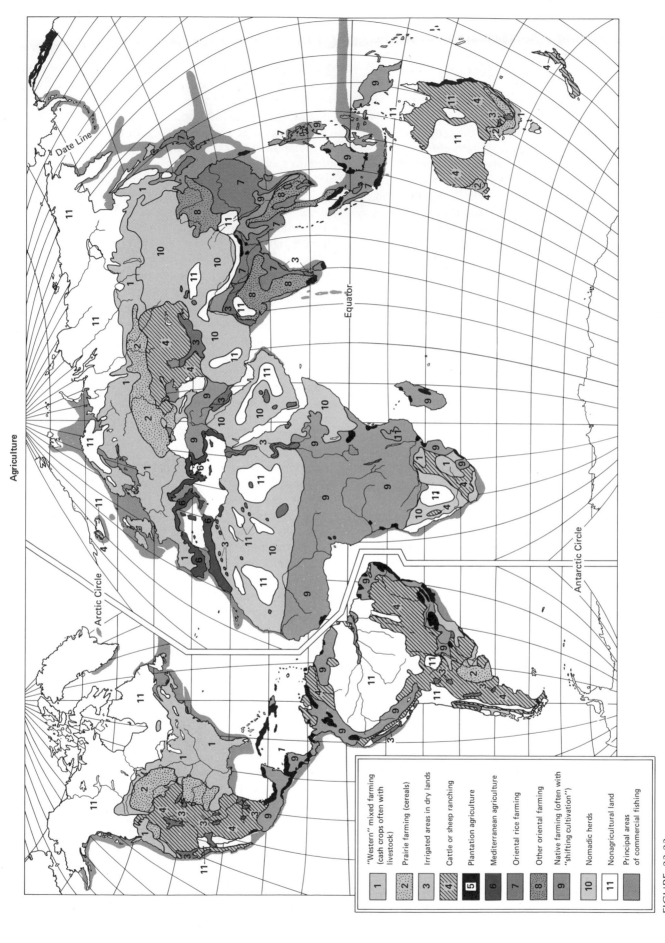

FIGURE 23.23
Agricultural products of the world.

Legend:

1 "Western" mixed farming (cash crops often with livestock)
2 Prairie farming (cereals)
3 Irrigated areas in dry lands
4 Cattle or sheep ranching
5 Plantation agriculture
6 Mediterranean agriculture
7 Oriental rice farming
8 Other oriental farming
9 Native farming (often with "shifting cultivation")
10 Nomadic herds
11 Nonagricultural land
Principal areas of commercial fishing

Date Line

Arctic Circle

Equator

Antarctic Circle

## Vegetation

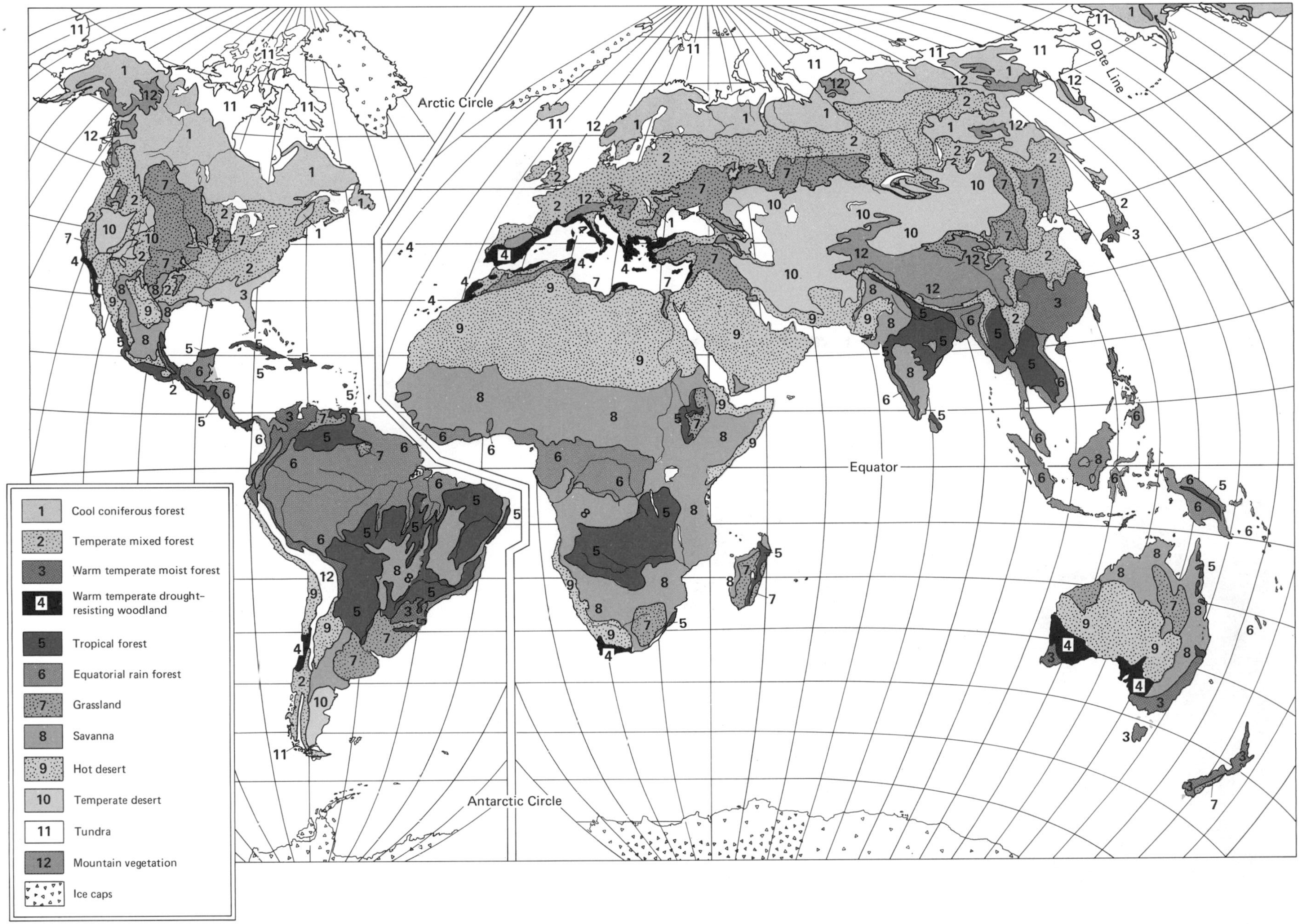
Arctic Circle
Date Line
Equator
Antarctic Circle
1 Cool coniferous forest
2 Temperate mixed forest
3 Warm temperate moist forest
4 Warm temperate drought-resisting woodland
5 Tropical forest
6 Equatorial rain forest
7 Grassland
8 Savanna
9 Hot desert
10 Temperate desert
11 Tundra
12 Mountain vegetation
Ice caps

FIGURE 23.24
The natural vegetation of the world.

FIGURE 23.25
Comparison of the Kit Fox with the Arctic Fox. Notice particularly the difference in the ears.

THE MARK OF CLIMATE ON THE EARTH

# PROBLEMS

**23-1** What is the difference between weathering and erosion?

**23-2** How does the rate of weathering depend on temperature and precipitation?

**23-3** Explain how frost heaves and pingos are formed.

**23-4** Explain why desert landscapes tend to be rather angular.

**23-5** Describe typical landforms of limestone in (a) arid, (b) cold humid, and (c) warm humid climates.

**23-6** Why are soils in Ar climates so low in nutrients?

**23-7** Why are soils in Ar and Aw climates enriched in iron and aluminium?

**23-8** Why are soils in B climates enriched in salts, silicon, and calcium?

**23-9** Explain the layering in the Dc and E soils.

**23-10** Explain why tundra soils can often be burned after they have been dried.

**23-11** Explain how rice and wheat can be grown in the same place even though they need completely different climate conditions.

**23-12** What is the potential danger in irrigating the soils of a region over prolonged periods?

**23-13** Make a list of animals that are widespread over the globe. Have these animals evolved differently as a result of climate differences (a) in arid versus humid regions or (b) in polar versus tropical regions?

# CLIMATES
# OF THE PAST

The ancient Greeks (yes, them again) were the first people to write intelligibly about fossils. In the sixth century B.C., some of the Greek philosophers commented that fossilized sea shells had been found high in the mountains, far from the seas in the deepest parts of rock quarries. They reasoned (as did Leonardo da Vinci 2000 years later) that these mountainous regions had once been covered by the sea, and that all animals on land had once descended from primitive fishes.

But people had dug up fossils long before the ancient Greeks wrote about them. Charms and necklaces, fashioned out of fossils by cave dwellers, have been found by modern fossil hunters. Stories of giants may well have been inspired by the discovery of fossil elephant bones. In fact, the fossil skulls of elephants may have prompted the creation of the myth of the cyclops (the one-eyed giant in *The Odyssey*), because elephant skulls have one large central opening for the trunk, which can be mistaken for an eye socket.

Throughout the 1700s more and more people became interested in fossils. This was the Age of Reason — although from some of the ridiculous theories on fossils you might not believe it. Along the way to making fossil collecting a science, there were some notable and even humorous scandals. One group of practical jokers bore a grudge against a stodgy paleontologist, Professor Beringer. They manufactured fossillike rocks with God's name written in Hebrew letters or imprints of frogs having

sexual intercourse. They then buried these fakes where the gullible "expert," Beringer, was most likely to search. Even after the hoax was exposed, Beringer persisted in his belief that these rocks contained divine imprints, much to the discredit of the entire field of paleontology.

But the mounting evidence soon pointed the way to a rational explanation of fossils, and by 1800 serious fossil collectors were beginning to realize that fossils were definitely the remains or impressions left in stone of animals and plants that had died long before any biblical flood.

Their great antiquity was not the only intriguing feature of the fossils. In many cases the fossils indicated that the environment had also drastically changed. Knowledge of fossils was soon combined with the expanding knowledge of biology and geology. William Smith, a British surveyor, made detailed observations of the different rock layers throughout England, and he was the first to realize that there was a close connection between rock layers and fossils. Each rock layer had its own characteristic fossils (and minerals) no matter where it appeared, in England or in Wales.

It was already deduced that the lower rock layers were laid down earlier as sediments and that the higher layers then accumulated on them. Using this fact, Smith was able to write a history of life on earth. Because the lower rock layers always contained simpler and more primitive forms of life, the theories of evolution evolved naturally.

Some of the fossils found in England belonged to

tropical plants and animals, implying that England once had a tropical climate. Then, in the early 1800s, geological research showed that in a more recent age an ice sheet covered the British Isles. As Goethe put it, "For all that ice we need cold. I would wager that an epoch of great cold passed over Europe, at any rate." The history of the earth was apparently becoming a history of great climatic changes as well.

This chapter is devoted to the subject of climate changes that have taken place on earth. We will first look at some of the evidence used to infer these past climates. This evidence relies on present-day knowledge of the relationships that exist among climate and landforms, soils, and plant and animal life, so that you must be familiar with the material of the last two chapters.

We will then examine some of the climates of the past, concluding with a few of the climate theories that are debated today.

# THE EARTH AS A CLIMATE RECORD BOOK

The *Guinness Book of World Records* may be useful for giving recent climate and weather records, but it is worthless for climates of the distant past. As you will soon see, various records kept by humankind have given us some picture of the climate of selected regions as far back as 3000 B.C. That is when the Egyptians began keeping records of the annual flooding of the Nile River. This provides us with an indirect but reliable measure of the past rainfall of tropical East Africa, which is north of the equator and the principal source region of the Nile. Before that, humans were silent witnesses to the climate, and we must look to the earth itself as the only climate record book of more ancient climates.

You must now become experts in reading the secrets of the earth. First, the history of the earth is conveniently divided into time periods, as shown in Table 24.1. Earlier in the book you read that the age of the earth is approximately 4.6 billion years. How do we know this? The dating of rocks and other objects is a somewhat recent subject. Radioactive elements, such as uranium, decay to other elements, such as lead, at a constant rate. In 1 billion years, half of the uranium decays. In the next billion years, half of the remaining uranium decays, so that only one-fourth of the original uranium remains after 2 billion years. All we must know is the original uranium content of the rock to determine its age. There are only certain rocks for which we know the original composition.

TABLE 24.1

| Geologic Succession | | | | Approximate Age $10^6$ years |
|---|---|---|---|---|
| Eon | Era | System or Period | Series or Epoch | |
| Phanerozoic | Cenozoic | Quaternary | Recent | |
| | | | Pleistocene | 3 |
| | | Neocene | Pliocene | |
| | | | Miocene | 22 |
| | | Paleocene | Oligocene | |
| | | | Eocene | |
| | | | Paleocene | |
| | Mesozoic | Cretaceous | | 62 |
| | | | | 130 |
| | | Jurassic | | 180 |
| | | Triassic | | 230 |
| | Paleozoic | Permian | | 280 |
| | | Carboniferous: Pennsylvanian | | |
| | | Carboniferous: Mississippian | | 340 |
| | | Devonian | | 400 |
| | | Silurian | | 450 |
| | | Ordovician | | 500 |
| | | Cambrian | | 570 |
| | | Ediacarian | | 640 |
| Precambrian | Proterozoic | Upper | | 950 |
| | | | | 1350 |
| | | Middle | | 1650 |
| | | | | 1800 |
| | | Lower | | 2600 |
| | | Archean | | 3600 |
| | | No record | | 4700 |

Geologists tell us there are three types of rocks—**igneous**, **metamorphic**, and **sedimentary**. Igneous rocks often form by refreezing or recrystallization of **magma** (molten rock such as volcanic lava). Sedimentary rocks are produced when sands or muds are deposited in thick layers and then buried, compressed, and cemented together. Metamorphic rocks are produced when great pressures and temperatures deep in the earth alter or distort the rocks, but do not quite melt them. Granite and basalt are examples of igneous rocks. Sandstone and limestone are examples of sedimentary rocks. Marble is the metamorphic rock produced from limestone and mica schist, a metamorphic rock often produced from sandstone (see Figure 24.1).

Only igneous rocks give a good idea of the original chemical composition, and therefore only these rocks can be dated directly. Unfortunately, igneous rocks seldom contain fossils. Most fossils are found in sedimentary rocks because, when a plant or animal dies, it is most likely to get buried and preserved in sand or mud. The only time an igneous rock can contain fossils is when living things get buried alive in lava or volcanic ash.

Fortunately, however, we *can* use igneous rocks to estimate the age of sedimentary and metamorphic rocks. Since the newer sediments accumulate on top of the older sediments and rocks (they certainly couldn't accumulate under them), the deeper you dig the older the rocks get. There is at least one important exception to this rule.

Magma (molten rock) sometimes forces its way up from the heated depths of the earth through cracks in the rock. When this magma reaches the earth's surface it produces volcanoes, but it often forms underground channels that never reach the surface. These channels, known as **igneous intrusions**, are certainly younger than the rocks that surround them. By using such information, geologists have been able to piece together a remarkably detailed and accurate calendar of earth history (see Figure 24.2).

Life may have begun more than 3.5 billion years ago in the Precambrian Era, but it remained primitive for well over 2 billion years. Then, suddenly, at the beginning of the Paleozoic Era (in the Cambrian Period), much more advanced forms of fossils appeared in the rocks. There was a continuous progression of life throughout the Paleozoic. At first, life was restricted to the sea but, during the Silurian, the world's first forests appeared. Then, in the Devonian, the first amphibians crawled out onto the land.

At the end of the Paleozoic there were wholesale extinctions and, by the beginning of the Mesozoic Era, new forms of life appeared. This was the time of the dinosaurs. Then, at the end of the Mesozoic, they too disappeared and mammals have dominated the earth ever since—in the Cenozoic Era.

Now let's consider the fossil and geological evidence that is used to infer the past climates. In the sections that follow, not only are the various indicators presented, but also the possible ambiguities. Therefore, when paleoclimatologists reconstruct the climates of the past, they never rely on one form of evidence alone, but use many in conjunction.

### Evidence of Warm Climates

Near the Connecticut River around Hartford there is a distinctive layer of red rocks. Red coloration in rocks generally indicates the presence of iron oxides, which accumulate in tropical soils. The red rock layer near Hartford was probably made from laterites that were subsequently buried, compressed, and cemented into rock.

Geologists have pointed out that the red coloration sometimes occurs thousands of years after the sediments are buried. Nevertheless, red rock layers, known as **redbeds** are usually a reliable indicator of a once tropical climate. And, indeed, the climate of Hartford *was* tropical when these Triassic redbeds were laid down.

Limestone deposits generally form in tropical waters from shells and coral. Calcium carbonate, the main mineral in limestone, is very soluble in cold water, and so it tends to dissolve rather than accumulate there. Besides, almost all present coral reefs are restricted to shallow tropical waters, because corals need warmth (average temperature at least 25.5°C) and sunlight. Corals grow much more rapidly as the temperature increases, so that in places the coral reefs show annual and even daily growth rings.

**Note:** The growth rings of coral from the Mesozoic Era indicate that each year was then almost 400 days long, but each day was only 22 hours long because the earth rotated faster. The earth's rotation rate has been slowing over time because of the tidal drag exerted by the moon.

Notice on the world map, Figure 24.3, that coral reefs are most restricted on the western coasts of South America and Africa. Why? Corals cannot survive near these coasts because of the upwelling of cold water. Ancient coral reefs, such as the rock formation now containing Carlsbad Caverns in New Mexico, were all built by corals that are now extinct. We therefore must *assume* that these ancient corals also needed warmth to grow. This assumption is quite reasonable since there is so much other corroborating evidence of warmth.

It has always amazed me that delicate things such as leaves have left their imprints in the rocks. Originally, the leaves dropped onto mud and, when more mud and leaves piled on top, an imprint was left in the mud. This was preserved when the mud was buried deep and hardened.

There is a far greater variety of plants and animals in tropical climates than in the colder climates; therefore, great variety in the fossil record is another indicator of a warm climate. Tropical plants generally also have very large leaves. These leaves tend to be **"entire margined"** (i.e., have no side lobes; as shown in Figure 24.4). In tropical forests today over 80% of the plants have entire-margined leaves, whereas this is true of less than 33% of the plants in the midlatitudes.

FIGURE 24.1
Various rocks including (a) granite, (b) basalt, (c) sandstone,
(d) limestone, (e) schist, and (f) marble.

(f)

FIGURE 24.2
Reconstruction of a sample history of rock layers. Extensive erosion took place after 2 and 3. Generally the oldest rocks are on the bottom, but the young igneous intrusions may cut through older rocks.

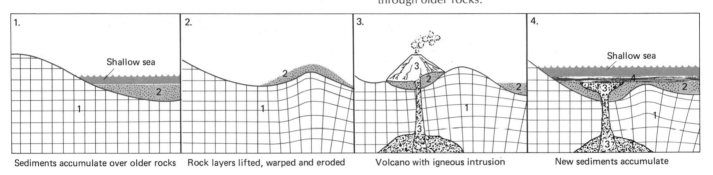

| 1. Shallow sea | 2. | 3. | 4. Shallow sea |
|---|---|---|---|
| Sediments accumulate over older rocks | Rock layers lifted, warped and eroded | Volcano with igneous intrusion | New sediments accumulate |

FIGURE 24.3
Current distribution of world's rock-forming corals. The inner shaded region shows where corals are abundant, while the heavy line outlines their poleward limits. Compare this with Figures 16.13 and 16.14 to see the relationship with ocean temperatures.

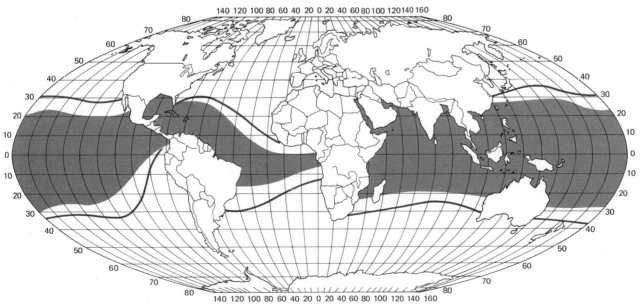

THE EARTH AS A CLIMATE RECORD BOOK 451

(a)

(b)

FIGURE 24.4
Entire-margined (a) and nonentire-margined (b) leaves. Tropical plants tend to have entire-margined leaves.

Not only do leaves tend to be larger in tropical climates, but so do insects. Large reptiles are also restricted to warm climates since they are cold-blooded.

It had long been assumed that the dinosaurs were cold-blooded, but now there is some convincing evidence that they were at least partially warm-blooded. If so, they would not have been restricted to warm climates. But even if the evidence concerning the dinosaurs is ambiguous, there were many other reptiles such as crocodiles that were widespread during the Mesozoic. We *know* these other reptiles were cold-blooded, because their cold-blooded descendants are alive today.

### Evidence of Cold Climates

The landforms created by glaciers constitute the principal form of evidence for colder climates. We also know that cold conditions lower the tree line and the snow line. The **snow line** is the height above which snow remains on the ground year round. The snow line is found at about 5000 m above sea level at the equator, at about 5500 m above sea level in the dry subtropics, and generally gets lower as

you approach the poles, where snow can remain on the ground even at sea level (see Figure 24.5).

The presence of former snow lines is indicated by **corries**—which are basically dug out areas on hills and mountain slopes that take the appearance of huge thumb-print impressions. However, some caution must be applied when using the snow line as an indicator of colder conditions, since wetter conditions can also lower the snow line. This is why the snow line is higher in the subtropics than it is near the equator, where there is adequate precipitation to maintain it. Usually, the temperature is the dominant factor that determines the snow line.

Fossils of beavers have been found in the Siberian tundra. Since beavers live in forests, this implies that the climate was once warmer. But the greatest of all Siberian fossil finds has been the mammoth. Several mammoths have been found preserved in the ice of the permafrost. As soon as these mammoths have been exposed by melting ice, they have been eaten by wolves, dogs, and even people. Mammoth steaks were once served at a dinner for a few fossil hunters, and the evening was a huge success—the steaks had been aged to perfection.

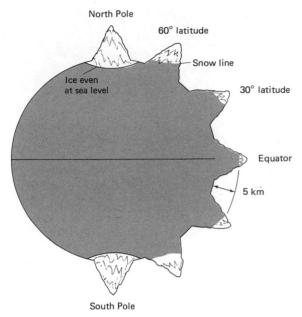

FIGURE 24.5
The snow line as a function of latitude. The snow line generally gets lower as you approach the poles, but it is highest in the dry subtropics.

The mammoth carcasses surprised many scientists. They were covered with thick, wooly hair, completely unlike all living elephants, and consequently were well adapted to the cold. This provides us with one more example of the great care needed in interpreting fossil information. Remember that animals and plants can either adapt to changing climate conditions, or two closely related breeds can live in completely different climate regions. In this case, some people have concluded that the mammoths actually lived in warm conditions, because that is where the present-day elephants are found — but they couldn't be more incorrect. The mammoths actually lived during frigid conditions, and the ones that have been preserved most likely fell down slippery slopes during blizzards and into marshes or ice-covered waters which thereafter froze.

Scientists now devote great time and effort to studying the oozing sediments that accumulate on the ocean floor. From mammoths we turn to a more humble creature, the one-celled shell-building foraminifera (foram), known as *Globigerina pachyderma*. When one of these forams dies, its shell slowly drops to the ocean floor. Those shells that do not dissolve accumulate in the sediments and can later be analyzed. In fact, scientists have found that the *Globigerina pachyderma* is something of a fossil "thermostat." When the temperature of the sea surface in April

(for the NH) is warmer than 7.2°C, the shells of the pachyderma coil to the right (dextrally) whereas, if the temperature lies below 7.2°C, the shells coil to the left (sinistrally), as shown in Figure 24.6. There are also a number of other similar living "thermostats."

One outstanding characteristic found among the plant fossils is the fact that **conifers** (cone bearers or evergreens) dominate in cold climates. Spruce trees are often able to withstand the most severe winter conditions and are found at the northern border of the boreal forests. Once the climate becomes somewhat milder, the spruce trees are replaced by the larger pines and firs.

## Evidence of Dry Climates
Since dry climates are characterized by an excess of evaporation over precipitation, evaporite deposits in the form of rock salt, potash salts, and gypsum can build up to surprising depths. These deposits often remain relatively unchanged even after the climate of the region changes.

There are many fossilized imprints of animal tracks (as in Figure 24.7) and footprints, and even impressions from raindrops (**rain pits**). Oddly enough, these tend to indicate a dry climate (although they can also form in the mud near riverbanks and lake shores). An animal slogging through the mud after a rare rainstorm in the desert makes tracks that the sun has time to bake and harden. The same tracks made in a humid climate would be washed away by the next rainstorm long before they could harden. Perhaps primitive people were aware of this when they constructed the large desert figures and "runways" in Chile and Peru, which some people have interpreted as the work of visitors from outer space.

Since sand tends to accumulate and blow around the

FIGURE 24.6
Shells of *Globigerina pachyderma*. (a) Dextral coiling indicates warmer temperatures and (b) sinistral coiling indicates colder temperatures.

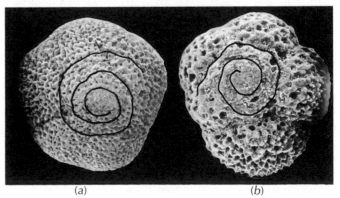

(a)                              (b)

FIGURE 24.7
Fossilized animal tracks in desert sandstone.

FIGURE 24.8
Fossilized ripples in sandstone. Guess which way the wind was blowing?

deserts, sandstone is often an indicator of former desert conditions. Sand dunes and other characteristic landforms result. An important form of evidence is given by ripple marks in the sandstone (see Figure 24.8). Ripple marks are also made in sand and mud that lies under running water, but underwater ripples are always much steeper than ripples made in the air. The ripples can also be used to tell the direction of the prevailing wind, because the steeper side of the ripple faces downwind.

### Evidence of Humid Climates

Coal deposits are the fossilized remains of ancient forests. Many people tend to think that coal was produced in the steamy jungles, but we have already seen that dead plant matter rots so quickly in the tropical rain forest that litter seldom accumulates. On the other hand, organic remains tend to accumulate in the boreal forest and in the peat bogs of colder climates.

Coal deposits therefore indicate a moist climate. Very often they probably formed in swamps where the lack of oxygen in the murky waters did not allow the remains to decompose too rapidly, or perhaps they even formed in swamps of the tropical rain forest. In order to determine whether a particular coal deposit formed in warm or cool conditions, you must inspect the type of fossils contained in the coal.

There are several salt lakes (such as Mono Lake) in the deserts of Nevada and Utah. These are the pitiful remnants of several giant lakes that once covered much of the land when the climate was wetter. When the lakes were larger their water levels were naturally higher, and in many places they created beaches that are now left high and dry, far above lake level. Some ancient beach terraces can now be seen surrounding the Great Salt Lake (as is shown in Figure 24.9) and the Dead Sea, and these inform us that the climate was once wetter. But such information might also indicate that the climate was cooler. The present low level of these lakes is not only the result of the scanty precipitation, but also a high evaporation rate. And you know that when the temperature drops, so does the evaporation rate.

Leaves growing in moist climates often have pointed ends called **drip points** (see Figure 24.10). This allows the excess moisture to drop off the leaves more easily (overcoming the surface-tension effect). Leaves are also broader than they are in dry climates, where a broad leaf would allow too much water loss.

### Other Climate Indicators

Outside of the rainy tropics there are always some significant seasonal changes. Therefore, plant growth outside the tropics is also seasonal. One manifestation of seasonal

454 CLIMATES OF THE PAST

FIGURE 24.9
Ancient beach terraces of Lake Bonneville (now the Great Salt Lake), Utah.

growth is the appearance of **tree rings**. (The light-colored part of each ring generally indicates growth during the spring and early summer, while the darker, denser wood indicates the slower growth at the summer's end.) In those seasons that the tree grows rapidly the tree rings are widely spaced, whereas in years of poor growth they are narrow (see Figure 24.11).

In dry climates growth is usually limited by the water supply. The trees grow quite well in rainy years, but in dry

FIGURE 24.10
Drip points on leaves indicate a wet climate.

FIGURE 24.11
Tree rings showing years of plenty, alternating with lean (neither too dry nor too cold) years.

years the growth is stunted. Thus, it is possible to piece together a record of rainfall from tree rings. At first you might think that such a record could not go back very far, but by using the Bristlecone Pine trees, which sometimes live almost 5000 years, we have an annual climate record for almost 10,000 years in the American Southwest.

Trees living at higher altitudes and latitudes such as the dwarf pine are more limited by cold weather than by dry conditions, because the temperature barely rises to the point where vegetation can grow. A summer that is as little as 1°C cooler than normal may cut the annual growth in half.

Tiny pollen grains often provide us with the best of all climate indicators. The microscopic pollen grains of each different species of plant are distinguishable from all others (see Figure 24.12). Pollen grains are carried by the winds (as people who suffer from hay fever might suspect) and the rivers and deposited in sediments where they are preserved. The pollen record is often so detailed that **palynologists** (pollen analysts) are actually able to estimate the average annual temperature and precipitation of the ancient climate!

FIGURE 24.12
Microscopic view showing that pollens from different plants can easily be identified.

Pollen analysis has its difficulties, too. Pollen can be blown or carried enormous distances before it is deposited in entirely different climate regions. Fortunately, palynologists are usually able to distinguish distant from local pollen merely on the basis of numbers—the distant pollen being less common.

## Isotope Techniques

Oxygen has two less common isotopes, $O^{17}$ and $O^{18}$, in addition to the normal $O^{16}$. These are not radioactive isotopes, but they do have a property that actually makes them behave like thermometers in fossil shells.

Almost exactly 1000 out of every 500,000 oxygen atoms in the ocean are $O^{18}$ atoms. Since $O^{18}$ is slightly heavier than $O^{16}$, the former tends to "settle" out of the water (i.e., crystallize) more rapidly into shells ($CaCO_3$). Thus, out of every 500,000 oxygen atoms, the shells contain about 1025 of $O^{18}$.

The essential point for us is that the "settling" out rate depends slightly on temperature. Therefore, when a shell forms in 0°C water it contains 1026 of $O^{18}$, whereas if it forms in 25°C water it contains only 1022 of $O^{18}$ (out of every 500,000 oxygen atoms). That is a difference of 4 atoms out of 500,000, but the measuring techniques have become so sensitive that they are reliable to 1°C or one atom difference out of every 6 million atoms!

A dramatic illustration of the $O^{18}$ technique was published in the early 1950s. An extinct shell-producing animal known as a *belemnite* used to be widespread in the shallow seas of the Jurassic. This animal built one cigar-shaped shell during its lifetime, building it progressively outward. Therefore, scientists were actually able to piece together a brief biography of one such belemnite. It was born in the fall and lived through three summers and four winters before dying in the spring of its fourth year (see Figure 24.13). By using the $O^{18}$ technique, we know the temperatures for a few seasons about 150 million years ago.

Even this revolutionary technique must be used with care. During ice ages so much water is evaporated from the ocean and accumulated on the ice sheets that the ratio of $O^{18}$ to $O^{16}$ in the oceans is affected. The heavier $O^{18}$ molecules do not evaporate as readily as the lighter $O^{16}$ molecules, and so the ocean becomes relatively enriched in $O^{18}$ at the peaks of the ice ages. Therefore, even seashells forming in tropical waters at this time are quite enriched in $O^{18}$. Scientists are now well aware of this added complexity and are careful to take it into account.

The $O^{18}$ concentrations are also affected by the salinity of the water. Thus the $O^{18}$ technique provides information about the temperature and salinity of the ocean. Unfortu-

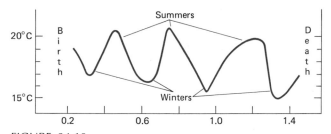

FIGURE 24.13
The biography of a belemnite, 150 million years ago.

nately, this means we have two unknowns—but only one equation. This requires using another isotope, such as $C^{13}$, that responds less to temperature changes and more to salinity changes than $O^{18}$.

# HELP FROM OUR ANCESTORS

The records of our ancestors also contain many indirect clues about the climate in the days before weather instruments were available.

Some of the climate clues are contained in myths or sagas, but these are not always easy to interpret. One notable exception to this is the information that can be gleaned from the Viking sagas. These provide us with a running record by which we can date some of the major changes in the sea ice and plant life over a number of centuries during the Middle Ages.

In Japan, records have been kept of the date of the festival of the cherry blossoms. The cherry blossoms come out earlier when the weather of the early spring is warmer than usual. Thus we have a record of the springtime temperatures in Japan that stretches back almost 1000 years.

A number of agricultural records that have been kept in Europe since the Middle Ages have been used to infer the climate. Because wheat and wine are sensitive to temperature, their crop histories provide valuable information on past climates. Wheat is a grain that is especially sensitive to weather conditions in the spring and early summer. When the wheat crop suffers, the spring and early summer probably have been both hot and dry. The grapes used for wine are also quite sensitive to the average temperature conditions over the entire summer. The cooler the summer, the later the wine is harvested. The latest wine harvest ever recorded was in the year 1816—How is that for proof! (See the discussion of the year without a summer later in this chapter.)

This is but a small sample of the many historical forms of climate information that are available; there is still much that can be done in this field.

# CLIMATE CHANGE AND CONTINENTAL DRIFT

Over the ages the continents have actually moved across the surface of the earth! When you look at a map of the world you can't help noticing that Africa and South America seem to fit together like two pieces of a jigsaw puzzle. People have probably noticed this fit ever since the first maps of the world were printed in the early 1500s. However, it was not until 1911 when Alfred Wegener began his detailed investigations on **continental drift** (now often called **plate tectonics**) that the idea received serious consideration. At first Wegener won many converts, but soon his idea fell out of favor. Unfortunately by the time he had died on the ice of Greenland, he had been temporarily discredited. Then, in the mid-1960s, new and indisputable evidence in favor of continental drift was dredged up from the ocean floors.

It may surprise you to learn that climate evidence provided Wegener (who, as you have seen in Chapter 12, was a brilliant meteorologist) with his strongest arguments in favor of continental drift. When he tried to make sense out of the world climate patterns of the late Carboniferous and early Permian periods, Wegener realized that it demanded that the continents be in much different positions than they are today. Let's examine the evidence.

Moraines from the Carboniferous Period have been found in many places. They occurred in eastern South America, in most of Africa south of the equator, in Australia, and even in India! At the same time, the greatest tropical rain forests in the earth's history (so far as we can tell) were creating the world's greatest coal deposits in the United States, Europe, and China. Also, evaporite deposits dating from this time were found both north and south of the coal deposits in America and Europe.

Now, how can glaciers be growing both north and south of the equator at the very same time that tropical rain forests were growing in the midlatitudes? A global climate pattern must be logical, and this enigma suggested that the continents were located elsewhere during the Permo-Carboniferous Period.

As you can see from the world map (Figure 24.14), at that time most of the continents were grouped together around Antarctica. Even India was wedged between Africa and Australia, and South America and Africa were together. All of these combined formed the giant continent called *Pangaea* (or Gondwandaland). During the Mesozoic, Pangaea began to split apart, and most of the continents drifted northward. The fastest moving of all the continents was India, which finally smashed into the rest of Asia where it is today. The Himalaya Mountains resulted from this collision!

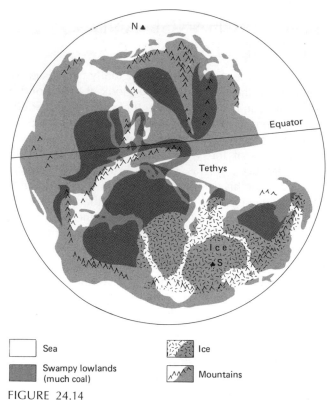

Sea

Swampy lowlands
(much coal)

Ice

Mountains

FIGURE 24.14
The world map during the Carboniferous Period.

# CLIMATES OF THE PAST

This section includes only the briefest outline of the earth's climate history. This record is by no means complete. Generally, the further back we go in time, the larger are the gaps in our knowledge and the more uncertain the information.

During the entire Precambrian Period the fossil record is too primitive to be useful to paleoclimatologists. No evaporite deposits have been found from this time either. There is only some suggestive evidence of ice ages. Both of these ice ages occurred in North America near the Great Lakes. The first occurred roughly 2 billion years ago, and the second 1 billion years ago.

In the early part of the Cambrian there is evidence of another ice age. This ice age seems to have been rather widespread, and ancient moraines have been found in Greenland, Scotland, Scandinavia, and China in the NH and in South America, southern Africa, and Australia in the SH. The sudden blossoming of life in the Cambrian seems to have taken place just after this ice age.

During the Ordovician Period there was also a small ice age. The Sahara was then located at the South Pole and was partly ice covered.

By the Devonian North America, Europe, China, and Australia had become tropical. Extensive coral reefs from the Devonian have been found in all these places. Evaporite deposits from this time have also been found in North America and Europe. The Devonian would have to be classified as a warm, dry time from the bulk of the evidence, although there is some evidence of glaciation at the southern tip of Africa and in South America.

The climate became much wetter during the Carboniferous Period, and the widespread ice age and tropical rain forests of this time have already been mentioned. Then, as the ice began to recede in the Permian, great deserts began to expand. The world's greatest evaporite deposits come from Permian times. The dry times continued into the Triassic Period when the climate began to become moderate.

During the severe times whole species became extinct, and new species of plants and animals replaced them. Surprisingly, it was during the Permian that the first mammallike reptiles appeared. However, as the climate eased these protomammals lost their advantages for survival, and reptiles took over as the dominant life form.

The entire Mesozoic was a warm period. There is no evidence of ice anywhere on earth. It was a time when mountain ranges were worn low by erosion and low-lying land alternated with shallow seas. This was the Age of the

As the continents moved northward they crossed different climate zones. Imagine a continent starting from the equator and gradually moving toward the North Pole, as in Figure 24.15. At first it will have a tropical rainy climate, but then as it moves through the subtropics dry conditions will prevail. As the continent continues northward it moves into the stormy middle latitudes and, finally, it arrives at the pole glazed with an ice cover.

As the continent drifts northward, animals and plants would have to migrate toward its south end, simply to remain in the same climate belt. Otherwise they would have to adjust to the changing climate conditions. If they couldn't do either, they would become extinct.

Continental drift can also tell us something about the climate of the future. Right now, Australia is moving northward at the rate of several centimeters per year. A large part of Australia is now desert because it is centered in the subtropics. Assuming it will continue to move at the same rate, 50 million years from now Australia should lie directly over the equator, and it will then have a tropical rain forest climate. To verify this, we'll simply have to wait and see.

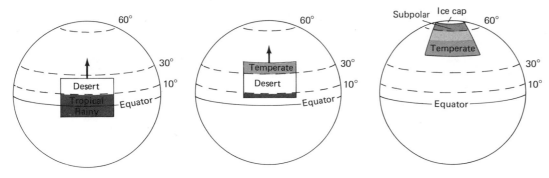

FIGURE 24.15
Climate changes experienced by a continent moving from equator to pole.

Dinosaur, who ruled the earth for 150 million years. Their enormous fossilized bones are now exhibited in natural history museums and are a "must" on anyone's list of things to see. They speak of times that are irretrievably lost.

The dinosaurs lived under the protection of Mesozoic warmth and, for the most part, adequate moisture. Tropical and subtropical conditions prevailed almost worldwide (although much evidence from Antarctica now lies buried by ice), and even Greenland had subtropical weather. Then, toward the end of the Mesozoic a new mountain-building era began and the shallow seas began to disappear. The weather grew somewhat cooler and drier as the continents of the NH continued moving northward.

One of the great mysteries of the earth's history is why the dinosaurs disappeared. Some people have pointed to the cooling and drying trends at the end of the Cretaceous Period, and so it has been pictured in the movie, *Fantasia*. But, to be quite honest, no one knows the answer. The fact is that the climate on earth did not change radically until much later. Perhaps there were some disastrous cold waves or droughts but if they took place they are not apparent in the geologic record. Perhaps the dinosaurs had evolved to the point where they had become too overspecialized to a specific environment and the slight climate changes were among the factors which helped them become extinct. In any case, they ruled the earth about 100 times longer than *We* have thus far. Will we last as long?

The fossil record tells us much more about the climates of the Cenozoic Era because we are finally dealing with plant and animal species that still exist today. The Cenozoic has been a time of more or less continuous cooling. We can observe this cooling trend in the Cenozoic by looking at the average temperatures of the Pacific coast of North America (see Figure 24.16).

At the beginning of the Cenozoic much of the United States was still located in subtropical latitudes, but it drifted slowly northward and soon reached the midlatitudes. Inland from the Pacific there was also a drying trend. We also know that 20 million years ago Utah was largely forested (judging from the pollen records), but that within the last 10 million years the forests dwindled and were replaced first by grassland and finally by desert.

At least 10 million years ago glaciers began forming on mountains of the NH. Then, perhaps 2 to 4 million years ago, it became so cold that ice began to form on the lowlands and in the ocean near the North Pole. This ice began to spread rapidly within the last 1 million years. In North America it spread from two centers—one in the Canadian Rockies and the other in the Laurentian Shield of eastern Canada. These two ice sheets merged several times so that at their maximum extent they covered almost all of Canada, southern Alaska, and the northern United States, with a layer anywhere from 1 to 3 km thick (see Figure 24.17).

There have been ten or more *major* advances and retreats of the continental ice sheets during the past million years, which at the maximum extent covered the cities of Montreal, Chicago, Detroit, Cleveland, and half of New York City! During these advances the climate was much cooler, and tundra and boreal forest spread across the middle of the United States.

Europe and Asia also felt the effects of these ice ages, as you have already seen in the last chapter. Glaciers spread from the Alps and from Scandinavia, as well as from several other centers in Siberia. These also advanced and retreated a number of times. Strange as it may seem, eastern Siberia and northern Alaska were almost free of ice during

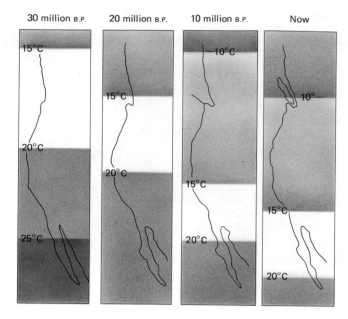

FIGURE 24.16
The cooling trend over the past 30 million years on the west coast of North America. (B.P. means *before present*.)

FIGURE 24.17
Temperatures in the North Atlantic and the thickness (in meters) of the ice sheet over North America in 18,000 B.P.

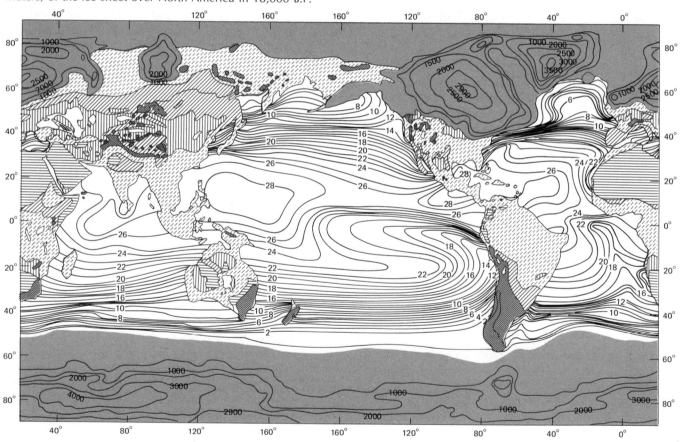

the height of the ice ages, probably because it was too dry there for glaciers to accumulate.

The SH also felt the effects of the ice ages. Glaciers spread from the Andes Mountains and, believe it or not, they also were slightly larger on Antarctica at those times.

In fact, ice had appeared on Antarctica perhaps 40 million years ago or even earlier — far longer than its stay in the NH. Indeed, most of Antarctica was ice covered by about 10 million years ago, and it has remained that way ever since.

During the height of the ice ages the worldwide temperature fell by about 5°C. The largest temperature drops occurred over land in the midlatitudes where it was approximately 10°C cooler! Near the equator, most places only suffered a cooling of about 1 to 2°C, so that these remained tropical.

These temperature changes led to a far larger north–south temperature gradient in the midlatitudes, which was also located much closer to the equator than it is now. The larger temperature gradient also probably led to much stronger winds than we have today on earth. Nevertheless, these were dry winds because there was less evaporation; hence, there was less precipitation with the lower ocean temperatures during the height of the Ice Age. Surprisingly to many, the Ice Age was a relatively dry period.

Some of the climate changes that occurred in the tropical and subtropical latitudes during the Ice Age follow directly from the temperature changes. For instance, the midlatitude storm belt probably moved closer to the equator because the region of largest temperature contrast was also closer to the equator. Thus, the poleward margins of all the subtropical deserts had at least as much rain as now, and the lower temperatures so reduced evaporation that they temporarily became grasslands.

On the other hand, the ITCZ was less free to wander and was more closely restricted to the vicinity of the equator. Therefore, the tropical rain forests were not as extensive as they are today and, in places, the deserts spread as much as 500 km closer to the equator than they are now. There are many places today in which jungle has been found growing over sand dunes that were blown there during the last Ice Age.

The last great Ice Age began rather abruptly about 75,000 years ago. After several minor variations the ice reached its maximum extent about 18,000 years ago, when it covered an area 6 times larger than the continental United States or twice as large as the U.S.S.R. Sea level was almost 100 m lower than now, so that large parts of the continental shelf today under water were then dry land.

For some reason, suddenly the ice began to melt rapidly and retreat. Most of Sweden was ice-free 11,000 years ago, but then the climate turned colder for the next 1000 years and the ice readvanced to the southern edge of Sweden — this was the last fling for the Ice Age and the ice began its final retreat (up to this time). The melting of the ice was so rapid (see Figure 24.18) that today's rivers seem little more than creeks when compared to those post-Ice Age rivers.

Between 5000 and 8000 years ago the weather was distinctly warmer. Temperatures in the midlatitudes were 2.5°C higher than they are today, and forests were able to grow where there is only tundra now. The glaciers also retreated farther poleward then, and as a result sea level was several meters higher.

During this warm time (which is sometimes called the *climatic optimum*) the Sahara became a grassland and all the other deserts of the world shrunk. Animals and humans freely roamed across the Sahara, with lions and elephants common as far north as the Mediterranean Sea.

After this, the climate began to cool and became sharply drier in the subtropics, where Egyptian records point to the constant threat of hungry peoples pressing in from the west. Nevertheless, it seems that it still was somewhat wetter than it is today. The famous astronomer, Ptolemy, also kept weather records for Alexandria, Egypt, during part of A.D. second century. These records show that the Cs climate was not so well developed at the time. Rain occurred all year round, with winter rain coming mostly from lows and summer rain from thunderstorms. Today there is more winter rain than there used to be, but almost no summer rain. In line with this, now there is a noticeable tendency for extremely hot days to occur both before and after summer (due to the Scirocco or Khamsin), but in Ptolemy's day hot days seem to have occurred mainly during summer.

In the waning days of the Roman Empire, the climate took a definite turn for the worse, and the winters grew colder. The snow line in the Alps also descended far enough to obstruct travel (see Figure 24.19). Then, the climate improved for a few hundred years. Between A.D. 900 and 1200 temperatures rose a degree or so higher than they are today. Polar ice retreated enough for the Vikings to sail freely across the Atlantic, which was much less stormy than it is now. During this time Greenland was colonized and America was discovered by Europeans.

After 1200, the climate began to cool and disastrous harvests with famine and plague became much more common. The years from about 1400 to 1850 have been termed the Little Ice Age. After 1200, sea ice advanced and helped to isolate the dying settlement on Greenland.

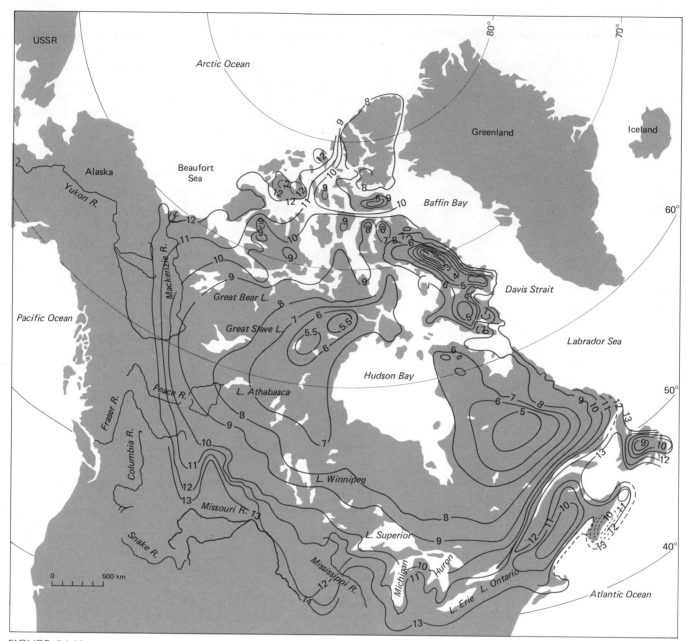

FIGURE 24.18
The retreat of the ice sheet on North America. Contours give the position of the edge of the ice in thousands of years before present (B.P.).

After 1570, glaciers in the Alps advanced to their greatest extent since the Great Ice Age 10,000 years earlier, and temperatures in Europe were 1 to 2°C cooler than they are today. On a number of occasions the Thames River was completely frozen over and festivals were held on the ice in London. (There were some tragedies when this ice wasn't thick enough to support everyone.) By the way, the Thames hasn't frozen over in the last 100 years.

Although the coldest winters and the greatest glacial advance occurred right around 1700, perhaps the culmination of the Little Ice Age came in 1816, which has come to be known as the year without a summer. Snow fell in northern New England from June 6–8, accumulating to 50 cm in places. Frosts occurred there in July and again in September, with ruinous results for the harvest. This is one of the important factors that led to a large-scale exodus of

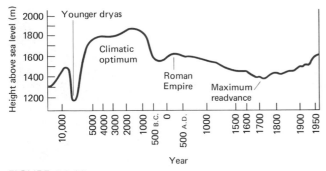

FIGURE 24.19
The changing snow line in Norway during the last 12,000 years. Generally, for every 100 m the snow line drops, the temperature drops 0.5°C.

farmers from New England across the Appalachian Mountains to Ohio. Conditions were no better in Europe, as you will see in the next chapter.

From 1850 until about 1940 temperatures rose worldwide and glaciers and sea ice retreated (see Figure 24.20). The dust bowl of the 1930s came at the culmination of this warming trend, with the hottest, driest summers in American history recorded for the Great Plains. Again, the weather reinforced the effects of an economic depression.

Then, just as a series of scientific papers were published to present evidence that industrialization was responsible for this warming trend, the climate began to cool again. This cooling trend has been most noticeable at high latitudes where the sea ice has spread southward again. However, the cooling trend seems to have leveled off since about 1970. Where we go from here is anyone's guess.

# A FEW COMMENTS ABOUT CLIMATE CHANGE

The last Ice Age both started and ended with astonishing rapidity (geologically speaking), switching on and off perhaps in only a few hundred years. Although some recent evidence seems to indicate that one hemisphere may have kindled the Ice Age changes, the fact remains that these climate changes were worldwide. Furthermore, the climate changes were so great during the Ice Age that even a warm Ice Age year in the midlatitudes was far colder than a cold year is today.

We cannot speak with such certainty about many of the smaller climate changes that have occurred since then. From year to year there is a natural variability in the weather, but only of a degree or so in the middle and high latitudes. Does this allow us to say that there has been a climate change when a few years in a row the average is about a degree less than the previous few years?

Many of the so-called climate changes that have occurred in historical times have not been worldwide. For instance, when the Little Ice Age was producing its coldest winters in Europe during the 1690s, western Siberia was experiencing unusually warm weather. This should bring to mind the fact that the waves in the upper troposphere can bring persistent northwest winds with cold weather to one place and at the very same time bring persistent southwest winds with warm weather to another place. Similarly, at the same time that Europe was experiencing its climatic optimum in the twelfth century, Japan was having a series of unusually cold springs (as we know from the dates of the cherry blossom festivals).

Finally, the scanty evidence we have from the SH indicates that during the Little Ice Age the climate changes in the two hemispheres were mostly out of phase. Thus, the Little Ice Age does not appear to have been a worldwide cooling, although it certainly was significant in North America, the North Atlantic Ocean, and in Europe.

Most major changes in climate are bound to take a long time (historically speaking) and last a long time because of the enormous heat (or cold) stored in the oceans and ice caps. For example, if sunlight suddenly became 10% more intense and *all* this extra heat were retained on earth, it would take over 100 years to melt the ice caps and warm the ocean by 10°C. In short, we must be careful before we can say with some certainty that a string of unusual weather means that the climate has changed.

# CLIMATE THEORIES

This section contains a sample of the many climate theories that have been proposed. Several of these have already been discussed. The first theory we discussed was the positive feedback mechanism of temperature and ice (see Chapter 8). Briefly reviewing the theory, remember that colder temperatures cause more ice to freeze and this increases the albedo, which in turn leads to even colder temperatures, and so on. This feedback mechanism is almost universally accepted by meteorologists as a vital factor in any climate theory. What is distinctly lacking from this theory is the triggering mechanism that would start the chain reaction in the first place. Could it simply be started by one freakishly cold, snowy winter, or would it require a fundamental change such as a change of sunlight or a change in the global-scale winds?

Continental drift is a second important ingredient in the very long-term climate theories. However, continental

FIGURE 24.20
The retreat of the Argentière glacier during the twentieth century.

drift cannot be used to explain the more recent climate variations, because the continents have hardly moved during the past million years, during which time there have been at least four major ice ages, as well as many smaller climatic variations.

One of the more straightforward theories asserts that many of the climate changes on earth have simply been the result of changes in the intensity of sunlight. However, no significant change of sunlight intensity has ever been observed since people began measuring the solar constant. Furthermore, astronomers are not yet certain if there has been much change in the solar constant, although astrophysical theory indicates that there has been a slow increase of perhaps 30% over the past 4 billion years. In short, it must be said that at present we have no way of verifying or disproving solar-change climate theories.

One related theory now undergoing something of a renaissance is the sunspot-cycle climate theory. This theory holds that many climate changes can be ascribed to the changing number of sunspots which appear on the face of the sun. The logic behind the theory is that there may be a small increase (less than 1%) in the heat coming

from the sun when there are more sunspots than usual.

One of the chief pieces of evidence being used in support of this theory is the fact that during the period 1645 to 1723 there were almost no sunspots; this corresponds to the height of the Little Ice Age.

However, this theory will have to wait many years for a verification. Only now do we have instruments in space and above the atmosphere measuring the direct solar intensity. These measurements will eventually confirm or deny if the sun's intensity changes with the sunspot number. But even if the solar constant does vary with the sunspot cycle, it may be some time before a coherent theory is formulated to explain the confusing mass of climate evidence. Indeed, there are climatic features that seem to vary with the sunspot cycle, but at most the sunspot cycle produces different weather changes in different places with no well-defined pattern as of yet.

One of the most popular climate theories (this doesn't make it correct) is that many climate variations are due to the variations in the earth's orbit around the sun. Several aspects of the earth's orbit are not quite the same each year but actually change slowly. Three factors are considered important in instigating climate changes.

**1.** Right now the earth is closest to the sun on January 1 and farthest from the sun on July 1. In about 11,000 years we will be closest to the sun on July 1, and in 22,000 years things will return to their present conditions. This is called the **precession of the equinoxes** and it has a 22,000-year cycle.

This factor is important because it is now making the NH winters milder and the summers cooler than normal, whereas the opposite is true in the SH. This *seems* to imply that any climate changes produced by this factor would occur out of phase in each hemisphere.

**2.** The distance from the earth to the sun now varies between 147 and 152 million km. This is just about as small as the variation gets, but it can be as large as 142.5 to 156.5 million km. This factor is known as the **eccentricity** of the earth's orbit. The eccentricity varies with a period of approximately 95,000 years.

This factor also affects the severity of the seasons, producing opposite effects in the two hemispheres at any given time. Thus, when the SH is having cold winters and warm summers (i.e., extreme temperature variations), the NH is having warm winters and cool summers (i.e., small annual temperature variations).

**3.** The earth's axis now tilts 23½ degrees toward or away from the sun. The tilt can be as much as 24¼ or as little as 21¾. The tilt is called the **obliquity of the ecliptic** by astronomers and has a cycle of 42,000 years (see Figure 24.21).

The larger the tilt, the higher the sun can get during summer in the polar skies. A large tilt thus leads to smaller temperature differences in the summer hemisphere and also warms the poles. In addition, a larger tilt increases the area of total darkness during the polar winter. Unlike the first two factors, this produces climate changes that are in phase in both hemispheres.

To test the orbital theory, the first step is to see if the climate record shows any of these periodicities. Sediment cores are obtained from the ocean floors and analyzed in detail by using the $O^{18}$ technique, observing the coiling of *Globigerina pachyderma,* and so on, and then the climate record is reconstructed from these cores. There has been some strong evidence supporting the orbital climate theory (the latest work as of 1978 claims to find an especially strong 100,000-year cycle), but much work still needs to be done. First of all, no one has yet shown precisely how the orbital changes should affect the climate. For instance, in order to produce an ice age is it more important to have cold winters to freeze more ice, or is it more important to have cool summers so that less ice melts? The latest choice is that an ice age is helped by cold summers in the NH, but cold winters in the SH—but no one alive today knows the answer to that question. The orbital climate theory still awaits testing by the large

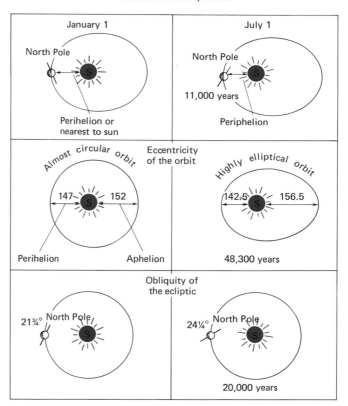

FIGURE 24.21
The variations of the earth's orbit.

GCMs. For what it's worth, the orbital theory has the earth now heading directly back to ice-age conditions.

Geologists have shown that ice ages occurred only during times when there has been active mountain building on earth. Mountains provide a "seed" area for glaciers, since the weather is cooler and often wetter in the mountains than in the lowlands. Once the glaciers have formed, the ice-albedo feedback mechanism can begin working.

Some scientists feel that ice ages can be started during times of extensive volcanic eruptions. Explosive eruptions of volcanoes such as Krakatoa in 1883 can fill the stratosphere with fine dust which takes several years to fall to the ground. The dust reduces the sunlight reaching the ground, presumably increasing the global albedo. According to some advocates of the volcanic ice age theory, this can reduce worldwide temperatures enough to initiate an ice age.

Recently the influence of volcanic dust on climate has been studied in detail. The results show that large eruptions have a small cooling effect on the climate, which lasts only a year or two after the eruption. Thus, for example, the year without a summer, 1816, may have been the result of the truly enormous eruption of Tambora in 1815.

However, there is no evidence that volcanic eruptions have ever produced any long-lasting climate changes. The only way for volcanic eruptions to do so would be if the level of volcanic activity was at least 10 times greater than it has been over the last century.

One of the popular theories during the 1950s was the carbon dioxide theory. According to this theory the warming trend that culminated about 1940 could be attributed to the increase of $CO_2$ that is the result of human activities. It is a fact that $CO_2$ has increased 15% since 1890, which in turn has caused an increase in the greenhouse effect. But late in the 1950s the climate statistics were finally updated and, lo and behold, they showed the warming trend had ended around 1940 and had been replaced by a cooling trend! Obviously, some other factor must be more important than the $CO_2$ effect. Results from the recent GCMs have shown that doubling the $CO_2$ will lead to a warming of roughly 2°C, even when the ice-albedo feedback mechanism is taken into account (see Figure 24.22).

There is now a resurgence of interest in the $CO_2$ climate theory because of the energy crisis and our increasing need to burn coal. Pound for pound, coal produces more than twice as much $CO_2$ as oil when burned, because oil is not pure carbon and also produces much $H_2O$ when burned.

But if the climate is now cooling, are we still to blame? Some ecologically minded people would answer, yes. Now we are told, even though the $CO_2$ we stuff into the atmosphere has an effect, it is the aerosols we produce that have the greater climatic effect. They block the sunlight and cool the climate. Scientists admit that, although they think that particles cause a net cooling by increasing the albedo, they are aware that the particles also produce a slight increase in the greenhouse effect as well. Thus, particles may not be such efficient coolers of the atmosphere after all. Then add to this the fact that most aerosols are produced naturally, and humans are at least partly off the hook as the culprits.

In the past few years one more feedback mechanism has been proposed. As the desert spreads, the albedo increases—first, because soil albedo increases as it dries and, second, because light-colored soils and sand become exposed, thus replacing the much darker plant debris. This leads to a local cooling effect that enhances sinking motion and dry conditions. The drier conditions in turn lead to an even further spread of the desert. This theory was perhaps inspired by the drought in the Sahel. It

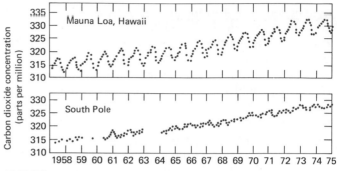

FIGURE 24.22

The recent increase in $CO_2$ at Hawaii and at the South Pole. The seasonal variation is caused by the growth of vegetation in the NH, which reduces the $CO_2$ to its lowest point in August of each year.

supposes that the increased desertification restricts the northward wandering of the ITCZ.

This theory (like all other climate theories) is also under attack. Opponents point out that the decreased precipitation in the newly formed deserts leads to lower evaporational cooling at the surface, which largely offsets the cooling due to increased albedo. A recent study of the drought years in the Sahel has shown that the ITCZ wandered as far north as in other years, but the intensity of atmospheric circulation throughout Africa for those years was weaker and precipitation was lower everywhere on the continent.

**Evaluating the Climate Theories**

We are now beginning to critically evaluate many of the climate theories by simulating them on the computer. The large GCMs are continually being improved and have become enormously complex. Not only are they able to simulate the behavior of the atmosphere, but they now are beginning to include some of the interactions of the atmosphere with the surface below (by incorporating the albedo-feedback mechanism) or with the ocean (by incorporating the effects of ocean currents).

Unfortunately, the GCMs are still subject to many artificialities. For example, they cannot directly predict cloudiness or precipitation produced by cumulus and cumulonimbus clouds, because the distance between the grid points is many times greater than the size of the largest cumulonimbus clouds. As a result, they must make an artificial prediction of the cumuliform cloudiness, based on the sunlight intensity, the temperature lapse rate, and the humidity profile. This prediction must be based on climatological statistics, and one can naturally question how accurate this approach is when applied to completely different climate conditions such as prevailed 18,000 years ago.

There is another important drawback to the GCMs—they are very time-consuming. For instance, if we want to test the orbital theory of climate, we must run the computer through several orbital cycles; this amounts to about 500,000 years of weather. It now takes the fastest computer about 5 minutes to run through one day of weather, even when the interactions with the surface and ocean are not computed. At that rate it would take 2000 years of computer time to test the theory! But even if the computations were not so time-consuming, the GCMs are so complex that it is very difficult to analyze causes and effects.

For these reasons, a number of simpler numerical climate models have been developed. These models start by using the Stefan-Boltzmann equation. They artificially (without computing actual winds) transport heat from the warmer tropics to the cooler poles. They also simulate the greenhouse effect and incorporate other effects, such as a variable albedo. As one example of how the models work, suppose we compute that the temperature at a place drops below about $-1$ or $-2°C$; with this change the albedo will then be increased to allow for extra snow. But now that there is a new albedo, the temperature must be recalculated, and changing the temperature at one point will have an effect on all the surrounding points. Thus you can see that even with these simpler models, an enormous number of calculations must be made before a consistent picture of climate is produced.

A typical experiment run on one of these models is to change the solar constant to see what effect this has on the global climate. The early versions of these models were unrealistically sensitive to any changes in solar intensity and indicated that changing the solar constant by as little as $±2\%$ would completely melt the ice caps or freeze the whole globe. More recent versions of those models now indicate that the solar constant would have to change by at least $±10\%$ to produce such drastic changes.

Thus, as the climate models become progessively more realistic and computers become faster, we nurture the hope that within this century we may begin to learn what makes the climate change. Armed with this knowledge, we may be better able to feed all the people on earth.

# PROBLEMS

**24-1** Explain why, in general, the lower the rock layer the older it is. Give exceptions to this rule.

**24-2** Why is it not possible to date sedimentary rocks by radioactive decay?

**24-3** Relate redbeds to the lateritic soils of the Aw climates.

**24-4** Take a trip to the greenhouse of the nearest botanical gardens and see for yourself how leaves vary with climate.

**24-5** In a particular region spruce and pine trees have been replaced by oak trees. What change occurred in the climate?

**24-6** Why are fossilized animal tracks more likely to have been produced in arid climates or climates with a long dry season?

**24-7** Why is coal an indicator of a formerly humid climate rather than a formerly warm climate?

**24-8** What are beach terraces, and what do they indicate about past climates?

**24-9** Comment on the following statement. Since it is relatively dry in the Mediterranean area, if the Straits of Gibraltar were dammed up the Mediterranean Sea would eventually evaporate, leaving huge salt deposits.

**24-10** How can an analysis of the oxygen isotopes in sea shells tell you the temperature they were formed at?

**24-11** Think of some human activity whose records might be useful for inferring past climates. (Use an example not given in this book.)

**24-12** What is continental drift, and what is its relationship to climate change?

**24-13** Israel is currently moving south at about 1.5 centimeter per year, and Jordan is moving north at the same rate. Comment on the expected, long-term changes of the climate in these two countries.

**24-14** Judging from the surface temperatures of the sea at the height of the last ice age (Figure 24.17), do you think hurricanes were more or less frequent than they are now? Why?

**24-15** Why was worldwide precipitation lower at the height of the ice age?

**24-16** (a) In how many years will we be closest to the sun during the NH summer? (b) In how many years will the earth get *both* farthest from and closest to the sun? (c) How often (in years) does the sun appear highest in the sky?

**24-17** Explain three different climate theories. Which of these do you find most convincing?

**24-18** Over the past five years have you noticed any significant climate changes.

**24-19** Comment on our ability to change the climate.

# 25

# THE MARK OF CLIMATE ON THE HUMAN RACE

It was the summer of 1816 or, rather, what should have been summer. Lord Byron, Poldini, and Percy Shelley and his wife, Mary, had come to Switzerland to enjoy nature. As Mary Shelley tells us, the weather altered their plans.

But it proved a wet, ungenial summer, and incessant rains often confined us for days to the house. Some volumes of ghost stories . . . fell into our hands.

"We will each write a ghost story," said Lord Byron . . . .

On the morrow I announced that I had *thought of a story*. I began that day with the words, "It was on a dreary night in November."

Mary Shelley, Introduction to *Frankenstein*

And so the tale, *Frankenstein* was created during the "year without a summer." Were it not for the bad weather, Mary Shelley may never have written this horror classic.

There were other repercussions. The glaciers of the Alps became enlarged in the decade around 1816 and then advanced down the valleys, threatening a number of Swiss villages with imminent burial in ice. It is no mere coincidence that the theory of the Great Ice Ages first took seed during this minor but impressive decade of glacial advance. What had greatly impressed the imaginative mind of Mary Shelley (for one chapter of Frankenstein takes place on the Des Bois glacier near Chamonix which Mary had seen moving) also impressed the supposedly more sober minds of naturalists such as Ignaz Venetz, Charpentier, and Agassiz.

The final consequence of the year without a summer is that it was also a year without any appreciable harvest. This resulted in inflation, starvation, and social unrest among a people recently ravaged by the Napoleonic Wars.

The purpose of this chapter is to examine some of the ways in which our history and activities have been influenced by the weather and climate. Of course, only a small number of the innumerable examples can be included here. We will look at how historical events have been shaped or altered by significant weather events or climatic changes, and how human health and the arts have been influenced by the weather. As in all matters touching human affairs, the weather is just one of many factors. Thus it is rarely possible to say with certainty exactly what the role of the weather was in a particular instance. And with this brief word of caution behind us, let's begin.

# WEATHER, CLIMATE, AND HUMAN HEALTH

During the spring of 1965 an American soldier serving overseas in Germany threw his one-year-old daughter to her death from a second-story window. Since this soldier did not live on an army base, he was tried by the local German civilian court. For this brutal act the man was actually pardoned. The reason? The German court acknowledged temporary insanity caused by the foehn.

Of course, this is an extreme example of the effects that weather can have on us, but perhaps not so extreme as it first appears to be. Diseases and mortality are quite closely related to the weather. Toward the end of every summer in the years before Salk's vaccine, mothers would not let their children go to the beach for fear of polio. And this was done with good reason; the incidence of polio always peaked in late summer.

Mortality tables from Chicago show that until about 1890 more people died during the summer but, for a while after that, the death rate was highest during the winter (see Figure 25.1). In this case, the infectious and parasitic summer diseases were chiefly eliminated by draining the swamps, installing adequate sewage systems, and cleaning the water supply. However, the respiratory diseases of winter were not proportionately reduced.

Every autumn outside the tropics we await the annual winter onslaught of influenza. Bronchial troubles (not caused by summer allergies) are also far more common during the winter, because of the stress of cold air on the lungs.

Deaths from heart attacks and heart disease increase when the body is subjected to great thermal stress (i.e., when it is too hot or too cold, or when the temperature changes sharply). Thus, in the cooler climates heart disease is highest in winter, while in the hotter climates it peaks during the hottest times of the year. Therefore, it is wise for people with heart disease to go south during winter and go north during summer.

People with old injuries or wounds and people with arthritis often forecast the weather with surprising skill. We are not exactly sure what causes their joints to throb with the approach of bad weather. Perhaps it is the falling barometer or the changing temperature. Or, perhaps it is the minute pressure variations from gravity waves or from the changes in the electric field (known as **sferics**). In any case their bodies are subject to an added stress they recognize as being related to changes in the weather. Indeed, it has been determined that the dramatic weather changes associated with frontal passages (especially warm fronts) bring the unhealthiest type of weather.

The notorious weather effects of the foehn and other

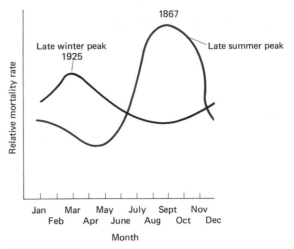

FIGURE 25.1
The seasonal variation of mortality rates in Chicago for 1867 and 1925. Originally, there was a sharp peak during the late summer. The winter peak is no longer so pronounced.

hot winds are well documented. They include depression, headaches, insomnia, and respiratory trouble, and have been connected with certain chemical changes in the body such as extremely reduced excretion of adrenalin in the urine. Often it is the temperature rise rather than the warmth itself that produces the ill effect. Recent research indicates that much of the discomfort at such times is due to the fact that the atmospheric electric field has reversed. One strange finding is that people gradually seem to lose their resiliency and, as time goes on, suffer more than ever from the hot winds. Newcomers to such lands tend to be much less affected.

Weather effects also show up in the statistics of accidents. Once again, frontal passages seem to have the worst effect on people, since it is during these times that the incidence of accidents increases dramatically. You might be tempted to argue that "of course there will be more accidents during bad weather such as fog simply because visibility is obstructed." But indoor accidents are also more common during frontal weather! During such weather, human reaction time slows down and work complaints and accidents rise about 45%, much of which is apparently due to biological changes produced by the weather changes.

If you take a trip to the tropics, you must be immunized against a number of diseases such as cholera and yellow fever, which are often thought of as tropical diseases. One of the most notorious of the so-called tropical diseases is malaria (from the Latin for *bad air*). It is impossible to underestimate the effect that malaria has had on humankind and civilization over the ages.

Carried by the anopheles mosquito which breeds in

still, warm waters it has decimated populations primarily in tropical and subtropical regions. The mosquito itself can breed when the average temperature for the warmest month reaches 16°C, which it does, say, in London. Fortunately, however, the parasite that produces the malaria does not become active until the average temperature reaches 19°C. Thus, only a climate change can bring malaria into England for an extended stay.

Malaria seems to have been unknown in ancient Greece before the Persian Wars, but it certainly spread with great rapidity from one city-state to another a generation later when Greek fought Greek during the Peloponnesian War. Malaria also became a scourge to the ancient Romans about the second century B.C. The Romans realized that the source of their troubles came from the swamps, although they thought that it was the cool, moist air that produced the malarial chills. Thus, truly Herculean efforts went into draining the swamps, and this kept the problem under control to some degree. In fact, many historians agree that malaria played no small role in the eventual demise of the Greek and Roman cultures. And once the civilizations crumbled, the swamps filled up again and the malarial problem became worse than ever. Malaria has no doubt also greatly inhibited civilizing efforts throughout the tropical parts of the world for thousands of years.

### Locusts and the Weather

Another weather-related scourge is the infamous plague of locusts. Locusts are "merely" mature grasshoppers that have bred too profusely and then migrated across the countryside, moving with the winds and eating everything green in their paths.

The desert locust (*Schistocerca gregaria*) of biblical fame is the most notable of all the locusts. This resident of the desert fringes of Africa and Asia is a large insect with a wingspan of 10 to 15 cm. Large swarms of up to 100 billion locusts can cover 1000 km². Each locust eats its weight in food every day, so that a large swarm can consume 100 million kg/day. There is no telling how many people have starved to death after locust hoards have flown by.

The desert locust breeds best on the fringe of the desert, laying its eggs in loose, moist sand in regions where the annual precipitation averages between 8 and 40 cm. In wetter climates the locust has so many natural enemies that it rarely can breed with any degree of success. Even people eat locusts, which are considered a delicacy in many cultures.

The locust population booms in the rains that follow a particularly dry year. During the dry year the number of natural enemies of the locust dwindles sharply. Then, in the rains that follow, the locust can breed with amazing rapidity. However, young hoppers do not swarm until the rains have stopped, after which they are forced to congregate on the few remaining green areas. Close packing has some strange effect on them, sending them through several molts in which they grow successively larger. Finally, they emerge as mature locusts and take off, flying with the winds. They have been known to cover over 3500 km in a single month.

Since the locust swarms generally move with the winds, meteorologists are able to make locust forecasts. There are several forecasting tricks. Locusts tend to converge in low pressure areas because that is where the low-level winds converge. It takes no great imagination, therefore, for any meteorologist to conclude that locusts will tend to converge in the ITCZ. Since the trade winds blowing into the ITCZ are quite reliable and change regularly with the seasons, the movement of locusts can be predicted days, and even weeks, in advance. This is of great help to spraying efforts.

The normal winds keep the desert locust away from Egypt (see Figures 16.10 and 16.11), but when the Khamsin sets in not only is the weather unbearable but the locusts may be sucked up into Egypt by the south and southeast winds, just as in biblical times.

## WEATHER, CLIMATE, AND HISTORY

In the sixth century B.C., the Persian Empire was an expanding colossus, stretching eastward to India and westward to the Mediterranean Sea. But Darius I was not satisfied. He wanted to expand even farther westward. He attacked Greece, but his huge army was defeated at the famous battle of Marathon. Darius returned to Persia and died there before he could launch another attack.

Themistocles had played no part in the battle of Marathon (much to his own shame), but he would soon play a great role. He had the feeling that the Persians would eventually return and knew that the ultimate Greek victory would have to come at sea. As leader of Athens, he then greatly expanded and trained the Athenian navy, preparing them for the next invasion.

The next invasion did come. Xerxes became ruler of Persia and sought revenge. Herodotus tells us that Xerxes assembled an army of over 2.3 million men, which grew by another 300,000 men by the time it reached Greek soil. The Persian navy consisted of over 1200 large ships, and the invading force was so large that it was said to dry up rivers when it stopped to drink and to bankrupt each town that tried to feed it for only one day! (Modern estimates for this army are placed at only 200,000.)

After a preliminary sea battle Themistocles assembled the entire Greek navy in the narrow strait between the island of Salamis and the mainland (see Figure 25.2). When the Persian fleet approached, it was so vast that it terrified the Greeks. Themistocles knew he could win the battle, so he actually sent a messenger to Xerxes advising him to cut off the Greeks' only escape route. Xerxes was surprised, but followed this advice. The surrounded Greeks had no choice but to fight.

The great battle of Salamis was fought the next day. Themistocles had chosen his battlefield well, since in the narrow strait the Persians could not make use of their full numbers. Furthermore, Themistocles also chose the time of the battle, and this is where the weather enters.

As Themistocles had fixed upon the most advantageous place, so, with no less sagacity, he chose the best time for fighting; for he would not run the prows of his galleys against the Persians, nor begin the fight till the time of day was come, when there regularly blows in a *fresh breeze from the open sea* [emphasis added], and brings with it a strong swell into the channel; which was no great inconvenience to the Greek ships, which were low built and little above the water, but did much hurt to the Persians, which had high sterns and lofty decks and were heavy and cumbrous in their movements, as it presented them broadside to

FIGURE 25.2
The battle of Salamis and the sea breeze.

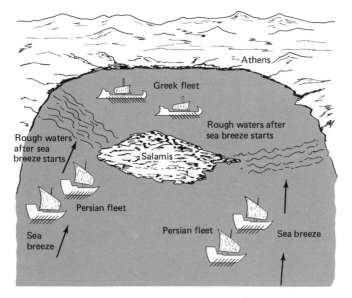

the quick charges of the Greeks, who kept their eyes upon the motions of Themistocles, as their best example. . . .

Plutarch, *Lives of the Noble Greeks and Romans*

The Persians were routed in this deciding battle of the war and, soon after, Xerxes retreated from Greece, never to come again. Thus, the outcome of one of the major battles in history was altered by using the sea breeze to advantage.

There is another notable invasion that was defeated largely by the weather. The great Kublai Khan decided to add Japan to his enormous realm. After one attack failed (largely because of a storm) he assembled an even larger army and navy which reached Japan. This enormous armada consisted of Korean and South Chinese ships, and involved over 140,000 men.

The invading army was a motley crew (just as the Persian army) and never did advance very far into Japan. But the final outcome of the invasion was decided on August 15, 1291 when a typhoon approached from the southeast. The invaders prepared to flee and boarded ships, but the ships were driven back by the wind and the rising waters. There (perhaps in the Imari Gulf) many ships collided and sank, and thousands drowned. The remaining soldiers who were not able to get on board in time largely dispersed during the storm and then were slain or enslaved by the Japanese Samurai. Thus ended the invasion of Japan.

Of course, you all know what the notorious Russian winters did to the unprepared invading armies of both Napoleon and Hitler. Any reading of military history will show no end of examples of how the weather changed the outcome of important battles.

There is one war in which the role of the weather has sometimes been overplayed. That was the defeat of the Spanish Armada by England. The first significant action took place early in 1588, when Francis Drake secretly sailed into Cadiz harbor and destroyed the fleet that the Spanish were preparing for the attack. At great cost the Spanish rebuilt their fleet and set sail only to encounter a fierce storm. But this storm did far less damage than Drake had done, and in another two months they were ready to sail again.

This time the Spaniards reached the coast of England with favorable southwest winds. The tide was also favorable for the Spanish cause. If they had sailed up into Plymouth harbor to the narrow waters, they would easily have destroyed the English. Instead, they hesitated for hours to allow the last stragglers to catch up. The tide changed by this time, allowing the British to get out into the open waters where they had a fighting chance. The Spanish never regained the offensive.

During the deciding battle a few weeks later the English used the wind to load eight of their ships with combustibles, set fire to them, and let them drift toward the Spanish Armada. At the sight of the approaching tinderboxes the Armada panicked and were routed. But by far the worst damage they received was after the battle when the crippled ships were wrecked by bad weather. Twenty ships sank near the coasts of Ireland and Scotland, and when the sick and starving sailors were able to reach shore, they were slaughtered by the local people.

Weather and climate have always influenced the food supply of nations, and thus history. The tree ring record provides us with a picture of the rainfall in the southwest United States for several thousand years. This record shows that there was an extensive and severe drought that occurred there from A.D. 1276 to 1299.

Shortly after A.D. 1000, a rather advanced American Indian civilization had sprung up in the southwest United States. Irrigation was developed to foster a thriving agriculture on the mesa tops. Pueblos were built on the mesa tops at first, but these were soon abandoned in favor of the famous cliff dwellings of Mesa Verde and Chaco Canyon, which were built on cliffs below the mesa tops.

Then, in 1276 the drought struck. The Indians were able to *weather* the situation for a time, but gradually the water table fell too low; by 1300 the cliff dwellings were abandoned throughout the entire area as the Indians left in search of water.

Society was less mobile in Europe. In fact, it was bound by ancient, repressive patterns when the famine of 1788–89 struck France. A hot, dry spring in 1788 combined with the antiquated and inefficient farming methods shriveled the grain harvest so that as much as 25% of the grain had to be saved for the next year's planting. And, to make matters worse, on July 13, 1788, a monstrous hailstorm totally devastated the crops over a long, wide swath, destroying several hundred villages in its path.

As we know from the history of the time, the French government was incompetent and insensitive to the masses who had always lived on the verge of starvation. The poor harvest was disastrous to them. Bread riots broke out beginning late in 1788, and the fury of the people was further ignited by the unfounded rumor (started by Rousseau) that their queen, Marie Antoinette, had said, "If they have no bread, let them eat cake." Food shortages are always most acute just before the harvest, especially following a crop disaster. So it was in 1789. On July 14, shortly before the summer harvest, the Bastille was stormed.

There are many other occasions when famines altered history. The biblical "seven lean years" probably describes the years when the Nile did not rise high enough to irrigate the lands (due to insufficient rains in tropical Africa). The poor harvests of 1816 and the Great Famine in Ireland in the late 1840s have already been described; both led to unrest and mass emigration. The reign of Louis XIV, so glorious in its early years, was later marked by wars, plague, and a famine in the 1690s that was directly attributable to the poor weather. During the last 30 years of the seventeenth century the population of France actually fell by 25%!

So far we have looked at how individual events were affected or changed by the weather. But the effect of climate is far more pervasive. Is it a mere coincidence that civilization sprang up in the millenia following the Ice Age? Great climate changes in the past were also associated with wholesale extinctions and the emergence of new, more advanced forms of life. Drastically changing environments demand drastic adaptations for all living creatures. Those unable to adapt perish.

People do not change willingly. It is said that Admiral Peary once asked one of his Eskimo guides, "Of what are you thinking," and the Eskimo answered, "I do not think, I have plenty of meat." On the other hand, when conditions are too severe, there is no leisure time to create a civilization.

Those who are awed by their surroundings do not think of change no matter how miserable their condition. When our mode of life is so precarious as to make it patent that we cannot control the circumstances of our existence, we tend to stick to the proven and familiar . . . . Fisherfolk, nomads and farmers who have to contend with the willful elements . . . all fear change . . . . There is thus a conservatism of the destitute as profound as the conservatism of the privileged.

Eric Hoffer, *The True Believer*

Will Durant expresses the preconditions of civilization in meteorological terms.

Certain factors condition civilization, and may encourage or impede it . . . . Civilization is an interlude between Ice Ages: at any time the current of glaciation may rise again, cover with ice and stone the works of man, and reduce life to some narrow segment of the earth . . . .

The heat of the tropics, and the innumerable parasites that infest them, are hostile to civilization; lethargy and disease and a precocious maturity and decay, divert the energies from those inessentials of life that make civilization . . . . Rain is necessary . . . ; the unintelligible whim of the elements may condemn to desiccation regions that once flourished with empire, like Ninevah and Babylon.

W. Durant and A. Durant, "Our Oriental Heritage," *The Story of Civilization*

In short, the conditions for civilization are quite precarious. They can neither be too harsh nor too easy, and there must be enough of a change in the environment to spur or perhaps necessitate ingenuity.

When does civilization begin? Often it is said to begin once agriculture has been established. Some time before 7000 B.C., it seems that agriculture began independently in the Mexican highlands and in Mesopotamia (where there was also animal herding). But what do these two regions have in common?

Both of these regions are subtropical and border on high mountain ranges so that there is a wide range of climate zones within a relatively short distance. At that time many people were seminomadic and followed the migrating animals and the ripening plants. In regions where the migrations covered great distances, there could not have been much time for leisure. People who travel long distances must travel light and cannot easily bring along their inventions. But in these regions of Mexico and Mesopotamia the migrations covered very short distances, so there was probably much more time for leisure and ingenuity. The common element in both regions was the climate.

# CLIMATE AND THE ARTS

We have now traveled through space and walked on the same moon that our ancestors may have revered. The voyages we have taken in rockets were taken in the minds of people once they had leisure to think and dream. Yes, despite the dictum that the only true study of mankind is man (Pope), we find that creative artists in all fields have payed homage to nature and have used the weather to reinforce the dramatic impact of their creations.

### Meteorology and Literature

Early literature often has an epic quality to it. Perhaps it is because people in ancient civilizations lived closer to nature and were more exposed to the elements then we are. In fact, in early literature there are many more references to nature then in most of our modern literature.

The Bible is full of meteorological wonders and disasters, several of which have already been mentioned. Indeed, the very language of the Old Testament is suffused with the feeling of the outdoors.

15. A fountain of gardens, a well of living waters, and streams from Lebanon.

16. Awake, O north wind; and come, thou south; blow upon my garden, that the spices thereof may flow out. Let my beloved come into his garden, and eat his pleasant fruits.

*King James Bible*, Song of Solomon, Chapter 4

The "epic" attitude sees nature as an almost personal force directed by the gods. Aware of the strength and the wonder of nature, humans learned to pray and, according to mythology, to occasionally please the gods, who could change the winds or the course of rivers.

In *The Iliad*, Achilles was almost drowned by the river, Xanthus, but his prayers roused the gods to burn the river. Later, Achilles again prayed for divine help when the funeral pyre of his slain friend, Patroclus, refused to burn.

There was some delay with the body of Patroclus also; the pyre refused to kindle. But a remedy suggested itself to the swift and excellent Achilles. Standing clear of the pyre, he prayed and offered splendid offerings to the two winds, Boreas of the north and Zephyr of the Western Gale . . . .

The two winds rose uproariously, driving the clouds before them . . . . When they came to the deep soiled land of Troy, they fell upon the funeral pyre and the fire blazed up with a terrific roar.

Homer, *The Illiad*

Even in those days there were cold fronts!

Occasionally in epic literature it is even possible to exert a brief but direct influence over the elements. Thus, during a raging thunderstorm in *Moby Dick*, Captain Ahab quelled a mutiny among his crew by snuffing out Saint Elmo's fire, which had lit up his harpoon.

. . . from the keen steel barb [of Ahab's harpoon] there now came a levelled flame of pale, forked fire . . . .

Snatching the burning harpoon, Ahab waved it like a torch among them; swearing to transfix with it the first sailor that but cast loose a rope's end . . . Ahab spoke again.

"All your oaths to hunt the White Whale are as binding as mine; and heart, soul, and body, lungs and life, old Ahab is bound. And that ye may know to what tune this heart beats; look ye here; thus I blow out the last fear!" And with one blast of his breath he extinguished the flame.

Herman Melville, *Moby Dick*

But Ahab's "victory" was temporary. He did not triumph over nature, but merely over his frightened crew. And soon the forces of nature—as embodied in the white whale—did triumph.

A modern epic with tragic overtones, Rolvaag's *Giants in the Earth* portrays the inner and outer struggles of immigrants on the Great Plains. From the first line in the book we know that the weather will be a dominant, omnipresent force.

Bright, clear sky over a plain so wide that the rim of the heavens cut down on it around the entire horizon . . . . Bright, clear sky, today, tomorrow, and for all time to come.

And sun! And still more sun! It set the heaven afire every morning; it grew with the day to quivering golden light—then softened into all the shades of red and purple as evening fell.

Per Hansa and his wife and children had come from a fishing village in Norway to the South Dakota territory. They struggled through locust-plagued summers and bitter, long winters to achieve prosperity. But when you contend with nature constantly you cannot always expect to succeed.

They say it rained forty days and forty nights once in the old days, and that was terrible; but during the winter of 1880–1881 it snowed twice forty days; that was more terrible. Day and night the snow fell. From the 15th of October when it began until the middle of April. . . . And all winter the sun stayed in his house [halo] . . . . One would leap out of bed. . . and start to go out— only to find that someone was holding the door. It wouldn't budge an inch. An immovable monster lay close outside . . . . Once outside, he found himself standing in an immense flour bin, out of which whirled the whiteness, a solid cloud. Then he had to dig his way down to the house again. And tunnels had to be burrowed from house to barn . . . .

O.E. Rolvaag, *Giants in the Earth*

Finally, one of the blizzards in the severe winter of 1880–81 claimed Per Hansa as it, in reality, claimed many others on the northern Great Plains.

There are also many references to the weather in tragedy, but the emphasis is different. In true tragedy the weather or the elements hardly ever cause the hero's downfall (in contrast with the epic viewpoint); instead, weather often sets the background mood. There may or may not be bright interludes, but the dominant mood set by the weather is one of darkness, bleakness, and storminess.

One of the best examples of the role of weather in tragedy comes from William Shakespeare's *King Lear*. Much of the action of this play takes place during a raging storm. King Lear, half senile and furious that he has bequeathed his kingdom to unloving daughters, goes raging about in the storm as if it were the cause of his troubles.

**Kent.**  Who's there besides foul weather?

**Gentleman.**  One minded like the weather, most unquietly.

**Kent.**  I know you. Where's the king?

**Gentleman.**  Contending with the fretful elements;
Bids the wind blow the earth into the sea
Or swell the curled waters 'bove the main
That things might change or cease; tears his white hair
Which the impetuous blasts, with eyeless rage,
Catch in their fury, and make nothing of
Strives in his little world of man to out-scorn
The to-and-fro-conflicting wind and rain. . . .

William Shakespeare, King Lear, Act III, Scene I

All of King Lear's fury and energy is misdirected. He is helpless against his true problems, which have nothing to do with the elements; the weather's "eyeless rage" merely emphasizes Lear's helplessness.

The gothic also uses nature and the elements as a backdrop. *Wuthering Heights* by Emily Bronte is a novel of wild, unfulfilled passion. The setting for the novel is appropriately the wild, windswept heath of the English coastline. Here, the weather plays a large role in the tempestuous lives and deaths of the would-be lovers, Heathcliff and Catherine.

The following evening was very wet—indeed it poured down, till day-dawn; and as I took my morning walk round the house, I observed the master's window swinging open, and the rain driving straight in . . . .

Mr. Heathcliff was there—laid on his back. His eyes met mine so keen and fierce, I started; and then he seemed to smile.

I could not think him dead—but his face and throat were washed with rain; the bed clothes dripped and he was perfectly still . . . .

Emily Bronte, *Wuthering Heights*

In *Wuthering Heights,* as in most gothics the weather plays an even greater role in setting the mood of overwhelming evil and terror that most tragedies. Very little action in gothic horror stories takes place during the light of day. The worst is always reserved for the night when the evil spirits run free.

There are few stories more lurid than *Dracula* by Bram Stoker. As Jonathan Harker, who unwittingly set Dracula loose in England, was riding innocently to Count Dracula's castle he felt a mood of darkness that prevailed over the landscape.

There were dark, rolling clouds overhead, and in the air the heavy, oppressive sense of thunder. It seemed as though the mountain range had separated two atmospheres, and that now we had got into the thunderous one.

And, again, at the end of the book nature plays a predominant role when there is a frenzied race with time to kill Dracula before sunset.

The sun was almost down on the mountain top and the shadow of the whole group fell upon the snow. I saw the count lying within the box upon the earth . . . . He was deathly pale, just like a waxen image, and the red eyes glared with a horrible vindictive look which I knew so well.

As I looked, the eyes saw the sinking sun, and the look of hate in them turned to triumph.

But, on the instant, came the sweep and flash of Jonathan's great knife. I shrieked as I saw it shear through the throat whilst at the same moment Mr. Morris' bowie knife plunged into the heart. . . the whole body crumpled into dust and passed from our sight. . . .

The sun was now right down upon the mountain top, and the red gleams fell upon my face, so that it was bathed in rosy light. . . .

The curse had passed away.

Bram Stoker, *Dracula*

There are also innumerable poems that paint word pictures or images of the weather. For example, the classic by Shelley, "Ode to the West Wind," sees the cold, strong west wind as an untamed spirit that must be free but, despite its uncompromising appearance, "Oh Wind, if Winter comes, can Spring be far behind?"

Robert Frost's poetry is permeated with the weather. We have become indoor people—Frost tells us to return outdoors to celebrate life.

*Come with rain, O loud Southwester!*
*Bring the singer, bring the nester,*
*Give the buried flower a dream*
*Make the settled snowbank stream . . .*
*Burst into my narrow stall;*
*Swing the picture on the wall;*
*Run the rattling pages o'er;*
*Scatter poems on the floor;*
*Turn the poet out of door.*

Robert Frost, "To the Thawing Wind," *The Poems of Robert Frost*

## Meteorology and the Art of Painting

Painters have always watched the sky. Sensitive to the beauty and power of the atmosphere, they are some of the best students of meteorology. A careful study of paintings will show that many artists, with the exception of most so-called "modern" artists, represent the weather with a knowledgeable eye attuned to detail, although with some artistic bias.

A meteorologist, Hans Neuberger, examined over 12,000 paintings for their meteorological impact. He then classified the paintings according to the nationality of the painter and measured the blueness of the sky, the cloudiness, and the visibility in them. He found that the Dutch and especially the British paintings tended to depict pale blue, cloudy skies with hazy air, whereas the Italian and Spanish paintings tended to have the clearest, deepest blue skies with the best visibility. And these findings stand even though many painters, such as Van Gogh, moved around and sometimes changed painting styles completely. One of the conclusions of the study was, naturally enough, that the artists accurately portrayed their meteorological environments. Neuberger also found some evidence for the Little Ice Age in the paintings between 1550 and 1850!

During the later Middle Ages there was a tremendous revival of art in Europe, but the themes were predominantly religious. In the paintings of these times, the sky was rarely emphasized—all of nature was still an unconquered enemy. Indeed, in paintings and manuscript illuminations before 1400, when the sky was depicted it often appears as a solid blue or gold background or even as a geometrical mosaic pattern.

But the world was changing in the 14th century. The forests were being cleared and commerce was rapidly expanding. The attitude toward Nature was changing. Thus, Petrarch climbed a mountain and told all Europe about its grandeur and even the theologians of the time began to see in the world of Nature further evidence of the magnificence of God.

The world of Art was set for a revolution. This came in Flanders shortly after 1400 and largely at the hands of Jan Van Eyck (1390?–1441). It is in his paintings that we can see for the first time a realistic sky. Consider his *Crucifixion* (c 1425/30) hanging in the Metropolitan Museum of Art. Realistic puffy and flat based cumulus clouds 'float' across the sky while higher up long cirrus streamers can be seen. Van Eyck also included effects known as **atmospheric perspective.** Thus the sky color grades from deep blue near the zenith to whitish as you near the horizon while the distant mountains fade somewhat and assume a bluish tinge we now know is due to scattered light from the intervening atmosphere.

In Renaissance Italy, painters were also devoting increased attention to the sky. Andrea Mantegna's (1431–1506) *Calvary* (1456–9) shown in Figure 25.3 represents one of the early attempts to portray the sky realistically. In it the deep blue sky whitens toward the horizon and is filled with striking clouds. There is however something wrong with the clouds. They are puffy and flat based like

FIGURE 25.3
The *Calvary* of Mantegna. The clouds are neither cumulus nor cirrus.

cumulus but are rather small and far too squashed vertically. Furthermore, they taper off to fibrous streaky endings on each side more like cirrus clouds.

However, the errors made by Mantegna were soon corrected by his brother-in-law, Giovanni Bellini (1431–1516). Consider his *Agony in the Garden* (c 1465). This shows a breathtaking dawn sky with cumulus clouds whose level but corrugated bottoms lie bathed in the pink light of the rising sun.

In the second half of the sixteenth century, Breughel popularized the landscape painting without reference to religious or classical themes. In many of his paintings he showed the common people of Holland at work and at play. And his mostly cloudy skies typify the Dutch winter weather. The many Dutch landscape painters who followed Breughel (e.g., Van Goyen, van Ruisdael, Cuyp, and Rembrandt) tend to show skies largely filled with very realistic, billowy cumulus clouds whose bases are quite low to the ground (see Figure 25.4). Indeed, such skies are common in Holland.

It is in the nineteenth century that we find many of the greatest paintings of the sky. In the last half of the eight-

eenth century Jean Jacques Rousseau attacked the spirit of the Age of Reason in his writings and replaced a God of Reason with a God of Nature. This changed emphasis caught on in all the arts, and after about 1800 artists, writers, and composers became "Romantic" and learned to love nature as never before.

Two British artists, John Constable (1776–1837) and Joseph M. W. Turner (1775–1851) were leaders in this "back to nature" movement of art. They began to paint their exquisite outdoor scenes about 1800. Both artists were consciously influenced by Luke Howard's classification of clouds, and Constable even made an extensive cloud study consisting of numerous sketches and paintings of all types of clouds and of single and double rainbows.

Constable's painting *Salisbury Cathedral from the Meadows* (1831) shows the cathedral strategically placed under the protection of a rainbow, while the sky contains well-developed cumulus clouds with dark bases and sunlit sides. Sunlight reaches the ground in places, as it must for a rainbow to form. Other parts of the ground lie in dark shadows, and the contrasts of light and dark add to

FIGURE 25.4

*View of Haarlem* by Jacob van Ruisdael. The sixteenth and seventeenth century Dutch painters knew their clouds.

the effectiveness and beauty of the painting. There is only one meteorological error; Constable did not make the sky under the rainbow brighter than the sky above it. (Rainbows have always given artists troubles).

J. M. W. Turner was an early forerunner of the Impressionists with his very brightly illuminated scenes. He, too, studied the clouds, and his painting, *The Grand Canal, Venice* (1835), is extremely bright with a deep blue sky near the zenith and white cirriform clouds in the background. This is one of the few paintings that have cirrus clouds—for some reason they have never been a favorite among artists (see Figure 25.5).

But Turner did not use cirrus clouds in his painting, *Snowstorm, Hannibal and His Army Crossing the Alps* (1812). In this painting, Turner used the contrasts of light and dark quite effectively to give an almost cataclysmic impression (see Figure 25.6). The snow is just beginning to fall from dark, swirling clouds on the right side of the picture, while the sky still shows some summerlike pale blue on the left. A "watery" sun shines feebly through the leading edge of the storm cloud.

Turner's snowstorm actually looks more like a thunderstorm than a winter blizzard and, indeed, the painting was inspired in 1810 when Turner was staying in York-

shire and sketched a thunderstorm as it passed overhead. " 'There,' said he, 'Hawley, in two years you will see this again and call it Hannibal crossing the Alps!' " (John Walker, *Turner*). Since Hannibal crossed the Alps in September, it is possible that his troops encountered a thunderstorm which produced snow.

Another painting which employs a very similar effect is the fourth in a series of five paintings by Thomas Cole (1801-1848) entitled, *The Course of Empire*. The series represents an allegory of the evolution of civilization from the savage state to abandoned ruins. The entire series takes place in one day (meteorologically speaking).

The first of the series, *The Savage State* (Figure 25.7a) shows primitive man still partially shrouded in the mists of early morning. The sky has all the colors of the sunrise. The second depicts *The Pastoral State* (Figure 25.7b)—the mists and fog of morning are almost completely "burned off," except for a few thin veneers against the hillside. The third shows the *Consummation of Empire* (Figure 25.7c); it has as its setting the hazy summer sky of early afternoon. Look carefully in the background over the tranquil waters and you can see cumulus congestus clouds billowing up. That is the one threatening note in this picture (unfortunately, it does not show up well in reprints). The fourth shows *The Destruction of Empire,* with civilization being ravaged by war and a consuming fire. Its background shows a dark swirling thunderstorm raging overhead. The fifth and final painting shows overgrown and abandoned ruins, illuminated by the moonlight shortly after sunset. And, true to life, the clouds are the stratocumulus or altocumulus clouds that often form from the dissipated cumulonimbus clouds of late afternoon.

The American landscape painters of the nineteenth century had an astonishingly broad knowledge of meteorology and all other aspects of nature. Jasper Cropsey expresses their philosophy in this passage.

The axe of our civilization is busy with our old forests . . . . Yankee enterprise has little sympathy with the picturesque, and it behooves our artists to rescue from its grasp the little that is left, before it is ever too late.

John Wilmerding, *Audubon, Homer, Whistler and 19th Century America*

With these painters the sky becomes real. But like all painters of the landscape, these meteorological artists had certain biases.

You will almost never see a completely clear sky or a uniformly gray overcast. The great majority of paintings of daytime scenes show cumulus clouds. Even when the artist intends to indicate a cloudy day, there will always be

FIGURE 25.5
*The Grand Canal, Venice* by J. M. W. Turner.

some break in the clouds that allows the sunlight either to strike one part of the ground or at least to illuminate the side of a cloud.

When sunset or sunrise scenes are depicted, the artist will often exaggerate the golden or reddish colors. The favorite sunset and sunrise clouds are the altocumulus and stratocumulus. This is based partly on fact, since these are common at the end of the day as a reminder of daytime cumulus. It is also based on aesthetics. The most beautiful sunsets and sunrises are seen when altocumulus clouds are present. The sun then illuminates the bottoms of the clouds, causing them to glow with incredible pinks and reds, which contrast with the blue sky. If you are going to paint a sunset, why not paint a beautiful one.

In the latter part of the nineteenth century, art became more and more abstract. Nature was still painted, but it is often a transformed nature that appears in the paintings of the Impressionists. (Even so, because of their concern with light, the Impressionists have produced some of the best paintings of skies filled with multiple cloud layers.) Van Gogh represents one extreme of Impressionism.

Van Gogh was Dutch and his early paintings are studies in somber colors. Then he moved to Paris and was astounded by the use of color and light in the paintings of the Impressionists. But Van Gogh did not find his subject until he saw the blazing blue skies of southern France where the Mistral blows. Never it seems has anyone been more affected by a change of climate. Van Gogh's later paintings are literally explosions of bright colors burned onto the paper by the intense Mediterranean sun. But the swirling heavy brush strokes for which he is famous show that Van Gogh was an interpretor of nature, not a mere imitator and you could not expect him to give his cumulus clouds flat bases.

Since the turn of the twentieth century, art has taken different paths in which nature, for the most part, plays no major role. Sometimes, as with the Surrealists, if a view of nature does emerge it is a remarkably detailed view that

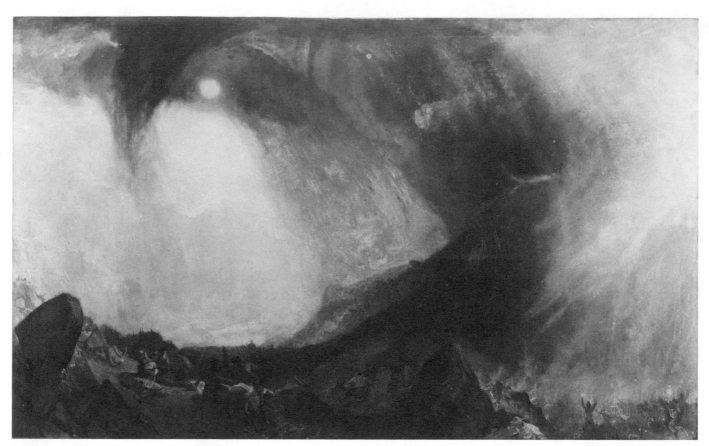

FIGURE 25.6
*Snowstorm, Hannibal and His Army Crossing the Alps* by J. M. W. Turner.

FIGURE 25.7
*The Course of Empire* by Thomas Cole. (a) *The Savage State;* (b) *The Pastoral State;* (c) *The Consummation of Empire.*

(a)

THE MARK OF CLIMATE ON THE HUMAN RACE

(b)

(c)

attempts to show some flagrant violation of a law of nature. In Chapter 2 Magritte's *The Empire of Light* represents a good example of this technique.

## Meteorology and Music

Meteorology has also played a role in music. In some cases there are definite sounds that the composer can imitate—the whistling of the wind, the roar of thunder, the patter of raindrops, and so on. In other cases the composer merely intends to set a mood to music, as in the *Spring* Symphony of Robert Schumann. Schumann said that the symphony was produced out of his feeling of expectant joy at the coming of spring and that we should not listen for specific sound effects.

Antonio Vivaldi was a prolific composer. For almost 200 years most of his music lay buried in vaults and his fame rested largely on a series of concertos known as the *Four Seasons*. Vivaldi prefaced each "season" with a poem so that we would know exactly what his musical intentions were. Some of his meteorological sound effects are truly remarkable.

During the *Summer* Concerto there is a full-scale thunderstorm. Before the thunderstorm arrives the music is very quiet, although the mood of serenity is disturbed several times by distant rumblings of thunder. Then, quite suddenly, the quiet is broken by the full fury of the storm. Thunder is simulated by the bass fiddles, while the driving rain is indicated by rapidly repeating descending scales on the violins. Ascending scales imitate the whistling, gusty winds. As in reality, the thunderstorm (the sound volume) gradually tapers off and the scene becomes peaceful once again. This amazingly realistic musical thunderstorm proved to be the forerunner of many other musical thunderstorms.

Vivaldi's *Winter* Concerto also has many meteorological sound effects. It begins with a musical picture of people stamping their feet from the cold; here Vivaldi uses a one note march. Then the gentle, continuous patter of splashing raindrops is simulated by plucked broken chords on the violin. Finally, a series of intertwined ascending and descending scales played by the violins simulates the swirling fall of snow.

As with the painters, the composers of the nineteenth century also became more interested in nature. The musical vision grew more "Romantic," and pastoral scenes became more commonplace.

Beethoven's Sixth Symphony, the *Pastoral* Symphony, contains one of music's most famous thunderstorms, woven into a musical paean to the pastoral beauty of summer. Much of the symphony has been used in the movie, *Fantasia*, during which there is an outdoor celebration by nymphs and satyrs. Their playful revel is broken up by the intense thunderstorm, but soon the earth is at peace again. As with Vivaldi (and nature), Beethoven's storm comes on rather suddenly and then gradually tapers off. Beethoven used basically the same musical devices to mimic the thunderstorm as Vivaldi but, in addition, drums were used for the thunder.

One more famous thunderstorm scene occurs in Rossini's *William Tell Overture* (the finale of which is familiar to most people as the "Theme from the Lone Ranger"). Rossini uses drums for distant thunder and cymbals for nearby thunder. Shortly before his storm strikes in full force, there is a light pitter-patter of the first few raindrops, reminiscent of Vivaldi's winter rain, except that Rossini uses flutes to play the staccato notes. The thunderstorm then tapers off with a few more pitter-patters, and the theme switches to one of chirping birds (a theme containing many trills), which has such a feeling of pastoral serenity that you may forget the grass is still wet.

Richard Wagner was a master of portraying both the placidness and the fury of nature. His storm scene from the Prelude to *The Flying Dutchman* portrays not only the storm itself, but also the turbulent, mountainous seas it produces. Waves are often portrayed by a repeating theme consisting of ascending and descending notes, in which the highest note is usually played loudest.

Wagner's *Die Walkurie* also contains a storm scene in the Prelude that is unmatched for its unrelieved feeling of ominous darkness. This is no summer thunderstorm—but an extratropical cyclone of hurricane proportions. Such pagan music could truly presage the twilight of the gods. And who hasn't heard the music of the screaming winds in the *Ride of the Walkurie* (often played on college radio stations as exam time approaches)?

As with painting, music has ceased to be a vehicle for imitating nature. Twentieth-century classical music is now concerned with other problems. Now and then, we will hear popular songs such as "Somewhere Over the Rainbow" and "Both Sides Now" that deal with feelings inspired by the weather.

In a way, Joni Mitchell's "Both Sides Now" is particularly relevant. The music is light and fluffy and has a tinkling, almost fairy-tale quality to it. The lyrics describe how clouds appear both from a plane and from the ground. The words tell much about the times we live in. For it is only within this century that we have been able to see clouds from both sides. Our knowledge of the natural world has increased so much that this would have been almost inconceivable 100 years ago. We now can use nature's laws to change the course of our lives and even the face of the earth. If we could go back in time and tell our ancestors of the technological advances we have made, they would all begin to write Utopias.

And yet these advances are not greeted with a sense of joy. Instead, there is a feeling prevalent that science and technology are responsible for the hollowness of our lives — and add no *real* spirit or knowledge. We are repeatedly told that now, having finally seen clouds from both sides, we *really* don't know clouds at all.

Perhaps our expectations are too high. Science cannot explain *ultimate* truth. The job of science is to explain the phenomena of nature in terms of a few, basic, general laws, so that events in the universe are more predictable and understandable. The job of technology is to use these basic laws to our advantage to make our lives more comfortable. The advances in science and technology cannot make our lives more or less meaningful.

In the end, the ultimate mystery and beauty of the universe still lie before us.

## PROBLEMS

**25-1** Perform the following experiment. Find some people who claim to feel changes in weather. Have them telephone you when they wake up in the morning and give you their forecast for the following day. See if they are as accurate as the official weather forecasts.

**25-2** How much would summer temperatures have to rise in England for malaria to become a problem there?

**25-3** Find specific historical examples of how malaria had a suppressing effect on civilizations or societies.

**25-4** What is the relationship between the appearance of the locusts in Egypt and the Khamsin?

**25-5** Find examples (not given in this book) of how weather either changed the course of history or at least strongly influenced it.

**25-6** List some literary works that have used weather events as part of their stories.

**25-7** Rank some of the famous painters according to their ability to paint clouds. (You should refer to Chapter 2 and start looking closely at the sky again.) Which of these painters do you feel looked at the sky carefully and knowingly?

**25-8** Compile a list of all musical pieces in chronological order that "describe" a) thunderstorms and b) the weather of the seasons. Then evaluate each composer for originality and for "accuracy" in portraying the mood of the weather.

# APPENDIXES

# APPENDIX 1
# Weather Symbols

| | | | | |
|---|---|---|---|---|
| **00** Cloud development NOT observed or NOT observable during past hour. | **01** Clouds generally dissolving or becoming less developed during past hour. | **02** State of sky on the whole unchanged during past hour. | **03** Clouds generally forming or developing during past hour. | **04** Visibility reduced by smoke. |
| **05** Dry haze. | **06** Widespread dust in suspension in the air, NOT raised by wind, at time of observation. | **07** Dust of sand raised by wind at time of observation. | **08** Well developed dust devil(s) within past hour. | **09** Duststorm or sandstorm within sight of or at station during past hour. |
| **10** Light fog. | **11** Patches of shallow fog at station, NOT deeper than 6 feet on land. | **12** More or less continuous shallow fog at station, NOT deeper than 6 feet on land. | **13** Lightning visible, no thunder heard. | **14** Precipitation within sight, but NOT reaching the ground at station. |
| **15** Precipitation within sight, reaching the ground but distant from station. | **16** Precipitation within sight, reaching the ground, near to but NOT at station. | **17** Thunder heard, but no precipitation at the station. | **18** Squall(s) within sight during past hour. | **19** Funnel cloud(s) within sight during past hour. |
| **20** Drizzle (NOT freezing and NOT falling as showers) during past hour, but NOT at time of ob. | **21** Rain (NOT freezing and NOT falling as showers during past hr., but NOT at time of ob. | **22** Snow (NOT falling as showers) during past hr., but NOT at time of ob. | **23** Rain and snow (NOT falling as showers) during past hour, but NOT at time of observation. | **24** Freezing drizzle or freezing rain (NOT falling as showers) during past hour, but NOT at time of observation. |
| **25** Showers of rain during past hour but NOT at time of observation. | **26** Showers of snow, or of rain and snow, during past hour, but NOT at time of observation. | **27** Showers of hail, or of hail and rain, during past hour, but NOT at time of observation. | **28** Fog during past hour, but NOT at time of ob. | **29** Thunderstorm (with or without precipitation) during past hour, but NOT at time of ob. |
| **30** Slight or moderate duststorm or sandstorm, has decreased during past hour. | **31** Slight or moderate duststorm or sandstorm, no appreciable change during past hour. | **32** Slight or moderate duststorm or sandstorm, has increased during past hour. | **33** Severe duststorm or sandstorm, has decreased during past hr. | **34** Severe dust storm or sandstorm, no appreciable change during past hour. |
| **35** Severe duststorm or sandstorm, has increased during past hr. | **36** Slight or moderate drifting snow, generally low. | **37** Heavy drifting snow, generally low. | **38** Slight or moderate drifting snow, generally high. | **39** Heavy drifting snow, generally high. |
| **40** Fog at distance at time of ob., but NOT at station during past hour. | **41** Fog in patches. | **42** Fog, sky discernible, has become thinner during past hour. | **43** Fog, sky NOT discernible, has become thinner during past hour. | **44** Fog, sky discernible, no appreciable change during past hour. |
| **45** Fog, sky NOT discernible, no appreciable change during past hour. | **46** Fog, sky discernible, has begun or become thicker during past hr. | **47** Fog, sky NOT discernible, has begun or become thicker during past hour. | **48** Fog, depositing rime, sky discernible. | **49** Fog, depositing rime, sky NOT discernible. |

| 50 | 51 | 52 | 53 | 54 |
|---|---|---|---|---|
| Intermittent drizzle (NOT freezing) slight at time of observation. | Continuous drizzle (NOT freezing) slight at time of observation. | Intermittent drizzle (NOT freezing), moderate at time of ob. | Continuous drizzle (NOT freezing) moderate at time of ob. | Intermittent drizzle (NOT freezing), thick at time of observation. |
| 55 | 56 | 57 | 58 | 59 |
| Continuous drizzle (NOT freezing), thick at time of observation. | Slight freezing drizzle. | Moderate or thick freezing drizzle. | Drizzle and rain slight. | Drizzle and rain, moderate or heavy. |
| 60 | 61 | 62 | 63 | 64 |
| Intermittent rain (NOT freezing), slight at time of observation. | Continuous rain (NOT freezing), slight at time of observation. | Intermittent rain (NOT freezing), moderate at time of ob. | Continuous rain (NOT freezing), moderate at time of observation. | Intermittent rain (NOT freezing), heavy at time of observation. |
| 65 | 66 | 67 | 68 | 69 |
| Continuous rain (NOT freezing), heavy at time of observation. | Slight freezing rain. | Moderate or heavy freezing rain. | Rain or drizzle and snow, slight. | Rain or drizzle and snow, mod. or heavy. |
| 70 | 71 | 72 | 73 | 74 |
| Intermittent fall of snowflakes, slight at time of observation. | Continuous fall of snowflakes, slight at time of observation. | Intermittent fall of snowflakes, moderate at time of observation. | Continuous fall of snowflakes, moderate at time of observation. | Intermittent fall of snowflakes, heavy at time of observation. |
| 75 | 76 | 77 | 78 | 79 |
| Continuous fall of snowflakes, heavy at time of observation. | Ice needles (with or without fog) | Granular snow (with or without fog). | Isolated starlike snow crystals (with or without fog). | Ice pellets (sleet, U.S. definition). |
| 80 | 81 | 82 | 83 | 84 |
| Slight rain shower(s). | Moderate or heavy rain shower(s). | Violent rain shower(s). | Slight shower(s) of rain and snow mixed. | Moderate or heavy shower(s) of rain and snow mixed. |
| 85 | 86 | 87 | 88 | 89 |
| Slight snow shower(s). | Moderate or heavy snow shower(s). | Slight shower(s) of soft or small hail with or without rain or rain and snow mixed. | Moderate or heavy shower(s) of soft or small hail with or without rain or rain and snow mixed. | Slight shower(s) of hail,[††] with or without rain or rain and snow mixed, not associated with thunder. |
| 90 | 91 | 92 | 93 | 94 |
| Moderate or heavy shower(s) of hail,[††] with or without rain or rain and snow mixed, not associated with thunder. | Slight rain at time of ob., thunderstorm during past hour, but NOT at time of observation. | Moderate or heavy rain at time of ob., thunderstorm during past hour, but NOT at time of observation. | Slight snow or rain and snow mixed or hail[†] at time of ob., thunderstorm during past hour, but not at time of ob. | Mod. or heavy snow, or rain and snow mixed or hail[†] at time of ob., thunderstorm during past hour, but NOT at time of observation. |
| 95 | 96 | 97 | 98 | 99 |
| Slight or mod. thunderstorm without hail[†] but with rain and or snow at time of ob. | Slight or mod. thunderstorm, with hail[†] at time of observation. | Heavy thunderstorm, without hail[†] but with rain and or snow at time of observation. | Thunderstorm combined with duststorm or sandstorm at time of ob. | Heavy thunderstorm with hail[†] at time of ob. |

† refers to "hail" only

†† refers to "soft hail" "small hail" and "hail"

| W — Past Weather | N — Total amount all clouds | a — Barometer characteristics |
|---|---|---|
| **0** Clear or few clouds. | **0** No clouds. | **0** Rising then falling. Now higher than 3 hours ago. |
| **1** Partly cloudy (scattered) or variable sky. | **1** Less than one-tenth or one-tenth. | **1** Rising, then steady; or rising, then rising more slowly. Now higher than 3 hours ago. |
| **2** Cloudy (broken or overcast. | **2** Two or three-tenths. | **2** Rising steadily, or unsteady. Now higher than, 3 hours ago. |
| **3** Sandstorm, or duststorm, or drifting or blowing snow. | **3** Four-tenths. | **3** Falling or steady, then rising, or rising, then rising more quickly. Now higher than, 3 hours ago. |
| **4** Fog, or smoke, or thick dust haze. | **4** Five-tenths. | **4** Steady. Same as 3 hours ago. |
| **5** Drizzle. | **5** Six-tenths. | **5** Falling, then rising. Same or lower than 3 hours ago. |
| **6** Rain. | **6** Seven- or eight-tenths. | **6** Falling, then steady; or falling, then falling more slowly. Now lower than 3 hours ago. |
| **7** Snow, or rain and snow mixed, or ice pellets (sleet). | **7** Nine-tenths or overcast with openings. | **7** Falling steadily, or unsteady. Now lower than 3 hours ago. |
| **8** Shower(s). | **8** Completely overcast. | **8** Steady or rising, then falling; or falling, then falling more quickly. Now lower than 3 hours ago. |
| **9** Thunderstorm, with or without precipitation. | **9** Sky obscured. | |

# APPENDIX 2
## Definitions of Climatic Symbols and Boundaries

## DEFINITIONS OF CLIMATIC SYMBOLS AND BOUNDARIES[1]

$A$ = killing frost absent; in marine areas, cold month over 18°C

    r (rainy) = 10 to 12 months wet; 0 to 2 months dry

    $w$ = winter (low-sun period) dry; more than 2 months dry

    $s$ = summer (high-sun period) dry; rare in $A$ climates

$B$ = evaporation exceeds precipitation. Boundary,

$$R = 2T + 40 - 0.5\ PW$$

    *where* $R$ = rainfall, in centimeters

    $T$ = temperature, in °C

    $PW$ = percent annual rainfall in winter half of year[2]

    Desert/steppe boundary is

$$R = T + 20 - 0.25\ PW$$

or half the amount of the steppe/humid boundary

$W$ (German *Wüste*) = desert or arid

---

[1] Based on Trewartha's modification of Köppen's climate classification system. From G. T. Trewartha *An Introduction to Climate* (4th ed.) (new York: McGraw-Hill), 1968.

[2] This single formula represents an approximation of the three more elaborate Köppen formulas used to calculate the boundary of dry climates.

    $S$ = steppe or semiarid

    $h$ = hot; 8 months or more with average temperature over 10°C

    $k$ (German *kalt*) = cold; fewer than 8 months average temperature above 10°C

    $s$ = summer dry

    $w$ = winter dry

    $n$ = frequent fog

$C$ = 8 to 12 months over 10°C; coolest month below 18°C

    $a$ = hot summer; warmest month over 22°C

    $b$ = cool summer; warmest month below 22°C

    $f$ = no dry season; difference between driest and wettest month less than required for $s$ and $w$; driest month of summer more than 3 cm

    $s$ = summer dry; at least three times as much rain in winter half of year as in summer half of year: driest summer month less than 3 cm; annual total under 90 cm

    $w$ = winter dry; at least ten times as much rain in summer as in winter

$D$ = 4 to 7 months inclusive over 10°C

    $o$ = oceanic or marine; cold month over 0°C (to 2°C) in some locations inland

    $c$ = continental; cold month under 0°C (to 2°C)

    $a$ = same as in $C$

    $b$ = same as in $C$

$f$ = same as in C
$s$ = same as in C
$w$ = same as in C
$E$ = 1 to 3 months inclusive over 10°C
$F$ = all months below 10°C
  $t$ = tundra; warmest month between 0 and 10°C
  $i$ = ice cap; all months below 0°C
$A/C$ boundary = equatorial limits of freeze; in marine locations the isotherm of 18°C, for the coolest month
$C/D$ boundary = 8 months 10°C
$D/E$ boundary = 4 months 10°C
$E/F$ boundary = 10°C for warmest month
$t/i$ boundary in $F$ climates = 0°C for warmest month
$B/A$, $B/C$, $B/D$, $B/E$ boundary = evaporation equals precipitation
$BS/BW$ boundary = one-half the $B/A$, $B/C$, $B/D$, $B/E$ boundary
$h/k$ boundary in dry climates = same as $C/D$
$Do/Dc$ boundary = 0°C (to 2°C) for coolest month

# BIBLIOGRAPHY

In writing this book, I have drawn on a large number of sources. Most of the books listed in this bibliography not only proved invaluable to me, but are also good as references if you should wish to do further reading.

A red entry number indicates that the book or journal is written at an advanced level.

Every day there are new discoveries in the atmospheric sciences so that books become all too quickly outdated. To keep up to date you must always have an eye on the journals. Some of the *technical* journals are:

1. *Agricultural Meteorology,* American Elsevier, New York.
2. *Climatic Change,* D. Reidel, Boston.
3. *Icarus* (for all the planets and their atmospheres), American Astronomical Society, Academic Press, New York.
4. *Journal of Applied Meteorology,* available from the American Meteorological Society (abbreviated AMS), Boston.
5. *Journal of Atmospheric and Terrestrial Physics* (for the upper atmosphere), Pergamon Press, Elmsford, N.Y.
6. *Journal of Geophysical Research,* American Geophysical Union, Washington, D.C.
7. *Journal of the Atmospheric Sciences,* AMS, Boston.
8. *Monthly Weather Review,* AMS, Boston.
9. *Quarterly Journal of the Royal Meteorological Society,* available from the Royal Meteorological Society (abbreviated RMS), Bracknell, Berks, England.
10. *Quaternary Research* (for recent climate changes), University of Washington Press, Seattle, Wash.
11. *Reviews of Geophysics and Space Physics,* American Geophysical Union.
12. *Tellus,* publication of Swedish Geophysical Society, Stockholm.

Journals written at a slightly lower level include:

13. *Bulletin of the American Meteorological Society,* AMS, Boston.
14. *Science,* American Association for the Advancement of Science.
15. *Scientific American,* Scientific American, Inc., New York.

Finally, there are the popular journals. These are easy to read once you have read the relevant sections in this book. The journals dealing with interesting meteorological phenomena are:

16. *Marine Observer.* British Meteorological Office.
17. *Meteorological Magazine,* British Meteorological Office.
18. *Weather,* RMS, Bracknell, Berks, England.
19. *Weatherwise,* AMS, Boston.

For astronomy and phenomena such as the aurora there is:

20. *Sky and Telescope,* Sky Publishing Corp., Cambridge, Mass.

The more general, popular journals include:

21. *Natural History.*

**22.** *Smithsonian.*

Now for the textbooks. Many of the topics in this book are discussed at a more advanced level in this excellent, up-to-date textbook:

**23.** Wallace, J. M., and P. V. Hobbs, 1977: *Atmospheric Science: An Introductory Survey.* Academic Press, New York.

Another advanced general text, which is better for its "weather" descriptions, is:

**24.** Byers, H., 1974: *General Meteorology.* McGraw-Hill, New York.

Textbooks that stress the mathematical aspects of meteorology are:

**25.** Haltiner, G. J., and F. L. Martin, 1957: *Dynamical and Physical Meteorology.* McGraw-Hill, New York.
**26.** Hess, S. L., 1959: *Introduction to Theoretical Meteorology.* Holt, Rinehart & Winston, New York.
**27.** Houghton, J. T., 1977: *The Physics of Atmospheres.* Cambridge University Press, New York.

References 23 to 27 can be consulted for further information on most topics in this book (up to Chapter 21); they will not be mentioned again in the chapter-by-chapter breakdown that follows unless their presentation is outstanding or unique.

## Chapter 1

The account of the hurricane of 1938 was largely taken from:

**28.** McCarthy, J., 1969: *Hurricane.* McGraw-Hill, New York.

There are only a small number of books on the early history of meteorology. The best of these is:

**29.** Shaw, Sir N., 1926: *Manual of Meteorology, Vol. 1.*
but it is unfortunately out of print. Disappointingly, the only recent book is not as complete:

**30.** Frisinger, H. H., 1977: *The History of Meteorology to 1800.* AMS, Boston.

There are a number of excellent books on more specialized topics:

**31.** Middleton, W., 1964: *History of the Barometer.* John Hopkins, Baltimore.
**32.** ———, 1966: *History of the Thermometer and its Use in Meteorology.* John Hopkins, Baltimore.
**33.** ———, 1968: *History of the Theories of Rain and Other Forms of Precipitation.* University of Chicago, Chicago.

**34.** ———, 1969: *Invention of the Meteorological Instruments,* John Hopkins, Baltimore.

You should be forewarned that it is usually best to know some meteorology before trying to read about the history of meteorology.

## Chapter 2

The standard cloud atlas is:

**35.** *International Cloud Atlas,* 1956: World Meteorological Organization, Geneva, Switzerland.

Unfortunately, there are very few books devoted to the subject of cloud forms. The best by far is:

**36.** Scorer, R. S., 1972: *Clouds of the World.* David & Charles, N. Pomfret, Vt.

## Chapter 3

Almost everything that has been seen in the atmosphere is described in the excellent book:

**37.** Minnaert, M., 1954: *The Nature of Light and Color in the Open Air.* Dover Publications, New York.

Mathematical treatments of Atmospheric Optics appear in the classic:

**38.** Humphreys, W. J., 1940: *Physics of the Air.* Dover Publications, New York.

and more recently with some corrections in:

**39.** Tricker, R. A. R., 1970: *Introduction to Meteorological Optics.* American Elsevier, New York.

Many historical aspects presented in this chapter were taken from:

**40.** Boyer, C., 1959: *Rainbow: From Myth to Mathematics.* Yoseloff, New York.

Two classic books on ice crystals, complete with many beautiful photographs, are:

**41.** Bentley, W. A., and W. J. Humphreys, 1931: *Snow Crystals.* McGraw-Hill, New York.
**42.** Nakaya, U., 1954: *Snow Crystals.* Harvard University Press, Cambridge, Mass.

A more modern treatment with fewer pictures is presented in the comprehensive book:

43. Hobbs, P. U., 1975: *Ice Physics*. Oxford University Press, New York.

**Chapter 4**

There are a large number of elementary books on astronomy. Two of the good ones are:

44. Abell, G., 1969: *Exploration of the Universe*. Holt, Rinehart & Winston, New York.
45. Jastrow, R., and M. Thompson, 1974: *Astronomy: Fundamentals and Frontiers*. Wiley, New York.

A detailed treatment of sunlight (particularly a description of how it strikes the ground) is presented in:

46. Sellers, W. D., 1965: *Physical Climatology*. University of Chicago, Chicago.

**Chapter 5**

The classic in a highly mathematical field, this book describes the composition of air as acting like a bunch of billiard balls:

47. Jeans, Sir J., 1940: *An Introduction to the Kinetic Theory of Gases*. Cambridge University Press, New York.

The concepts of pressure, temperature, and energy are developed in many physics textbooks at a fairly elementary level. Two basic books are:

48. Krauskopf, G. B., and A. Beiser, 1973: *The Physical Universe*. McGraw-Hill, New York.
49. Williams, G. A., 1976: *Elementary Physics*, McGraw-Hill, New York.

The wind-chill factor is discussed in the charming, popular book:

50. Landsberg, H. E., 1969: *Weather and Health*. Anchor/Doubleday, New York.

**Chapter 6**

Reference 48 provides a good introduction to chemistry, which is a subject of this chapter. The chemistry of our atmosphere and the atmospheres of the other planets is discussed at an advanced level in:

51. McEwan, M. J., and L. F. Philips, 1975: *Chemistry of the Atmosphere*. Edward Arnold, London.

A popular book on the atmospheres of the other planets is:

52. Ohring, G., 1966: *Weather on the Planets*. Anchor/Doubleday, New York.

but this is more of historical interest because the entire field of planetary atmospheres is so new. An entire issue of *Scientific American* (September 1976) was devoted to the solar system and is an important reference for Chapter 6. New results come in almost every day, and it is necessary to read the journals to keep up to date.

**Chapter 7**

At a basic level, the field of modern physics is well presented in References 48 and 49, and at a more advanced level in:

53. Eisberg, R. M., 1976: *Fundamentals of Modern Physics*. Wiley, New York.

Most of the phenomena of the upper atmosphere (i.e., above the troposphere) are discussed with a popular approach in:

54. Craig, R. A., 1968: *The Edge of Space*. Anchor/Doubleday, New York.
55. Ratcliffe, J. F., 1970: *Sun, Earth and Radio*. McGraw-Hill, New York.

and at an advanced level in:

56. Ratcliffe, J. A., 1972: *An Introduction to the Ionosphere and Magnetosphere*. Cambridge University Press, New York.
57. Rishbeth, H., and O. K. Garriet, 1969: *Introduction to Ionospheric Physics*. Academic Press, New York.

Sunlight and skin color are discussed in Reference 50 and in:

58. Claiborne, R., 1970: *Climate, Man and History*. Norton, New York.

Pictures of aurorae are often shown in *Sky and Telescope* (Reference 20).

**Chapter 8**

The history of the thermometer is discussed in Reference 32. References 116 to 120 provided information for the discussion of worldwide temperatures in Chapter 22.

Radiation, heat, and temperature are discussed in great detail in the comprehensive work:

59. Kondratyev, K., 1969: *Radiation in the Atmosphere*. Academic Press, New York.

The subjects of radiation, heat, and temperature are also presented, with an emphasis on climate, in Reference 46 and in:

60. Budyko, M. I., 1974: *Climate and Life*. Academic Press, New York.

Other references dealing with climate theories are listed under Chapter 23.

**Chapter 9**

Much of the material in this chapter was taken from a gem of a book:

61. Geiger, R., 1965: *The Climate Near the Ground*. Harvard University Press, Cambridge, Mass.

Another good book with some newer information is:

62. Yoshino, M. M., 1975: *Climate in a Small Area, An Introduction to Local Meteorology*. International Scholarly Book Services, Forest Grove, Oregon.

References 46 and 60 are also useful from a technical point of view.

Climate in the city is also discussed in Reference 50 and, because of the inevitable connection with pollution, in the references listed under Chapter 14.

**Chapter 10**

Water vapor and its effects in the atmosphere are discussed in detail at an advanced level in:

63. Iribarne, J. V., and W. L. Godson, 1973: *Atmospheric Thermodynamics*, D. Reidel, Boston.

The hydrologic cycle and the flow of water on and under the ground are discussed in most introductory geology textbooks, one of which is:

64. Press, F., and R. Siever, 1974: *Earth*, W. H. Freeman, San Francisco.

Floods and droughts are discussed in a number of popular weather books such as:

65. Bryson, R. A., and T. J. Murray, 1977: *Climates of Hunger*. University of Wisconsin Press, Madison, Wis.
66. Hoyt, W. G., and W. B. Langbein, 1955: *Floods*, Princeton University Press, Princeton, N.J.
67. Lane, F. W., 1965: *The Elements Rage*.
68. Tannehill, I. R., 1946: *Drought: Its Causes and Effects*. Princeton University Press, Princeton, N.J.

**Chapter 11**

Fog is discussed in great detail in Reference 24 and also in:

69. Petterssen, S. 1956: *Weather Analysis and Forecasting* (Volumes I and II). McGraw-Hill, New York.

Clearing fog is a form of weather modification and is discussed in all the references listed under Chapter 12 (except Reference 74).

**Chapter 12**

Books dealing with precipitation statistics are listed in the references under Chapter 22. Precipitation processes are discussed in a number of books. A popular approach to this subject can be found in:

70. Blanchard, D. C., 1967: *From Raindrops to Volcanoes*. Anchor/Doubleday, New York.

At a somewhat more technical level there is:

71. Mason, B. J., 1975: *Clouds, Rain and Rainmaking*. Cambridge University Press, New York.

The advanced level books include the comprehensive:

72. Mason, B. J., 1972: *The Physics of Clouds*. Oxford University Press, New York.
73. Pruppacher, H. R., and J. D. Klett, 1978: *Microphysics of Clouds and Precipitation*. D. Reidel, Boston.

A book devoted entirely to hailstones is:

74. Gokhale, N. R., 1976: *Hailstorms and Hailstone Growth*, State University of New York Press, Albany, N.Y.

Weather modification (especially cloud seeding) is discussed in References 71 to 73 and also in the popular books:

75. Battan, L. J., 1969: *Harvesting the Clouds*. Anchor/Doubleday, New York.
76. Halacy, D. S., Jr., 1968: *The Weather Changers*. Harper & Row, New York.

**Chapter 13**

One of the more complete presentations of the thermodynamic diagram is given in Reference 63, an advanced level book. Many practical applications of the thermodynamic diagram are presented in Reference 69.

Representative soundings for many different climate regions are discussed in:

77. Crowe, P. R., 1972: *Concepts in Climatology*. St. Martin's Press, New York.

**Chapter 14**

Air pollution is another subject in which significant new findings are constantly announced; once again, you will have to read the journals to keep up to date. Two popular books on the subject are:

**78.** Battan, L. J., 1966: *The Unclean Sky.* Anchor/Doubleday, New York.

**79.** Scorer, R. S, 1968: *Air Pollution.* Pergamon Press, New York.

Reference 79 has a large collection of color prints of smoke plumes. A more advanced treatment of air pollution and its effects is given in:

**80.** Williamson, S. J., 1973: *Fundamentals of Air Pollution.* Addison-Wesley, Reading, Mass.

The discussion of atmospheric aerosols given in Reference 23 is also particularly good.

## Chapter 15

The principal source for the section on windmills is:

**81.** *Wind Energy Hearing,* presented before the Subcommittee on Energy of the Committee on Science and Astronautics. U.S. House of Representatives. May 21, 1974. No. 49.

Historical aspects of windmills, watermills, and sailboats can also be found in:

**82.** Durant, W., and A. Durant, *The Story of Civilization* (11 volumes). Simon & Schuster, New York.

**83.** Sarton, G. *History of Science* (2 volumes). Harvard University Press, Cambridge, Mass.

Newton's laws of motion are discussed at a rather basic level in References 48 and 49, as well as:

**84.** Sears, F. W., and M. Zemansky, 1976: *University Physics,* Addison-Wesley, Reading, Mass.

A good reference at a more advanced level is:

**85.** Holton, J. P., 1972: *An Introduction to Dynamic Meteorology.* Academic Press, New York.

Ocean currents and upwelling (and many other topics in oceanography) are treated at an advanced level in the fine book:

**86.** Neumann, G., and W. Pierson, 1966: *Principles of Physical Oceanography.* Prentice-Hall, Englewood Cliffs, N.J.

There is, of course, the popular book:

**87.** Carson, R., 1961: *The Sea Around Us.* Oxford University Press, New York.

## Chapter 16

For a detailed account of Columbus' voyages (and life) refer to:

**88.** Morrison, S. E., 1942: *Admiral of the Ocean Sea.* Little, Brown, Boston.

A history of views on large-scale winds, as well as a superb presentation of the general circulation of the atmosphere, appears in:

**89.** Lorenz, E., 1967: *The Nature and Theory of the General Circulation of the Atmosphere.* World Meteorological Organization, Geneva, Switzerland.

Wind and weather systems at all scales are treated in the indispensable reference:

**90.** Palmen, E., and C. W. Newton, 1969: *Atmospheric Circulation Systems.* Academic Press, New York.

The jet streams are discussed in:

**91.** Reiter, E. R., 1963: *Jet Stream Meteorology.* University of Chicago Press, Chicago.

and, with a more popular approach, there is:

**92.** Reiter E. R., 1966: *Jet Streams.* Anchor/Doubleday, New York.

## Chapter 17

Air masses are discussed at length in many climatology books, as well as in references 69 and 90 which are good references for most of the material in this chapter (except for the last section). There is also the monumental book:

**93.** Godske, C. L., T. Bergeron, J. Bjerknes, and R. C. Bundgaard, 1957: *Dynamic Meteorology and Weather Forecasting,* AMS, Boston.

Information for the case studies comes from:

**94.** Daily Weather Maps, Department of Commerce, NOAA, Washington, D.C.

**95.** Storm Data, Department of Commerce, NOAA, Washington, D.C.

as well as from newspapers and news magazines. Highly recommended to "weather nuts," the following two popular books provided the information on the extreme weather conditions discussed in this chapter:

96. Ludlam, D., 1976: *The Country Journal New England Weather Book,* Houghton Mifflin, New York.
97. ——, 1967: *Early American Winters: 1604–1870.* AMS, Boston.

## Chapter 18

Popular books on hurricanes include:

98. Dunn, G. E., and B. I. Miller, 1960: *Atlantic Hurricanes.* Louisiana State University Press, Baton Rouge, La.
99. Ludlam, D. 1963: *Early American Hurricanes, 1492–1970.* AMS, Boston.
100. Tannehill, I. R., 1944: *Hurricanes.* Princeton University Press, Princeton, N.J.

In addition to Reference 90 and many recent findings on hurricanes and tropical meteorology, which appear in the scientific journals, there is also the old standby:

101. Riehl, H., 1954: *Tropical Meteorology.* McGraw-Hill, New York.

Information on storm surges can be found in Reference 86.

## Chapter 19

There are many popular books on thunderstorms, hailstorms, and tornadoes, some of which are:

102. Flora, S., 1953: *Tornadoes of the United States.* University of Oklahoma Press, Norman, Okla.
103. Flora, S., 1956: *Hailstorms of the United States.* University of Oklahoma Press, Norman, Okla.
104. Ludlam, D., 1970: *Early American Tornadoes, 1586–1870.* AMS, Boston.
105. Weems, J., 1977: *The Tornado.* Doubleday, New York.

The structure of air-mass thunderstorms is described in:

106. Byers, H. R. (director), 1949: *The Thunderstorm: Report of the Thunderstorm Project.* U.S. Department of Commerce, Washington, D.C.

Severe hailstorms are described in:

107. Foote, G. B., and C. A. Knight, Eds., 1977: "Hail: A Review of Hail Science and Hail Suppression." *Meteorological Monographs,* Vol. 16, No. 38, AMS, Boston.

Most of the information, however, is only available in the scientific journals because of the rather recent introduction of doppler radar.
For a popular approach to weather radar there is:

108. Battan, L. J., 1968: *Radar Observes the Weather.* Anchor/Doubleday, New York.

A more technical approach to weather radar is found in:

109. Battan, L. J., 1973: *Radar Observation of the Atmosphere.* University of Chicago Press, Chicago.

Thunderstorms and atmospheric stability are treated in some detail in Reference 69, in which several relatively simple forecasting techniques for severe thunderstorms, hail, and tornadoes are presented.
Lighting and thunder are beautifully described in:

110. Schonland, B. F. J., 1964: *The Flight of Thunderbolts.* Clarendon Press, Oxford.

A more advanced book on the subject is:

111. Uman, M. A., 1969: *Lightning.* McGraw-Hill, New York.

## Chapter 20

Many features of small-scale winds are described in References 61, 62, and 116. For information on the bora there is:

112. Yashino, M. M., Ed., 1976: *Local Wind Bora.* International Scholarly Book Service, Inc.

Besides these sources, the bulk of the information in this chapter can be found in the journals.

## Chapter 21

The classic in the field of numerical weather prediction is:

113. Richardson, L. F., 1922: *Weather Prediction by Numerical Process.* Dover Publications, New York.

Two more recent books on the same subject, written after the computer was invented, are:

114. Haltiner, G. J., 1971: *Numerical Weather Prediction.* Wiley, New York.
115. Thompson, P. D., 1961: *Numerical Weather Analysis and Prediction.* Macmillan, New York.

A brief but good presentation of temperature determination from satellites is given in Reference 27. However, the most useful information on satellite meteorology is available only in Government Technical Reports or in the journals. There is an urgent need for a basic book covering the materials of this chapter.

## Chapter 22

Perhaps the most charming book describing climates all over the world is:

116. Kendrew, W. G., 1961: *Climates of the Continents.*

The climate classification system used in this book is taken from:

117. Trewartha, G. T., 1968: *An Introduction to Climate.* McGraw-Hill, New York.

A basic climate book, especially informative on the dependence of vegetation on climate, is:

118. Rumney, G. R., 1968: *Climatology and the World's Climate.* Macmillan, New York.

A good discussion of the regional anomalies of climate is presented in:

119. Trewartha, G. T., 1962: *The Earth's Problem Climates.* University of Wisconsin Press, Madison, Wis.

To date the most comprehensive treatment of the world's climates is given in the 14-volume set:

120. *World Survey of Climatology.* American Elsevier, New York.

## Chapter 23

Two good basic references for the entire chapter are:

121. Oliver, J. E., 1973: *Climate and Man's Environment: An Introduction to Applied Climatology.* Wiley, New York.
122. Strahler, A., and A. Strahler, 1977: *Principles of Modern Geography.* Wiley, New York.

Reference 121 is an excellent reference for Chapter 24 as well.
Another basic geology text (besides Reference 64), good on landforms, is:

123. Putnam and collaborators, 1977: *Geology.* Oxford University Press, New York.

The subject of landforms is treated at length in:

124. Garner, H. F., 1974: *The Origin of Landscapes.* Oxford University Press, New York.

Landforms and soils are discussed at a rather advanced level in:

125. Ollier, C. D., 1975: *Weathering.* Longman, New York.

For a basic book on world soils refer to:

126. Bridges, E. M., 1970: *World Soils.* Cambridge University Press, New York.

Climate, agricultural products, and livestock are discussed in many books on economic geography such as:

127. Hodder, B. K., and R. K. Lee, 1974: *Economic Geography.* St. Martin's Press, New York.

## Chapter 24

Perhaps the classic book on climates of past ages is:

128. Schwarzbach, M., 1974: *Climates of the Past.* Van Nostrand, New York.

Almost every aspect of climate is discussed in the monumental, two-volume work:

129. Lamb, H. H., 1972: *Climate: Present, Past and Future. Vol. 1.* Methuen, London.
130. ———, 1978: *Climate: Present, Past and Future, Vol. 2. Climatic History and the Future.* Harper & Row, New York.

Volume 2 is especially pertinent to the material of this chapter. Another good reference for most of the material in this chapter is Reference 121.
For those who have no background in geology, it is best to look at appropriate sections of References 48, 64, or 123 to learn about the nature of rocks.
A book that reconstructs the climates of the past 1000 years, mostly from historical records, is the charming and intriguing:

131. Ladurie, E. L. R., 1971: *Times of Feast, Times of Famine: A History of Climate Since the Year 1000.* Doubleday, New York.

The subject of continental drift (now called plate tectonics) is discussed in References 48, 64, and 123, but an especially good presentation is found in:

132. Takeuchi, H., S. Uyeda, and H. Kanamori, 1967: *Debate About the Earth.* Freeman Cooper,

Graphic presentations of this subject have appeared in a series of articles in *Scientific American.*
Books dealing entirely with climate theories and climate changes include:

133. Budyko, M. I., 1977: *Climatic Changes.* American Geophysical Union, Washington, D.C.

And with an emphasis on the potential impact of humans on climate:

134. *Inadvertent Climate Modification.* Report of the Study of Man's Impact of Climate (SMIC) 1971, The M.I.T Press, Cambridge, Mass.

Because climate change is now an "in" topic, there has been a glut of popular books. Some of these are too eager to forecast doom (either another ice age or else a torrid earth is supposed to be imminent) and must be taken with a "grain of salt." They include Reference 65 and:

**135.** Calder, M., 1975: *The Weather Machine*. The Viking Press, New York.
**136.** Gribbin, J., 1976: *Forecasts, Famine and Freezes: The World's Climate and Man's Future*. Walker & Company, New York.
**137.** Hays, J. D., 1977: *Our Changing Climate*. Atheneum Publishers, New York.
**138.** Mosley, F., and A. E. Wright, 1975: *Ice Ages: Ancient and Modern*. Seel House.
**139.** Ponte, L., 1976: *The Cooling*. Prentice-Hall, Englewood Cliffs, N.J.

**Chapter 25**

For climate and human health, refer to References 50 and 122, which are good for starters, and there is now:

**140.** Sulman, F. G., 1977: *Health, Weather and Climate*. S. Karger.

Information on locusts and the weather in this chapter was based on two principle sources. The first is a comprehensive work on locusts:

**141.** Uvarov, Sir B., and collaborators, 1977: *Grasshoppers and Locusts, Vol. 2*. Centre for Overseas Pest Research.

An interesting meteorological study, with many weather maps showing the movement of locusts, can be found in:

**142.** Rainey, R. C., 1963: *Meteorology and the Migration of the Desert Locusts*. WMO No. 138, T.P. 64.

There are a number of sources that discuss the interrelationship of weather, climate, and history. References 58 and 131 are especially recommended, and in Reference 82 much information can also be found. However, we are still waiting for a good book on weather and history.
One book which looks into the future is:

**143.** Schneider, S. H., and L. E. Meisirow, 1976: *The Genesis Strategy: Climate and Global Survival*. Plenum, New York.

There is also a need for books that discuss the relationship between meteorology and the arts. Some literature, such as:

**144.** Conrad, J., *Typhoon*.
**145.** Stewart, G. R., 1947: *Storm*. Modern Library, New York.

deal directly with weather phenomena, but there are innumerable examples in which weather forms a background. Throughout the book I have tried to work literary references into the fabric of the discussion; there are many other references for you to discover.
A good classic book on art history is:

**146.** Janson, H. W., 1977: *History of Art*. Prentice-Hall, Englewood Cliffs, N.J.

Besides this, there are many art books containing prints of the world's masterpieces from which you can see for yourself how artists have represented the sky. One in particular is:

**147.** Stechow, W., 1966: *Dutch Landscape Painters of the 17th Century*. Phaidon, New York.

Landscape artists were quite conscious of what they were doing; to verify this fact, read some of the quotes by American landscape artists in:

**148.** McCoubrey, J. W., 1965: *American Art: 1700–1960*. Prentice-Hall, Englewood Cliffs, N.J.

To gain some idea of the role of weather in music, simply look through an anthology of musical themes, such as:

**149.** Ewen, D., 1959: *Encyclopedia of Concert Music*. Hill and Wang, N.Y.

# GLOSSARY

It has been my policy to avoid stressing technical terms as much as possible throughout the book. The glossary therefore contains many terms that are not mentioned in the chapters, but that you may read elsewhere and may not understand. For a more complete glossary you should (as I have here) refer to the *Glossary of Meteorology* (1959) by the American Meteorological Society.

**Absolute humidity.**   The density of water vapor in the air.
**Absolute instability.**   The state of a column of air when it has a superadiabatic lapse rate.
**Absolute stability.**   The state of a column of air when it has a lapse rate less than the moist adiabatic lapse rate.
**Absolute zero.**   The zero point on the Kelvin temperature scale where all molecular motion ceases. The temperature 0°K corresponds to −273.16°C.
**Absorption.**   The process in which incident radiation is retained by a substance.
**Acceleration.**   The rate of change of velocity with time.
**Accretion (coalescence).**   The growth of a precipitation particle by collision of a frozen particle with a supercooled liquid droplet that freezes on contact.
**Adiabatic.**   Without change of heat.

**Advection.**   Transfer of atmospheric properties by horizontal winds.
**Aerosol.**   A solid or liquid particle in the air (usually excluding *hydrometeors*).
**Airglow.**   The quasisteady radiant emission from the upper atmosphere.
**Air mass.**   A widespread body of air of somewhat uniform temperature, humidity, and stability properties.
**Air parcel.**   An imaginary blob of air that has uniform properties and that is subjected to lifting or *subsidence* without appreciably affecting the surroundings.
**Aitken nuclei.**   Aerosols less than 0.2 microns in diameter.
**Albedo.**   The percentage of light reflected from a body.
**Aliasing.**   The introduction of error into the analysis of cyclic data by discrete sampling at too great an interval.
**Alluvial fan.**   Sediment usually laid "fanwise" by a rapidly flowing stream as it enters an open valley.
**Alpine gardens.**   Tundra in the high mountains.
**Altimeter.**   A device for measuring altitude; often a barometer with a height scale.
**Altocumulus.**   A middle-level cloud often with a wavy or cellular appearance. Individual cloud elements subtend an angle between 1 and 5° when seen high in the sky.
**Altostratus.**   A middle-level cloud forming a sheetlike layer. A "watery" sun or moon can often be seen through this cloud.

**Anabatic wind.** An upslope wind (opposite of *katabatic wind*).

**Anemometer.** An instrument designed to measure wind speed.

**Aneroid.** Not wet; applied to a barometer that contains no liquid.

**Anticyclonic.** Having a sense of rotation about the vertical opposite to that of the earth's rotation—that is, clockwise in the Northern Hemisphere, counterclockwise in the Southern Hemisphere, and undefined at the equator.

**Aphelion.** The point on the orbit of the earth that is furthest from the sun (opposite of *perihelion*).

**Aquifer.** A rock layer through which water can flow or percolate.

**Arc of Lowitz.** An arc of light extending downward from the parhelia to the small halo.

**Arctic Circle.** An imaginary line at latitude 66°32′ N on which the sun does not set on June 21 and does not rise on December 21.

**Arcus (roll cloud).** A horizontal roll-shaped cloud situated on the lower front part, or in advance, of the main cloud.

**Arroyo (wadi).** A streambed or gully in the desert.

**Atmosphere.** The envelope of air surrounding the earth and bound to it more or less by virtue of the earth's gravitation.

**Aurora.** The sporadic radiant emission from the upper atmosphere over middle and high latitudes; often spectacular at night.

**Autumn (fall).** The season beginning with the autumnal equinox and ending with the winter solstice; this occurs from September 21 to December 21 in the Northern Hemisphere and March 21 to June 21 in the Southern Hemisphere.

**Azimuth.** The angle of a point in the horizontal from north measured clockwise; thus north is 0°, east 90°.

**Backing.** A counterclockwise turning of the wind direction.

**Badlands.** A landscape characterized by gullies.

**Banner cloud.** A cloud plume often observed extending downwind from isolated mountain peaks even on clear days.

**Baroclinic.** Of the state where the air density varies from one place to another at the same atmospheric pressure.

**Baroclinic instability.** A form of instability arising from an excessively large meridional temperature gradient (and hence *thermal wind*); as a result of this instability, many extratropical cyclones and anticyclones develop.

**Barograph.** A recording barometer.

**Barometer.** An instrument that measures atmospheric pressure.

**Barotropic.** Of the state of the atmosphere where air density depends only on the pressure.

**Beach terrace.** A terrace in the landscape that in the geological past was a beach.

**Beaufort Wind Scale.** A system for estimating the wind speed by observing the state of the sea.

**Billow cloud.** Broad, nearly parallel, lines of cloud that are oriented at right angles to the wind direction.

**Black body.** An object that absorbs all radiation incident on it and radiates radiation at a maximum rate for a given temperature.

**Blizzard.** A storm with winds above about 30 knots, with visibility less than 150 meters because of snow.

**Blocking.** The obstructing, on a large scale, of the normal west-to-east progress of migratory cyclones and anticyclones.

**Boiling point.** The temperature at which the saturation vapor pressure equals the surrounding pressure. Boiling is the *rapid* change from liquid to gas (as opposed to the slower evaporation).

**Boreal forest (taiga).** The forested region that adjoins the tundra and consists largely of conifers.

**Breeze.** A gentle wind—less than 27 knots.

**Bright band.** The enhanced radar echo of snow as it melts to rain.

**Bronchial tree.** The passageways for air leading into the lungs.

**Brownian motion.** The random movements of small aerosols due to the impacts of molecules in air or a liquid.

**Buoyancy.** That property of an object that enables it to rise through a fluid. An object is buoyant when it has a smaller density than that of the surrounding fluid.

**Calm.** The absence of wind.

**Canopy.** The crown or leafy area of trees in a forest.

**Cap cloud.** An approximately stationary or standing cloud on, or hovering above, an isolated mountain peak.

**Ceiling.** The height of the lowest cloud layer that covers at least six-tenths of the sky.

**Cenozoic.** The Age of Recent Life starting about 60 million years ago.

**Centrifugal force.** The apparent force in a rotating system that deflects masses radially outward from the axis of rotation.

**Centripetal acceleration.** The acceleration of a particle moving in a curved path due to the change in direction.

**Chaparral.** A low, scrubby form of wooded vegetation found in the Southwest United States.

**Chernozem.** A soil type that is characteristic of cool or temperate semiarid climates.

**Chinook.** The name given to the foehn on the eastern side of the Rocky Mountains.

**Circumhorizontal arc.** An arc of light parallel to the horizon and below the sun (or moon) that is produced by the refraction of light through ice crystals.

**Circumscribed halo.** The halo phenomena seen just outside the 22° halo when the sun is about 45° above the horizon. This is a form of the upper- and lower-tangent arcs, produced by the refraction of light through ice crystals.

**Circumzenithal arc.** A circular arc of light around the zenith point and above the sun that is produced by the refraction of light through ice crystals.

**Cirrocumulus.** A cloud type composed of very small elements (each subtending an angle of less than 1°) in the form of grains, ripples, and so on.

**Cirrostratus.** A cloud type consisting of a whitish veil that may totally cover the sky and that often produces halo phenomena.

**Cirrus.** A cloud type composed of ice crystals and detached delicate filaments, patches, or bands.

**Clay.** A fine-textured, sedimentary, or residual deposit.

**Clear air turbulence (CAT).** Turbulence encountered by aircraft when flying through airspace devoid of clouds.

**Climate.** The long-term manifestations of the weather.

**Closed low.** A low that may be completely encircled by an isobar or contour line.

**Cloud.** A hydrometeor consisting of a visible aggregate of minute water and/or ice particles in the atmosphere above the earth's surface (*fog* by contrast touches the ground).

**Cloud base.** The level of the cloud bottom.

**Cloudburst.** An unofficial, or unpredicated, sudden, heavy rain that falls at a rate greater than 100 millimeters per hour.

**Cloud droplet.** A particle of liquid water within a cloud that is less than 200 microns in diameter.

**Cloud seeding.** Any technique of adding to a cloud certain particles with the intent to alter that cloud's natural evolution.

**Coalescence.** The merging of two water drops into a single larger drop.

**Col (saddle point).** The point of lowest pressure between two highs or the highest pressure between two lows.

**Cold front.** Any nonoccluded front that moves so that the colder air replaces the warmer air.

**Cold wave.** A rapid fall in temperature within 24 hours that requires substantially increased protection to agriculture and other commerce affected by the cold.

**Condensation.** The physical process by which a vapor becomes a liquid or solid (opposite of *evaporation*).

**Condensation nucleus.** A particle either liquid or solid upon which condensation of water vapor begins in the atmosphere.

**Condensation trail (contrail).** A cloudlike streamer frequently observed to form behind aircraft flying in clear, cold, humid air.

**Conditional instability.** The state of a column of air when its lapse rate is less than dry adiabatic but greater than moist adiabatic.

**Conditional instability of the second kind (CISK).** The process whereby clouds and rising air are self-enhancing by latent heat release, even though the lapse rate is less than moist adiabatic.

**Conduction.** The transfer of energy within and through a substance by means of internal particle or molecular activity and without any net external motion.

**Conifer.** Any of the cone-bearing evergreen trees or shrubs.

**Constant pressure chart.** The synoptic chart for any surface of constant pressure that usually contains plotted data (such as height of the surface, temperature, humidity, and wind) and analyses of their distribution. Most common are the 850 mb, 700 mb, 500 mb, 300 mb, and 200 mb charts.

**Continental air.** A type of air whose characteristics are developed over a large land area and hence has characteristically low-moisture content.

**Continental drift.** The phenomenon of the moving of entire continents relative to one another across the surface of the earth. Now often referred to as *plate tectonics*.

**Continental shelf.** The zone around the continents extending from the low water mark seaward to where there is a marked increase in slope to greater depths.

**Contour line.** A line with a constant value of some variable, often height.

**Convection.** Mass motions (predominantly vertical in meteorology) within a fluid resulting in the transport and mixing of the properties (such as heat) of that fluid.

**Convective instability (potential instability).** The state of an unsaturated layer or column of air whose equivalent potential temperature decreases with height.

**Convergence.** The contraction of a vector field (such as the wind).

**Coral reef.** A ridge of rock consisting of skeletons of corals at or near the surface of the water.

**Coriolis acceleration.** An acceleration of a particle moving in a rotating system.

**Coriolis force.** The apparent force on moving particles in a rotating system; used to account for the coriolis acceleration.

**Corona.** A set of one or more colored rings of small radii surrounding the disk of the sun or moon when veiled by a thin cloud and due to diffraction of light by cloud droplets.

**Corrie.** A semiamphitheater-shaped depression on a mountainside that is caused by the action of ice.

**Crepuscular rays.** Alternating lighter and darker bands of light that appear to diverge in fanlike array from chinks in the clouds or from the sun's position at about twilight.

**Cross section.** A graphic representation of a vertical surface in the atmosphere along a given horizontal line or path—like a slice through a cake.

**Crystal.** A solid with a regular periodic internal structure.

**Cumulonimbus.** An exceptionally dense and vertically developed cloud that often has an anvil-shaped top with a generally mountainous appearance; frequently these clouds produce thunderstorms.

**Cumulus.** Clouds comprised of individual, detached elements that are usually dense, with well-defined outlines; often resembling cauliflower on top with flat bases.

**Current.** Any movement of a quantity, such as electricity or water, by which there is a net transport of that quantity.

**Cutoff high.** A warm high that has become displaced poleward and out of the basically westerly current; often acts as a blocking high.

**Cutoff low.** A cold low that has become displaced equatorward out of the basically westerly current.

**Cycle.** One complete and consecutive set of all the changes that occur in a recurrent action or phenomenon, ending with all conditions as they were at the start.

**Cyclogenesis.** Any development or strengthening of cyclonic circulation in the atmosphere (opposite of *cyclolysis*).

**Cyclone.** A system with cyclonic circulation, that is, counterclockwise in the Northern Hemisphere and clockwise in the Southern Hemisphere.

**Cyclostrophic wind.** The horizontal wind velocity for which the centripetal acceleration balances the horizontal pressure force.

**Dawn.** The first appearance of light in the eastern sky before sunrise.

**Deciduous.** Having the characteristic of shedding leaves seasonally.

**Declination.**  The angular distance measured by drawing an angle from a point on the equator at noon, to the center of the earth, and to the sun.

**Deepening.**  A decrease in the central pressure of a pressure system or the decrease in the central height (on a constant pressure chart); usually referred to a low (opposite of *filling*).

**Degree.**  (a). A unit of temperature. (b). A unit of angular measure equaling 1/360 part of a circle.

**Degree day.**  The number of degrees that the mean daily temperature falls below or above a certain standard (often 65°F).

**Deliquescence.**  The slight wetting of a surface due to condensation.

**Dendrochronology.**  The analysis of the annual growth of tree rings.

**Density.**  The ratio of the mass of any substance to the volume it occupies.

**Depression.**  An area of low pressure—a low or trough. Usually applied to the early stage of development of a tropical cyclone.

**Desert.**  A region of extreme aridity.

**Dew.**  Water that has condensed on grass and other objects near the ground when their temperatures have fallen below the dew point temperature of the surface air.

**Dew point (dew point temperature).**  The temperature to which a given parcel of air must be cooled at constant pressure and constant water vapor content in order for saturation to occur.

**Diffraction.**  The process by which the direction of radiation is changed so that it spreads into the geometric shadow region of an opaque or refractive object.

**Diffusion.**  The exchange of fluid parcels (or molecules) between regions in space on a scale too small to be treated directly by Newton's laws of motion.

**Discharge.**  (a). The rate of flow past a point in a stream. (b). The flow of electricity in a gas, which results in the emission of radiation.

**Dispersion.**  The process by which radiation is separated into its component wavelengths.

**Dissolve.**  To pass into solution.

**Disturbance.**  Any agitation or disruption of a steady state, loosely used for a somewhat weak, low pressure area.

**Diurnal.**  Daily, especially with regard to daily cycles.

**Divergence.**  The expansion of a vector field (opposite of *convergence*).

**D-layer.**  The lowest layer of the ionosphere.

**Dog days.**  The period of greatest heat in summer.

**Doldrums (equatorial calms).**  A nautical term for the equatorial trough, with special reference to the light and variable nature of the winds.

**Doppler effect.**  The change of frequency with which waves are observed when the receiver and the energy source are in motion relative to one another.

**Doppler radar.**  A radar that determines from the Doppler effect the radial velocity of a target.

**Downdraft.**  Rapidly downward moving air, especially in a thunderstorm.

**Downstream (downwind).**  The direction toward which the fluid is flowing.

**Drag.**  The frictional impedance offered by air to the motion of bodies passing through it.

**Drip point.**  The sharp point on leaves that enables water to drip off easily.

**Drizzle.**  Drops between 0.2 and 0.5 millimeters in diameter that produce a precipitation rate almost inevitably less than 1 millimeter per hour.

**Drought.**  A period of abnormally dry weather sufficiently prolonged for the lack of water to cause a serious hydrologic imbalance (i.e., crop damage) in the affected area.

**Dry adiabat.**  A line of constant potential temperature on a thermodynamic diagram.

**Dry adiabatic lapse rate.**  The rate of decrease of temperature with height of a parcel of dry air lifted adiabatically through an atmosphere in hydrostatic equilibrium.

**Dry air.**  Air containing no water vapor. (Sometimes used to denote unsaturated air.)

**Dry bulb temperature.**  A fancy term for air temperature.

**Dune.**  A sand hill or ridge formed by the wind.

**Dust devil.**  A rapidly rotating column of air rendered visible by dust, sand, and debris picked up from the ground.

**Dust storm.**  A frequently severe weather condition characterized by strong winds and dust-filled air over an extensive area in which visibility is less than 1 kilometer.

**Easterlies.**  Any wind with a component from the east.

**Easterly wave.**  A migratory wavelike disturbance in the tropical easterlies.

**Eccentricity.**  The degree of deviation from circularity.

**Echo.**  In radar, the return signal from the target.

**Ecliptic.**  The circle that is the apparent annual path of the sun around the earth.

**Eddy.**  A "glob" of fluid within the fluid mass that has a certain integrity and life history of its own.

**Ekman spiral.**  The graphic representation of the way in which theoretical wind-driven ocean currents or frictionally influenced winds in the atmosphere vary with altitude or depth.

**Electric field.**  A region wherein any charged particle would experience an electrical force.

**Electromagnetic radiation.**  Energy propagating through space or material media in the form of an advancing disturbance in electric and magnetic fields.

**Electron.**  The subatomic particle that possesses the smallest possible negative electric charge.

**Element.**  A substance that cannot be broken down into simpler components by ordinary chemical means.

**Elevation angle.**  The vertical angle above the horizon.

**Emissivity.**  The ratio of the rate at which a body emits radiation to the black body rate.

**Energy.**  Any quantity with the dimensions mass × length$^2$ ÷ time$^2$.

**Entire margined.**  A leaf that is neither toothed nor lobed.

**Entrainment.**  The mixing of environmental air into a preexisting organized air current.

**Environmental lapse rate.**  The rate of decrease of temperature with elevation.

**Epiphyte.** A plant that grows attached to the stem or leaves of another plant but is not a parasite.

**Equator.** The great imaginary circle on the earth's surface that is at latitude 0° and is equidistant from the poles.

**Equilibrium.** Any state of a system that does not undergo change or acceleration.

**Equinox.** Either of the two days each year when the sun is overhead at the equator at noon.

**Equivalent potential temperature.** The potential temperature of air that has been cooled by lifting until all its water vapor has condensed.

**Erosion.** The movement of soil or rock by the action of the sea, the wind, or any other natural process.

**Escape velocity.** The vertical speed at which a particle must start (neglecting friction) in order to escape from the gravitational field of a planet or star.

**Estuary.** The portion of a river that is affected by tides.

**Eucalyptus.** Any of a type of aromatic evergreen trees.

**Evaporation.** The opposite of condensation.

**Evapotranspiration.** The total evaporation from plants and the ground.

**Exosphere.** The outermost or topmost portion of the atmosphere.

**Extratropical.** Outside the belt of tropical easterlies.

**Extratropical cyclones.** A term usually referring to the migratory frontal cyclones of middle and high latitudes, as opposed to tropical cyclones (or hurricanes).

**Eye.** The eye of the storm is the clear central region with light winds.

**Fall wind.** A strong, cold, downslope wind.

**Fata morgana.** A complex mirage that is characterized by a multiple distortion of images, with magnification which makes cliffs and cottages appear as fantastic as castles.

**Feedback mechanism.** A process in which changes in a situation react either to enhance the change (positive feedback) or to reduce the change (negative feedback). A chain reaction is an extreme example of a positive feedback mechanism.

**Ferrel cell.** The weak meridional cellular motion with rising air at about 60° and sinking air at 30°.

**Firn.** Old snow that has become granular and compacted.

**Flare.** A bright eruption from the sun's chromosphere.

**Flash flood.** A flood that rises and falls quite rapidly with little or no advance warning.

**Floe.** A piece of sea ice of any size.

**Flood.** The condition that occurs when a body of water overflows its confines or accumulates over low-lying areas.

**Flood stage.** The height above which a stream overflows its natural banks and begins to cause damage.

**Foehn.** A warm dry wind on the lee (downwind) side of a mountain range, particularly in the Alps.

**Foehn wall.** The steep leeward boundary of flat cumuliform clouds that are formed on the peaks of mountains during foehn conditions.

**Fog.** A cloud in contact with the ground, reducing the visibility to below 1 kilometer.

**Forest wind.** A light breeze blowing from forests toward open country on clear calm days.

**Fossil.** Any remains, impression, or trace of an animal or plant of a former geologic age.

**Fragmentation.** The process by which an ice crystal or raindrop breaks up into a large number of fragments, each of which can now serve as a new center of growth.

**Freezing.** The phase transition of a substance passing from liquid to solid state.

**Freezing nucleus.** Any particle that will initiate the growth of an ice crystal about itself.

**Freezing rain.** Rain that freezes on contact with a solid object.

**Frequency.** The rate of recurrence of an event in periodic motion.

**Friction.** The mechanical resistive force offered by one medium or body to the relative motion of another in contact with it.

**Front.** An interface or transitional zone between two air masses of different density.

**Frontal cyclone.** Any cyclone associated with a front.

**Frontogenesis.** The initial formation of a front or frontal zone (opposite of *frontolysis*).

**Frost (hoarfrost).** A deposit of ice crystals formed by the direct sublimation on objects.

**Frost heaving.** The lifting of a surface by the internal action of frost.

**Funnel cloud.** The popular name for a tuba or tornado cloud.

**Galaxy.** A distinct aggregate of stars.

**Gale.** A wind of 28 to 47 (or according to the Beaufort Scale, 55) knots.

**Gas.** The state of matter in which individual molecules or atoms move about freely and randomly, generally at a great distance compared to their size.

**General circulation.** The complete statistical description of atmospheric motions over the earth, usually referring to those wind systems extending over a large portion of the earth.

**General Circulation Model (GCM).** Any of the large computer models used to simulate the general circulation, often with the goal of predicting weather or climate.

**Geologic time.** Time considered in terms of the vast geologic past and divided into eras, which are subdivided into periods and further subdivided into epochs.

**Geology.** The study of the earth.

**Geomorphology.** The study of the form and surface configuration of the earth's surface.

**Geostrophic.** Of the balance between the horizontal pressure-gradient force and the Coriolis force.

**Glacial erratic.** A rock or boulder transported by a glacier.

**Glacier.** A mass of land ice.

**Glory.** A ring or series of rings of colored light seen directly around the shadow of the observer as it appears on a cloud of water droplets. Most often seen around the shadow of aircraft.

**Gradient.** The space rate of decrease of a function.

**Gradient wind.** Wind for which the Coriolis acceleration and centripetal acceleration balance the pressure-gradient force.

**Gradient wind level.** Height above which friction is negligible for the wind.

**Graupel.**  Snow pellets.

**Gravitation.**  The acceleration produced by the mutual attraction of two masses.

**Gravity.**  The force imparted by the earth to a mass at rest relative to the earth. (This includes the effects of the earth's rotation.)

**Gravity wave.**  A wave disturbance in which buoyancy acts as the restoring force.

**Green flash.**  A brilliant green coloration of the sun's upper limb just as it lies above the horizon.

**Greenhouse effect.**  The heating effect exerted by the atmosphere by virtue of absorption and reemission of infrared radiation.

**Growing season.**  The period of the year when the temperature is sufficiently high to allow plant growth.

**Gust.**  A sudden, brief increase in the wind speed (generally less than 20 seconds).

**Guttation.**  The process by which plants expel liquid water from uninjured leaves in excess of transpiration.

**Hadley cell.**  The direct (warmer air rising, colder air sinking) meridional cellular motion, producing rising air near the equator, sinking air near 30° latitude, and the trade winds in each hemisphere.

**Hail.**  Precipitation in the form of balls or lumps of ice, having a diameter of 5 millimeters or more. Anything smaller than this is called ice pellets, snow pellets, or graupel.

**Halcyon days.**  A period of fine weather.

**Half-life.**  The time required for a radioactive element to be reduced to one-half of its original amount.

**Halo.**  Any of a large class of atmospheric optical phenomena that appear as colored or whitish rings and arcs about the sun and moon when seen through an ice crystal cloud. Coloration is produced by refraction and dispersion, whitish luminosity by reflection.

**Haze.**  Fine dust or salt particles dispersed through a portion of the atmosphere and markedly diminishing visibility.

**Heat.**  A form of energy transferred between systems by virtue of a difference in temperature and existing only in the process of energy transformation.

**Heat capacity.**  The ratio of the heat absorbed (or released) by a system to the corresponding temperature rise (or fall).

**Heat lightning.**  The luminosity observed from ordinary lightning that is too far away for its thunder to be heard.

**Heat wave.**  A period of abnormally and uncomfortably hot and usually humid weather.

**Heiligenschein.**  A diffuse white ring of light that surrounds the shadow of an observer's head on a dew-covered lawn.

**High.**  An area of relatively high pressure on a constant height chart or an area of relatively greater height on a constant pressure chart.

**Hoarfrost.**  Frost.

**Horizon.**  The actual lower boundary of the observed sky, especially level ground or level water.

**Horizontal pressure gradient.**  The horizontal component of the pressure gradient that is useful in the calculation of forces on the winds.

**Horse latitudes.**  The belts of latitude over the oceans, at approximately 30 to 35° north and south, that are characterized by light winds and hot, dry weather.

**Humidity.**  A measure of the water-vapor content of the air.

**Humus.**  The dark organic material in soils that is produced by the decay of vegetable or animal matter.

**Hurricane.**  A severe tropical cyclone in the North Atlantic Ocean, Caribbean Sea, Gulf of Mexico, or Eastern North Pacific Ocean in which the maximum surface wind exceeds 65 knots.

**Hydrodynamics.**  The study of fluid (liquid or gas) motion.

**Hydrologic cycle.**  The composite picture of the interchange of water substance among the earth, the atmosphere, and the sea.

**Hydrology.**  The scientific study of the waters of the earth (distinct from oceanography).

**Hydrometeor.**  Any product of condensation or sublimation of atmospheric water vapor; also any water or ice particles blown by the wind from the earth's surface.

**Hydrosphere.**  The water portion of the earth.

**Hydrostatic equilibrium.**  The state in which there is complete balance between the force of gravity and the vertical component of the pressure-gradient force.

**Hygrometer.**  An instrument that measures the water vapor content of the air.

**Hygroscopic.**  Pertaining to a marked ability to accelerate or increase the condensation of water vapor.

**Ice.**  The solid form of water substance.

**Ice age.**  A major interval of geologic time during which extensive ice sheets formed over many parts of the world.

**Iceberg.**  A mass of land ice that has broken away from the land and floats in the sea.

**Ice cap.**  A perennial cover of ice and snow over an extensive portion of the earth's surface.

**Ice pellets (sleet).**  A type of precipitation consisting of transparent or translucent pellets of ice 5 millimeters or less in diameter.

**Ice polygons.**  The polygonal-patterned appearance of the ground in flat tundra regions that is produced by repeated freezing of water and replacement of soil by ice.

**Ice wedge.**  A vertical wedge-shaped block of ice in the ground that is formed during repeated freezings of water in which ice replaces soil.

**Icing.**  Any deposit or coating of ice on an object that is caused by impingement and freezing of liquid hydrometeors (also called *riming*).

**Ideal gas.**  A gas that obeys the ideal gas equation.

**Igneous rocks.**  Rocks produced by solidification of magma or other more complicated crystal-forming processes.

**Index of refraction.**  A measure of the amount of refraction: the ratio of the speed of light in a vacuum to that in a substance.

**Indian summer.**  A period in mid or late autumn of abnormally warm weather.

**Indirect cell.**  A closed circulation in a vertical plane in which the rising motion occurs at lower potential temperature than the descending motion; characteristic of the Ferrel cell.

**Induction.**  In static electricity, the process of electrical charging by inducing an electrical field.

**Inertia.**  The property of matter by which it retains its velocity if not acted on by a force.

**Inferior mirage.** A mirage seen below the true position of the object (opposite of *superior mirage*).

**Infrared radiation.** Electromagnetic radiation lying in the approximate wavelength interval, 0.74 to 1000 microns.

**Insolation.** Solar radiation received at the earth's surface.

**Instability.** A state of precarious balance in a system such that tiny perturbations in the system will amplify and destroy the initial balance.

**Interference.** The vector addition of two or more waves.

**Internal energy.** A measure of the molecular activity of a system.

**Intertropical convergence zone (ITCZ).** The dividing line between the trade winds of each hemisphere.

**Inversion.** A departure from the normal increase or decrease with altitude of the value of an atmospheric property—most commonly an increase of temperature with altitude.

**Ion.** Any electrically charged submicroscopic particle.

**Ionosphere.** The atmospheric layer that is characterized by a high density of ions.

**Iridescent clouds.** Clouds that exhibit brilliant spots or borders of color, usually red and green, observed up to about 30° from the sun.

**Irrigation.** The artificial routing of water over land to promote the growth of crops.

**Isentrope.** A line of equal or constant potential temperature.

**Isobar.** A line of equal or constant pressure.

**Isobaric surface.** A surface of constant pressure.

**Isotach.** A line of equal or constant wind speed.

**Isotherm.** A line of equal or constant temperature.

**Isotope.** One of two or more chemical species (often atoms) having the same atomic number but different atomic mass.

**Jet stream.** Relatively strong winds that are concentrated within a narrow air current in the atmosphere.

**Karst.** An area of limestome formations that are characterized by sinks, ravines, and underground streams.

**Katabatic wind.** Any wind blowing down an incline (opposite of *anabatic wind*) and usually cold.

**Kelvin (absolute) temperature scale.** The temperature scale for which a given volume of gas has a temperature proportioned to the pressure and at which all molecular motion ceases at 0°.

**Khamsin.** A dry, dusty, and generally hot desert wind occurring mainly in Egypt.

**Kinetic energy.** The energy that a body possesses as a consequence of its motion, which is defined as $\frac{1}{2} mv^2$.

**Kinetic theory.** The derivation of the bulk properties of fluids from the properties and interactions of their molecules.

**Lake breeze.** A wind similar to the sea breeze but blowing from the surface of a lake.

**Land and sea breeze.** The complete cycle of diurnal local winds that occurs on sea coasts due to differences in surface temperatures of the land and sea. The land breeze blows from land to sea, usually at night; the sea breeze blows from sea to land.

**Lapse rate.** The decrease of an atmospheric variable (most often temperature) with height.

**Large scale.** A scale in which the curvature of the earth is not negligible.

**Latent heat.** The heat released or absorbed per unit mass by a system in a change of phase.

**Laterite.** A residual clay that is formed under tropical climate conditions.

**Leaching.** The removal of mineral salts from the soil by the action of percolating water.

**Lee side.** The downstream or downwind side.

**Lee wave.** Any wave disturbance caused by, and usually stationary with respect to, some barrier in the fluid flow.

**Lenticular.** Lens shaped.

**Liana.** A woody vinelike climber that is typical of tropical forests.

**Lichen.** A large group of composite plants that consist of an alga and fungus; often seen covering rocks under the most forbidding climate conditions.

**Lifting condensation level (LCL).** The level at which a parcel of air lifted dry adiabatically would become saturated.

**Light.** Visible electromagnetic radiation (about 0.4 to 0.75 microns in wavelength).

**Lightning.** Any of the various forms of visible electrical discharge produced by thunderstorms.

**Liquid.** A state of matter in which the shape depends on the container, but the volume is independent of the container.

**Local winds.** Winds that, over a small area, differ from those that would be appropriate to the general pressure distribution.

**Loess.** A wind-borne clay consisting of fine particles that originate in arid regions.

**Looming.** A mirage effect enabling objects normally below the horizon to be seen.

**Low.** An area of low pressure (compare *high*).

**Lunar.** Referring to the moon.

**Mackeral sky.** A sky with considerable cirrocumulus or small-element attocumulus clouds which resemble the scales on a mackeral.

**Magma.** Molten rock.

**Magnetic field lines.** Imaginary lines in a region containing a magnetic field that are so drawn that a piece of iron would tend to move along them.

**Mamma.** Hanging pouchlike protuberances on the undersurface of a cloud.

**Maquis.** *Chaparral* vegetation (in Europe).

**Maritime air.** A type of air whose characteristics are developed over an extensive water surface and therefore have high humidity content in at least its lower levels.

**Mean sea level.** The average height of the sea surface.

**Melting level.** The altitude at which ice crystals and snowflakes melt as they descend through the atmosphere.

**Meridional.** Along a meridian; northerly or southerly.

**Mesometeorology.** That portion of the science of meteorology concerned with the study of phenomena that range in size from about 1 to 100 kilometers.

**Mesosphere.** The atmospheric shell between about 20 and 80 kilometers above sea level.

**Mesozoic.** The geological era that extends from about 225 to 65 million years ago.

**Metamorphic rocks.** Rocks that have been structurally changed, usually by great heat and pressure.

**Meteor.** Anything in the air.

**Meteorite.** That portion of a relatively large meteoroid that survives its passage through the atmosphere and reaches the earth's surface.

**Meteoroid.** Any of the countless solid particles in interplanetary space that when encountering the atmosphere produces a meteor.

**Meteorology.** The study dealing with the phenomena of the atmosphere.

**Microclimate.** The fine climatic structure of the airspace from the ground to that height whose characteristics are independent of the peculiarities of the surface immediately below.

**Microwave radiation.** Electromagnetic radiation with a wavelength that ranges from about 0.01 to 100 centimeters.

**Mineral.** A naturally occurring substance of more or less definite chemical composition and physical properties.

**Mirage.** A refraction phenomenon in which an image of some object is made to appear displaced from its true position.

**Mist.** A hydrometeor that in all its characteristics is intermediate between haze and fog and produces a thin, greyish veil over the landscape.

**Mixing ratio.** The ratio of the mass of water vapor to the mass of dry air; often approximately equal to the *specific humidity*.

**Mock sun.** Another expression for *parhelion*.

**Moist air.** Air that is a mixture of dry air with any amount of water vapor; often used to refer to air with a high relative humidity.

**Molecule.** The smallest part of an element or compound that exhibits all the chemical properties of that substance.

**Momentum.** That property of a particle that is given by the product of its mass, *m*, with its velocity, *v*.

**Monsoon.** Seasonal winds; often applied to the winds over the Arabian Sea and India.

**Moonbow.** A rainbow formed by light from the moon.

**Moraine.** A ridge, mound, or irregular mass of boulders, rock, and sand that is transported by a glacier.

**Mountain and valley winds.** A system of diurnal winds that blows downhill at night (mountain winds) and uphill by day (valley winds).

**Mud cracks.** Cracks formed on the surface of the ground and produced by the shrinking of mud during drying.

**Muggy.** Humid, warm, uncomfortable.

**Nacreous clouds.** Clouds rarely seen, which form at about 20 to 30 kilometers above the earth (also called mother-of-pearl clouds).

**Nadir.** The point vertically downward, that is, directly opposite the zenith.

**Neutron.** A subatomic particle with no electric charge and with a mass slightly greater than a proton.

**Nimbostratus.** A gray-colored, often dark cloud that is frequently rendered diffuse by falling precipitation particles and not accompanied by lightning, thunder, or hail.

**Noctilucent clouds.** Clouds that form at altitudes roughly between 75 and 90 kilometers.

**Nucleation.** Any process in which a phase change of a substance, when moving to a more condensed state, is initiated at a certain locus.

**Nucleus.** (a) A particle of any kind about which condensation occurs. (b) The central, positively charged core of an atom.

**Numerical weather prediction.** The forecasting of the weather by the finite difference solution of the governing fundamental equations of the air.

**Obliquity of the ecliptic.** The angle between the plane of the earth's orbit (ecliptic) and the plane of the earth's equator.

**Obscuration.** A designation for sky cover when the sky is completely hidden by surface-based obscuring phenomena (such as fog).

**Occluded cyclone.** Any cyclone within which there has developed an occluded front.

**Occluded front.** A rather complicated front ideally composed of a combination of a cold front and a warm front.

**Orographic.** Of or caused by mountains.

**Overcast.** Descriptive of a cloudy sky cover that encompasses 95% or more of sky.

**Paleontology.** The study of life in the geologic past.

**Paleoclimate.** The climate of a time period in the geologic past.

**Palynology.** The study of pollen.

**Parallax.** The change in apparent position of a nearby object compared to a remote reference point.

**Parcel.** See *air parcel*.

**Parhelic circle.** An arc of light at the same elevation angle as the sun, which is caused by the reflection of light by ice crystals.

**Parhelion.** Either of the two colored luminous spots of light at the same elevation angle as the sun and somewhat outside the 22° halo, which is caused by the refraction of light by ice crystals. (Also called *mock sun* and *sun dog*.)

**Peak gust.** The highest "instantaneous" wind speed recorded during a specific period.

**Peat.** A highly organic soil that is more than 50% combustible, is composed of partially decayed vegetable matter, and is found in marshy or damp regions.

**Period.** The time interval between the passage of a given *phase* of a wave.

**Permafrost.** A layer of soil or bedrock that has been continuously frozen for at least a few years.

**Perturbation.** Any departure from an assumed steady state.

**Phase.** For any type of periodic phemomena, a point in the period to which the phenomena has advanced relative to a given initial point.

**Photon.** The elementary quantity of radiant energy.

**Photosphere.** The intensely bright portion of the sun that is visible to the naked eye.

**Pileus.** An accessory cloud in the form of a cap, hood, or scarf that occurs above, or is attached to, a cumuliform cloud, which often pierces it.

**Pingo.** A large frost mound whose life is more than one year.

**Planetary wave.** See *Rossby wave*.

**Plate tectonics.** See *continental drift*.

**Pluvial.** Pertaining to rain or a geologic time when it was rainy.

**Podzol.** The typical ashlike, usually acidic, soil that is characteristic of subpolar regions.

**Point discharge.** A gaseous electrical discharge from a pointed conductor.

**Point source.** A single point that emits radiation, pollution, or some other substance.

**Polar front.** The semipermanent, semicontinuous front that separates the air masses of tropical and polar origin.

**Potential energy.** The energy that a body possesses as a consequence of its position in the field of gravity.

**Potential temperature.** The temperature of an air parcel brought dry adiabatically to a pressure of 1000 millibars.

**Prairie.** A flat, gently undulating, grassy plain that is generally treeless.

**Precipitable water.** The total depth of water produced by condensing all water vapor from a vertical air column.

**Precipitation.** Any or all forms of water particles, whether liquid or solid, that fall from the atmosphere and reach the ground.

**Pressure.** Force divided by the area. (Magnitude only.)

**Prevailing wind.** The wind direction that is most frequently observed during a given period.

**Prognostication.** Forecast.

**Proton.** A positively charged subatomic particle having a mass slightly less than a neutron but about 1846 times greater than an electron.

**Pseudoadiabat (moist adiabat).** See *saturation adiabat*.

**Psychrometer.** A type of hygrometer that uses the principles of evaporation.

**Radar.** An electronic instrument used to detect objects by their property of scattering radio waves.

**Radiation.** The process by which wave energy (principally electromagnetic radiation) is propagated.

**Radioactivity.** The property of an element to disintegrate spontaneously into simpler elements; this process is accompanied by the emission of particles and energy.

**Radiosonde.** A balloon-borne instrument package for the simultaneous measurement and transmission of meteorological data.

**Rainbow.** A luminous circular arc of light produced by the refraction and reflection of light from a sheet of raindrops.

**Raindrop.** A drop of water of more than 0.5 millimeters in diameter that is falling through the atmosphere.

**Rain forest.** A forest that grows in a region of heavy annual precipitation. The term *jungle* is often used for a tropical rain forest.

**Rain pit.** Impressions left on mud or sand that are produced by the impact of raindrops.

**Rain shadow.** The region on the leeward side of a mountain range where precipitation is noticeably reduced.

**Ray.** An elemental path of radiated energy; the direction of wave motion.

**Redbed.** A layer of reddish sedimentary rock often formed from laterites.

**Reflection.** The process whereby a surface of discontinuity sends back a portion of incident radiation.

**Refraction.** The process in which the direction of energy propagation is changed as a result of a change of medium or a change in the density of a single medium.

**Relative humidity.** The ratio of the actual vapor pressure of the air to the saturation vapor pressure or, almost equivalently, the ratio of the actual mixing ratio of the air to the saturation mixing ratio. (You choose—they are quite close.)

**Ridge (wedge).** An elongated area of high pressure (opposite of *trough*).

**Rime.** A white or milky granular deposit of ice formed by the rapid freezing of supercooled water drops as they impinge on an exposed object.

**Rock.** An aggregate of minerals. A type of music.

**Roll cloud.** See *arcus*.

**Rossby wave (planetary wave).** Any of the large (several thousand kilometers wide) patterns seen on upper-air charts in which the winds meander along as they circuit the globe.

**Rotor cloud.** A turbulent altocumulus-type cloud formation that is formed in the lee of some large mountain ridges in which the air rotates around an axis parallel to the ridge.

**Runaway-greenhouse effect.** A greenhouse effect with a positive feedback that leads to far greater temperatures than in its absence.

**Runoff.** The water that ultimately reaches stream channels.

**Saint Elmo's fire.** A luminous, and often audible, electrical discharge emitted from objects (especially pointed ones) usually near the end of a thunderstorm.

**Salinity.** A measure of the quantity of dissolved salts in sea water.

**Sandstorm.** A strong wind carrying sand through the air; most of these particles are between 0.08 and 1 millimeter in diameter.

**Saprophyte.** An organism that obtains its food from dead organic matter.

**Saturation.** The condition in which the pressure of any fluid constituent is equal to its maximum value, such that any increase will lead to condensation.

**Saturation (moist) adiabatic lapse rate.** The rate of decrease in temperature with height of a saturated air parcel lifted through an atmosphere in hydrostatic equilibrium.

**Savannah.** A tropical or subtropical region of grassland and other drought-resistant vegetation.

**Scattering.** The process by which small particles diffuse a portion of the incident radiation in all directions.

**Scintillation.** The generic term for rapid variations in apparent position, brightness, or color of a distant luminous object viewed through the atmosphere.

**Scud.** Ragged low clouds, especially when such clouds are moving rapidly beneath a layer of nimbostratus.

**Sea breeze.** See *land breeze*.

**Sediment.** Material carried in suspension by air or water, or the deposits of airborne or water-borne materials.

**Sedimentary rocks.** Rocks formed by the compaction and concretization of sediments.

**Sferics (atmospherics).** The radio-frequency electromagnetic radiations that originate from lightning.

**Shear.** The variation of a vector field along a given direction of space. (See *wind shear.*)

**Shearing instability.** The complex process by which perturbations on a basic steady wind or current field grow and distort the flow as a result of the shear.

**Shimmer.** Scintillation of objects within the atmosphere.

**Shower.** Precipitation from a cumuliform cloud.

**Sidereal day.** The time (23 hours, 56 minutes, 4 seconds) it takes for the earth to rotate 360° around its axis and face the same point in space.

**Sinkhole.** A steep, walled depression in the ground that is caused by the collapse of the roof of a cave.

**Sky cover.** The fraction of the sky that is covered by clouds or other obscuring phenomena.

**Sleet.** See *ice pellets.*

**Smog.** Air rendered visible by pollution.

**Snow.** Precipitation composed of white or translucent ice crystals, which are usually in complex form.

**Snow banner.** Snow being blown from a mountain crest.

**Snow cover.** The arcal extent of snow-covered ground.

**Snow line.** Generally the outer boundary (often in elevation) of the permanent snow cover.

**Snow pellets.** Precipitation consisting of white, opaque, approximately round or conical ice particles that have a snowlike structure and are about 2 to 5 millimeters in diameter.

**Snow roller.** A cylindrical spiral of snow that is formed by wind when it starts rolling its own snowballs.

**Soil.** The portion of the earth's surface that consists of disintegrated rock and humus.

**Solar constant.** The rate at which solar radiation is received from outside the earth's atmosphere at the earth's mean distance from the sun.

**Solifluction.** The rapid creep or downslope flow of saturated soil.

**Solstice.** Either of the two points on the sun's apparent annual path where it is displaced farthest north (June 21) or south (December 21).

**Sonic boom.** The explosive sound produced by any object moving faster than the speed of sound.

**Sound.** A mechanical disturbance propagated through a material medium by virtue of its compressibility. The advancing disturbance is called a *sound wave* and it travels at the speed of sound, which is governed by the nature of the medium.

**Sounding.** Any penetration of the natural environment for scientific observation.

**Specific heat.** The heat capacity of a system divided by its mass.

**Specific humidity.** The ratio of the mass of water vapor in an air parcel to its total mass.

**Spectrum.** The entire range of wavelengths of any wave phenomena (e.g., electromagnetic radiation).

**Spiral arms.** Spiral-shaped aggregates of stars in a galaxy.

**Spiral bands.** Spiral-shaped aggregates of cumulonimbus clouds in a hurricane.

**Spring.** The period extending from the vernal equinox to the summer solstice.

**Squall.** A strong wind characterized by a sudden onset and termination, and lasting only a few minutes.

**Squall line.** Any nonfrontal line or narrow band of active thunderstorms.

**Stability.** Generally used to indicate *static stability* (opposite of *instability*).

**Standing wave.** A wave that is stationary with respect to the medium in which it is embedded.

**Static stability.** The stability of an atmosphere in hydrostatic equilibrium with respect to vertical displacements, usually of air parcels.

**Steam.** In general, denotes water vapor but, in popular terms, the visible condensation of water vapor as it passes from a warmer to a colder environment.

**Steering.** A loosely used term for the influence of the atmospheric winds on the motion of weather systems.

**Steppe.** An area of grass-covered and generally treeless plains with a semiarid climate that forms a broad belt from southeastern Europe through southwestern U.S.S.R.

**Stooping.** Opposite of *towering.*

**Storm.** Any disturbed state of the atmosphere—a strong indication of severe weather.

**Storm surge (storm tide).** An abnormal rise of the sea along a shore as the result of a storm.

**Storm track.** The path taken by the center of a low.

**Storm wind.** A wind whose speed is between 56 and 63 knots.

**Stratified fluid.** A fluid having potential density variations along the axis of gravity, usually implying *static stability.*

**Stratocumulus.** A cloud type that is predominantly stratiform, but with elements in the form of roll clouds (round blobs, etc.), subtending between 5 and 30° of arc.

**Stratosphere.** The atmospheric shell above the troposphere and below the mesosphere that is generally characterized by a temperature increase with height.

**Stratus.** A cloud type in the form of a gray layer, with a somewhat uniform base, whose precipitation is most commonly drizzle.

**Stream flow.** The water flowing in a stream channel.

**Subatomic particle.** Any particle of less than atomic mass, for example a proton, neutron, or electron.

**Sublimation.** The transition of a substance from the solid phase directly to the vapor phase.

**Subpolar.** Having a latitude between about 50 and 70°.

**Subsidence.** A descending motion of air.

**Subtropical.** Having a latitude between about 20 and 35°.

**Summer solstice.** For either hemisphere, the solstice at which the sun is above that hemisphere, marking the beginning of summer.

**Sun cross.** A rare halo phenomenon in which bands of white light intersect over the sun at right angles.

**Sun drawing water.** The popular phrase for crepuscular rays.

**Sun pillar.** A luminous streak of light, vertically above or below the sun when the sun is near the horizon, that is produced by the reflection of light from ice crystals.

**Sunspot.** A relatively dark area on the surface of the sun. Sunspots come in cycles of approximately 11 years.

**Superadiabatic lapse rate.** A lapse rate greater than the dry adiabatic lapse rate and hence unstable.

**Supercooling.** The reduction in the temperature of any liquid below its melting point without freezing.

**Superior mirage.** The mirage of an object that is formed above its true position.

**Supernumerary rainbows.** A set of dimly colored rainbow arcs that are mainly inside the primary rainbow and are produced by diffraction.

**Supersaturation.** The condition existing when the relative humidity is greater than 100%.

**Surface chart.** An analyzed synoptic chart of weather observations made at ground level.

**Surface tension.** A phenomenon peculiar to the surface of liquids that is caused by a strong attraction toward the interior of the liquid of the molecules at its surface.

**Swell.** Ocean waves that have traveled out of the region in which they were generated and are characterized by regularity.

**Synergy.** The combination of two biological processes to produce a far more severe effect.

**Synoptic chart.** Any chart or map presenting data and analyses of the state of the atmosphere over a wide area.

**Taiga (boreal forest).** The open northern part of the boreal forest that consists mainly of coniferous trees growing in a rich floor of lichen.

**Tangent arcs.** The generic term for several types of halo arcs that are tangent to other halos.

**Temperate zone.** In general, the region between 23½ and 66½° latitude.

**Temperature.** A measure of the translational molecular kinetic energy; in general, a measure of the degrees of hotness or coldness.

**Tendency.** The rate of change of a quantity with time at a given point in space.

**Terminal velocity.** The falling speed at which the drag on an object balances the gravitational force.

**Terrestrial.** Of the earth.

**Thermal.** (a). Pertaining to temperature or heat. (b). A relatively small rising parcel of warm air.

**Thermal wind.** The difference in the geostrophic wind at two different heights that is produced by horizontally varying temperature.

**Thermocline.** A layer of water with an anomalously large vertical temperature gradient.

**Thermodynamic (adiabatic) diagram.** Any chart or graph representing values of pressure, density, temperature, water vapor, and functions thereof, such that the equations for adiabatic and saturation-adiabatic processes are satisfied.

**Thermometer.** An instrument for measuring temperature. A recording thermometer is called a thermograph.

**Thermosphere.** The atmospheric layer extending from the top of the mesosphere to outer space.

**Thickness.** Usually the vertical depth between two surfaces of constant pressure.

**Thunder.** The sound emitted by rapidly expanding gases along the channel of a lightning discharge.

**Thunderstorm.** A local storm invariably produced by a cumulonimbus cloud and always accompanied by lightning and thunder.

**Tide.** The periodic rising and falling of the earth's oceans and atmosphere.

**Tongue.** A pronounced protrusion or extension of air with a property markedly different from that of the surroundings.

**Topography.** The disposition of the major natural features and the manufactured physical features of the earth's surface.

**Tornado.** A violently rotating column of air pendant from a cumulonimbus cloud and nearly always observable as a funnel cloud or *tuba*.

**Towering.** A vertical magnification of a looming mirage (opposite of *stooping*).

**Trace.** A precipitation amount of less than 0.1 millimeter.

**Trade winds.** The wind system, found in most of the tropics, that blows from the subtropical highs toward the equatorial trough.

**Transverse.** At right angles or normal.

**Tree line.** (a). The poleward or altitudinal limit of tree growth. (b). The botanical boundary between tundra and taiga.

**Tree ring.** The ringlike band of alternating dark and light wood in a tree trunk showing annual or seasonal growth.

**Tropical air.** A type of air whose characteristics (of high-surface temperature, etc.) are developed over low latitudes.

**Tropical cyclone.** A general term for a cyclone that originates over the tropical oceans. These cyclones are classified as follows: winds up to 34 knots, tropical depression; winds from 35 to 64 knots, tropical storm; winds 65 knots or greater, hurricane or typhoon (with some other regional names).

**Tropopause.** The boundary between the troposphere and stratosphere, usually characterized by an abrupt change of lapse rate.

**Troposphere.** The lowest layer of the atmosphere that extends to 10 or 20 kilometers and in which most weather phenomena occur.

**Trough.** An elongated area of relatively low pressure (opposite of *ridge*); or the low point of a wave.

**Tsunami.** An ocean wave that is produced by the landslide of a submarine earthquake or volcanic eruption, which can be truly enormous at coastlines.

**Tuba.** A cloud column or an inverted cloud cone pendant from a cloud base that when it reaches the ground is a tornado or waterspout. Popularly called a *funnel cloud*.

**Tundra.** Treeless plains that lie poleward of, or above, the tree line with a layer of permafrost throughout most of the subsoil.

**Turbidity.** Any condition of the atmosphere that reduces its visibility.

**Turbulence.** A state of fluid flow in which the instantaneous velocities are so irregular that only their average and statistical properties can be subjected to analysis and prediction.

**Twilight.** The period of incomplete darkness, following sunset and preceding sunrise, when the sun is less than 6°, 12°, or 18° (for civil, nautical, and astronomical twilight, respectively).

**Typhoon.** A hurricane occurring in the region of the western Pacific Ocean.

**Ultraviolet radiation.** Electromagnetic radiation in the approximate wavelength interval from 0.00150 to 0.4 microns.

**Umbrella.** A device that is essential whenever the weather forecast is sunny.

**Updraft.** A strong vertically ascending wind, often in a thunderstorm.

**Upper-air chart.** A synoptic chart of meteorological conditions at any level at or above 850 millibars.

**Upstream (upwind).** The direction from which a fluid is moving.

**Upwelling.** The rising of water (often cold) toward the surface from the depths.

**Vacuum.** Empty space.

**Valley wind.** See *mountain and valley winds.*

**Vapor.** Any substance existing in the gaseous state at a temperature below its boiling point.

**Vapor pressure.** The pressure exerted by the molecules of a given vapor.

**Varves.** The annual layers of sediment deposited by meltwater from glaciers.

**Vector.** Any quantity that has both magnitude and direction, such as the wind velocity.

**Veering.** A clockwise turning of the wind direction, often with time or altitude (opposite of *backing*).

**Velocity.** The time rate of change of position, including magnitude (speed) and direction.

**Virga.** Wisps or streaks of water or ice particles that fall from a cloud but evaporate before reaching the earth's surface.

**Virtual temperature.** The temperature of dry air, with the same density and pressure as moist air.

**Viscosity.** The molecular property of a fluid similar to friction.

**Visibility.** The greatest distance in a given direction at which it is just possible to see and identify with the unaided eye.

**Volcano.** A mountain that is formed by expelling molten lava.

**Vortex.** Any flow possessing *vorticity,* especially when marked swirling motion is apparent.

**Vorticity.** A vector measure of the local rotation rate in a fluid flow.

**Wadi.** See *arroyo.*

**Warm front.** Any nonoccluded front that moves in such a way that the warmer air replaces the colder air.

**Warm sector.** That area of a wave cyclone where the warm air is found.

**Water.** Dihydrogen oxide, $H_2O$.

**Water spout.** A tornado or a lesser whirlwind over water.

**Water table.** The surface defined by the upper limit of unconfined groundwater.

**Watery sun (moon).** The faded image of the sun or moon that is visible through some altocumulus or altostratus clouds.

**Wave cyclone.** A cyclone that forms and moves along a front.

**Wave length.** The mean distance between successive maxima or minima of a wave.

**Weather.** The state of the atmosphere, primarily with respect to its effects on life and human activity.

**Weathering.** The mechanical, chemical, or biological action of the atmosphere, hydrometeors, and aerosols on the form, color, or constitution of exposed material (to be distinguished from *erosion*).

**Westerlies.** The dominant west-to-east motion of the atmosphere that is centered over the middle latitudes of both hemispheres.

**Wet bulb temperature.** The temperature produced by evaporational cooling to the point of saturation.

**Whiteout.** An optical phenomenon (mainly of the polar regions) that is produced when a uniform cloud cover forms over an unbroken snow cover; with these conditions all sense of depth and orientation is lost, engulfed in a uniformly white glow.

**Wind.** Air in motion relative to the surface of the earth.

**Wind-chill index.** The cooling effect of any combination of temperature and wind.

**Window.** A wavelength band of infrared radiation that largely penetrates the atmosphere.

**Wind shear.** The local variation of the wind vector in a given direction, mainly perpendicular to the wind itself.

**Winter.** The period extending from the winter solstice to the vernal equinox when the sun is in the opposite hemisphere.

**Work.** A form of energy that arises from the motion of a system against a force and exists only in the process of energy conversion.

**X rays.** Electromagnetic radiation with wavelengths between about 0.00001 and 0.00150 microns.

**Zenith.** The point vertically overhead.

**Zenith angle.** The angle from the zenith.

**Zonal.** Easterly or westerly (as opposed to *meridional*).

# HINTS AND SOLUTIONS FOR SELECTED PROBLEMS

## CHAPTER 1

**1-4** Think about the temperature of the air in the flame.

**1-5** Counterclockwise and outward.

**1-6** Think of how the pressure of the escaping air changes.

**1-7** No, can you see air.

**1-9** Think of how rising or sinking air relates to rain. Then check by looking at Figure 12.1

**1-11** 2°C; 2°C; 1003 mb; overcast; continuous light rain; wind from the east; 10 knots.

**1-12** Think of where it rains harder—to the north.

**1-13** 3°C. Check this estimate with the actual temperature in Figure 17.32.

**1-14** 0.03 mb/km.

## CHAPTER 2

**2-1** Take your textbook along to verify your observations.

**2-2** Air must rise to a certain level before condensation can occur, but how much higher it rises is quite variable.

**2-3** Think of what vertical motions are induced around large cumulonimbus clouds.

**2-7** Check Figures 16.13 and 16.14 to find cold coastal ocean currents similar to those off the California coast; then verify your answer with Figure 11.2.

**2-9** A watery sun; a halo.

**2-10** Check the sizes of the individual cloud puffs. See definitions in the glossary.

**2-11** The precipitation is roughly 400 km away and will arrive within 12 to 24 hours.

**2-13** Wind differences at different levels cause the streamers to tilt. Usually the streamer top is located where the winds are fastest and are farthest downwind.

**2-15** Waves typically form downstream from an obstacle.

**2-16** Relate this to cloud streets. Birds soar where air rises.

**2-18** Cumulus clouds form when sunlight produces rising hot air from the ground. At higher levels the sunlight will often strike clouds, heating them without producing rising motion. Thus they evaporate.

**2-19** If the answer isn't apparent, check with Figure 17.28 or the discussion of this weather situation in Chapter 17.

**2-20** At night the top of the cloud layer cools off by radiating its heat out to space.

# CHAPTER 3

**3-3** About 0.22 microns.

**3-7** Look at Figure 3.8.

**3-8** In such a case the air temperature near the ground is rapidly modified by the underlying surface.

**3-11** Yes, and they are called *moon rainbows*. Colors are usually not seen since these rainbows are rarely bright enough.

**3-12** Think first of where the sun is in the sky at noon during the summer.

**3-16** Keep in mind that the rainbow ray is the most deflected of all the light rays.

**3-17** Keep in mind that thunderstorms in the tropics move from east to west. Then redraw Figure 3.14 accordingly.

**3-20** The light near the horizon always comes from a greater effective distance. Check Figure 3.24.

**3-21** Keep in mind the discussion of diffraction and the fact that blue has a shorter wavelength than red.

**3-22** A corona.

**3-23** Cirrostratus clouds are composed mainly of ice crystals, whereas altostratus and altocumulus clouds often consist mainly of droplets.

**3-24** Crepuscular rays are easier to see because you are looking toward the sun.

# CHAPTER 4

**4-2** Refer to Figures 4.5 and 4.6

**4-3** Basically because the earth is larger than the moon. Refer to Figure 4.1.

**4-4** Light passing entirely through the earth's atmosphere is refracted into the shadow and is red because the blue has been depleted by scattering, as in Figure 3.24.

**4-5** Closer objects always appear to be moving faster.

**4-7** Refer to Figure 4.8.

**4-8** Use Figures 4.7 and 4.8 to help.

**4-10** Again, use Figures 4.7 and 4.8.

**4-11** The sun rises just about 5:00 A.M. and sets just about 7:00 P.M., so daylight lasts 14 hours.

**4-13** (a) 1.92 cal/(cm²)(min); (b) 0.91, assuming the solar constant is 2.0. Use the sine law.

**4-15** Jupiter. This is found by multiplying the direct solar intensity on each planet times the planet's area. Venus takes second place.

**4-16** 47.5 million km. Use the inverse square law.

# CHAPTER 5

**5-3** No, the water can't rise more than about 10 m. All the giant can do is make a barometer.

**5-4** At the Dead Sea.

**5-5** As you rise through the atmosphere there is less air above you.

**5-6** (a) 50 watts; (b) 100 m/s; (c) 239 food calories (kilogram calories).

**5-7** 77°F; 298°K.

**5-8** (a) −10°C, 263°K; (b) −125°C, −193°F.

**5-9** (a) 9,600°K; (b) 154,000°K.

**5-10** (a) 1,680°K; (b) 27,000°K. Certainly the average molecule will not escape from Mercury, but there would be a significant number of molecules fast enough to escape so that in a few million years Mercury would lose any atmosphere it obtained.

**5-12** Your ears "pop" from the change of air pressure outside.

**5-14** You cannot answer this question using only the ideal gas equation because it has two unknowns: the temperature and the volume. You will be able to answer this question after reading Chapter 13.

**5-15** Yes, the final pressure would reach 3000 mb.

**5-16** It *halves*. This is an example of a law known as *Boyle's law*.

**5-17** It *doubles*. This is an example of a law known as *Charles' law*.

**5-19** Use Table 5.2. Neither choice is too good, but (b) is better.

**5-20** It takes 5.4 times as much heat to boil water than to heat it from 0 to 100°C. Therefore you have 54 minutes.

**5-24** 300 minutes.

**5-25** The shortage of oxygen, the result of much reduced air pressure at high altitudes.

# CHAPTER 6

**6-1** You may stop whenever you get tired.
**6-2** Oxygen and silicon.
**6-6** They have been eroded away by the actions of the atmosphere—wind, rain, freezing, and so on.
**6-8** Photosynthesis.
**6-9** On earth most of the $CO_2$ is locked up in the ocean or in limestone rocks; on Venus it is too hot for this to happen.

# CHAPTER 7

**7-1** First think of their wavelength (Figure 3.3), and then think of Einstein's law.
**7-2** The 0.56 micron light is green and is produced by the oxygen; the 0.66 micron light is red.
**7-3** 0.66 microns, which corresponds to red. Did you realize the connection with Problem 7-2?
**7-4** The concentration of molecules is so small that it is rare to get three molecules colliding at once.
**7-5** Fortunately the radiation, which can dissociate oxygen molecules, doesn't reach sea level.
**7-6** We don't need it to breathe; we need it as a shield against ultraviolet radiation.
**7-7** Remember that pressure halves every 5.5 km you rise.
**7-11** The warm air near the ground at the equator rises so high that by the time it stops rising it has cooled enormously.
**7-12** Use the fact that temperature decreases roughly 6° per km. Then you have to go up about 5.0 km to reach subfreezing daytime temperatures.
**7-15** The appearances of sunspots and solar flares indicate particles are being ejected from the sun. These take about three days to reach earth.
**7-16** (b), (e), and (f). Did you ever hear of these places?
**7-17** More energetic electrons can penetrate closer to sea level. Then use Einstein's law.
**7-19** Extrapolating from Figure 6.6, you can deduce that 1987 will most likely be a year with few sunspots. Now what do you think this implies about auroral activity?

# CHAPTER 8

**8-1** Think of what happens to mercury below −39°C.
**8-4** Blue. First use Wien's law. Then find the color from Figure 3.3.

**8-5** According to Wien's law or Planck's law a higher percentage of a hot star's light would be ultraviolet, which is not too healthy.
**8-6** You have not been given enough information in this book to answer this question. Certainly, by applying Wien's law, the object should glow red by about 4000°K, but in fact, using Planck's law, there is enough visible radiation for an object to glow by about 1500°K.
**8-7** 16 times. Use the Stefan-Boltzmann law.
**8-8** The balance temperature is inversely proportional to the square root of the distance. Find this answer by combining the inverse square law and the Stefan-Boltzmann law.
**8-9** 27°C. First use the Stefan-Boltzmann law to obtain 300°K; then switch to degrees centigrade.
**8-10** See Figure 8.5
**8-12** 232°K according to the inverse square law and the Stefan-Boltzmann law. This is the temperature of the Venus cloud top and it shows how very strong the greenhouse effect is on Venus.
**8-14** It surely would not make the climate cooler.
**8-15** The Stefan-Boltzmann law gives us an answer of 246°K. Adding 30°K you get 276°K; this is just enough to melt the ice if there weren't much cooling due to the winds. Moreover, at lower temperatures the greenhouse effect would not be as strong; see Problem 8-18. Thus, it appears the ice would not melt if it reached the equator.
**8-18** Since the amount of water vapor the air *can* hold increases with temperature, it is reasonable to assume that the actual vapor content also increases with temperature. Thus, the strength of the greenhouse effect due to water vapor most likely increases with temperature.
**8-20** This will happen if the cloud bottom is warmer than the ground, since the cloud then radiates more heat downward than the ground radiates upward.
**8-21** Actual temperatures often rise 2°C or more per hour near the ground during the morning hours.
**8-22** Think of which hemisphere has a larger percentage of land.
**8-23** It takes time for the heat to build up.
**8-24** Can you see the relation to the last question? Now explain why the coldest time of day is shortly after dawn and not midnight.
**8-25** The earth's surface also gets heat radiated from the atmosphere. This question is another way of rephrasing the greenhouse effect.
**8-26** Look at Figure 8-14.

# CHAPTER 9

**9-3** Relate this to Figure 9.2.

**9-5** At ground level wind speeds are generally greater during the day when the warm air near the ground that has been slowed by friction readily rises and exchanges with colder, heavier air above that is moving faster.

**9-6** The reasoning here is similar to that of Problem 9-5. On clear nights it gets colder near the ground, so the air is less able to mix with the faster air above, and the wind speed gradually slows down.

**9-7** On a windy clear day the warm air near the ground is swept up and mixed more efficiently with the colder air above.

**9-8** A glance at Figure 9-5 should help. Also, there is much cooling by evaporation in bogs.

**9-10** The more efficiently heat is conducted through the ground, the less temperature will rise when the sun is up and the less it will fall when the sun is down. Thus moderation is a golden rule.

**9-13** Superimpose a hill on Figure 4.7 or Figure 4.11; then qualitatively apply the sine law of light intensity.

**9-14** New trees cannot grow without enough sunlight.

**9-15** In the city rainwater runs down into the sewers and later cannot reevaporate through the pavement. Thus, in the country there is more cooling by evaporation.

**9-16** Compare the nighttime temperatures of downtown Cincinnati with those of the surrounding countryside from the ground up to 300 m. Use Figure 9.12.

**9-17** Think about the vertical motion of warm air.

**9-18** Extra pollutants, extra clouds.

**9-19** The extra pollution and reduced sunlight certainly *do not* help.

# CHAPTER 10

**10-1** The humid air is lighter (less dense!). This is simply because molecules of water vapor (molecular weight, 18) are lighter than the average air molecule (average molecular weight, 29) they replace.

**10-2** No.

**10-3** $W = 5.11\%_0$; $W_s = 20.44\%_0$. This information comes from Table 10.1.

**10-4** See Example 10.2, page 145.

**10-5** (a) $T_v = 313.6°K = 40.6°C$; (b) $T_v = 310°K = 37°C$.

Since (a) has a higher virtual temperature it is lighter.

**10-7** 110%

**10-8** The droplet will evaporate and the crystal will grow. With respect to ice, RH = 115%.

**10-9** Yes.

**10-10** $T_d = 5°C$. I hope you didn't waste any time trying to find $T$—you can't.

**10-11** $T_w$.

**10-12** On a hot, humid day. $T_d$.

**10-13** *Deathly* hot and humid.

**10-14** (a) $T_w = 16.5°C$; (b) $T_w = 16.5°C$. Both should feel equally cold while you are wet. See Table 10.2b.

**10-15** $T_w = -4°C$. The water on the windshield got cooled to the wet bulb temperature and froze. Of course, this only happened when the car was moving with respect to the air. See Table 10.2b.

**10-16** As the rain fell into the dry air some of it evaporated, cooling the air and lowering its temperature. Eventually it became cold enough to snow. From Table 10.2a the RH = 20%; from Table 10.2b the estimated $T_w = -4°C$.

**10-17** (a) THI = 87; (b) THI = 92. Both are oppressive but (b) is far worse.

**10-18** See the answer to Problem 10-19.

**10-19** The boiling point there is only about 87°C.

**10-20** At −15°C, because the difference between the saturated vapor density over water and over ice is greatest at this temperature.

**10-21** 23.05 g/m³. Extreme values of liquid water rarely exceed 5 g/m³, except in the most intense thunderstorms where the value may reach 10 g/m³. For stratus clouds the typical liquid water content varies between 0.1 and 1.0 g/m³.

**10-22** Proceed as in Example 10.10. About 3.2 times as much rain is needed at 30°C than at 10°C.

**10-23** Where would the water come from in the Sahara?

**10-24** (a) When water evaporates almost all the salt remains behind. (b) Over 2000 mm.

**10-25** During droughts the water table lowers.

**10-26** According to Figure 10.8, about 110/250 = 44%.

# CHAPTER 11

**11-3** Even though fog reflects much sunlight, quite a bit still penetrates to ground level, heating the ground and the air nearest it.

**11-4** Radiation fog forms best on clear nights.

**11-5** Since the albedo of fog is higher than that of land, the ground temperature rises most slowly where fog lasts the longest. This of course suppresses cumulus clouds.

**11-6** The cloudy afternoon implies that the temperature was not able to rise too much above the dew point, but it can quickly cool at night if it clears.

**11-7** The humid, Pacific air rapidly cools under calm conditions in winter, and once fog forms it reflects much sunlight, thereby reinforcing itself. Notice the large high pressure area with high winds and clear skies in Figure 17.28. See if you can find the central valley of California in the satellite picture of Figure 17.30.

**11-10** They get much of their water from the fog, which is most common right along the coast.

**11-11** Steam fog.

**11-12** Proceed as in Example 11.1, assuming that breath has $T = 35°C$, RH = 100%. Mixing equal masses of your breath with outside air produces in (a) an RH equal to 173% and in (b) an RH equal to 86%. Thus, condensation is produced only in (a) and is always more likely the colder it gets.

**11-13** Over the Grand Banks the air from the northwest can be very cold during winter.

**11-15** Look at Figure 11.10.

**11-16** This is related to upslope fog.

# CHAPTER 12

**12-1** See Figure 12.1. The Andes Mountains force the winds to rise on the western side where it rains and to sink on the eastern side where desert conditions prevail.

**12-3** Check Figures 12.2a and 12.2b.

**12-4** Size.

**12-5** The many salt nuclei in the air grow slightly when the RH is high, but not necessarily 100%.

**12-6** See the comment for Problem 12-5.

**12-7** Think of the effects of surface tension.

**12-8** Think of the effects of surface tension.

**12-10** Surface tension can no longer hold them together once they are subject to any impact or disturbance. In fact, the breakup of larger drops (and ice crystals) greatly increases the rate at which raindrops are produced in a cloud.

**12-12** Growth by condensation can only occur when the air is slightly supersaturated.

**12-13** Rain forms in the clouds, not at ground level.

**12-14** No nuclei—no condensation; too many nuclei—

droplets remain too small.

**12-15** They don't have too many nuclei.

**12-17** If there are too many freezing nuclei, graupel might remain small.

**12-18** Such altocumulus clouds consist of supercooled water droplets, and in some regions ice crystals are nucleated and grow at the expense of the droplets.

**12-19** Refer back to Chapter 6.

**12-20** Slightly more than 200 microns in diameter for an average drop. In fact, because of the statistical nature of coalescence, a small percentage of droplets (or crystals) grow faster by coalescence than by condensation at about 50 microns in diameter.

**12-22** Look at Table 12.6.

**12-24** This is a terminal velocity.

**12-26** It can make a ship top heavy.

**12-28** When the ground is above freezing it is unlikely for falling rain to freeze on it. This is possible, however; see Problem 10-15.

# CHAPTER 13

**13-1** (b), by 4°.

**13-2** 30°C at sea level and −30°C at 6 km. Normally mixing air is thought to equalize temperatures, but in the atmosphere mixing air equalizes potential temperatures.

**13-3** 1.5 km. See Example 13.1.

**13-7** (a) 11‰; (b) 28‰; (c) 39%; (d) 20°C; (e) 293°K; (f) 334; (g) 800 mb; (h) 1.9 km; (i) −21°C; (j) 1.8‰; (k) 328°K; (l) 334°K.

**13-8** Latent heat of condensation was released but, by definition, this doesn't affect the equivalent potential temperature.

**13-9** The final mixing ratio is lower because of condensation. Presumably much of the difference reached the ground as rain.

**13-10** When you blow out, but constrict your mouth, the air inside your lungs gets pressurized; however, constant contact with the body keeps its temperature from rising inside you. On the way out the pressure falls and so does temperature. (You can develop an excess pressure of over 100 mb in your lungs.) This example shows why people who always keep their mouths wide open are said to be full of hot air.

**13-12** Starting from the 1000 to 900 mb layer, the stability characteristics are in order: C,S,U,S,C,S,U,S. Inversions exist between 900 and 800 mb and also between 300 and 200 mb.

**13-13** A strong wind will produce mixing with air from above. Now look at Problem 13-2.

**13-16** 27.5°C.

# CHAPTER 14

**14-1** Think of how the change in lapse rate affects the dispersal of pollutants aloft.

**14-2** "Fizzed up" from the sea, the salt is easily blown inland; the larger particles settle quickly.

**14-4** Look at Figure 14.5.

**14-5** Think of a corroding by-product of coal burning.

**14-8** Ozone is produced when sunlight acts on certain pollutants.

**14-9** Think of the intensity of sunlight in the city.

**14-10** Think of inversions.

**14-11** The lower the sun in the sky, the longer the path through the atmosphere.

# CHAPTER 15

**15-2** Winds tend to be weak in summer.

**15-4** Think of how a rocket moves, or blow up a balloon and then release it.

**15-6** 4 m/s².

**15-7** Almost 24 seconds, and he will be falling at 68 m/s.

**15-8** (a) $a_c = 0.05$, $C = 0.005$; (b) $a_c = 2.5$, $C = 0.00005$ (all units m/s²).

**15-9** Your weight, because it varies with the gravity of the planet.

**15-10** 0.1 pascals per meter.

**15-11** $(3.5)(10)^{-3}$ pascals per meter; $(2.9)(10)^{-3}$ meters per square second.

**15-16** Follow the approach used in Figure 15.20.

**15-17** Follow the approach used in Figure 15.20.

**15-20** 29 m/s. Notice that the actual wind is about 10 m/s (20 knots), crossing the isobars at an angle of approximately 30° toward lower pressure.

**15-21** (a) $(3.4)(10)^{-4}$ m/m; (b) 34 m/s. The actual wind is 70 knots or 35 m/s. Thus, the geostrophic wind approximation is quite accurate in speed (and in direction).

**15-22** True, the geostrophic wind speed varies inversely with the sine of the latitude.

**15-23** Friction and the geostrophic wind always cause

the formula to overestimate the wind in a low.

**15-25** As you know from Problems 9-5 and 9-6, the wind speed is greater when it is less stable because then there is freer exchange between air at the ground and air above. Since the air above blows almost parallel to the isobars, more mixing implies that the air at the ground will then blow at a smaller angle toward low pressure.

**15-26** Relate this to the solution for 15-25, keeping in mind that the wind speed generally increases with height.

# CHAPTER 16

**16-3** This should be obvious from Figure 16.7.

**16-4** One of the factors is the unbroken expanse of ocean (which has less friction) in the SH.

**16-5** As the wind crosses the equator the direction of the Coriolis force changes.

**16-6** The equatorial countercurrents.

**16-7** The currents reverse because the winds reverse.

**16-8** These coastal regions are dominated by the same basic wind system—the stable side of the subtropical oceanic highs.

**16-9** Think of how the Coriolis force depends on latitude.

**16-10** The southern island of New Zealand lies in the westerly wind zone, but the northern island is situated near the border of easterlies and westerlies.

**16-11** The westerlies are located farther north in July.

**16-12** Madagascar is located in the trade wind belt.

**16-13** 4 hours and 15 minutes from Los Angeles, 6 hours and 10 minutes from New York.

**16-14** It went with the winds. Icing is not as much a problem in August as in winter.

**16-16** Cold highs and warm lows weaken with height.

**16-17** (a) It should be getting warmer, as you can see from Figure 16.24. (b) This eliminates the turning of the wind due to frictional effects alone.

**16-18** Simply interchange right and left.

**16-19** (a) Westerlies, because it is generally colder at the poles than at the equator. (b) Easterlies, because it is generally warmest at the poles. In the thermosphere the reverse is true, as is implied by the temperatures at the mesopause. All these winds are quite strong.

**16-21** Winds are weaker when the air is stable near the ground, as it often is in January (because of the low sun).

# CHAPTER 17

**17-1** Look at Figure 17.2 and think about the wind speeds there.

**17-2** The coldest air is against the ground, so it doesn't easily pass over the mountains. If a strong wind does blow it over, this increases mixing with air above, which is potentially much warmer. The result is much like what happens in Problem 13-2.

**17-3** Think of what the sun is doing in each season.

**17-4** Over the subpolar oceans there is a large stretch of almost uniformly cold water.

**17-5** Look at Figure 17.1.

**17-6** The surface air is dry, and there is often an inversion above.

**17-7** Ottawa is warmer. Look at the wind directions on each side of the front.

**17-8** 30 km/h (which is rather slow). Warm air comes from the south, cold air from the north.

**17-9** The slope is 0.02

**17-12** Remember that the cold front is a boundary line showing the leading edge of cold air.

**17-14** The air comes from the ocean and therefore is rarely below freezing, but is often humid; when it is forced upslope showers result.

**17-15** The air aloft, which is blowing from the south above the frontal surface, is often quite warm because it frequently has just been in contact with the relatively warm surface of the Atlantic Ocean to the south.

**17-18** (a) Clear day ahead. True, but later in the day the surface wind became northeast.
(b) Bad weather ahead. True.
(c) Clear day ahead. True.
(d) Bad weather continues. True, but later in the day the surface wind became northwest.
(e) Slow change for the worse. This is not quite true. The practical rule is not very useful in the warm sector, because the winds are more or less parallel at all heights and because it cannot predict the approach of a cold front or of showers and thundershowers.

**17-19** First, because of the ice-skater effect east of the Appalachians and, second, because of the large temperature contrast between the warm Gulf Stream waters and the cold land.

**17-20** The explanation is similar to that for Problem 17-19, except the large temperature contrasts in this case result from the ease with which warm Gulf air can meet cold Canadian air in the flat Midwest.

**17-21** Radiation fog. It is early morning and almost calm in the center of a large clear high pressure area.

**17-22** No chance for warming with such strong northwest winds aloft.

# CHAPTER 18

**18-5** These hurricanes must move over cold waters, which kill them.

**18-8** The overall winds are easterly, so the tropical weather systems are steered that way.

**18-10** 4.5°C. This is the amount needed to compensate for the adiabatic cooling, as determined from the thermodynamic diagram, Figure 13.4.

**18-12** Recall what happens to warm lows with increasing height.

**18-14** They lose their source of energy.

# CHAPTER 19

**19-2** About one minute.

**19-6** Thunderstorms in Florida rarely produce hail, but Wyoming thunderstorms often do, as is shown in Figure 12.10. However, the most severe thunderstorms occur over the Great Plains, as is suggested by Figure 19.7, so this is where tornadoes and *large* hail (but not hail in general) are most common.

**19-8** SSI = −4, which indicates severe thunderstorms. But these did not occur showing that (a) you cannot reliably compute SSI by using surface temperature instead of 850 mb temperature and that (b) the large-scale weather pattern should also be favorable for producing thunderstorms.

**19-9** The anvil of the thundercloud is composed of cirrus or cirrostratus clouds and is blown rapidly downwind by the strong winds near the tropopause.

**19-10** (a) 2°C/km; (b) −2°C/km; (c) 8°C/km. Thus, in general, rising motion makes the lapse rate less stable, and sinking air makes it more stable. The destabilization of rising air is enhanced if the bottom of the layer is saturated and the top unsaturated.

**19-11** Too short a wavelength is scattered by cloud droplets as well as by raindrops; too long a wavelength is not scattered by raindrops.

**19-12** 65,000 feet.

**19-19** Not if you want to stay alive.

# CHAPTER 20

**20-1** This is true if the sea breeze rolls in during the late morning hours.

**20-3** The sea breeze circulation is restricted to the lower troposphere.

**20-4** As the air moves inland it warms and the fog evaporates.

**20-6** You can see from the winds at inland stations that the general winds are opposing the sea breeze in New Jersey, but actually aid it on Long Island and Connecticut.

**20-7** This was simply a strong downslope breeze that was the result of sinking, cold air from the plateau at night.

**20-9** (a) 4°C; (b) 16°C. The more stable the air, the greater the potential warming of the chinook.

**20-12** Northerly winds blow almost constantly, as they have for thousands of years. They are known as the Etesian winds and effectively kill the Khamsin.

**20-13** In July the Etesian winds of Tel Aviv blow from the Mediterranean Sea and, since they are remarkably persistent, no extraordinary heat from the Khamsin is ever produced in July.

**20-15** 50 m/s.

**20-16** 330 m/s.

**20-17** Think of how the stronger winds at upper levels (which steer the thunderstorms) will refract the sound.

**20-18** 5 km.

# CHAPTER 21

**21-2** The grid points are too far apart.

**21-3** (a) infrared; (b) visible: (c) infrared; (d) visible when distinguishing from bare earth, but infrared when distinguishing from thick clouds; (e) infrared; (f) infrared.

**21-5** Forecaster A is better. Why even hire forecaster B?

**21-7** Absolute zero.

# CHAPTER 22

**22-1** (a) savannah; (b) forest; (c) forest; (d) tundra.

**22-2** (b), (d), and (f).

**22-3** Think of the prevailing wind direction in the tropics and above 40° latitude.

**22-4** (b) This climate has persistent westerlies that, because of the Andes Mountains, blow downslope and remain dry even though the ocean is nearby. Lows do form but invariably farther east. (d) The eastward bulge of South America extends far into the subtropical high pressure area of the South Atlantic Ocean. This suppresses thunderstorm activity, keeping the ITCZ away. Then, the southeast winds around the high blow downslope after passing over the coastal mountains (where they do produce precipitation). (f) During the normal rainy season (July) north of the equator, the winds in the Horn of Africa are basically downslope and have lost their moisture over the highlands of Africa and Madagascar. During January the winds here have a long trajectory over the Arabian Desert and therefore are quite dry.

**22-5** 27°C; 21°C.

**22-6** (a) June and July; (b) December; (c) it is the rainy, humid season.

**22-7** Technically it shouldn't be, but the neighboring regions do have more precipitation.

**22-9** Consider the prevailing wind direction.

**22-11** July. January is the rainy season.

**22-13** A single storm can easily exceed the average annual precipitation by many times.

**22-16** (a) January; (b) July.

**22-18** The zone of prevailing westerlies drifts southward hitting San Francisco first. When it drifts back north it leaves Los Angeles first.

**22-19** Look at Figure 22.2.

**22-21** In Cs winters there is a great deal of sunshine.

**22-22** Because of the strength of the subtropical high, upwelling is strongest in July, but it weakens in September.

**22-24** Look at the orientation of the coastline and then at the winds. In addition, the ocean is warmer in autumn than winter or spring; some precipitation also comes from hurricanes.

**22-26** The Hudson Bay, an important moisture source, freezes over by mid-December.

**22-29** There is too much ocean at the appropriate latitudes.

**22-30** (a) 3 km; (b) 1.5 km. Think of how high you must rise for the warmest months to have a temperature less than 10°C. In extensive highlands the cooling rate may be less than 6°C/km.

**22-31** (77) BSh; (78) Dca; (79) H; (80) Aw; (81) BWk; (82) Aw; (83) Dcb; (84) Cs; (85) Aw; (86) Cf; (87) BWh; (88) Cf; (89) Ar; (90) Cf.

**22-33** (a), (d), and (e).

**22-34** (a) and (f).

# CHAPTER 23

**23-5** In arid regions limestone features are resistant and stand out sharply. In cold, humid regions they lead to caves and are rapidly worn away. In warm, humid regions they tend to stand out, as in Figures 23.14 and 23.15.

**23-10** The organic material does not disintegrate rapidly in such cold conditions; therefore, the soil is composed largely of dead plant material in many places.

**23-11** They aren't grown in the same season.

**23-12** If not flushed, the soil will become progressively saltier.

# CHAPTER 24

**24-5** This indicates a warming trend.

**24-7** Usually plant matter disintengrates more rapidly than it grows in warm climates.

**24-9** This actually happened about 6 million years ago; as a result, huge salt deposits lie under the Mediterranean Sea today.

**24-13** Israel is moving closer to the BWh zone, and Jordan is moving away from it. Thus, the rainfall in Israel will decrease while Jordan's will increase.

**24-14** Colder tropical waters reduced the probability of hurricanes, but the water was still warm enough to spawn some hurricanes.

**24-15** Because of the dependence of the vapor-holding capacity on temperature, a reduced temperature implies a reduced precipitation.

**24-16** (a) 11,000; (b) 48,300; (c) 40,000.

**24-18** I hope not—there haven't been any.

# CHAPTER 25

**25-2** At least 2°C.

# PHOTO CREDITS

**Chapter 1**

| | |
|---|---|
| Opener | Bruce Roberts/Photo Researchers |
| 1-1 | Adler's Photo Service |
| 1-2 | Library of Congress |
| 1-3, 1-10 & 1-14 | NOAA |

**Chapter 2**

| | |
|---|---|
| Opener | U.S. Department of Commerce, Weather Bureau |
| 2-1 | Courtesy Los Alamos Scientific Laboratory |
| 2-2 | Stanley Gedzelman |
| 2-3 | NOAA |
| 2-7 | Louis J. Battan, *Fundamentals of Meteorology,* Prentice-Hall, Inc., 1979 |
| 2-8 | NASA |
| 2-9 | NOAA |
| 2-11 | Robert Gedzelman |
| 2-13 | Stanley Gedzelman |
| 2-14 | Robert Gedzelman |
| 2-15 | NOAA |
| 2-16 | Robert Gedzelman |
| 2-17 & 2-18 | NOAA |
| 2-19 | U.S. Forest Service |
| 2-21 | NOAA |
| 2-22 | Courtesy Brookhaven National Laboratory |
| 2-24 | Stanley Gedzelman |
| 2-25 | Brooks Martner, University of Wyoming |
| 2-28 | Stanley Gedzelman |
| 2-29 | Paul E. Branstine/American Meteorological Society |
| 2-30 | Stanley Gedzelman |
| 2-31 | J. M. Walker |
| 2-32, 2-33 & 2-34 | NASA |
| 2-35 | NOAA |

**Chapter 3**

| | |
|---|---|
| Opener | Stanley Gedzelman |
| 3-12 | Scott Davis |
| 3-18 | Vincent J. Schaefer |
| 3-20 | Emil Schulthess |
| 3-26 | R. A. R. Tricker |

**Chapter 4**

| | |
|---|---|
| Opener | NASA |
| 4-14, 4-15 & 4-16 | NASA |

**Chapter 5**

| | |
|---|---|
| Opener | Leonard Lee Rue III/National Audubon Society-Photo Researchers |

**Chapter 6**

| | |
|---|---|
| Opener | NASA |
| 6-5 | Hale Observatories |
| 6-8 & 6-9 | NASA |
| 6-10 | Hale Observatories |
| 6-12 | Lawrence Pringle/Photo Researchers |

**Chapter 7**

| | |
|---|---|
| Opener | NASA |
| 7-9 | Nancy Lockspeiser/Stock, Boston |
| 7-14 | Kathy Bendo |

**Chapter 8**

| | |
|---|---|
| Opener | U.S. Navy |
| 8-1 | Cincinnati Post, Photo by Alex Burrows |

**Chapter 9**

| | |
|---|---|
| Opener | Grant Heilman |
| 9-3 | Florida Department of Citrus |
| 9-10 | Hubertus Kanus/Rapho-Photo Researchers |

**Chapter 10**

| | |
|---|---|
| Opener | Alain Nogues/Sygma |
| 10-1 | Lawrence Pringle/Photo Researchers |
| 10-2 | Ivan Massar/Black Star |
| 10-9 | Jerome Wyckoff |

| 10-11 | Wide World |
| 10-12 | NASA |
| 10-14 | Courtesy ESSA |
| 10-15 | NASA |
| 10-16 | FAO |

## Chapter 11

| Opener | Margaret Durrance/Photo Researchers |
| 11-1 | Radio Times Hulton Picture Library |
| 11-4 | Eugene Anthony/Black Star |
| 11-5, 11-7, 11-9 & 11-11 | NOAA |
| 11-13 | F. B. Grunzweig/Photo Researchers |
| 11-16 | Courtesy Air Force Geophysics Laboratory |

## Chapter 12

| Opener | NOAA |
| 12-5 | Lawrence Pringle/Photo Researchers |
| 12-11 | Courtesy Nancy and Charles Knight/National Center for Atmospheric Research |
| 12-13 | A. T. & T. Photo Center |

## Chapter 13

| Opener | George Rodger, Life Magazine © 1944 Time, Inc. |
| 13-13 | NOAA |

## Chapter 14

| Opener | United Nations |
| 14-1 | Courtesy Dr. H. B. Kettlewell |
| 14-4 | NASA |
| 14-5 | Ralph Turcotte, Beverly Times |
| 14-6 | Brookhaven National Laboratory |
| 14-13 | Photo by Charles E. Grover, Courtesy EPA |

## Chapter 15

| Opener | U.S. Coast Guard |
| 15-1 | Courtesy Belfort Instrument Company |
| 15-4 | © Ginger Chih, 1978/Peter Arnold |

## Chapter 16

| Opener | Photo supplied by Professor Richard L. Pfeffer, Director, Geophysical Fluid Dynamics Institute, Florida State University |
| 16-4 & 16-5 | National Climatic Center, NOAA |
| 16-6, 16-8 & 16-18 | NOAA |

## Chapter 17

| Opener | Deborah Collings/Stock, Boston |
| 17-9, 17-12 | NOAA |
| 17-13 | Dr. Ken Dewey/NOAA |
| 17-17, 17-30 | NOAA |

## Chapter 18

| Opener | Milton Adams/Sygma |
| 18-1 | Chauncey T. Hinman |
| 18-5, 18-13, 18-14 & 18-16 | NOAA |
| 18-19 | Wide World |

## Chapter 19

| Opener | ESSA |
| 19-2 | Steve A. Tegtmeier |
| 19-3 | James Meyer |
| 19-4 | Federal Aeronautics Administration |
| 19-5 | T. Theodore Fujita, University of Chicago |
| 19-11 & 19-15 | NOAA |
| 19-16 | Courtesy Pauline M. Austin, Massachusetts Institute of Technology |
| 19-17 | National Weather Service |
| 19-22 | NOAA |
| 19-25 | E. Phillip Krider, Institute of Atmosphere Physics, University of Arizona |

| | |
|---|---|
| 25-4 | *View of Harlem* by Jacob van Ruisdael Kunsthaus Zurich, fondation Professor Dr. L. Ruzicka |
| 25-5 | J. M. W. Turner, *The Grand Canal, Venice,* 1835. The Metropolitan Museum of Art, New York City |
| 25-6 | J. M. W. Turner, *Snowstorm, Hannibal and his Army Crossing the Alps,* The Tate Gallery, London |
| 25-7 | Thomas Cole, *The Course of Empire.* Courtesy, New York Historical Society, New York City |

# COLOR PHOTO CREDITS

| | |
|---|---|
| Plate 1 | Robert Gedzelman |
| Plate 2 | Stanley Gedzelman |
| Plate 3 | Robert Gedzelman |
| Plate 4 | NASA |
| Plate 6 | Christian Grund |
| Plates 7 & 8 | Stanley Gedzelman |
| Plate 9 | © Alistair B. Fraser 1971 |
| Plate 10 | NASA |
| Plate 11 | Ned Haines/Photo Researchers |
| Plates 12 & 13 | NASA |
| Plate 14 | W. H. Hodge/Peter Arnold |

# INDEX

534 INDEX

Wet bulb temperature, 144, 147-149, 199-202, 394
Wheat, 440-441, 457
Wien, Wilhelm (1864-1928), 119-121, 124
"William Tell Overture," 482
Wind chill, 84-85, 399
Windmills, 229, 232, 270-271
Winds, 1, 4, 229-271, 276-308, 359-373, 482
  and air pollution, 215-216, 217-220, 221
  aloft, 96, 110, 229, 237, 248-249, 265-269, 289, 294-308, 314-316, 319, 340-342, 371, 378
  cause of, 2, 5-8, 80, 128, 245-246
  in city, 139-141
  and climate, 118, 262-265, 387-418
  and cloud shapes, 21, 33-38
  cyclone scale, 276-308, 363, 365
  diurnal variations of, 366-367
  and evaporation, 149, 152-153
  and fog formation, 166-170
  forces on, 234-250
  in forest, 139
  geostrophic, 246-248, 267-269
  local scale, 359-373
  global scale, 253-271
  and locusts, 471
  gradient, 248, 267
  and ocean currents, 262-265
  in hurricanes, 1, 311, 312
  of past, 461
  power from, 232-234
  seasonal variations of, 255-256
  shear of, 35, 133
  speeds of, 229-232
  thermal, 269, 379
  in tornadoes, 230, 332
  trade, 253-255, 259, 314-315, 471
  and viscosity, 85
Wine, 457
*Wuthering Heights,* 475

Xerxes, 471-472

Year without a summer, 457, 462-463, 466, 469
Young, Thomas, 45, 51, 57

Zonda, 365. *See also* Chinook
Zone, of possible life, 122-123
  of silence, 110, 370